GRADE
2

Everyday Mathematics®

The University of Chicago School Mathematics Project

TEACHER'S LESSON GUIDE
VOLUME 2

Mc
Graw
Hill
Education

Bothell, WA • Chicago, IL • Columbus, OH • New York, NY

The University of Chicago School Mathematics Project

Max Bell, Director, *Everyday Mathematics* First Edition; James McBride, Director, *Everyday Mathematics* Second Edition; Andy Isaacs, Director, *Everyday Mathematics* Third and Fourth Editions; Amy Dillard, Associate Director, *Everyday Mathematics* Third Edition; Rachel Malpass McCall, Associate Director, *Everyday Mathematics* CCSS and Fourth Editions; Mary Ellen Dairyko, Associate Director, *Everyday Mathematics* Fourth Edition

Authors
Max Bell, Jean Bell, John Bretzlauf, Amy Dillard, Robert Hartfield, Andy Isaacs, James McBride, Cheryl G. Moran, Kathleen Pitvorec, Peter Saecker

Fourth Edition Grade 2 Team Leader
Cheryl G. Moran

Writers
Camille Bourisaw, Mary Ellen Dairyko, Gina Garza-Kling, Rebecca Williams Maxcy, Kathryn M. Rich

Open Response Team
Catherine R. Kelso, Leader; Steve Hinds

Differentiation Team
Ava Belisle-Chatterjee, Leader; Jean Marie Capper

Digital Development Team
Carla Agard-Strickland, Leader; John Benson, Gregory Berns-Leone, Juan Camilo Acevedo

Virtual Learning Community
Meg Schleppenbach Bates, Cheryl G. Moran, Margaret Sharkey

Technical Art
Diana Barrie, Senior Artist; Cherry Inthalangsy

UCSMP Editorial
Don Reneau, Senior Editor; Rachel Jacobs, Kristen Pasmore, Luke Whalen

Field Test Coordination
Denise A. Porter

Field Test Teachers
Kristin Collins, Debbie Crowley, Brooke Fordice, Callie Huggins, Luke Larmee, Jaclyn McNamee, Vibha Sanghvi, Brook Triplett

Contributors
William B. Baker, John Benson, James Flanders, Lila K. S. Goldstein, Funda Gönülateş, Lorraine M. Males, John P. Smith III, Kathleen Clark, Patti Satz, Penny Williams

Center for Elementary Mathematics and Science Education Administration
Martin Gartzman, Executive Director; Jose J. Fragoso, Jr., Meri B. Fohran, Regina Littleton, Laurie K. Thrasher

External Reviewers
The *Everyday Mathematics* authors gratefully acknowledge the work of the many scholars and teachers who reviewed plans for this edition. All decisions regarding the content and pedagogy of *Everyday Mathematics* were made by the authors and do not necessarily reflect the views of those listed below.

Elizabeth Babcock, California Academy of Sciences; Arthur J. Baroody, University of Illinois at Urbana-Champaign and University of Denver; Dawn Berk, University of Delaware; Diane J. Briars, Pittsburgh, Pennsylvania; Kathryn B. Chval, University of Missouri–Columbia; Kathleen Cramer, University of Minnesota; Ethan Danahy, Tufts University; Tom de Boor, Grunwald Associates; Louis V. DiBello, University of Illinois at Chicago; Corey Drake, Michigan State University; David Foster, Silicon Valley Mathematics Initiative; Funda Gönülateş, Michigan State University; M. Kathleen Heid, Pennsylvania State University; Natalie Jakucyn, Glenbrook South High School, Glenview, IL; Richard G. Kron, University of Chicago; Richard Lehrer, Vanderbilt University; Susan C. Levine, University of Chicago; Lorraine M. Males, University of Nebraska-Lincoln; Dr. George Mehler, Temple University and Central Bucks School District, Pennsylvania; Kenny Huy Nguyen, North Carolina State University; Mark Oreglia, University of Chicago; Sandra Overcash, Virginia Beach City Public Schools, Virginia; Raedy M. Ping, University of Chicago; Kevin L. Polk, Aveniros LLC; Sarah R. Powell, University of Texas at Austin; Janine T. Remillard, University of Pennsylvania; John P. Smith III, Michigan State University; Mary Kay Stein, University of Pittsburgh; Dale Truding, Arlington Heights District 25, Arlington Heights, Illinois; Judith S. Zawojewski, Illinois Institute of Technology

Note
Many people have contributed to the creation of *Everyday Mathematics*. Visit http://everydaymath.uchicago.edu/authors/ for biographical sketches of *Everyday Mathematics* Fourth Edition staff and copyright pages from earlier editions.

www.everydaymath.com

Send all inquiries to:
McGraw-Hill Education
8787 Orion Place
Columbus, OH 43240

ISBN: 978-0-02-140995-2
MHID: 0-02-140995-1

Printed in the United States of America.

1 2 3 4 5 6 7 8 9 WEB 19 18 17 16 15 14

Contents

Volume 2

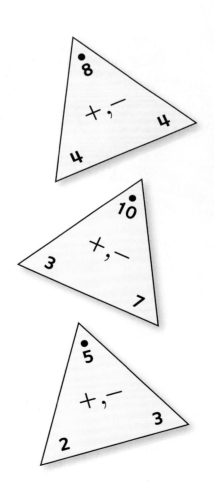

(t)Mark Steinmetz

A digital clock

(t)©iStockphoto.com/ImageGap, (b)©United States Mint image

arm span

Contents   **vii**

Four equal groups of eggs

(t)©iStockphoto.com/fotosipsak, (b)Mazer Creative Services

Unit 5 Organizer
Addition and Subtraction

In this unit, children review addition and subtraction problems in the context of money and number stories. They learn strategies for mentally adding and subtracting 10 and 100. Children's learning will focus on three clusters of the Common Core's content standards, as well as in-depth work on two of the Mathematical Practices.

CCSS Standards for Mathematical Content

Domain	Cluster
Operations and Algebraic Thinking	Represent and solve problems involving addition and subtraction.
Number and Operations in Base Ten	Use place value understanding and properties of operations to add and subtract.
Measurement and Data	Work with time and money.

Because the standards within each domain can be broad, *Everyday Mathematics* has unpacked each standard into Goals for Mathematical Content GMC . For a complete list of Standards and Goals, see page EM1.

For an overview of the CCSS domains, standards, and mastery expectations in this unit, see the **Spiral Trace** on pages 438–439. See the **Mathematical Background** (pages 440–442) for a discussion of the following key topics:

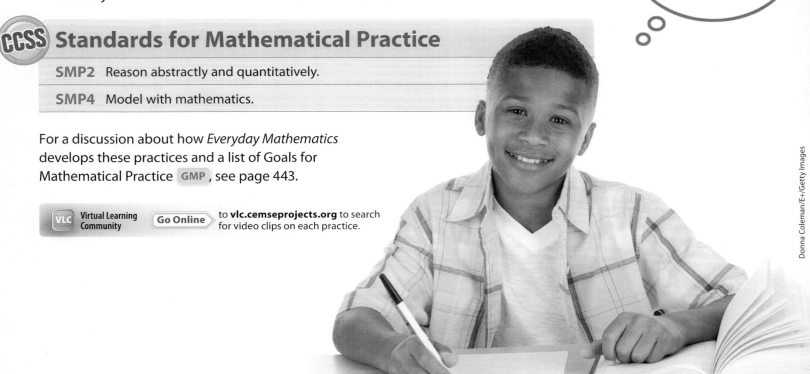

34 plus 10 is 44, plus 10 more is 54.

- Fact Power
- Money
- Open Number Lines
- Number Stories and Number Models

CCSS Standards for Mathematical Practice

SMP2   Reason abstractly and quantitatively.

SMP4   Model with mathematics.

For a discussion about how *Everyday Mathematics* develops these practices and a list of Goals for Mathematical Practice GMP , see page 443.

VLC **Virtual Learning Community**   Go Online to **vlc.cemseprojects.org** to search for video clips on each practice.

Donna Coleman/E+/Getty Images

Go Digital with these tools at **connectED.mheducation.com**

ePresentations   Student Learning   Facts Workshop   eToolkit   Professional   Home   Spiral Tracker   Assessment   English Learners   Differentiation
                 Center   Game   Development Connections   and Reporting Support   Support

Contents

Lesson and Overview	Page	CCSS Common Core State Standards*
5-1 **Playing *Beat the Calculator*** Children play *Beat the Calculator* to develop fact power by using mental strategies to add two 1-digit numbers.	444	2.OA.2 SMP5, SMP6
5-2 **Using Coins to Buy Things** Children review coin equivalencies and make different combinations of coins for the same amount of money.	450	2.NBT.2, 2.MD.8 SMP1, SMP2, SMP3
5-3 **Counting Up with Money** Children find coin combinations to pay for items and make change by counting up.	456	2.NBT.2, 2.NBT.5, 2.NBT.7, 2.MD.8 SMP1, SMP5
5-4 **Coin Calculations** Children make purchases and practice making change.	462	2.NBT.2, 2.NBT.7, 2.MD.8 SMP2, SMP5
5-5 (Explorations) **Exploring Arrays, Time, and Shapes** Children make arrays, match clock faces to digital notation, and construct shapes on geoboards.	468	2.OA.4, 2.MD.7, 2.G.1 SMP2, SMP7
5-6 **Mentally Adding and Subtracting 10 and 100** Children develop strategies for mentally adding and subtracting 10 and 100.	474	2.NBT.2, 2.NBT.5, 2.NBT.7, 2.NBT.8, 2.NBT.9 SMP1, SMP7, SMP8
5-7 **Open Number Lines** Children use open number lines as a tool for solving number stories.	480	2.OA.1, 2.NBT.5, 2.NBT.7, 2.NBT.8, 2.MD.6 SMP1, SMP2
5-8 **Change-to-More Number Stories** Children solve change-to-more number stories.	486	2.OA.1, 2.NBT.5, 2.NBT.7 SMP1, SMP2, SMP4
5-9 **Parts-and-Total Number Stories** Children solve parts-and-total number stories.	494	2.OA.1, 2.OA.2, 2.NBT.5, 2.NBT.7 SMP1, SMP2, SMP4
5-10 **Change Number Stories** Children solve change number stories involving temperature.	500	2.OA.1, 2.NBT.2, 2.NBT.5, 2.NBT.7 SMP1, SMP4
2-Day Lesson **5-11** (Open Response) **Adding Multidigit Numbers** **Day 1:** Children complete an open response problem by solving an addition problem using two different strategies. **Day 2:** The class discusses selected strategies, and children revise their work.	506	2.NBT.5, 2.NBT.9, 2.MD.8 SMP1, SMP4, SMP5
2-Day Lesson **5-12** (Assessment) **Unit 5 Progress Check** **Day 1:** Administer the Unit Assessments. **Day 2:** Administer the Open Response Assessment.	514	

*The standards listed here are addressed in the **Focus** of each lesson. For all the standards in a lesson, see the Lesson Opener.

Unit 5 Materials

VLC Virtual Learning Community

See how *Everyday Mathematics* teachers organize materials.
Search "Classroom Tours" at **vlc.cemseprojects.org**.

Lesson	Math Masters	Activity Cards	Manipulative Kit	Other Materials
5-1	pp. 118–119; *Assessment Handbook*, pp. 98–99		calculator; per group: 4 each of number cards 0–9; inch ruler	slate; per group: large paper triangle (optional); dominoes, number grid (optional); scissors
5-2	pp. 120–122; TA7; G10; G22	64–65	toolkit coins; per group: one $1 bill; centimeter ruler	Number-Grid Poster (optional); Class Number Scroll (optional); Class Data Pad; per group: sheet of paper labeled "Bank", large paper clip; scissors
5-3	pp. 121; 123–125	66–67	toolkit coins; per group: 4 each of number cards 0–10; per partnership: two 6-sided dice	slate; scissors
5-4	pp. 126–127; TA3; TA24; G19–G20	68	toolkit coins and bills; per partnership: base-10 blocks (10 longs and 30 cubes), 4 each of number cards 0–9; calculator (optional)	slate
5-5	pp. 128–129; per group: 2 copies of p. 130; 131–132; TA25; G18	69–71	per partnership: two 6-sided dice, 36 centimeter cubes; geoboards; rubber bands; per group: 4 each of number cards 0–10; pattern blocks	slate; envelope; scissors; straightedge; Pattern-Block Template
5-6	pp. 133–134; G23; G24 (optional)	72–73	Class Number Line; calculator; per partnership: die (see Before You Begin); centimeter ruler; 4 each of number cards 0–9; two 6-sided dice	slate; Number-Grid Poster; Class Number Scroll; per partnership: paper clip; crayons
5-7	pp. 135–136; *Assessment Handbook*, pp. 98–99	74	Quick Look Cards 96, 108, and 117; Class Number Line; 4 each of number cards 0–9; per group: calculator	slate; per group: large paper triangle (optional); number grid
5-8	pp. 137–139; TA7; TA26 (optional)	75	centimeter ruler; inch ruler	slate; shopping bag; number line
5-9	pp. 140–142; TA7; TA27 (optional)	76	toolkit coins (optional); counters	slate; number grid (optional); Fact Triangles; craft sticks
5-10	pp. 143–147; TA26 (optional); G7–G8		4 each of number cards 0–9; centimeter ruler	slate; Class Thermometer Poster; glue or tape; number line
5-11	pp. 148–150; TA5		base-10 blocks; $10 and $1 toolkit bills	slate; Standards for Mathematical Practice Poster; number grid; Guidelines for Discussions Poster; selected samples of children's work; children's work from Day 1; colored pencils (optional)
5-12	pp. 151–154; *Assessment Handbook*, pp. 31–36		toolkit coins	

Go Online to download all Quick Look Cards.

Problem Solving Professional Development

Everyday Mathematics emphasizes equally all three of the Common Core's dimensions of **rigor:** conceptual understanding, procedural skill and fluency, and applications. Math Messages, other daily work, Explorations, and Open Response tasks provide many opportunities for children to apply what they know to solve problems.

▶ Math Message

Math Messages require children to solve a problem they have not been shown how to solve. Math Messages provide almost daily opportunities for problem solving.

▶ Daily Work

Journal pages, Home Links, Writing/Reasoning prompts, and Differentiation Options often require children to solve problems in mathematical contexts and real-life situations. **Minute Math+** offers Number Stories for transition times and spare moments throughout the day. See Routine 6, pages 38–43.

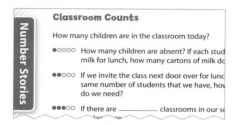

▶ Explorations

In Exploration A of Lesson 5-5, children make arrays with centimeter cubes and write number models to express the total number of cubes in the array as a sum of equal addends. In Exploration C, children make shapes on geoboards with a specified number of sides.

▶ Open Response and Reengagement

In Lesson 5-11, children show two ways they could determine the total cost of two items. Children can use tools, paper-and-pencil strategies, or mental methods. The reengagement discussion on Day 2 might focus on different ways that the same tool could be used, or different ways of representing the same strategy. Solving a problem in more than one way is the focus mathematical practice for the lesson. GMP1.5   When children use multiple methods to solve the same problem, they deepen their understanding of the mathematics by thinking it through in a different way. They also gain confidence in their problem-solving abilities as they use a second method to confirm that their answers are correct.

VLC Virtual Learning Community   **Go Online** to watch an Open Response and Reengagement lesson in action. Search "Open Response" at **vlc.cemseprojects.org**.

▶ Open Response Assessment

In Progress Check Lesson 5-12, children find possible coin combinations for 75¢ and explain how they know one of the combinations equals 75¢. GMP3.1

Look for GMP1.1–1.6 markers, which indicate opportunities for children to engage in SMP1: "Make sense of problems and persevere in solving them." Children also become better problem solvers as they engage in all of the CCSS Mathematical Practices. The yellow GMP markers throughout the lessons indicate places where you can emphasize the Mathematical Practices and develop children's problem-solving skills.

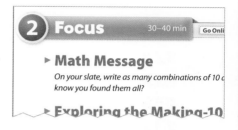

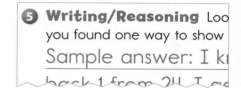

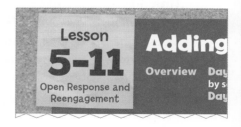

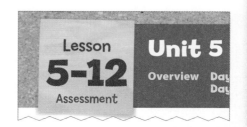

Assessment and Differentiation

See pages xxii–xxv to learn about a comprehensive online system for recording, monitoring, and reporting children's progress using core program assessments.

  to **vlc.cemseprojects.org** for tools and ideas related to assessment and differentiation from *Everyday Mathematics* teachers.

✓ Ongoing Assessment

In addition to frequent informal opportunities for "kid watching," every lesson (except Explorations) offers an **Assessment Check-In** to gauge children's performance on one or more of the standards addressed in that lesson.

Lesson	Task Description	CCSS Common Core State Standards
5-1	Demonstrate fluency with addition facts.	2.OA.2
5-2	Show two different combinations of coins for the same amount.	2.NBT.2, 2.MD.8, SMP1, SMP2
5-3	Select coins or bills with a value greater than the cost of the item.	2.NBT.2, 2.NBT.7, 2.MD.8, SMP5
5-4	Use toolkit coins to show exact change.	2.MD.8
5-6	Use a mental strategy to add 10 to and subtract 10 from any 2-digit number and 3-digit multiple of 10.	2.NBT.5, 2.NBT.8
5-7	Mentally add 10s and 1s.	2.NBT.5, 2.NBT.7, 2.NBT.8
5-8	Solve change-to-more number stories.	2.OA.1, 2.NBT.5, 2.NBT.7
5-9	Solve parts-and-total number stories.	2.OA.1, 2.NBT.5, 2.NBT.7
5-10	Solve change number stories.	2.OA.1, 2.NBT.5, 2.NBT.7
5-11	Find the total cost of two items.	2.NBT.5, SMP1

▶ Periodic Assessment

Unit 5 Progress Check This assessment focuses on the CCSS domains of *Operations and Algebraic Thinking, Number and Operations in Base Ten,* and *Measurement and Data.* It also contains an Open Response Assessment to test children's ability to explain how different coin combinations have the same value. GMP3.1

> **NOTE** Odd-numbered units include an **Open Response Assessment.** Even-numbered units include a **Cumulative Assessment.**

▶ Unit 5 Differentiation Activities Differentiation Support ELL English Learners Support

Differentiation Options Every regular lesson provides **Readiness, Enrichment, Extra Practice,** and **English Language Learners Support** activities that address the Focus standards of that lesson.

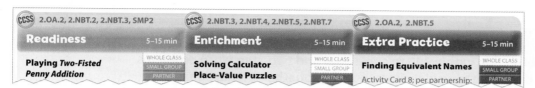

CCSS 2.OA.2, 2.NBT.2, 2.NBT.3, SMP2	CCSS 2.NBT.3, 2.NBT.4, 2.NBT.5, 2.NBT.7	CCSS 2.OA.2, 2.NBT.5
Readiness 5–15 min	**Enrichment** 5–15 min	**Extra Practice** 5–15 min
Playing *Two-Fisted Penny Addition*	Solving Calculator Place-Value Puzzles	Finding Equivalent Names

Activity Cards These activities, written to the children, enable you to differentiate Part 2 of the lesson through small-group work.

English Language Learners Activities and point-of-use support help children at different levels of English language proficiency succeed.

Differentiation Support Two online pages for most lessons provide suggestions for game modifications, ways to scaffold lessons for children who need additional support, and language development suggestions for Beginning, Intermediate, and Advanced English language learners.

Activity Card 67

Differentiation Support online pages

For **ongoing distributed practice,** see these activities:
- Mental Math and Fluency
- Differentiation Options: Extra Practice
- Part 3: Journal pages, Math Boxes, *Math Masters,* Home Links
- Print and online games

Ongoing Practice Differentiation Support

▶ Games

Games in *Everyday Mathematics* are an essential tool for practicing skills and developing strategic thinking.

Lesson	Game	Skills and Concepts	CCSS Common Core State Standards
5-1	*Beat the Calculator*	Practicing addition facts	2.OA.2
5-1	*Beat the Calculator* with Extended Facts	Applying knowledge of basic addition facts to compute extended facts	2.OA.2, 2.NBT.5
5-2	*Spinning for Money*	Making coin and bill exchanges	2.MD.8, SMP5
5-2	*Dime-Nickel-Penny Grab*	Finding the total value of various coin combinations	2.MD.8
5-3	*Salute!*	Practicing addition facts and finding missing addends	2.OA.2, SMP6
5-4	*Target*	Using base-10 blocks to model addition and subtraction	2.NBT.1, 2.NBT.3, 2.NBT.7, SMP2
5-5	*Addition Top-It*	Practicing addition facts	2.OA.2, 2.NBT.4
5-5	*Clock Concentration*	Matching times shown on clock faces to digital notation	2.MD.7
5-6	*Addition/Subtraction Spin*	Adding and subtracting 10 and 100 mentally	2.NBT.5, 2.NBT.7, 2.NBT.8, SMP8
5-7	*Beat the Calculator*	Practicing addition and subtraction facts	2.OA.2
5-10	*Number Top-It*	Comparing numbers	2.NBT.1, 2.NBT.3, 2.NBT.4, SMP7

VLC Virtual Learning Community **Go Online** to look for examples of *Everyday Mathematics* games at **vlc.cemseprojects.org.**

CCSS Spiral Trace: Skills, Concepts, and Applications

⭐ **Mastery Expectations** This Spiral Trace outlines instructional trajectories for key standards in Unit 5. For each standard, it highlights opportunities for Focus instruction, Warm Up and Practice activities, as well as formative and summative assessment. It describes the **degree of mastery**— as measured against the entire standard—expected at this point in the year.

Operations and Algebraic Thinking

2.OA.1   Use addition and subtraction within 100 to solve one- and two-step word problems involving situations of adding to, taking from, putting together, taking apart, and comparing, with unknowns in all positions, e.g., by using drawings and equations with a symbol for the unknown number to represent the problem.

3-5 Warm Up	3-7 through 3-9 Warm Up Focus Practice	4-11 Warm Up Practice	5-7 through 5-10 Warm Up Focus Practice	5-12 Progress Check	6-2 through 6-5 Warm Up Focus Practice	6-7 Practice	6-9 Focus	7-2 Focus

⭐ By the end of Unit 5, expect children to **add and subtract within 100 to solve one-step word problems involving situations of adding to, taking from, putting together, and taking apart, e.g., by using drawings to represent the problem.**

2.OA.2   Fluently add and subtract within 20 using mental strategies. By end of Grade 2, know from memory all sums of two one-digit numbers.

4-1 Practice	4-2 Practice	4-4 Warm Up Practice	4-11 Warm Up Focus	5-1 Focus Practice	5-3 Warm Up Practice	5-5 Practice	5-7 Warm Up Practice	5-9 Focus Practice	5-12 Progress Check	6-2 Warm Up Practice	6-8 Practice

⭐ By the end of Unit 5, expect children to **know doubles and combinations of 10, and apply strategies to solve all addition and subtraction facts.**

Number and Operations in Base Ten

2.NBT.2   Count within 1000; skip-count by 5s, 10s, and 100s.

2-2 Practice	2-4 Practice	2-8 Focus Practice	2-11 Warm Up Practice	2-12 Focus Practice	2-13 Progress Check	4-1 Warm Up	4-2 Warm Up Focus Practice	5-2 through 5-4 Warm Up Focus Practice	5-6 Warm Up Focus	5-10 Warm Up Focus

⭐ By the end of Unit 5, expect children to **count by 1s within 1,000 and skip count by 5s, 10s, and 100s.**

Spiral Tracker

Go to **connectED.mheducation.com** for comprehensive trajectories that show how in-depth mastery develops across the grade.

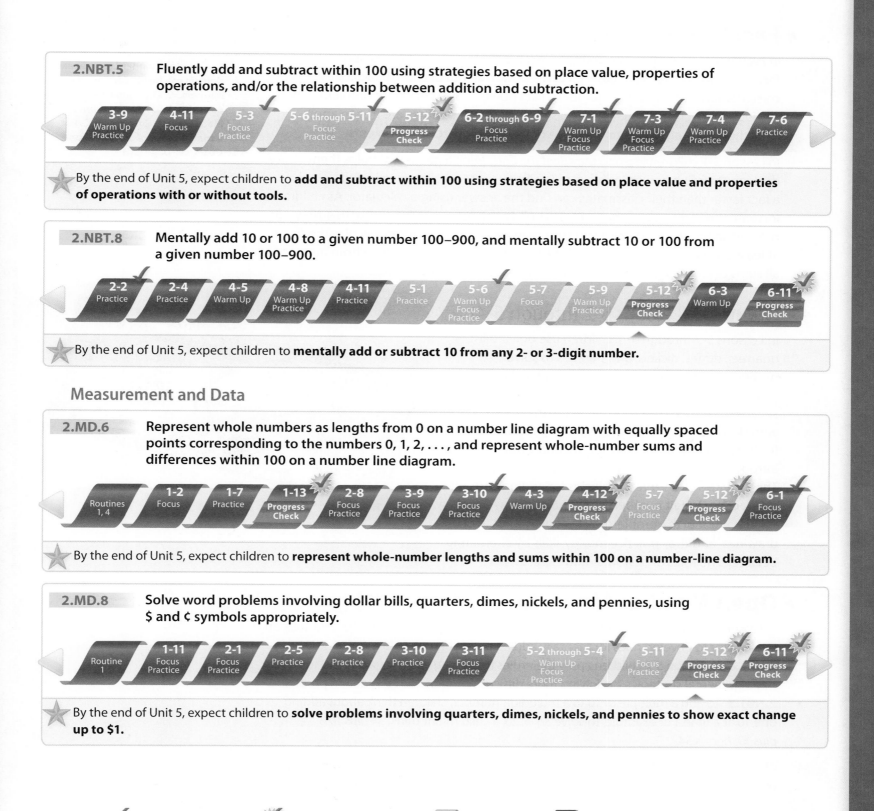

2.NBT.5 Fluently add and subtract within 100 using strategies based on place value, properties of operations, and/or the relationship between addition and subtraction.

| 3-9 Warm Up Practice | 4-11 Focus | 5-3 Focus Practice | 5-6 through 5-11 Focus Practice | 5-12 Progress Check | 6-2 through 6-9 Focus Practice | 7-1 Warm Up Focus Practice | 7-3 Warm Up Focus Practice | 7-4 Warm Up Practice | 7-6 Practice |

⭐ By the end of Unit 5, expect children to **add and subtract within 100 using strategies based on place value and properties of operations with or without tools.**

2.NBT.8 Mentally add 10 or 100 to a given number 100–900, and mentally subtract 10 or 100 from a given number 100–900.

| 2-2 Practice | 2-4 Practice | 4-5 Warm Up | 4-8 Warm Up Practice | 4-11 Practice | 5-1 Practice | 5-6 Warm Up Focus Practice | 5-7 Focus | 5-9 Warm Up Practice | 5-12 Progress Check | 6-3 Warm Up | 6-11 Progress Check |

⭐ By the end of Unit 5, expect children to **mentally add or subtract 10 from any 2- or 3-digit number.**

Measurement and Data

2.MD.6 Represent whole numbers as lengths from 0 on a number line diagram with equally spaced points corresponding to the numbers 0, 1, 2, . . . , and represent whole-number sums and differences within 100 on a number line diagram.

| Routines 1, 4 | 1-2 Focus | 1-7 Practice | 1-13 Progress Check | 2-8 Focus Practice | 3-9 Focus Practice | 3-10 Focus Practice | 4-3 Warm Up | 4-12 Progress Check | 5-7 Focus Practice | 5-12 Progress Check | 6-1 Focus Practice |

⭐ By the end of Unit 5, expect children to **represent whole-number lengths and sums within 100 on a number-line diagram.**

2.MD.8 Solve word problems involving dollar bills, quarters, dimes, nickels, and pennies, using $ and ¢ symbols appropriately.

| Routine 1 | 1-11 Focus Practice | 2-1 Focus Practice | 2-5 Practice | 2-8 Practice | 3-10 Practice | 3-11 Focus Practice | 5-2 through 5-4 Warm Up Focus Practice | 5-11 Focus Practice | 5-12 Progress Check | 6-11 Progress Check |

⭐ By the end of Unit 5, expect children to **solve problems involving quarters, dimes, nickels, and pennies to show exact change up to $1.**

Key ✓ = Assessment Check-In   ✹ = Progress Check Lesson   ▢ = Current Unit   ▢ = Previous or Upcoming Lessons

Mathematical Background: Content

 This discussion highlights the major content areas and the Common Core State Standards addressed in Unit 5. See the online Spiral Tracker for complete information about the learning trajectories for all standards.

▶ Fact Power (Lesson 5-1)

In *Everyday Mathematics,* the ability to recall basic facts instantly is called *fact power.* Instant recall of basic facts is a powerful tool because it facilitates computation with multidigit numbers. Fact power is also critical for problem solving. When children are able to easily recall their facts, they are free to apply all of their thinking power to the more challenging aspects of problems.

In Lesson 5-1 children learn to play *Beat the Calculator,* a game that helps them develop their fact power. In this game children try to recall the answer to a fact faster than their classmates can find the answer using a calculator. As children increase their fact power, they will also build confidence as they realize that they can provide an answer to a basic fact faster than a calculator. Children will play the game at least once per unit for the rest of the year to work toward knowing from memory all sums of two 1-digit numbers by the end of second grade. **2.OA.2**

Standards and Goals for Mathematical Content

Because the standards within each domain can be broad, *Everyday Mathematics* has unpacked each standard into Goals for Mathematical Content **GMC**. For a complete list of Standards and Goals, see page EM1.

▶ Money (Lessons 5-2 through 5-4)

In Lessons 5-2 through 5-4, children use dollar bills, quarters, dimes, nickels, and pennies to solve problems that involve buying items and making change. **2.MD.8** The activities in these lessons serve several purposes. They provide contextualized practice with skip counting, which helps to prepare children for the mental addition and subtraction strategies introduced in this unit and later units. **2.NBT.2** The activities also serve as vehicles for adding and subtracting 2-digit numbers. **2.NBT.5** Children model the problems with toolkit coins and drawings, and they practice money skills by alternating between the roles of customer and clerk. **2.NBT.7**

▶ Open Number Lines (Lessons 5-6 and 5-7)

In Lesson 5-6 children use patterns they identify when skip counting by 10s and 100s on a calculator to develop rules for mentally adding and subtracting 10 and 100. **2.NBT.2, 2.NBT.8** In Lesson 5-7 children apply their rules for adding and subtracting 10 to develop mental strategies for solving more difficult addition and subtraction problems. **2.NBT.5** To add a multiple of 10 to a multidigit number, children think about adding one 10 at a time. For example, to solve 34 + 20, a child might think: *34 plus 10 is 44, plus 10 more is 54.*

Children then extend this strategy to include adding numbers that are not multiples of 10. They add the tens in a number first and then continue by adding the ones.

Unit 5 Vocabulary

addition fact

array

change diagram

change-to-less number story

change-to-more number story

degree Fahrenheit (°F)

equivalencies

fact power

mental addition

mental subtraction

open number line

parts-and-total diagram

parts-and-total number story

thermometer

total

Open Number Lines *Continued*

For example, to add 34 + 23, a child might think: *34 plus 10 is 44, plus 10 more is 54. Then I still need to add 3 more. So, 54 plus 3 is 57.*

At this point in the year, most children should be able to fluently count forward by 10s and 1s, and most children should be able to perform in their heads the individual steps of the strategies described above. However, it can be difficult for children to keep track of the steps and remember how much they have added and how much more they have to go. In Lesson 5-7 children are introduced to *open number lines,* a new tool to help them record and keep track of the steps in their mental strategies.

Children have counted up and back on number lines to solve addition and subtraction problems since *First Grade Everyday Mathematics.* The number line provides a powerful representation of numbers that helps children visualize the way numbers increase when we count up and decrease when we count back. However, as children begin solving problems with larger numbers, it becomes cumbersome to draw fully labeled number lines. Not only can it be difficult to fit marks for every whole number needed to represent a computation on a typical sheet of paper, it is often impossible to determine the range of whole numbers before starting to solve the problem.

The use of open number lines resolves these issues. They look much like the number lines that you and children in your class are familiar with, but they show only the numbers that are needed to solve a particular problem with a particular strategy. Children picture the full number line in their minds, but they use an open number line to record only the numbers that are "stopping points" in their mental strategies. For example, consider the strategy described above for solving 34 + 23. The diagrams at the right and below show how children working with this strategy might use an open number line to record their steps. **2.MD.6**

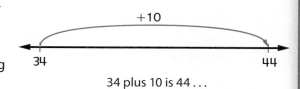

34 plus 10 is 44 . . .

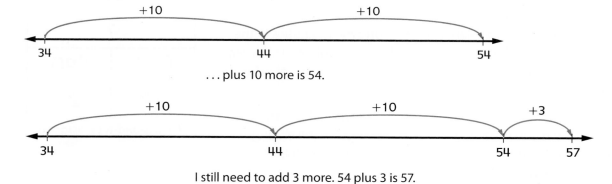

. . . plus 10 more is 54.

I still need to add 3 more. 54 plus 3 is 57.

Open Number Lines *Continued*

Open number lines are quick and easy to draw, and they provide children with a flexible problem-solving tool that they can use to record many different strategies. For example, open number lines can be helpful when children are counting up to solve subtraction problems. Instead of looking for where they land on a number line to determine the answer, children count the total number of hops taken to get from the smaller number to the larger number. The open number line below illustrates one mental strategy for solving $56 - 29$. By using open number lines, children are able to keep track of and reflect on their own thinking and explain their strategies to others. **2.NBT.9**

Starting at 29, hop 1 space to get to 30. Then hop 10 more to 40, 10 more to 50, and 6 more to 56. It took $1 + 10 + 10 + 6 = 27$ hops to get to 56, so $56 - 29 = 27$.

▶ Number Stories and Number Models

(Lessons 5-8 through 5-10)

There are two basic types of problem situations for addition: changing to more (or adding to) and putting together. Change-to-more number stories involve a start quantity that increases. This means the end quantity will be more than the start quantity. Parts-and-total number stories involve two or more parts that are combined to find a total. Change-to-more number stories are discussed in Lesson 5-8; parts-and-total number stories are discussed in Lesson 5-9. **2.OA.1**

Children use change diagrams and parts-and-total diagrams to organize the information in these stories. They fill in the numbers that they know, using a question mark to represent the number that they need to find. (*See margin.*) They start by filling in predrawn diagrams and then transition to making their own sketches of the diagrams. You can sketch diagrams to model the process for children and support this transition. Because these diagrams help children organize their thinking, they can be a helpful tool for problem solving. (See the section on Mathematical Practice 4 on page 443 for more information about filling in these diagrams.)

After filling in the diagrams, children write number models to represent the stories, again using a question mark for the unknown number. The number model for the change-to-more number story in the margin would be $40 + 20 = ?$, and the number model for the parts-and-total number story would be $12 + 23 = ?$ **2.OA.1** These number models represent and clarify the relationships among the quantities in the problems and may help some children decide how to solve the problems. However, children are not expected to formally manipulate the number models to solve for the unknown number.

In Lesson 5-10 children solve temperature-change stories. **2.OA.1** In these stories any number could be unknown: the starting number, the change, or the ending number. Change diagrams and number models can become important tools for organizing information and deciding on a solution strategy.

Amanda had $40. She earned $20 more mowing lawns. How much money does she have now?

A change diagram

There are 12 fourth graders and 23 third graders on a bus. How many children in all are on the bus?

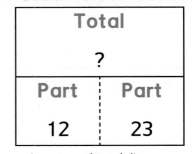

A parts-and-total diagram

Mathematical Background: Practices

 In Everyday Mathematics, *children learn the **content** of mathematics as they engage in the **practices** of mathematics. As such, the Standards for Mathematical Practice are embedded in children's everyday work, including hands-on activities, problem-solving tasks, discussions, and written work. Read here to see how Mathematical Practices 2 and 4 are emphasized in this unit.*

▶ Standard for Mathematical Practice 2

Mathematical Practice 2 states that mathematically proficient students have "the ability to *decontextualize*—to abstract a given situation and represent it symbolically." During Unit 5 children develop their ability to decontextualize by "making connections between representations." GMP2.3 In Lesson 5-7 children use open number lines to represent their strategies for solving number stories. (See pages 440–442 for more information on open number lines.) To make sense of open number lines, children connect them to their work with the concrete number lines they have used since kindergarten. As they transition from counting every number on concrete number lines to drawing simpler open number lines, children are further decontextualizing the situations in number stories and using more abstract representations.

In Lessons 5-8 through 5-10, children make connections between situation diagrams and number models. A situation diagram uses the numbers in the story along with a few words to show the relationships among the quantities, whereas a number model uses only numbers and mathematical symbols to show the relationships. As children use situation diagrams to write the more abstract number models, they are developing their ability to decontextualize.

▶ Standard for Mathematical Practice 4

Mathematical Practice 4 states that mathematically proficient students are "able to identify important quantities in a practical situation and map their relationships using such tools as diagrams." In Lessons 5-8 through 5-10, children use change diagrams and parts-and-total diagrams to "model real-world situations" presented in number stories. GMP4.1

Consider this temperature-change story: *When Peter got up, it was 35°F. At lunch time it was 13°F warmer. What was the temperature at lunch time?* The change diagram shown in the margin is a mathematical model of this problem because it shows the important quantities in the problem and how they are related. To fill in the diagram, children must ask themselves questions such as the following: *Where do I write the numbers that I know? Did the temperature change to more or change to less? What number do I need to find?* After they have filled out the diagram, children can "use [the] mathematical model to solve [the] problem" because the model makes it easier to see that the unknown number can be found by adding 13 to 35. GMP4.2

 Standards and Goals for Mathematical Practice

SMP2 Reason abstractly and quantitatively.

GMP2.1 Create mathematical representations using numbers, words, pictures, symbols, gestures, tables, graphs, and concrete objects.

GMP2.2 Make sense of the representations you and others use.

GMP2.3 Make connections between representations.

SMP4 Model with mathematics.

GMP4.1 Model real-world situations using graphs, drawings, tables, symbols, numbers, diagrams, and other representations.

GMP4.2 Use mathematical models to solve problems and answer questions.

Go Online to the *Implementation Guide* for more information about the Mathematical Practices.

For children's information on the Mathematical Practices, see *My Reference Book*, pages 1–22.

Change

Start		End
35	+ 13	?

Playing *Beat the Calculator*

Overview Children play *Beat the Calculator* to develop fact power by using mental strategies to add two 1-digit numbers.

▶ **Before You Begin**
Decide how you will distribute *Math Journal 2*. For *Beat the Calculator* in Part 2, you may want to provide large paper triangles with corners labeled Calculator, Caller, and Brain. (*See page 447.*)

▶ **Vocabulary**
addition fact · fact power

Common Core State Standards

Focus Cluster
Add and subtract within 20.

	Materials	
① Warm Up 15–20 min		
Mental Math and Fluency Children use <, >, and = to compare numbers.	slate	2.NBT.4
Daily Routines Children complete daily routines.	See pages 4–43.	See pages xiv–xvii.
② Focus 30–40 min		
Math Message Children write addition facts they know.		2.OA.2
Discussing Fact Power Children discuss how fact power can help them solve problems efficiently.	calculator	2.OA.2 SMP5, SMP6
Introducing *Beat the Calculator* Children race to add numbers mentally and with calculators.	per group: 4 each of number cards 0–9, calculator	2.OA.2
Playing *Beat the Calculator* **Game** Children practice addition facts.	*Assessment Handbook,* pp. 98–99; per group: 4 each of number cards 0–9, calculator, large paper triangle (optional)	2.OA.2
✓ **Assessment Check-In** See page 448.	*Assessment Handbook,* pp. 98–99	2.OA.2

CCSS 2.OA.2 **Spiral Snapshot**

GMC Know all sums of two 1-digit numbers automatically.

4-4 Practice	4-6 Warm Up	4-9 Warm Up Practice	5-1 Focus Practice	5-3 Practice	5-5 Practice	5-7 Warm Up Practice	5-9 Practice

Spiral Tracker   **Go Online**   to see how mastery develops for all standards within the grade.

③ Practice 15–20 min		
Solving Subtract-10 Number Stories Children solve subtraction number stories.	*Math Journal 2,* p. 103 and inside back cover (optional)	2.OA.1, 2.NBT.8
Math Boxes 5-1 Children practice and maintain skills.	*Math Journal 2,* p. 104; inch ruler	See page 449.
Home Link 5-1 **Homework** Children write sums for addition facts.	*Math Masters,* p. 119	2.OA.2

connectED.mheducation.com ▶

Plan your lessons online with these tools.

 ePresentations   Student Learning Center   Facts Workshop Game   eToolkit   Professional Development   Home Connections   Spiral Tracker   Assessment and Reporting   English Learners Support   Differentiation Support

Differentiation Options

RtI

CCSS 2.OA.2

Readiness   5–15 min

Reviewing Fact Inventories

WHOLE CLASS
SMALL GROUP
PARTNER
INDEPENDENT

Math Journal 1, pp. 22 and 94–97

Children get support for addition facts by using their Addition Facts Inventory Record on journal pages 94–97. Discuss strategies they might use to solve facts they don't yet know. Remind children to refer to their My Addition Facts Strategies table on journal page 22.

CCSS 2.OA.2, 2.NBT.5

Enrichment   5–15 min

Playing *Beat the Calculator* with Extended Facts

WHOLE CLASS
SMALL GROUP
PARTNER
INDEPENDENT

per group: 4 each of number cards 0–9, calculator, large paper triangle (optional)

Children apply their knowledge of basic facts to explore extended facts by playing an extended-facts version of *Beat the Calculator.* The caller draws two number cards and treats each number as a multiple of 10 (for example, 5 is treated as 50). If the caller draws a 3 and a 9, the problem is 30 + 90.

CCSS 2.OA.2, 2.NBT.5

Extra Practice   5–15 min

Making a Fact-Family Chain

WHOLE CLASS
SMALL GROUP
PARTNER
INDEPENDENT

sheet of paper, scissors, dominoes (optional)

For additional practice with fact families, children make a fact-family chain. See the Extra Practice activity in Lesson 3-3 for details.

English Language Learners Support

Beginning ELL Show children a picture of the Fact-Power Kids (*Math Masters,* page 118) to help them make the connection between developing physical power by doing pull-ups and developing fact power by practicing facts.

Go Online   ELL English Learners Support

Standards and Goals for
Mathematical Practice

SMP5 **Use appropriate tools strategically.**
GMP5.2 Use tools effectively and make sense
of your results.

SMP6 **Attend to precision.**
GMP6.4 Think about accuracy and efficiency
when you count, measure, and calculate.

1 Warm Up 15–20 min   Go Online   ePresentations   eToolkit

▶ Mental Math and Fluency

Have children use <, >, and = to compare numbers on their slates.
Leveled exercises:

●○○ 34 __<__ 43; 98 __>__ 93

●●○ 276 __<__ 286; 413 __<__ 431

●●● 620 __>__ 602; 701 __<__ 710

▶ Daily Routines

Have children complete daily routines.

2 Focus 30–40 min   Go Online   ePresentations   eToolkit

▶ Math Message

*Write five **addition facts** that you know.*

▶ Discussing Fact Power

| WHOLE CLASS | SMALL GROUP | PARTNER | INDEPENDENT |

Math Message Follow-Up Have several volunteers call out addition
facts with the sums as you record them. Record some horizontally with
the sum on the left, some horizontally with the sum on the right, and
some vertically.

Academic Language Development Discuss examples of good
classroom habits, such as pushing in your chair when leaving your seat.
Then ask children to name activities they do every day without thinking.
Explain that a *habit* is something we do automatically, or without
thinking, because we have done that activity over and over again. Point
out that learning facts is like developing a habit. In *Everyday Mathematics*
having good fact habits is called having *fact power*. Having fact power
makes it much easier to solve problems in mathematics because you
know facts automatically without having to figure them out.

Divide the class into two groups and pose an addition fact. Children in one group use calculators to find the sum. GMP5.2 Children in the other group find the sum mentally, without using calculators or any other tool.

Repeat the activity with other addition facts. Mix easy and challenging facts so that each group of children has the chance to find the sum first. Children are generally surprised to find that having **fact power**—knowing the facts automatically—makes it possible to beat the calculator. Discuss the idea that by developing fact power, children can solve facts more efficiently in their heads than they can with a calculator. GMP6.4

Tell children that they will use their brains and their calculators to practice addition facts.

► Introducing *Beat the Calculator*

| WHOLE CLASS | SMALL GROUP | PARTNER | INDEPENDENT |

Select three children to demonstrate the game.

Directions

1. One player is the Caller, another player is the Calculator, and a third player is the Brain.

2. The Caller places the deck of number cards (4 each of number cards 0–9) number-side down and draws two cards from the top. The Caller asks for the sum of the numbers.

3. The Calculator solves the problem with a calculator while the Brain solves it mentally. The Caller determines who gets the correct answer first.

4. The player who gets the correct answer first collects the two cards.

5. All three children rotate roles every five turns so that each player gets a turn in each role.

► Playing *Beat the Calculator*

Assessment Handbook, pp. 98–99

| WHOLE CLASS | SMALL GROUP | PARTNER | INDEPENDENT |

Divide the class into groups of three. Remind children to rotate roles every five turns.

Circulate as children play, offering guidance as needed. Emphasize that the purpose of the game is for children to help one another develop fact power.

> **NOTE** Some teachers provide children with a large paper triangle with the corners labeled for the three roles: Calculator, Caller, and Brain. Children rotate the triangle to keep track of role assignments. (*See margin.*)

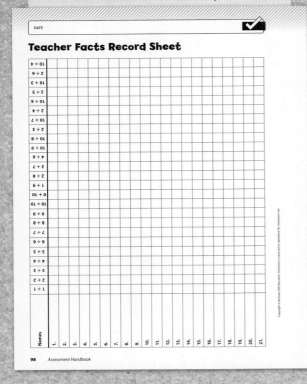

Observe

- Which children in the Brain role are automatic in their fact recall?
- Which children in the Brain role appear to be using efficient strategies to find the sums?

Discuss

- *For facts that you didn't automatically know, what strategies did you use to find the sums?*
- *What did you find easy about this game? What did you find challenging?*

Playing *Beat the Calculator* frequently can help children develop automaticity with basic facts.

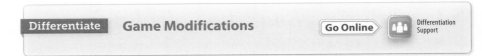

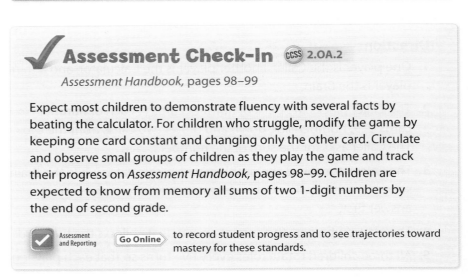

✓ Assessment Check-In CCSS 2.OA.2

Assessment Handbook, pages 98–99

Expect most children to demonstrate fluency with several facts by beating the calculator. For children who struggle, modify the game by keeping one card constant and changing only the other card. Circulate and observe small groups of children as they play the game and track their progress on *Assessment Handbook,* pages 98–99. Children are expected to know from memory all sums of two 1-digit numbers by the end of second grade.

Assessment and Reporting Go Online to record student progress and to see trajectories toward mastery for these standards.

Summarize Have children discuss ways to improve their fact power. Expect responses to include practicing with Fact Triangles and playing fact games.

Math Journal 2, p. 103

Solving Number Stories   Lesson 5-1

Solve. You may use the number grid on the inside back cover of your journal to help.

1. Justin brought 30 blueberries for a snack. He ate 10 of them. How many blueberries does he have left?

   Answer: __20__ blueberries

2. Davion's string was 48 inches long. He cut off 10 inches. How long is Davion's string now?

   Answer: __38__ inches

3. Haily picked up 41 shells on the beach. She picked up 10 more than her sister. How many did her sister pick up?

   Answer: __31__ shells

4. Rosa had 108 stamps in her collection. Of these, 78 stamps were from the United States. How many of her stamps were from other countries?

   Answer: __30__ stamps

5. The school library ordered 56 new animal books. Children in Ms. Tran's class checked out some of them. The library now has 46 new animal books. How many were checked out?

   Answer: __10__ books

2.OA.1, 2.NBT.8   one hundred three 103

③ Practice   15–20 min   Go Online

ePresentations   eToolkit   Home Connections

▶ Solving Subtract-10 Number Stories

Math Journal 2, p. 103

| WHOLE CLASS | SMALL GROUP | PARTNER | INDEPENDENT |

Children solve subtract-10 number stories. As needed, allow children to use the number grids on the inside back covers of their journals.

▶ Math Boxes 5-1 ✏️

Math Journal 2, p. 104

| WHOLE CLASS | SMALL GROUP | PARTNER | INDEPENDENT |

Mixed Practice Math Boxes 5-1 are paired with Math Boxes 5-3.

▶ Home Link 5-1

Math Masters, p. 119

Homework Children solve addition facts and complete a maze.

Planning Ahead

If you have not already done so, move the Temperature Routine forward by recording the daily temperature to the nearest degree. (For more details, see the Temperature Routine on page 30.) This will help prepare children for Lesson 5-10, where thermometers will be used as the context for change number stories.

Math Journal 2, p. 104

Math Boxes

Lesson 5-1
DATE

① Estimate the length of this line segment: Answers vary.

About _____ inches

Use your inch ruler to measure the length of the line segment.

About __2__ inches

② Write the number word for 98.

Ninety-eight

③ Use the digits 6, 1, and 8 to make the smallest number possible.

__168__

Use the same digits to make the largest possible number.

__861__

④ Count back by 100s.

900, 800, 700, 600

500, 400

⑤ **Writing/Reasoning** Explain how you made the largest possible number in Problem 3.

Sample answer: I put the biggest number in the hundreds place, the next-biggest number in the tens place, and the smallest number in the ones place.

① 2.MD.1, 2.MD.3 ② 2.NBT.3
③ 2.NBT.1, 2.NBT.3, 2.NBT.4 ④ 2.NBT.2

104 one hundred four ⑤ 2.NBT.1, 2.NBT.3, SMP7

Math Masters, p. 119

Solving Addition Facts

Home Link 5-1
NAME
DATE

Family Note

Today we continued working with addition facts. Children can develop number-fact reflexes the same way that they develop any other habit—by practicing them over and over. In *Everyday Mathematics* knowing facts automatically is called fact power. We discussed ways to develop fact power, such as practicing with Fact Triangles and playing fact games.

When your child has solved the addition facts below and is ready to draw the mouse's path through the maze, explain that the mouse can move up, down, left, right, or diagonally to find the cheese.

Please return this Home Link to school tomorrow.

Solve the facts. Then draw a path for the mouse to find the cheese. The mouse can go through only those boxes with a sum of 7.

2.OA.2
one hundred nineteen 119

Lesson 5-2

Using Coins to Buy Things

Overview Children review coin equivalencies and make different combinations of coins for the same amount of money.

▶ **Before You Begin**
For Part 2, prepare a Table of Equivalencies on the Class Data Pad (*see page 452*).

▶ **Vocabulary**
equivalencies

Common Core State Standards

Focus Clusters
- Understand place value.
- Work with time and money.

	Materials	
① Warm Up 15–20 min		
Mental Math and Fluency Children skip count by 5s, 10s, and 25s.	Number-Grid Poster (optional), Class Number Scroll (optional)	**2.NBT.2**
Daily Routines Children complete daily routines.	See pages 4–43.	See pages xiv–xvii.

② Focus 30–40 min		
Math Message Children find the total value of specified toolkit coins.	toolkit coins (10 pennies, 6 nickels, 6 dimes, and 4 quarters)	**2.NBT.2, 2.MD.8**
Finding the Total Children share strategies for counting their toolkit coins.	toolkit coins	**2.NBT.2, 2.MD.8** **SMP3**
Reviewing Money Equivalencies Children discuss equivalencies among coins and $1 bills.	*My Reference Book*, pp. 110–111; Class Data Pad (see Before You Begin); coins and one $1 bill (for demonstration)	**2.MD.8**
Buying and Selling Children show different ways to pay for items.	*Math Journal 2*, pp. 106–107; *Math Masters*, p. 121; toolkit coins	**2.NBT.2, 2.MD.8** **SMP1, SMP2**
✓ **Assessment Check-In** See page 454.	*Math Journal 2*, p. 107	**2.NBT.2, 2.MD. 8, SMP1, SMP2**

 2.MD.8 **Spiral Snapshot**

GMC Solve problems involving coins and bills.

| 2-5
Practice | 2-8
Practice | 3-10
Practice | 3-11
Focus
Practice | 5-2
Focus
Practice | 5-3
Focus
Practice | 5-4
Warm Up
Focus
Practice | 5-11
Focus
Practice |

 Spiral Tracker   **Go Online**   to see how mastery develops for all standards within the grade.

③ Practice 15–20 min		
Playing *Spinning for Money* **Game** Children practice coin and bill exchanges.	*Math Journal 2*, p. 108; *Math Masters*, p. G10; per group: pencil, large paper clip, one $1 bill, one sheet of paper labeled "Bank"; per player: 7 pennies, 5 nickels, 5 dimes, 4 quarters	**2.MD.8** **SMP5**
Math Boxes 5-2 Children practice and maintain skills.	*Math Journal 2*, p. 105; centimeter ruler	See page 455.
Home Link 5-2 **Homework** Children draw coins and bills to pay for items.	*Math Masters*, p. 122	**2.OA.2, 2.MD.8**

 **connectED.mheducation.com**

Plan your lessons online with these tools.

 ePresentations   Student Learning Center   Facts Workshop Game   eToolkit   Professional Development   Home Connections   Spiral Tracker   Assessment and Reporting   ELL English Learners Support   Differentiation Support

Differentiation Options RtI

Readiness   5–15 min

Dime-Nickel-Penny Grab

Activity Card 64;
Math Masters, p. G22;
toolkit coins (20 pennies,
8 nickels, and 10 dimes)

WHOLE CLASS
SMALL GROUP
PARTNER
INDEPENDENT

For support counting coins and
determining the total value of various coin
combinations, children grab coins and
count the amount. They record their work
on *Math Masters*, page G22.

Enrichment   5–15 min

Writing Number Stories with Money

Math Journal 2, p. 106;
Math Masters, p. TA7

WHOLE CLASS
SMALL GROUP
PARTNER
INDEPENDENT

To further explore making purchases
with exact change, children write number
stories on *Math Masters*, page TA7
involving items from Pine School's Fruit
and Vegetable Sale on journal page 106.
Encourage children to buy multiple items
and pay for them with $1 or more. Partners
exchange number stories to solve. You
might want to suggest that children make
the stories into a class book.

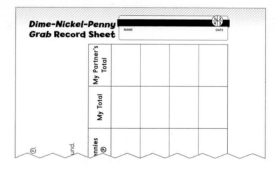

A Number Story

Unit

TA7

Extra Practice   5–15 min

Buying and Selling

Activity Card 65;
Math Masters, p. 120;
toolkit coins; scissors

WHOLE CLASS
SMALL GROUP
PARTNER
INDEPENDENT

For practice counting money, children
use their toolkit coins to buy fruit and
vegetables using exact change. GMP5.2

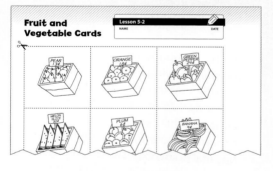

Fruit and Vegetable Cards

English Language Learners Support

Beginning ELL To prepare students for making trades with coins, demonstrate the
meaning of the term *trade* as "to exchange for something equivalent." Role-play making
trades with objects that look different but have equivalent values. For example, trade
10 cubes for 1 long or 10 longs for 1 flat. Alternatively, trade objects that look different but
are the same size, such as three single-subject notebooks for a three-subject notebook.

Go Online   ELL English Learners Support

SMP1 Make sense of problems and persevere in solving them.
   GMP1.5 Solve problems in more than one way.

SMP2 Reason abstractly and quantitatively.
   GMP2.1 Create mathematical representations using numbers, words, pictures, symbols, gestures, tables, graphs, and concrete objects.

SMP3 Construct viable arguments and critique the reasoning of others.
   GMP3.1 Make mathematical conjectures and arguments.

NOTE Throughout this and other lessons involving money, be sure to use different money notations, such as 25¢ and $0.25 or 100¢, $1, and $1.00. Although children are not yet expected to use the $ symbol and the decimal point, exposure to various money notations in class prepares children for exposure to them in everyday life. In addition, learning to use both $ and ¢ symbols correctly is a Grade 2 goal.

1 Warm Up   15–20 min   Go Online   ePresentations   eToolkit

▶ Mental Math and Fluency

Have children count by 5s, 10s, and 25s. Children may refer to the Number-Grid Poster or the Class Number Scroll as they count. *Leveled exercises:*

● ○ ○ Begin at 5. Count by 5s to 100. 5, 10, 15, 20, . . . , 100
         Begin at 20. Count by 10s to 100. 20, 30, 40, 50, . . . , 100
         Begin at 0. Count by 25s to 150. 0, 25, 50, 75, . . . , 150

● ● ○ Begin at 50. Count by 5s to 150. 50, 55, 60, 65, . . . , 150
         Begin at 140. Count by 10s to 250. 140, 150, 160, 170, . . . , 250
         Begin at 100. Count by 25s to 300. 100, 125, 150, 175, . . . , 300

● ● ● Begin at 115. Count by 5s to 200. 115, 120, 125, 130, . . . , 200
         Begin at 153. Count by 10s to 253. 153, 163, 173, 183, . . . , 253
         Begin at 175. Count by 25s to 350. 175, 200, 225, 250, . . . , 350

▶ Daily Routines

Have children complete daily routines. See the Planning Ahead note in Lesson 5-1 for an adjustment to the Temperature Routine that will prepare children for Lesson 5-10.

2 Focus   30–40 min   Go Online   ePresentations   eToolkit

▶ Math Message

Take 10 Ⓟ, 6 Ⓝ, 6 Ⓓ, and 4 Ⓠ from your toolkit. How much money is that?
200¢, $2, or $2.00

▶ Finding the Total

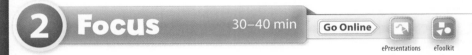

| WHOLE CLASS | SMALL GROUP | PARTNER | INDEPENDENT |

Math Message Follow-Up Ask children to share their strategies and justify their answer for finding the total value of the coins. GMP3.1
Model the following strategies:

• Count the coins in order of value, from largest to smallest. First count the 4 quarters by 25s to reach 100¢, or $1. Then count the 6 dimes by 10s to reach 160¢, or $1.60. Then count the 6 nickels by 5s to reach 190¢, or $1.90. Finally, count the 10 pennies by 1s for a total of 200¢, or $2.

• Make exchanges: 10 pennies for a dime, 4 quarters for a $1 bill, and so on.

▶ Reviewing Money Equivalencies

My Reference Book, pp. 110–111

| WHOLE CLASS | SMALL GROUP | PARTNER | INDEPENDENT |

With the class, read about money and **equivalencies** in *My Reference Book,* pages 110–111. Have children respond in unison to the following questions about the coins and the $1 bill.

Display a nickel. Ask: *What is this called?* A nickel *How much is it worth?* 5 cents Write *nickel* and *5 cents* on the Class Data Pad. Ask: *How much are 2 nickels worth?* 10 cents Repeat with a penny, a dime, a quarter, and a $1 bill.

Pose questions about coin equivalencies and record them in a Table of Equivalencies on the Class Data Pad or another display. (*See margin.*)

- *How many pennies would you trade for a nickel?* 5 pennies
- *How many pennies would you trade for a dime?* 10 pennies *For a quarter?* 25 pennies *For 1 dollar?* 100 pennies *For 2 dollars?* 200 pennies
- *How many nickels would you trade for a dime?* 2 nickels *For a quarter?* 5 nickels *For a dollar?* 20 nickels
- *How many dimes would you trade for a dollar?* 10 dimes

Explain that children will exchange coins while buying and selling items at Pine School's Fruit and Vegetable Sale.

▶ Buying and Selling

Math Journal 2, pp. 106–107; *Math Masters,* p. 121

| WHOLE CLASS | SMALL GROUP | PARTNER | INDEPENDENT |

Display the Pine School's Fruit and Vegetable Sale poster from *Math Masters,* page 121. Ask children to count out the coins they would use to pay for one pear. Partners check each other's coin combinations.

Ask volunteers to share the coin combinations they used. List their responses. GMP1.5 Here are four possible coin combinations for 13¢:

Ⓟ Ⓟ Ⓟ Ⓟ Ⓟ   Ⓓ Ⓟ   Ⓝ Ⓟ   Ⓝ Ⓟ Ⓟ
Ⓟ Ⓟ Ⓟ Ⓟ Ⓟ   Ⓟ   Ⓝ Ⓟ   Ⓟ Ⓟ Ⓟ
Ⓟ Ⓟ Ⓟ   Ⓟ   Ⓟ   Ⓟ Ⓟ Ⓟ

Ask children whether there are other ways they can pay for the pear. GMP1.5 Sample answer: I can pay for the pear with 1 dime and 1 nickel and get 2 pennies in change. Repeat the activity with other fruits and vegetables.

Table of Equivalencies

1 Ⓝ	=	5 Ⓟ
1 Ⓓ	=	10 Ⓟ
1 Ⓠ	=	25 Ⓟ
$1	=	100 Ⓟ

Math Masters, p. 121

Pine School's Fruit and Vegetable Sale

Lesson 5-2

one hundred twenty-one 121

Math Journal 2, p. 107

Buying Fruit and Vegetables at the Sale

Lesson 5-2
DATE

For Problems 1–2 follow these steps:
- Write the name and the cost of one item from journal page 106 that you want to buy.
- Draw the coins you can use to pay for it using Ⓟ, Ⓝ, Ⓓ, and Ⓠ.
- Draw another way to pay using a different combination of coins.

For Problems 3–4 write the names of two or more items you want to buy and how much they cost. Draw coins for the total amount of money you would spend. Then write the total. **Sample answers:**

I Bought	It Costs	I Paid	I Paid
Example: Orange	18¢	ⒹⓃ ⓅⓅⓅ	ⓃⓃⓅⓅ ⓅⓅⓅⓅ
1. Melon slice	30¢	⒬Ⓝ	ⒹⒹⓃ ⓅⓅⓅⓅⓅ
2. Lettuce	45¢	⒬ⓃⓃⓃⓃ	ⒹⒹⒹⒹⓃ
3. Apple	12¢	ⓃⓃⓅⓅ	ⒹⓅⓅⓅⓅⓅ ⓅⓅⓅ
Plum	6¢	ⓃⓅ	Total: 18¢
4. Corn	15¢	ⒹⓃ	ⒹⒹⒹ ⓃⓃⓅ
Banana	9¢	ⓃⓅⓅⓅⓅ	ⒹⒹⒹ ⓃⓃⓅ
Orange	18¢	ⒹⓃⓅⓅⓅ	Total: 42¢

2.NBT.2, 2.MD.8, SMP1, SMP2   one hundred seven   107

Math Journal 2, p. 108

Spinning for Money

Lesson 5-2
DATE

Materials	☐ *Spinning for Money* Spinner
	☐ pencil ☐ large paper clip
	☐ one sheet of paper labeled "Bank"
	☐ 7 pennies, 5 nickels, 5 dimes, and 4 quarters for each player
	☐ one $1 bill for the group
Players	2, 3, or 4
Skill	Exchange coins and dollar bills
Object of the Game	To be first to exchange for a $1 bill

How to Play

1. Each player puts 7 pennies, 5 nickels, 5 dimes, and 4 quarters into the bank. Each group puts one $1 bill into the bank.

2. Players take turns spinning the *Spinning for Money* Spinner and taking the coins shown by the spinner from the bank.

3. Whenever possible, players exchange coins for a single coin or bill of the same value. For example, a player could exchange 5 pennies for a nickel or 2 dimes and 1 nickel for a quarter.

4. The first player to exchange for a $1 bill wins.

Use a large paper clip and pencil to make a spinner.

108   one hundred eight   2.MD.8, SMP5

Then have partners take turns being customer and clerk at Pine School's Fruit and Vegetable Sale. The customer points to an item on journal page 106 and pays the exact amount with coins. The clerk checks that the customer has paid the correct amount. Partners each record four transactions on journal page 107, showing two possible combinations for each transaction. **GMP2.1, GMP1.5**

Summarize   Have volunteers share some of their answers and discuss the possible coin combinations. At the end of the activity, have children separate their coins and return them to their toolkits.

Differentiate   **Adjusting the Activity**

- To support children's thinking, have them count out the number of pennies for an item. Then have them make any possible exchanges for nickels, dimes, and finally quarters.

- To extend children's thinking, set parameters for coin combinations. For example, have them make and record coin combinations with the fewest possible coins or coin combinations without using nickels.

Go Online Differentiation Support

✓ Assessment Check-In   CCSS 2.NBT.2, 2.MD.8

Math Journal 2, p. 107

Expect most children to be successful with Problems 1–2. **GMP1.5, GMP2.1** For children who struggle showing two different coin combinations for the same amount, consider using the first suggestion in the Adjusting the Activity note. Because this is children's first exposure to combining two or more sets of coins to find the total, do not expect all children to be successful with Problems 3–4.

✓ Assessment and Reporting   Go Online ▷ to record student progress and to see trajectories toward mastery for these standards.

③ Practice   15–20 min   Go Online

ePresentations   eToolkit   Home Connections

▶ Playing *Spinning for Money*

Math Journal 2, p. 108; *Math Masters*, p. G10

WHOLE CLASS	**SMALL GROUP**	**PARTNER**	INDEPENDENT

Players put pennies, nickels, dimes, quarters, and a $1 bill into the bank. They take turns spinning the *Spinning for Money* spinner and taking the coins indicated by the spinner from the bank.

Whenever possible, players should exchange coins of a smaller denomination for a coin or bill of a larger domination that has the same value. GMP5.2 The first player to exchange for the $1 bill wins.

NOTE Ask children to write their names on the back of their *Spinning for Money* spinners. Collect the spinners for future use.

▶ Math Boxes 5-2

Math Journal 2, p. 105

| WHOLE CLASS | SMALL GROUP | PARTNER | INDEPENDENT |

Mixed Practice Math Boxes 5-2 are paired with Math Boxes 5-4.

▶ Home Link 5-2

Math Masters, p. 122

Homework Children use store advertisements to find items that cost less than $2. They draw coins and bills to show a way to pay for each item.

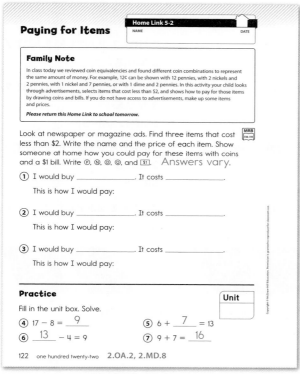

Math Masters, p. 122

Math Masters, p. G10

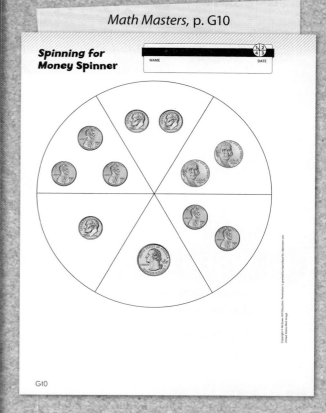

Math Masters, p. G10

Math Journal 2, p. 105

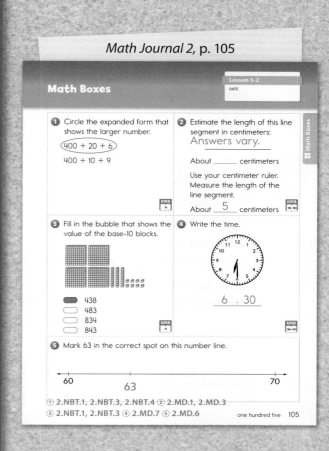

Counting Up with Money

Overview Children find coin combinations to pay for items and make change by counting up.

 Common Core State Standards

Focus Clusters
- Understand place value.
- Use place value understanding and properties of operations to add and subtract.
- Work with time and money.

1 Warm Up 15–20 min

Materials

	Materials	
Mental Math and Fluency Children solve number stories.	slate	2.OA.1, 2.OA.2
Daily Routines Children complete daily routines.	See pages 4–43.	See pages xiv–xvii.

2 Focus 30–40 min

Math Message Children draw coins to show the cost of an item.	slate	2.NBT.2, 2.NBT.5, 2.NBT.7, 2.MD.8
Making Change Children make change by counting up from the cost of an item to the amount paid.	*Math Journal 2*, p. 106; toolkit coins	2.NBT.2, 2.NBT.5, 2.NBT.7, 2.MD.8 SMP1, SMP5
Going Shopping Children make change by counting up in a shopping activity and record their transactions.	*Math Journal 2*, p. 109; toolkit coins	2.NBT.2, 2.NBT.5, 2.NBT.7, 2.MD.8 SMP5
✓ **Assessment Check-In** See page 460.	*Math Journal 2*, p. 109	2.NBT.2, 2.NBT.7, 2.MD.8, SMP5

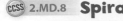 2.MD.8 **Spiral Snapshot**

GMC Solve problems involving coins and bills.

2-5 Practice	2-8 Practice	3-10 Practice	3-11 Focus Practice	5-2 Focus Practice	5-3 Focus Practice	5-4 Warm Up Focus Practice	5-11 Focus Practice

 Spiral Tracker 〈 Go Online 〉 to see how mastery develops for all standards within the grade.

3 Practice 15–20 min

Playing *Salute!* **Game** Children find missing addends.	*My Reference Book*, pp. 162–163; per group: 4 each of number cards 0–10	2.OA.2 SMP6
Math Boxes 5-3 Children practice and maintain skills.	*Math Journal 2*, p. 110; centimeter ruler	See page 461.
Home Link 5-3 **Homework** Children make change by counting up.	*Math Masters*, pp. 124–125	2.OA.2, 2.NBT.2, 2.NBT.5, 2.NBT.7, 2.MD.8

 connectED.mheducation.com

Plan your lessons online with these tools.

 ePresentations   Student Learning Center   Facts Workshop Game   eToolkit   Professional Development   Home Connections Spiral Tracker   Assessment and Reporting   English Learners Support Differentiation Support

Differentiation Options RtI

CCSS 2.NBT.2

Readiness — 5–15 min

Using Dice to Count Up

| WHOLE CLASS |
| SMALL GROUP |
| **PARTNER** |
| INDEPENDENT |

Activity Card 66,
per partnership: 2 dice

For practice counting up, partners use dice to generate numbers. After modeling the activity and assigning partnerships, circulate and ask children to count aloud to make sure they are using the counting-up strategy.

CCSS 2.NBT.2, 2.NBT.5, 2.NBT.7, 2.MD.8

Enrichment — 5–15 min

Solving a Coin Puzzle

| WHOLE CLASS |
| SMALL GROUP |
| PARTNER |
| **INDEPENDENT** |

Math Masters, p. 123

To apply their understanding of coin values and coin combinations, children solve coin puzzles on *Math Masters*, page 123. After they complete the page, children write their own coin puzzles on the back of the page.

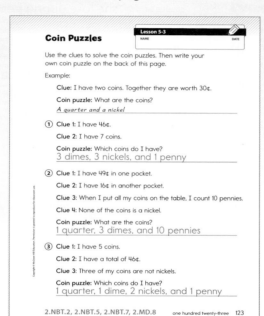

CCSS 2.NBT.2, 2.NBT.5, 2.NBT.7, 2.MD.8

Extra Practice — 5–15 min

Making Change by Counting Up

| WHOLE CLASS |
| SMALL GROUP |
| **PARTNER** |
| INDEPENDENT |

Activity Card 67;
Math Masters, p. 121; scissors;
toolkit coins

For additional practice making change by counting up, partners take turns buying and selling items at Pine School's Fruit and Vegetable Sale.

English Language Learners Support

Beginning ELL In this lesson children take turns being both a customer and a clerk to practice making change. To prepare them for acting in these roles, review polite phrases used in conversations between customers and clerks, such as the following:

- Thank you.
- How much is that?
- Here is your change.
- You're welcome.

- Please.
- Your bill is _____.
- Your change is _____.

 ELL English Learners Support

1 Warm Up 15–20 min [Go Online] ePresentations eToolkit

▶ Mental Math and Fluency

Pose number stories for children to solve on their slates. Have children share their strategies. *Leveled exercises:*

- ●○○ Tommy has passed 7 levels on his video game. To win, he must pass 11 levels in all. How many more levels must Tommy pass to win the game? 4 levels
- ●●○ Alex has some flowers in a vase. He takes out 7 flowers. Then there are 6 flowers left in the vase. How many flowers were in the vase before he took some out? 13 flowers
- ●●● Fay started her golf game with 14 golf balls. She lost some while playing. Now she has only 8 golf balls left. How many golf balls did she lose during her game? 6 golf balls

▶ Daily Routines

Have children complete daily routines. See the Planning Ahead note in Lesson 5-1 for an adjustment to the Temperature Routine that will prepare children for Lesson 5-10.

2 Focus 30–40 min [Go Online] ePresentations eToolkit

▶ Math Message

You want to buy a toy that costs 48¢. Which coins would you use to pay for it? Draw the coins on your slate. Use some or all of these symbols: Ⓟ, Ⓝ, Ⓓ, *and* Ⓠ.

▶ Making Change

Math Journal 2, p. 106

| WHOLE CLASS | SMALL GROUP | PARTNER | INDEPENDENT |

Math Message Follow-Up Ask several children to share their solutions. Explain that any group of coins that has a total value of 48¢ is a correct solution. Ask: *How can you find the fewest possible coins needed to buy the toy?* Sample answer: I start with the coin with the highest value, a quarter. I can use 1 quarter. Then I count on from 25¢ with the coin with the next highest value. I can use 2 dimes, so I count 25¢, 35¢, 45¢. I cannot use any nickels. So I count on from 45¢ with pennies: 45¢, 46¢, 47¢, 48¢. The fewest possible coins needed to buy the toy is six: 1 quarter, 2 dimes, and 3 pennies.

Math Journal 2, p. 106

Pine School's Fruit and Vegetable Sale Lesson 5-2 DATE

PEAR 13¢ ORANGE 18¢ GREEN PEPPER 24¢
MELON SLICE 30¢ PLUM 6¢ BANANA 9¢
APPLE 12¢ TOMATO 20¢ ONION 7¢
LETTUCE 45¢ CORN 15¢ CABBAGE 40¢

106 one hundred six

Point out that there are times when you do not have exact change when buying an item, so the clerk needs to give you back the correct amount of change. Tell children that in today's lesson they will learn how to make change by counting up to find the correct amount.

Have children turn to Pine School's Fruit and Vegetable Sale on journal page 106. Then pose the following problem. Say: *I am buying an orange. I give the clerk 2 dimes. How much change should the clerk give back?* Demonstrate how to make change by counting up:

- Start with the cost of the item: 18¢.
- Count up to the amount of money used to pay for the item: 20¢. Say "19, 20" while putting down 2 pennies, one at a time.
- Display the transaction as follows:

I Bought	It Cost	I Paid	My Change Was
an orange	18¢	ⒹⒹ	⒫⒫

Point out that in the sample problem, a child could have paid for the orange with a quarter. In that case the change could have been 1 nickel and 2 pennies or 7 pennies. Explain that usually a single purchase can be made and change can be given using various coin combinations.

Pose a few more similar problems. Have children help you determine the change and record the transactions in the display.

Have partners take turns being the customer and the clerk. The customer selects an item and pays for it by giving coins to the clerk with a total value greater than the price of the item. The clerk counts up to make change. GMP5.2 Have children share several transactions with the class. Add the transactions to the table and compare how they are alike and different. GMP1.6

Differentiate **Adjusting the Activity**

To help children who are struggling to make change, present examples where the clerk gives no more than 4 pennies in change. Then move on to making change with nickels, dimes, and quarters.

Guide children to write number models for transactions:

- You pay for a 13¢ pear with 2 dimes. Write the number model 13¢ + ? = 20¢ to show counting up to make the change. The cost is 13¢, and the amount you paid is 20¢, so the change is 7¢.

- You could show the same transaction with the number model 20¢ − 13¢ = ?

Go Online ▸ 👥 Differentiation Support

Academic Language Development

Children may be familiar with the phrase *make change* as meaning "to make an equal exchange"—for example, making change for a dollar by exchanging it for 4 quarters. Explain that *make change* also can mean "to give the difference between the price of an item and the value of the coins or bills used to pay for it." Use shopping simulations to demonstrate the meaning of *making change* or *giving change* as related to subtraction.

Differentiate Some children may choose combinations of coins that have a value significantly greater than the value of the item. It does not make sense to pay for a 7¢ onion with two quarters. Have children look at the value of the item and their coins. Ask: *Do you have one coin that will cover the cost? What combination of coins is worth only a little more than the price of the item?*

Go Online | Differentiation Support

▶ Going Shopping

Math Journal 2, p. 109

| WHOLE CLASS | SMALL GROUP | **PARTNER** | INDEPENDENT |

Partners continue the shopping activity. They take turns being the customer and the clerk. Each child records a few transactions as customer in their journals, using the following symbols: (P), (N), (D), (Q), and $1 .
GMP5.2

✓ **Assessment Check-In** CCSS 2.NBT.2, 2.NBT.7, 2.MD.8

Math Journal 2, p. 109

This lesson is an introduction to making change, so do not expect most children to do it successfully. Do expect most children to select coins or bills with a value greater than the cost of the item. **GMP5.2** Instruct children who select coins with a value *less than* the price of an object to compare the value of the coins they selected to the cost. Ask questions such as the following: *Is the amount you are paying for the item more or less than the cost of the item? Do you have enough to pay for the item? What coin(s) do you need to make more than the cost of the item?* For children who select coins with a value significantly greater than the price of an item, consider the suggestion in the Common Misconception note.

✓ Assessment and Reporting | **Go Online** to record student progress and to see trajectories toward mastery for these standards.

Summarize Invite partners to share the transactions they recorded on journal page 109. Then have children separate their coins and bills and return them to their toolkits.

③ Practice 15–20 min

Go Online

ePresentations · eToolkit · Home Connections

▶ Playing *Salute!*

My Reference Book, pp. 162–163

| WHOLE CLASS | **SMALL GROUP** | PARTNER | INDEPENDENT |

Have children play *Salute!* See Lesson 3-4 for detailed directions.

Observe

- How are children figuring out their numbers (the missing addends)?
- Do children understand the relationship between the numbers?

Discuss

- *How did you figure out the number on your card?*
- *When was it easy to figure out the number? When was it hard?* **GMP6.4**

Math Journal 2, p. 109

Shopping at Pine School's Fruit and Vegetable Sale

Lesson 5-3
DATE

Price per Item					
Pear	13¢	Melon slice	30¢	Lettuce	45¢
Orange	18¢	Apple	12¢	Green pepper	24¢
Banana	9¢	Tomato	20¢	Corn	15¢
Plum	6¢	Onion	7¢	Cabbage	40¢

Complete the table. Sample answers:

I Bought	It Costs	I Paid	I Got in Change
Example: One Melon slice	**30¢**	©©	**20¢**
1. One Pear	13¢	⒟⒟	7¢
2. One Apple	12¢	⒟Ⓝ	3¢
3. One Green pepper	24¢	©	1¢
Try This			
4. One Lettuce	45¢	$1	
and one Cabbage	40¢		15¢

2.NBT.2, 2.NBT.5, 2.NBT.7, 2.MD.8, SMP5   one hundred nine 109

▶ Math Boxes 5-3 ✏️

Math Journal 2, p. 110

WHOLE CLASS	SMALL GROUP	PARTNER	INDEPENDENT

Mixed Practice Math Boxes 5-3 are paired with Math Boxes 5-1.

▶ Home Link 5-3

Math Masters, pp. 124–125

Homework Children pretend to have a garage sale and price small items at 25¢ or less. They show someone at home how they make change by counting up.

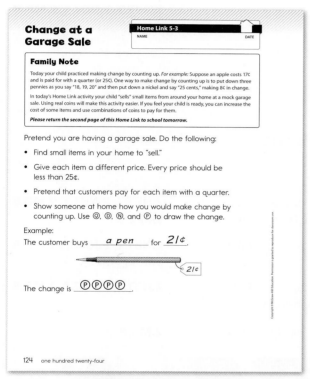

Math Masters, p. 124

Math Journal 2, p. 110

Math Boxes   Lesson 5-3
                                   DATE

① Estimate the length of this line segment:

About _____ centimeters

Use your centimeter ruler to measure the length of the line segment.

About __6__ centimeters

② Write the number word for 79.

Seventy-nine

③ Use the digits 7, 1, and 9 to make the smallest number possible.

179

Use the same digits to make the largest number possible.

971

④ Count back by 10s.

340, __330__ __320__ __310__ __300__ __290__

⑤ **Writing/Reasoning** Suppose you measure the line segment in Problem 1 in inches. Will the number of inches be more than or less than the number of centimeters? Explain.

Sample answer: The number will be less. Inches are longer than centimeters, so fewer of them are needed.

① 2.MD.1, 2.MD.3 ② 2.NBT.3
③ 2.NBT.1, 2.NBT.3, 2.NBT.4 ④ 2.NBT.2
⑤ 2.MD.1, 2.MD.2, SMP5

110 one hundred ten

Math Masters, p. 125

Change at a Garage Sale (continued)   Home Link 5-3
                                                 NAME   DATE

① The customer buys _____ for _____.   Answers vary.

The change is _____.

② The customer buys _____ for _____.

The change is _____.

③ The customer buys _____ for _____.

The change is _____.

④ The customer buys _____ for _____.

The change is _____.

Practice

Fill in the unit box. Solve.

⑤ 11 − __3__ = 8

⑥ 8 + __7__ = 15

⑦ __7__ + 7 = 14

⑧ 13 − 8 = __5__

Unit

2.OA.2, 2.NBT.2, 2.NBT.5, 2.NBT.7, 2.MD.8   one hundred twenty-five 125

Coin Calculations

Overview Children make purchases and practice making change.

▶ **Before You Begin**
For Part 2 decide how you will display the Vending Machine poster from *Math Masters,* page TA24.

 Common Core State Standards

Focus Clusters
- Understand place value.
- Use place value understanding and properties of operations to add and subtract.
- Work with time and money.

① Warm Up 15–20 min

	Materials	
Mental Math and Fluency Children show combinations of coins used to pay for items.	slate	2.NBT.2, 2.NBT.7, 2.MD.8
Daily Routines Children complete daily routines.	See pages 4–43.	See page xiv–xvii.

② Focus 30–40 min

Math Message Children prepare to discuss the milk and juice vending machine.	*Math Journal 2,* p. 1112	SMP2
Buying Items with Exact Change Children show coin combinations to pay for items.	*Math Journal 2,* p. 112; *Math Masters,* p. TA24; slate	2.NBT.2, 2.NBT.7, 2.MD.8 SMP2
Buying Items without Exact Change Children use strategies to make change.	*Math Journal 2,* p. 112; toolkit coins; slate	2.NBT.2, 2.NBT.7, 2.MD.8 SMP2, SMP5
Making Vending Machine Purchases Children buy items from a vending machine.	*Math Journal 2,* pp. 112–113; toolkit coins (optional)	2.NBT.2, 2.NBT.7, 2.MD.8
✓ **Assessment Check-In** See page 466.	*Math Journal 2,* pp. 112–113	2.MD.8

 2.MD.8 **Spiral Snapshot**

GMC Read and write monetary amounts.

| Routine 1 | 1-11 Focus Practice | 2-1 Focus | 3-11 Focus Practice | 5-2 Focus Practice | 5-4 Focus Practice | 5-11 Focus Practice | |

 **Spiral Tracker**   **Go Online** ▷ to see how mastery develops for all standards within the grade.

③ Practice 20–25 min

Playing *Target* to 50 **Game** Children model addition and subtraction using base-10 blocks.	per player: *Math Masters,* pp. G19 and G20; per partnership: base-10 blocks (10 longs and 30 cubes), 4 each of number cards 0–9	2.NBT.1, 2.NBT.3, 2.NBT.7 SMP2
Math Boxes 5-4 Children practice and maintain skills.	*Math Journal 2,* p. 111; centimeter ruler	See page 467.
Home Link 5-4 **Homework** Children practice making change.	*Math Masters,* p. 127	2.OA.2, 2.NBT.2, 2.NBT.7, 2.MD.8

connectED.mheducation.com

Plan your lessons online with these tools.

ePresentations | Student Learning Center | Facts Workshop Game | eToolkit | Professional Development | Home Connections | Spiral Tracker | Assessment and Reporting | English Learners Support | Differentiation Support

Differentiation Options

RtI

Readiness
5–15 min

WHOLE CLASS
SMALL GROUP
PARTNER
INDEPENDENT

Counting on the Number Grid with Coins

Math Masters, p. TA3; toolkit coins

To explore counting by 5s, 10s, and 25s using a visual model, children count up along with you as you place coins on a number grid. Start at 0 and count up by nickels, then dimes, and then quarters. As children become more comfortable counting by 5s, 10s, and 25s, start at a number other than 0. As children progress further, begin to count up by quarters, then switch to dimes, and then to nickels.

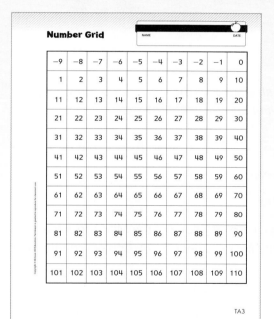

Number Grid

−9	−8	−7	−6	−5	−4	−3	−2	−1	0
1	2	3	4	5	6	7	8	9	10
11	12	13	14	15	16	17	18	19	20
21	22	23	24	25	26	27	28	29	30
31	32	33	34	35	36	37	38	39	40
41	42	43	44	45	46	47	48	49	50
51	52	53	54	55	56	57	58	59	60
61	62	63	64	65	66	67	68	69	70
71	72	73	74	75	76	77	78	79	80
81	82	83	84	85	86	87	88	89	90
91	92	93	94	95	96	97	98	99	100
101	102	103	104	105	106	107	108	109	110

TA3

Enrichment
5–15 min

WHOLE CLASS
SMALL GROUP
PARTNER
INDEPENDENT

Calculating the Value of a Name

Activity Card 68, paper, toolkit coins, calculator (optional)

To apply their understanding of coin values and coin combinations, children calculate the value of their own names. They find the value of each letter (based on a key) and then calculate the total. Then they use toolkit coins to represent the total value of their names.

Calculating the Value of a Name (68)

My name is Dan. It has the value of 19¢.

What You Need
paper
toolkit coins
calculator (optional)

What To Do
1. Make a key for the value of each letter in the alphabet. Give the letter a value based on its order in the alphabet. **Example:** A = 1¢, B = 2¢, C = 3¢, and so on.
2. Write your name. Use the key to write the value of each letter in your name.
3. Find the total value of your name. (You can use a calculator.)
4. Use toolkit coins to show the total value of your name with coins.

Talk About It
Compare the value of your name to the value of someone else's name.

More You Can Do
Find words with a total value of $1.00.

2.NBT.2, 2.NBT.7, 2.MD.8

Extra Practice
5–15 min

WHOLE CLASS
SMALL GROUP
PARTNER
INDEPENDENT

Practicing Making Change

Math Masters, p. 126; toolkit coins and bills

For additional practice making change, children complete *Math Masters,* page 126. Children select an item, show an amount used to pay for it, and show the amount of change.

Practicing Making Change | Lesson 5-4

Snack List

Applesauce	45¢	Popcorn	63¢
Banana	50¢	Raisins	43¢
Milk	86¢	Yogurt	70¢
Orange	62¢	Carrots	38¢

- Choose an item from the Snack List. Write the name in the table.
- Write the cost of the item.
- Use your toolkit money. Pay with coins or a $1 bill. Use $1, Q, D, N, and P to show how you pay.
- Count up to make change.
- For Problem 4, buy two items that cost less than $1 all together.

I Buy	It Costs	I Pay With	My Change
Example: applesauce	45 ¢	QQ	5¢
1.	¢		
2.	¢	Answers vary.	
3.	¢		
Try This 4. and	¢		
	¢	$1	

126 one hundred twenty-six   2.NBT.2, 2.NBT.7, 2.MD.8

English Language Learners Support

Beginning ELL If possible, use a toy vending machine (Kindergarten teachers may have one) to introduce the terms shown on the picture on journal page 112. Alternatively, make a large copy of *Math Masters,* page TA24. Show children how to push the *push* button, where to put in *$1 bills,* where to put in *coins,* and where the *change* comes out. Then provide practice with Total Physical Response commands. Say: *Push the button for chocolate milk. Put in a $1 bill. Put in 35 cents. Get the change.*

Go Online ELL English Learners Support

Standards and Goals for
Mathematical Practice

SMP2 **Reason abstractly and quantitatively.**
GMP2.1 Create mathematical representations using numbers, words, pictures, symbols, gestures, tables, graphs, and concrete objects.

GMP2.3 Make connections between representations.

SMP5 **Use appropriate tools strategically.**
GMP5.2 Use tools effectively and make sense of your results.

Math Journal 2, p. 112

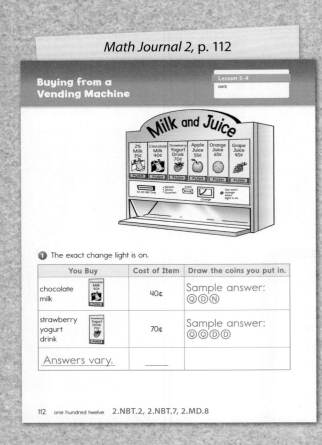

Buying from a Vending Machine	Lesson 5-4
	DATE

① The exact change light is on.

You Buy	Cost of Item	Draw the coins you put in.
chocolate milk	40¢	Sample answer: ⓆⒹⓃ
strawberry yogurt drink	70¢	Sample answer: ⓆⓆⒹ
Answers vary.		

1 **Warm Up** 15–20 min 〉Go Online ePresentations eToolkit

▶ Mental Math and Fluency

Children show how they could pay for each item by drawing Ⓝ, Ⓓ, and Ⓠ symbols on slates. *Leveled exercises:*

● ○ ○ A package of gum that costs 35¢ Sample answer: ⓆⒹ
● ● ○ An apple that costs 55¢ Sample answer: ⓆⓆⓃ
● ● ● A bottle of juice that costs 75¢ Sample answer: ⓆⓆⓆ

▶ Daily Routines

Have children complete daily routines. See the Planning Ahead note in Lesson 5-1 for an adjustment to the Temperature Routine that will prepare children for Lesson 5-10.

2 **Focus** 30–40 min 〉Go Online ePresentations eToolkit

▶ Math Message

Math Journal 2, p. 112

Turn to page 112 in your journal. What does the machine do? How does it work? Be ready to talk about your answers. GMP2.3

▶ Buying Items with Exact Change

Math Journal 2, p. 112; *Math Masters,* p. TA24

WHOLE CLASS	SMALL GROUP	PARTNER	INDEPENDENT

Math Message Follow-Up Display a copy of the Milk and Juice Vending Machine poster (*Math Masters,* page TA24) or have children turn to journal page 112. Ask:

• *What is this machine? How does it work?* Answers vary.
• *Which coins or bills can you use in the machine?* Nickels, dimes, quarters, and $1 bills
• *Can you buy something if you don't have the exact amount of coins or bills?* Yes, unless the "exact change" light is on. *What does the "exact change" light mean?* If it is on, the machine won't return change if the buyer puts in more money than an item costs.

Review the concept of making change: the buyer pays with coins or bills that add up to more than the cost of the item, and the vending machine gives back the money owed (the difference). Explain that in today's lesson children will buy items from the vending machine with and without exact change.

Academic Language Development *Exact change* may be confusing for some children because they may understand *change* as the coins or money you get in return when an item costs less than the amount of money tendered. Point out that *exact change* means that you must have the exact amount an item costs, and sometimes it can involve the use of bills.

Ask children to pretend that the exact change light is on and that they want to buy a can of grape juice. On their slates have them draw the coins they would use. **GMP2.1** Sample answers: 1 quarter and 2 dimes; 4 dimes and 1 nickel; 9 nickels Have children share their coin combinations as you display them. Ask: *How are these combinations the same?* Sample answer: They all show the same amount. *How are they different?* Sample answer: The combinations use different coins. Repeat for other items from the vending machine.

▶ **Buying Items without Exact Change**

Math Journal 2, p. 112

| WHOLE CLASS | SMALL GROUP | PARTNER | INDEPENDENT |

Ask children what happens when the exact change light is off. Sample answer: The buyer can put in more money than the cost of an item, and the machine will give change.

Ask children to pretend that they want to buy a carton of orange juice. Have them suggest various coin combinations they might use to pay with exact change. **GMP5.2** Sample answers: 2 quarters, 1 dime, and 1 nickel; 6 dimes and 1 nickel

Then ask children to pretend that they don't have the exact change to buy the juice. Have them take out 3 quarters from their toolkit coins. Ask: *What change would the machine give back?* Sample answer: 1 dime Have them show the transaction with coins. Ask: *Are there are other ways to pay for the juice without exact change?* Sample answers: Pay with a $1 bill and receive 35¢ change. Pay with 7 dimes and receive 1 nickel change. Have children act out these transactions with coins and record the change they get on slates. Record their responses. (*See next page.*) Repeat for other items as necessary. **GMP2.1**

Math Journal 2, p. 113

Buying from a Vending Machine (continued)

Lesson 5-4
DATE

② The exact change light is off.

You Buy	It Costs	Draw the coins or the $1 bill you put in.	How much change will you get?
2% milk	35¢	◎◎	15¢
chocolate milk	40¢	$1	60¢
Answers vary.	___		___
Answers vary.	___		___
Answers vary.	___		___

Try This

③ You want to buy a carton of orange juice with $1. Will you get back enough change to also buy a carton of 2% milk? Explain.

Sample answer: Yes. If I pay for the orange juice with $1, I will get 35¢ back, which is enough to buy a carton of milk.

2.NBT.2, 2.NBT.7, 2.MD.8   one hundred thirteen   113

Target Record Sheet

NAME   DATE

For each of your turns, record the number you make and the value you show with base-10 blocks on the *Target* Game Mat.

Turn	Number You Made	Value on the *Target* Game Mat
1		
2		
3		
4		
5		
6		
7		
8		
9		
10		

G19

Target Game Mat

NAME   DATE

Ones

Tens

Hundreds

G20

I Bought	I Paid	My Change Was
orange juice for 65¢	$1	QD	35¢
	QQQ	D	10¢
	DDDD DDD	N	5¢

▶ Making Vending Machine Purchases

Math Journal 2, pp. 112–113

| WHOLE CLASS | SMALL GROUP | **PARTNER** | INDEPENDENT |

Have partners work together to solve Problems 1–2 on journal pages 112–113. In Problem 1 the exact change light is on; in Problem 2 it is off. Children take turns buying items from the vending machine and checking each other's work. They may act out the transactions with toolkit coins if they wish.

Differentiate   Adjusting the Activity

Encourage children to write the value of each coin above the symbol before they calculate the total.

Go Online ▷   Differentiation Support

✓ **Assessment Check-In**   CCSS 2.MD.8

Math Journal 2, pp. 112–113

Expect most children to be successful using toolkit coins to show exact change in Problem 1 on journal page 112. Some children may be able to solve the problem without coins. Because this is an early exposure to buying items without exact change, expect that some children may need support with Problem 2.

✓ Assessment and Reporting   Go Online ▷   to record student progress and to see trajectories toward mastery for these standards.

Summarize Have children share their transactions for Problem 2 on journal page 113. At the end of the activity, have children separate their coins and return them to their toolkits.

3 Practice · 20–25 min

Go Online

ePresentations · eToolkit · Home Connections

▶ Playing *Target* to 50

Math Masters, pp. G19 and G20

| WHOLE CLASS | **SMALL GROUP** | **PARTNER** | INDEPENDENT |

Have children play *Target* to 50. See Lesson 4-7 for detailed directions. Children record their turns on the *Target* Record Sheet.

Observe

- Which children are correctly representing their numbers with base-10 blocks? **GMP2.1**
- Which children seem to have a strategy for deciding whether to make a 1- or a 2-digit number? To add or subtract their numbers?

Discuss

- *How did you decide whether to make a 1- or a 2-digit number? To add or subtract your number?* **GMP2.2**
- *How did you know when to make an exchange?* **GMP2.2**

▶ Math Boxes 5-4

Math Journal 2, p. 111

| WHOLE CLASS | **SMALL GROUP** | **PARTNER** | **INDEPENDENT** |

Mixed Practice Math Boxes 5-4 are paired with Math Boxes 5-2.

▶ Home Link 5-4

Math Masters, p. 127

Homework Children practice making change.

Math Journal 2, p. 111

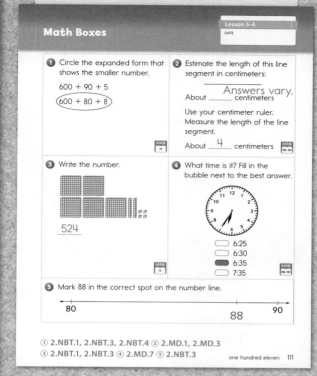

Math Boxes

Lesson 5-4
DATE

① Circle the expanded form that shows the smaller number.

600 + 90 + 5

(600 + 80 + 8)

② Estimate the length of this line segment in centimeters:

_____ Answers vary.

About _____ centimeters

Use your centimeter ruler. Measure the length of the line segment.

About __4__ centimeters

③ Write the number.

524

④ What time is it? Fill in the bubble next to the best answer.

○ 6:25
○ 6:30
● 6:35
○ 7:35

⑤ Mark 88 in the correct spot on the number line.

80 ———————————— 88 —— 90

① 2.NBT.1, 2.NBT.3, 2.NBT.4 ② 2.MD.1, 2.MD.3
③ 2.NBT.1, 2.NBT.3 ④ 2.MD.7 ⑤ 2.NBT.3   one hundred eleven 111

Math Masters, p. 127

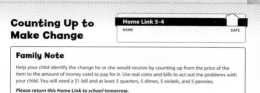

Counting Up to Make Change

Home Link 5-4
NAME
DATE

Family Note

Help your child identify the change he or she would receive by counting up from the price of the item to the amount of money used to pay for it. Use real coins and bills to act out the problems with your child. You will need a $1 bill and at least 3 quarters, 5 dimes, 5 nickels, and 5 pennies.

Please return this Home Link to school tomorrow.

Complete the table.

I Buy	It Costs	I Pay With	My Change
A box of raisins	70¢	◎◎◎	_5_ ¢
A box of crayons	65¢	$1	_35_ ¢
A pen	59¢	◎◎◎	_16_ ¢
An apple	45¢	◎◎◎◎◎	_5_ ¢
A notebook	73¢	◎◎◎◎Ⓝ	_2_ ¢
A ruler	48¢	$1	_52_ ¢
Answers vary.			___ ¢

Practice

Solve.

① 12 − _3_ = 9   ② 9 + _7_ = 16

③ _3_ + 8 = 11   ④ 14 − 8 = _6_

Unit

2.OA.2, 2.NBT.2, 2.NBT.7, 2.MD.8   one hundred twenty-seven 127

Exploring Arrays, Time, and Shapes

Overview Children make arrays, match clock faces to digital notation, and construct shapes on geoboards.

▶ **Before You Begin**

For the Math Message, display a 2-by-5 array (*see page 470*). For Exploration A you may want to make additional copies of *Math Masters,* page TA25. For Exploration B make two copies of *Math Masters,* page 130 on card stock for each small group.

▶ **Vocabulary**

array

CCSS **Common Core State Standards**

Focus Clusters
- Work with equal groups of objects to gain foundations for multiplication.
- Work with time and money.
- Reason with shapes and their attributes.

1 Warm Up 15–20 min

	Materials	
Mental Math and Fluency Children write numbers in expanded form.	slate	2.NBT.1, 2.NBT.3
Daily Routines Children complete daily routines.	See pages 4–43.	See pages xiv–xvii.

2 Focus 30–40 min

Math Message Children determine the number of dots in an array.		2.OA.4
Introducing Arrays Children skip count to find the number of dots in arrays and write addition number models to represent them.	slate	2.OA.4 SMP2
Exploration A: Making Arrays Children make arrays and write addition number models to represent them.	Activity Card 69; *Math Masters,* p. TA25; per partnership: two 6-sided dice, 36 centimeter cubes	2.OA.4 SMP2
Exploration B: Playing *Clock Concentration* Children match times shown on clock faces to digital notation.	Activity Card 70; *Math Journal 2,* p. 114; per group: 2 copies of *Math Masters,* p. 130 (see Before You Begin); envelope; scissors	2.MD.7
Exploration C: Making Shapes Children make shapes on geoboards.	Activity Card 71; *Math Masters,* p. 131; geoboard; rubber bands; straightedge	2.G.1 SMP7

3 Practice 15–20 min

Playing *Addition Top-It* **Game** Children practice addition facts.	*My Reference Book,* pp. 170–172; *Math Masters,* p. G18; per group: 4 each of number cards 0–10	2.OA.2, 2.NBT.4
Math Boxes 5-5 Children practice and maintain skills.	*Math Journal 2,* p. 115	See page 473.
Home Link 5-5 **Homework** Children match clock faces to digital notation.	*Math Masters,* p. 132	2.OA.2, 2.MD.7

Differentiation Options

RtI

CCSS 2.G.1, SMP7

Readiness
5–15 min

WHOLE CLASS
SMALL GROUP
PARTNER
INDEPENDENT

Identifying Pattern-Block Template Shapes

Math Masters, p. 128; Pattern-Block Template

To explore 3-, 4-, and 5-sided polygons using a concrete model, children use their Pattern-Block Templates to draw and identify shapes on *Math Masters,* page 128. When children have completed the page, encourage them to describe any similarities and differences among the shapes. **GMP7.1**

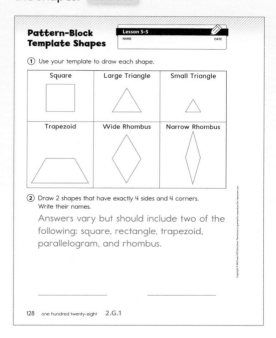

CCSS 2.G.1, SMP1

Enrichment
5–15 min

WHOLE CLASS
SMALL GROUP
PARTNER
INDEPENDENT

Working with Pattern-Block Puzzles

Math Masters, p. 129; pattern blocks; Pattern-Block Template

To extend their understanding of 2-dimensional shapes, children use pattern blocks to solve pattern-block puzzles on *Math Masters,* page 129. Invite children to compare their solutions. **GMP1.6**

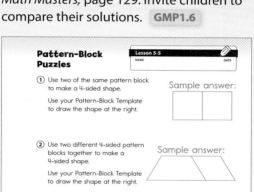

CCSS 2.MD.7

Extra Practice
5–15 min

WHOLE CLASS
SMALL GROUP
PARTNER
INDEPENDENT

Playing *Clock Concentration*

Activity Card 70; per group: 1 set of *Clock Concentration* Cards (*Math Masters,* p. 130)

For additional practice telling time to the nearest 5 minutes, children continue to play *Clock Concentration.*

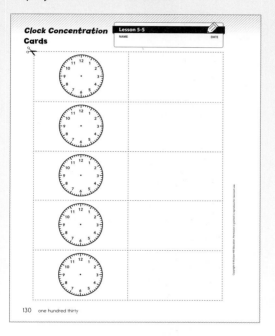

English Language Learners Support

Beginning ELL To prepare children to identify and describe shapes in terms of the number of sides, use Total Physical Response modeling and prompts to provide experiences hearing and using the terms _____-*sided shape* and *a shape with* _____ *sides. For example:* Point to a square, trace the sides with your finger, and say: *This is a 4-sided shape. It is a shape with 4 sides.* Then ask children to show you a 4-sided shape. Encourage children to use the term _____-*sided.* Continue with a variety of 3-, 4-, 5-, and 6-sided figures.

 Go Online   ELL English Learners Support

Standards and Goals for
Mathematical Practice

SMP2 **Reason abstractly and quantitatively.**

GMP2.1 Create mathematical representations using numbers, words, pictures, symbols, gestures, tables, graphs, and concrete objects.

GMP2.3 Make connections between representations.

SMP7 **Look for and make use of structure.**

GMP7.1 Look for mathematical structures such as categories, patterns, and properties.

1 Warm Up 15–20 min Go Online ePresentations eToolkit

▶ Mental Math and Fluency

Have children write numbers in expanded form on their slates.
Leveled exercises:

● ○ ○ **235** 200 + 30 + 5; **358** 300 + 50 + 8

● ● ○ **405** Possible answers: 400 + 5; 400 + 0 + 5; **399** 300 + 90 + 9

● ● ● **600** 600; **2,304** Possible answers: 2,000 + 300 + 4; 2,000 + 300 + 0 + 4

▶ Daily Routines

Have children complete daily routines. See the Planning Ahead note in Lesson 5-1 for an adjustment to the Temperature Routine that will prepare children for Lesson 5-10.

2 Focus 30–40 min Go Online ePresentations eToolkit

▶ Math Message

How many dots are there?

▶ Introducing Arrays

| WHOLE CLASS | SMALL GROUP | PARTNER | INDEPENDENT |

Math Message Follow-Up Explain that the arrangement of dots in rows and columns, as shown in the Math Message, is called an **array.** All of the rows have the same number of dots, and all of the columns have the same number of dots. Ask:

• *How many rows of dots are there?* 2

• *How many dots are in each row?* 5

• *How many dots are there in all?* 10

Have children share their strategies for finding the total number of dots. Guide them to conclude that skip counting by either 2s or 5s is more efficient than counting each individual dot. Arranging the dots in rows and columns, rather than leaving the dots scattered, makes it easier to skip count the dots.

Display the number sentence $5 + 5 = 10$. Explain that this addition number model uses equal addends to represent the array: 5 dots in the first row plus 5 dots in the second row is 10 dots in all. On slates, have children write a different addition number model that uses equal addends to represent the array. **GMP2.3** $2 + 2 + 2 + 2 + 2 = 10$

NOTE Because the number models express the total number of dots as a sum of equal addends, it doesn't matter whether children think of the array as 2 groups of 5 dots or 5 groups of 2 dots. Some children may write $1 + 1 + 1 + 1 + 1 + 1 + 1 + 1 + 1 + 1 = 10$, which expresses 10 as a sum of equal addends but does not reflect the 2-by-5 array structure. Arrays provide an important model for multiplication; working with them helps prepare children for later work with multiplication.

Display a 3-by-4 array. Have partners briefly discuss how they might skip count to find the total number of dots in the array. After a minute or two, have the class skip count by 4s and then by 3s. Then have children write addition number models for the array. **GMP2.1** $4 + 4 + 4 = 12$; $3 + 3 + 3 + 3 = 12$

Differentiate **Adjusting the Activity**

To support children who struggle with skip counting, illustrate the counts with hops on a number line.

Go Online ▸ 👥 Differentiation Support

After explaining the Explorations activities, assign groups to each one. Plan to spend most of your time with children working on Exploration A.

▶ Exploration A: Making Arrays

Activity Card 69; *Math Masters,* p. TA25

| WHOLE CLASS | **SMALL GROUP** | **PARTNER** | INDEPENDENT |

Partners complete the steps on Activity Card 69 to make arrays with centimeter cubes. They record at least five rectangular arrays on centimeter grid paper (*Math Masters,* page TA25). Below each array drawn, children write a number model to express the total number of cubes in the array as a sum of equal addends. **GMP2.1**

NOTE You may want to provide children with additional copies of the centimeter grid paper from *Math Masters,* page TA25.

Academic Language Development Children may initially think they are hearing "a ray" when they hear the term *array*. Have partners complete a 4-Square Graphic Organizer (*Math Masters,* page TA42) for the term *array*. They should illustrate an array, show examples (e.g., 2 rows of 6 eggs in an egg carton, 3 rows of 8 crayons in a box) and nonexamples, and write their own definition of the term.

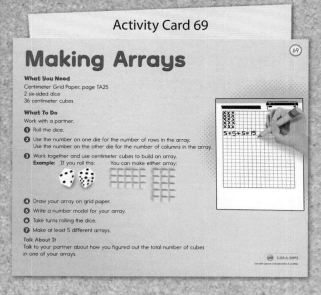

Activity Card 69

Making Arrays

What You Need
Centimeter Grid Paper, page TA25
2 six-sided dice
36 centimeter cubes

What To Do
Work with a partner.
1. Roll the dice.
2. Use the number on one die for the number of rows in the array. Use the number on the other die for the number of columns in the array.
3. Work together and use centimeter cubes to build an array.
   Example: If you roll this: You can make either array:
4. Draw your array on grid paper.
5. Write a number model for your array.
6. Take turns rolling the dice.
7. Make at least 5 different arrays.

Talk About It
Talk to your partner about how you figured out the total number of cubes in one of your arrays.

2.OA.A, SMP2

Math Journal 2, p. 114

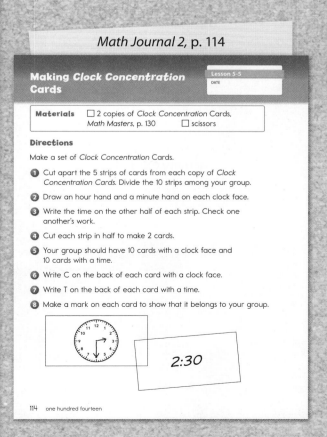

Making *Clock Concentration* Cards   Lesson 5-5
DATE

Materials   ☐ 2 copies of *Clock Concentration* Cards, *Math Masters,* p. 130   ☐ scissors

Directions

Make a set of *Clock Concentration* Cards.

1. Cut apart the 5 strips of cards from each copy of *Clock Concentration* Cards. Divide the 10 strips among your group.
2. Draw an hour hand and a minute hand on each clock face.
3. Write the time on the other half of each strip. Check one another's work.
4. Cut each strip in half to make 2 cards.
5. Your group should have 10 cards with a clock face and 10 cards with a time.
6. Write C on the back of each card with a clock face.
7. Write T on the back of each card with a time.
8. Make a mark on each card to show that it belongs to your group.

2:30

114   one hundred fourteen

Activity Card 70

Exploration B: Playing *Clock Concentration*

Activity Card 70; *Math Journal 2,* p. 114; *Math Masters,* p. 130

| WHOLE CLASS | **SMALL GROUP** | **PARTNER** | INDEPENDENT |

Groups follow directions on journal page 114 to make a set of *Clock Concentration* Cards from two copies of *Math Masters,* page 130. After the cards are made, have children follow the directions on Activity Card 70 to match times shown on a clock face to digital notation.

Differentiate   Adjusting the Activity

You may want to limit the number of cards in the set. For example, instead of ten Cs and ten Ts, use only five Cs and five Ts. After each round, change the set of cards to provide practice using all twenty cards. Alternatively, do the activity with all the cards faceup so children can focus on matching the clock faces to digital notation.

Go Online   Differentiation Support

Exploration C: Making Shapes

Activity Card 71; *Math Masters,* p. 131

| WHOLE CLASS | **SMALL GROUP** | **PARTNER** | INDEPENDENT |

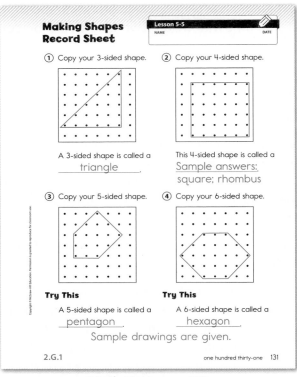

Math Masters, p. 131

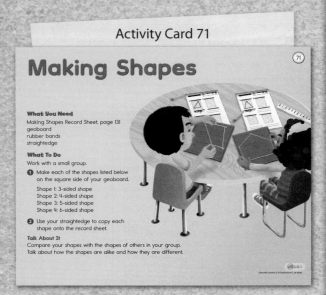

Activity Card 71

Children follow the directions on Activity Card 71 to make shapes with a specified number of sides on geoboards. They use a straightedge to copy each shape onto *Math Masters,* page 131 and then label each shape with its name.

When everyone in a group has completed the page, ask children to compare their shapes. Guide them to see that there are multiple names for 4-sided shapes. **GMP7.1**

Summarize Have children share the mathematics they engaged in for each exploration.

3 Practice 15–20 min

Go Online | ePresentations | eToolkit | Home Connections

▶ Playing *Addition Top-It*

My Reference Book, pp. 170–172; *Math Masters,* p. G18

| WHOLE CLASS | SMALL GROUP | PARTNER | INDEPENDENT |

Review the directions on *My Reference Book,* pages 170–172 as needed. Then have partners play *Addition Top-It* and record their number models on the *Addition Top-It* Record Sheet (*Math Masters,* page G18).

Observe

- What strategies are children using to determine the sums?
- Which children are using the correct comparison symbols on the *Addition Top-It* Record Sheet?

Discuss

- *How did you figure out the sums?*
- *How did you know which comparison symbol to write on the record sheet?*

▶ Math Boxes 5-5

Math Journal 2, p. 115

| WHOLE CLASS | SMALL GROUP | PARTNER | INDEPENDENT |

Mixed Practice Math Boxes 5-5 are paired with Math Boxes 5-7.

▶ Home Link 5-5

Math Masters, p. 132

Homework Children match clock faces to digital notation.

Math Journal 2, p. 115

Math Boxes
Lesson 5-5
DATE

1 In the number 300 there are
___3___ hundreds.
___0___ tens.
___0___ ones.
MRB 71

2 Write <, >, or =.
549 __<__ 595
378 __>__ 308
956 __>__ 856
MRB 74-71

3 Name something in the classroom that is about 1 foot long. Sample answer:
My Math Journal
MRB 103

4 Draw at least two ways to show 30¢ using Q, D, and N.
Sample answers:
Q N; D D D
MRB 118-271

5 **Writing/Reasoning** For Problem 3, how did you find something that was about 1 foot long?
Sample answer: I know that my foot-long ruler was about as long as my arm, and my journal is about as long as my arm.
MRB 103

① 2.NBT.1, 2.NBT.1b ② 2.NBT.4 ③ 2.MD.3, SMP6
④ 2.MD.8 ⑤ 2.MD.3, SMP6
one hundred fifteen 115

Math Masters, p. 132

Clock Faces and Digital Notation
Home Link 5-5
NAME
DATE

Family Note
Today your child played *Clock Concentration,* a game that involves matching clock faces to times in digital notation (such as 6:00 or 12:30). By the end of Grade 2, your child is expected to tell time to the nearest 5 minutes. By the end of Grade 3, your child will be expected to tell time to the nearest minute.
Please return this Home Link to school tomorrow.

Draw a line matching each clock face to a time. MRB 146-148

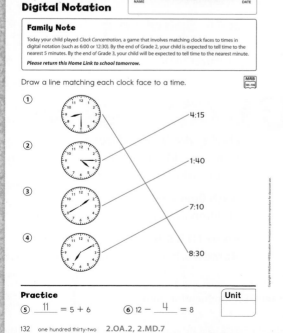

① ————— 4:15
② ————— 1:40
③ ————— 7:10
④ ————— 8:30

Practice
⑤ __11__ = 5 + 6   ⑥ 12 − __4__ = 8

Unit

132 one hundred thirty-two **2.OA.2, 2.MD.7**

Lesson 5-6

Mentally Adding and Subtracting 10 and 100

Overview Children develop strategies for mentally adding and subtracting 10 and 100.

▶ **Before You Begin**
Display a unit box and fill in a unit to use throughout the lesson. For Part 2 prepare one die per partnership by affixing stickers, one to each side, marked with + 10, + 10, − 10, + 100, + 100, − 100.

▶ **Vocabulary**
mental addition • mental subtraction

CCSS Common Core State Standards

Focus Clusters
• Understand place value.
• Use place value understanding and properties of operations to add and subtract.

1 Warm Up   15–20 min

	Materials	
Mental Math and Fluency Children count up and back by 10s on a number grid.	Number-Grid Poster	2.NBT.2, 2.NBT.8
Daily Routines Children complete daily routines.	See pages 4–43.	See pages xiv–xvii.

2 Focus   20–30 min

Math Message Children solve a + 10 addition problem.		2.NBT.5, 2.NBT.7, 2.NBT.8
Sharing Strategies Children share strategies for adding 10 to a 3-digit number.	Class Number Line, Number-Grid Poster, Class Number Scroll	2.NBT.5, 2.NBT.7, 2.NBT.8 SMP1
Adding and Subtracting 10 and 100 Children use calculator counts to develop rules for mentally adding and subtracting 10 and 100.	calculator, slate	2.NBT.2, 2.NBT.5, 2.NBT.7, 2.NBT.8, 2.NBT.9 SMP7, SMP8
✓ **Assessment Check-In** See page 478.		2.NBT.5, 2.NBT.8
Introducing *Addition/Subtraction Spin* Children learn to play *Addition/Subtraction Spin*.	*My Reference Book,* pp. 138–139; *Math Masters,* p. G23 and p. G24 (optional); per partnership: paper clip, pencil, die (see Before You Begin), calculator	2.NBT.5, 2.NBT.7, 2.NBT.8

CCSS 2.NBT.8 **Spiral Snapshot**

GMC Mentally add 100 to or subtract 100 from a given number.

| 4-11 Practice | 5-6 Focus Practice | 5-9 Warm Up Practice | 5-11 Warm Up Practice | 6-3 Warm Up | 7-6 Warm Up | 7-8 Warm Up | 7-9 Practice |

Spiral Tracker   **Go Online** to see how mastery develops for all standards within the grade.

3 Practice   15–20 min

Playing *Addition/Subtraction Spin* **Game** Children practice mentally adding and subtracting 10 and 100 with 3-digit numbers.	See above.	2.NBT.5, 2.NBT.7, 2.NBT.8 SMP8
Math Boxes 5-6 Children practice and maintain skills.	*Math Journal 2,* p. 116; centimeter ruler	See page 479.
Home Link 5-6 **Homework** Children mentally add and subtract 10 and 100.	*Math Masters,* p. 134	2.OA.1, 2.OA.2, 2.NBT.5, 2.NBT.7, 2.NBT.8

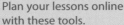
connectED.mheducation.com

Plan your lessons online with these tools.

ePresentations   Student Learning Center   Facts Workshop Game   eToolkit   Professional Development   Home Connections   Spiral Tracker   Assessment and Reporting   English Learners Support   Differentiation Support

Differentiation Options RtI

CCSS 2.NBT.2, 2.NBT.8, SMP7	CCSS 2.NBT.7, 2.NBT.8	CCSS 2.NBT.7, 2.NBT.8

Readiness   5–15 min

Counting by 10s on a Number Grid

WHOLE CLASS
SMALL GROUP
PARTNER
INDEPENDENT

Math Masters, p. 133; crayons

To explore patterns in counts by 10, children count up and back on a number grid. After they have completed the page, have children share patterns they see in the colored numbers. **GMP7.1** Sample answers: All of the yellow numbers have 0 in the ones place. All of the green numbers have 2 in the ones place.

Enrichment   5–15 min

Adding and Subtracting 10s and 100s

WHOLE CLASS
SMALL GROUP
PARTNER
INDEPENDENT

Activity Card 72,
4 each of number cards 0–9,
1 six-sided die, calculator, paper

To further explore mentally adding and subtracting 10s and 100s, children add more than one 10 or 100 to 3-digit numbers or subtract more than one 10 or 100 from 3-digit numbers. *For example:* A child might start with 236 and add three 10s: 236 plus 10 is 246, plus 10 more is 256, and plus 10 more is 266.

Extra Practice   5–15 min

Adding and Subtracting 10 and 100

WHOLE CLASS
SMALL GROUP
PARTNER
INDEPENDENT

Activity Card 73,
4 each of number cards 0–9,
2 six-sided dice, calculator, paper

To practice mentally adding and subtracting 10 and 100 with 3-digit numbers, children use number cards to generate 3-digit numbers and roll dice to determine what to add or subtract.

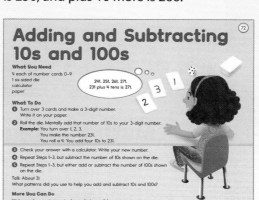

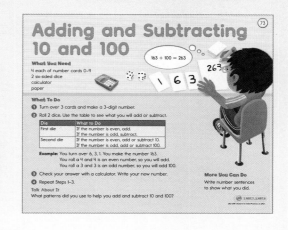

English Language Learners Support

Beginning ELL Using Total Physical Response prompts, model instructions for children, saying the following as you model: *Point to the number _____ on the number grid.* Use a teacher think-aloud to model counting on 10 more, saying the following as you model: *Circle the number that is 10 more.* Model with a number line and say: *Let's use a number line. Use your finger to hop from _____ to _____.* After demonstrating several times with different numbers, using both the number grid and the number line, have children respond to the same types of directions without teacher modeling.

Go Online | **ELL** English Learners Support

Academic Language Development

Use gestures, such as pointing to your head and closing your eyes, to accompany the use of *mental* and *mentally* to help children understand these terms to mean using only their minds or heads to find an answer. Use role plays to contrast adding and subtracting *mentally* with adding or subtracting using tools such as calculators or paper and pencil.

---

1 Warm Up   15–20 min   Go Online   ePresentations   eToolkit

▶ Mental Math and Fluency

Have children count up and back by 10s. As children count aloud, have a volunteer point to the counts on the Number-Grid Poster. *Leveled exercises:*

● ○ ○ Begin at 20. Count up by 10s past 100.   20, 30, 40, 50, 60, 70, 80, 90, 100, . . .
Begin at 90. Count back by 10s.   90, 80, 70, 60, 50, . . . , 0

● ● ○ Begin at 28. Count up by 10s past 100.   28, 38, 48, 58, 68, 78, 88, 98, 108, . . .
Begin at 82. Count back by 10s.   82, 72, 62, 52, 42, . . . , 2

● ● ● Begin at 156. Count up by 10s past 200.   156, 166, 176, 186, 196, 206, . . .
Begin at 209. Count back by 10s.   209, 199, 189, 179, 169, . . . , 9

▶ Daily Routines

Have children complete daily routines. See the Planning Ahead note in Lesson 5-1 for an adjustment to the Temperature Routine that will prepare children for Lesson 5-10.

---

2 Focus   20–30 min   Go Online   ePresentations   eToolkit

▶ Math Message

What is 120 + 10? Be prepared to explain how you found your answer.

▶ Sharing Strategies

Math Message Follow-Up Have children share how they solved 120 + 10. 130 Sample strategies may include the following:

- Use the Class Number Line. Start at 120. Count up 10. Land at 130.
- Use the Class Number Scroll. Find 120. Move down one row. Land at 130.
- Ignore the 100 in 120. Find 20 + 10 = 30. Add back in the 100 to get 130.
- Think about counting by 10s: 100, 110, 120, 130. Ten more than 120 is 130, so the answer is 130.

Ask children which strategy they like best and why. **GMP1.6** Sample answer: I like going down one row on the number scroll because it is fast.

Tell children that adding numbers in their heads in called **mental addition.** Subtracting numbers in their heads is called **mental subtraction.** Explain that children will use patterns to help them find rules for adding and subtracting 10 and 100 mentally.

▶ Adding and Subtracting 10 and 100

| WHOLE CLASS | SMALL GROUP | PARTNER | INDEPENDENT |

Review the procedure for skip counting on a calculator. (See Lesson 1-6 for detailed instructions.) Have children start at 221 and count up by 10s to 281. Record the counts as children read the counts aloud chorally. 221, 231, 241, 251, 261, 271, 281 Ask: *What patterns do you notice?* **GMP7.1** Sample answers: The ones digit is always 1. The tens digit goes up 1 in each count. The hundreds digit is always 2.

Have children continue the same count to 331 without clearing their calculators. Record the counts. 291, 301, 311, 321, 331 Ask: *What do you notice about the hundreds digit?* **GMP7.1** Sample answer: It stays the same until the tens digit is 9. When you count up by one more 10, the tens digit becomes 0, and the hundreds digit goes up 1. *Why do you think that happens?* Sample answer: When you have nine 10s and add one more 10, you get ten 10s, or 100. It is like trading base-10 blocks.

Next have children use their calculators to count back by 10s from 547 to 477. 547, 537, 527, 517, 507, 497, 487, 477 Ask: *What patterns do you notice?* **GMP7.1** Sample answers: The ones digit stays the same, and the tens digit goes down 1 at each count. The hundreds digit stays the same until the tens digit is 0. When I count back by one more 10, the hundreds digit goes down 1, and the tens digit changes to 9.

Ask: *How is counting by 10s similar to adding or subtracting 10s?* Sample answers: Adding 10 is like counting up 10. Subtracting 10 is like counting back 10. *How could we use these patterns to think of a rule for mentally adding or subtracting 10?* **GMP8.1** Sample answers: When you add 10, the tens digit goes up 1, and the other digits stay the same unless you have to make a new hundred. When you subtract 10, the tens digit goes down 1, and the other digits stay the same unless you have to trade a hundred to get more 10s.

Repeat the activity, with children using their calculators to count up and back by 100s from 3-digit numbers. Discuss any patterns and prompt children to identify rules for mentally adding and subtracting 100. **GMP8.1** Sample answers: When you add 100, the hundreds digit goes up 1, and the other digits stay the same. When you subtract 100, the hundreds digit goes down 1, and the other digits stay the same.

Common Misconception

Differentiate When counting up by 10 from a 3-digit number with 9 in the tens place, some children may change the tens digit to 0 without increasing the hundreds digit. For example, they may think the next count after 291 is 201. Have these children build the starting number with base-10 blocks. Have them add a long each time they count up by 10 and trade 10 longs for 1 flat at each hundreds transition.

Go Online Differentiation Support

Games

Addition/Subtraction Spin

Materials	☐ 1 Addition/Subtraction Spin Spinner
	☐ 1 paper clip
	☐ 1 pencil
	☐ 1 die marked with + 10, − 10, + 100, and − 100
	☐ 1 calculator
	☐ 2 sheets of paper
Players	2
Skill	Mentally adding and subtracting 10 and 100
Object of the Game	To have the larger total.

Directions

1. Players take turns being the "Spinner" and the "Checker."

2. The Spinner uses a pencil and a paper clip to make a spinner.

3. The Spinner spins the paper clip and writes the number that the paper clip points to. If the paper clip points to more than one number, the Spinner writes the smaller number.

MRB
138   one hundred thirty-eight

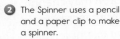
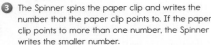

Addition/Subtraction Spin Spinners

NAME   DATE

G23

Dictate 2- and 3-digit numbers and have children apply the rules generated by the class to mentally add and subtract 10 and 100. They record their answers on slates. Include a variety of numbers, including 2- and 3-digit multiples of 10 and numbers that require a hundreds transition. *Suggestions:*

- Add 10 to 50 60; 46 56; 278 288; 460 470; and 598. 608
- Subtract 10 from 30 20; 78 68; 130 120; 433 423; and 703. 693
- Add 100 to 34 134; 600 700; 340 440; and 596. 696
- Subtract 100 from 500 400; 980 880; 156 56; and 432. 332

NOTE For now, work only with 3-digit numbers when adding and subtracting 100. It is not necessary to extend the activity to 4-digit numbers because children in second grade are expected to add and subtract only within 1,000.

✓ Assessment Check-In   CCSS 2.NBT.5, 2.NBT.8

Expect that most children will be able to use a mental strategy to add 10 to and subtract 10 from any 2-digit number and any 3-digit multiple of 10. If they struggle adding and subtracting 10 with 2-digit numbers, have them complete the Readiness activity. If they struggle adding and subtracting 10 with 3-digit multiples of 10, have them play *Addition/Subtraction Spin* using the bottom spinner on *Math Masters*, page G23. (*See below.*) Expect children to add and subtract 100 to multiples of 100. Children may also struggle adding and subtracting 10 with 3-digit numbers that are not multiples of 10, particularly for problems involving hundreds transitions, such as 198 + 10 = 208, or 208 − 10 = 198.

✓ Assessment and Reporting   **Go Online** to record student progress and to see trajectories toward mastery for these standards.

▶ Introducing *Addition/Subtraction Spin*

My Reference Book, pp. 138–139; *Math Masters*, p. G23

WHOLE CLASS   **SMALL GROUP**   PARTNER   INDEPENDENT

Read the directions for *Addition/Subtraction Spin* on *My Reference Book*, pages 138–139. Model the game by playing several rounds against the class before children play on their own.

The top spinner on *Math Masters*, page G23 is for practice mentally adding and subtracting 10 and 100 with any 3-digit number. The bottom spinner includes 3-digit multiples of 10 only. You can also use *Math Masters*, page G24 to create spinners with other numbers, as appropriate.

Differentiate   **Game Modifications**   **Go Online** Differentiation Support

Summarize Have partners explain to each other a rule they can use to add or subtract 10 or 100 as they play *Addition/Subtraction Spin*.

3 Practice 15–20 min

Go Online

ePresentations • eToolkit • Home Connections

▶ Playing *Addition/Subtraction Spin*

My Reference Book, pp. 138–139; *Math Masters*, p. G23

| WHOLE CLASS | SMALL GROUP | **PARTNER** | INDEPENDENT |

Have partners play several rounds of *Addition/Subtraction Spin*.

Observe

- Which children are engaged in the game?
- Which children need additional support to play the game?

Discuss

- *What shortcut or rule did you use to help you add or subtract?*
- *How do you know your rule works?* **GMP8.1**

Academic Language Development Provide sentence frames for children to use in their discussions of the game:

- I spun a _____.
- I added _____.
- My rule was _____.

▶ Math Boxes 5-6

Math Journal 2, p. 116

| WHOLE CLASS | **SMALL GROUP** | PARTNER | INDEPENDENT |

Mixed Practice Math Boxes 5-6 are paired with Math Boxes 5-10.

▶ Home Link 5-6

Math Masters, p. 134

Homework Children practice adding and subtracting 10 and 100.

Math Journal 2, p. 116

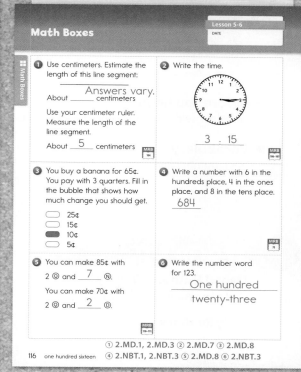

Math Masters, p. 134

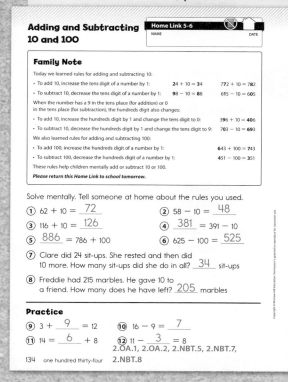

Open Number Lines

Overview Children use open number lines as a tool for solving number stories.

Common Core State Standards

Focus Clusters
- Represent and solve problems involving addition and subtraction.
- Use place value understanding and properties of operations to add and subtract.
- Relate addition and subtraction to length.

▶ **Before You Begin**
For Part 1 select and sequence Quick Look Cards 96, 108, and 117.

▶ **Vocabulary**
open number line

1 Warm Up 15–20 min

	Materials	
Mental Math and Fluency Children view Quick Look Cards.	Quick Look Cards 96, 108, and 117	2.OA.2
Daily Routines Children complete daily routines.	See pages 4–43.	See pages xiv–xvii.

2 Focus 20–30 min

Math Message Children solve a + 10 number story.	Class Number Line	2.OA.1, 2.NBT.5, 2.NBT.7, 2.NBT.8, 2.MD.6
Introducing Open Number Lines Children share strategies for adding 10 to a 2-digit number.	Class Number Line	2.OA.1, 2.NBT.5, 2.NBT.7, 2.NBT.8, 2.MD.6, SMP2
Using Open Number Lines Children use open number lines to solve number stories.	*Math Journal 2,* p. 117; slate	2.OA.1, 2.NBT.5, 2.NBT.7, 2.NBT.8, 2.MD.6, SMP1
✓ **Assessment Check-In** See page 484.	*Math Journal 2,* p. 117	2.NBT.5, 2.NBT.7, 2.NBT.8

CCSS 2.MD.6 Spiral Snapshot

GMC Represent sums and differences on a number-line diagram.

| 2-8
Focus
Practice | 2-11
Practice | 3-9
Focus
Practice | 3-10
Focus
Practice | 4-3
Warm Up | 5-7
Focus
Practice | 6-1
Practice | 6-3
Practice |

/// Spiral Tracker   **Go Online** to see how mastery develops for all standards within the grade.

3 Practice 15–20 min

Playing *Beat the Calculator* **Game** Children practice addition and subtraction facts.	*Assessment Handbook,* pp. 98–99; per group: 4 each of number cards 0–9, calculator, large paper triangle (optional)	2.OA.2
Math Boxes 5-7 Children practice and maintain skills.	*Math Journal 2,* p. 118	See page 485.
Home Link 5-7 **Homework** Children practice using open number lines.	*Math Masters,* p. 136	2.OA.1, 2.NBT.5, 2.NBT.7, 2.MD.6

connectED.mheducation.com

Plan your lessons online with these tools.

 ePresentations   Student Learning Center   Facts Workshop Game   eToolkit   Professional Development   Home Connections   Spiral Tracker   Assessment and Reporting   English Learners Support   Differentiation Support

Differentiation Options RtI

 CCSS 2.NBT.5, 2.NBT.8, SMP7

Readiness
5–15 min

WHOLE CLASS
SMALL GROUP
PARTNER
INDEPENDENT

Adding 10s and 1s

number grid (*Math Journal 2*, inside back cover)

For experience with mental addition strategies using a visual model, children count by 10s and 1s on a number grid. Refer children to the number grid on the inside back cover of their journals. Name a number and have children find it on the grid. Ask them to add one or more 10s to the number. For example, say: *Put your finger on 24. Add two 10s, or 20. Where does your finger land?* 44 *What pattern do you notice on the number grid when you add 10s?* GMP7.1 Sample answer: The ones digits stay the same, but the tens digits go up by 1 for each 10 I add.

When children are comfortable adding 10s, present problems that include adding both 10s and 1s. For example, say: *Put your finger on 42. Add 35.* 77 Pose additional problems as needed.

CCSS 2.OA.1, 2.NBT.5, 2.NBT.7, 2.NBT.8, 2.MD.6, SMP5

Enrichment
5–15 min

WHOLE CLASS
SMALL GROUP
PARTNER
INDEPENDENT

Using Open Number Lines with Larger Numbers

Math Masters, p. 135

To extend their work with open number lines, children use them to solve number stories involving larger numbers. Encourage children to think about how they could use open number lines to record bigger jumps, such as jumps of 100. GMP5.2

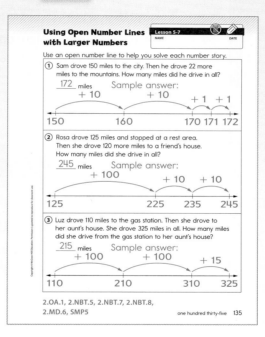

CCSS 2.NBT.5, 2.NBT.7, 2.NBT.8, 2.MD.6, SMP5

Extra Practice
5–15 min

WHOLE CLASS
SMALL GROUP
PARTNER
INDEPENDENT

Using Open Number Lines to Add

Activity Card 74, 4 each of number cards 1–9, paper

For practice using a visual model to add numbers, children solve addition problems using open number lines. GMP5.2

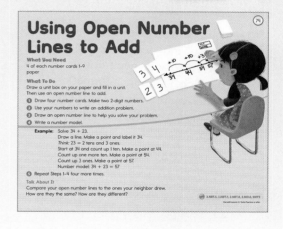

English Language Learners Support

Beginning ELL To help English language learners understand and use the terms *open number line* and *regular number line,* show examples of each, accompanied by teacher think-aloud statements such as the following: *The open number line has no numbers. The regular number line has numbers.* To provide practice differentiating between the two types of number lines, use a variety of Total Physical Response prompts, such as the following: *Show me the open number line. Show me the number line with no numbers. Draw a regular number line. Point to the numbers on the regular number line.*

Go Online ELL English Learners Support

Mathematical Practice

SMP1 **Make sense of problems and persevere in solving them.**

GMP1.2 Reflect on your thinking as you solve your problem.

GMP1.6 Compare the strategies you and others use.

SMP2 **Reason abstractly and quantitatively.**

GMP2.3 Make connections between representations.

Professional Development

Open number lines are powerful problem-solving tools that can help children record and keep track of their thinking. For more information about open number lines, see the Mathematical Background section of the Unit 5 Organizer.

Go Online Professional Development

Academic Language Development

The term *open number line* may be confusing for children because of their prior experiences with regular number lines and the word *open*. Have partners complete a 4-Square Graphic Organizer (*Math Masters*, page TA42) for the term to consolidate their understanding of its meaning and use. For more information, see the *Implementation Guide*.

---

1 Warm Up   15–20 min   Go Online
ePresentations   eToolkit

▶ Mental Math and Fluency

Flash Quick Look Cards 96, 108, and 117. Always allow children a second look and follow up by asking both *what* children saw and *how* they saw it. Asking such questions will allow a variety of strategies to emerge. *Leveled exercises:*

- ●○○ **Quick Look Card 96** Sample answer: I moved one row of 5 over to fill the first ten frame, and then there was 1 left. So that's 11.

- ●●○ **Quick Look Card 108** Sample answer: I moved 4 over to fill the first ten frame, and there were still 3 left. So that's 13.

- ●●● **Quick Look Card 117** Sample answer: I saw 2 empty spaces on one frame, so I moved 2 counters over to make 10. There were still 5 left, so that makes 15.

▶ Daily Routines

Have children complete daily routines. See the Planning Ahead note in Lesson 5-1 for an adjustment to the Temperature Routine that will prepare children for Lesson 5-10.

---

2 Focus   20–30 min   Go Online
ePresentations   eToolkit

▶ Math Message

Elyse has 23 buttons in her craft box. She buys a package of 10 more buttons. How many buttons does Elyse have now? 33 buttons *Use the Class Number Line to help you solve the number story.*

▶ Introducing Open Number Lines

| WHOLE CLASS | SMALL GROUP | PARTNER | INDEPENDENT |

Math Message Follow-Up Have volunteers share how they used the number line. Expect responses to include the following:

- I started at 23 and counted up 10 spaces to 33.
- I knew that 10 more than 23 is 33, so I started at 23 and hopped right to 33.

Ask children to recall the rule they found in the previous lesson for adding 10. When you add 10, the tens digit goes up by 1. Ask: *Do we need to count up by 1s to know where we will land on the number line?* Sample answer: No. We know we'll land on 33, so we can just jump 10 all at once. Display a number line to illustrate a jump of 10 from 23 to 33. (*See margin.*)

Introduce the **open number line** as a tool that allows us to quickly record our thinking when we use mental strategies to add or subtract. Beneath the first number line, display an open number line. (*See margin.*) Explain that this is a quick and easy way to show the jump of 10 that children calculated mentally. Ask: *How is the open number line like the other number lines we have used? How is it different?* GMP2.3 Sample answers: It is like a regular number line because the numbers get bigger as you go to the right. It is different from a regular number line because it doesn't show every tick mark, just the tick marks we need. *How might the open number line be helpful when we solve number stories?* Sample answers: It is quick and easy to draw. It can help us keep track of our steps.

Tell children that they will use open number lines to solve number stories and record their thinking.

▶ Using Open Number Lines

Math Journal 2, p. 117

WHOLE CLASS | SMALL GROUP | PARTNER | INDEPENDENT

Pose the following number story and have children draw open number lines on their slates to record their thinking: *Peter has 64 blocks in his toy box and 10 blocks on the table. How many blocks does he have in all?* 74 blocks Check that children draw open number lines similar to the one shown in the margin below.

Then have children work in partnerships to solve the following problem, recording their thinking on open number lines: *Peter has 64 blocks in his toy box and 30 blocks on the table. How many blocks does he have in all?* 94 blocks Have volunteers share strategies and display their open number lines. *Sample strategies:*

- I already knew that $64 + 10 = 74$. Because I have to add 30 this time, I made two more jumps of 10. So $74 + 10 = 84$, and $84 + 10 = 94$. The answer is 94.

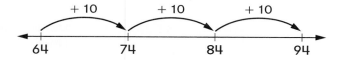

- I already knew that $64 + 10 = 74$. I know that $30 = 10 + 20$, so I have to jump 20 more. So I added $74 + 20$ to get 94.

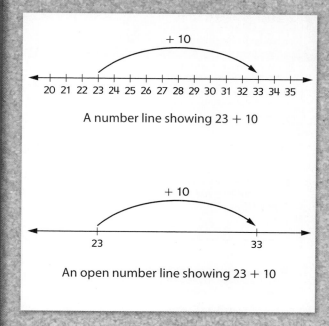

A number line showing $23 + 10$

An open number line showing $23 + 10$

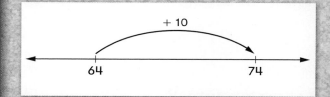

Using Open Number Lines to Solve Number Stories

Lesson 5-7
DATE

You are building towers with red and blue blocks. Use an open number line to help you solve each number story. Show your work.

① You build the first tower with 15 red blocks and 20 blue blocks. How many blocks did you use? __35__ blocks   Sample answer:

② You build the second tower with 37 red blocks and 32 blue blocks. How many blocks did you use? __69__ blocks   Sample answer:

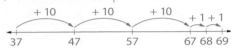

③ You build the third tower with 16 red blocks. You use 39 blocks in all. How many blue blocks did you use? __23__ blue blocks   Sample answer:

2.OA.1, 2.NBT.5, 2.NBT.7, 2.NBT.8, 2.MD.6   one hundred seventeen   117

Point out that drawing an open number line is a way to keep track of how much you have already added and how much you have left to add. **GMP1.2** Ask: *How could we use an open number line to solve the problem if Peter had 35 blocks on the table?* Sample answer: Just add another jump of 5, so you land at 99. To illustrate this strategy, add a jump of 5 to the previous open number line.

Pose a new story and have partners solve it using open number lines to record their thinking: *Sherry has 21 marbles. How many more does she need to have 57 marbles?* 36 marbles Have several children share their strategies and display them on the board. Sample strategies should include the following:

- I started from 21 and made a jump of 20 to get to 41. I knew I could make another jump of 10, which put me at 51. Then I had to jump 6 more to get to 57. So all together I hopped $20 + 10 + 6 = 36$.

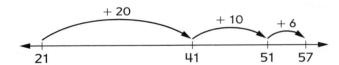

- I wanted to get to an easy number to start. So from 21 I went up 9 to 30. From there I made jumps of 10 to get to 40 and then to 50. I had to go 7 more to get to 57. So in all, I hopped $9 + 10 + 10 + 7$, which is 36.

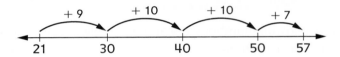

Highlight strategies that involve jumping to an easy number first, making a connection to the going-though-10 strategy for solving subtraction facts. Have children compare the different strategies. **GMP1.6** Sample answer: In the first strategy, jumps of 10 were made first. In the second strategy, the first jump was only 9 because that got us to an easy number. Guide the discussion to help children see that they can use open number lines to illustrate the different mental steps they might take to solve addition or subtraction problems.

Pose additional problems as needed. Then have children work in partnerships to complete journal page 117.

✓ Assessment Check-In   CCSS 2.NBT.5, 2.NBT.7, 2.NBT.8

Math Journal 2, p. 117

Expect that most children will be able to mentally add 10s and 1s to solve Problems 1–2 on journal page 117, with or without recording their steps on an open number line. If children struggle to mentally add 10s, complete or revisit the Readiness activity, paying particular attention to patterns in the tens digits on the number grid. Some children may be able to solve Problem 3. Practice with open number lines will continue throughout the year.

 Assessment and Reporting   Go Online   to record student progress and to see trajectories toward mastery for these standards.

Summarize Have children discuss with a partner whether using an open number line to solve number stories is easier or harder than using a regular number line. Sample answer: It is easier to draw an open number line because I don't have to mark every number.

3 Practice 15–20 min

Go Online

ePresentations eToolkit Home Connections

▶ Playing *Beat the Calculator*

Assessment Handbook, pp. 98–99

| WHOLE CLASS | **SMALL GROUP** | PARTNER | INDEPENDENT |

Have small groups play *Beat the Calculator,* as it was introduced in Lesson 5-1. As you circulate and observe, consider using *Assessment Handbook,* pages 98–99 to monitor children's progress with addition and subtraction facts. By the end of Grade 2, children are expected to know from memory all sums of two 1-digit numbers.

Observe

• Which facts do children know from memory?
• Which children need additional support to play the game?

Discuss

• *What strategies did you use to solve the facts you did not know?*
• *Why is it helpful to know addition and subtraction facts?*

▶ Math Boxes 5-7

Math Journal 2, p. 118

| WHOLE CLASS | SMALL GROUP | PARTNER | INDEPENDENT |

Mixed Practice Math Boxes 5-7 are paired with Math Boxes 5-5.

▶ Home Link 5-7

Math Masters, p. 136

Homework Children use open number lines to solve number stories.

Math Journal 2, p. 118

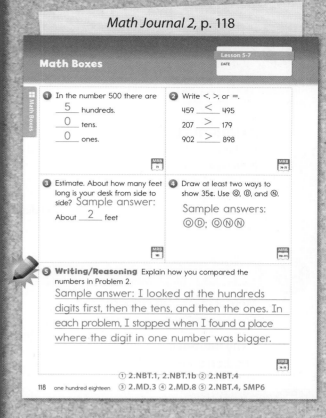

Math Masters, p. 136

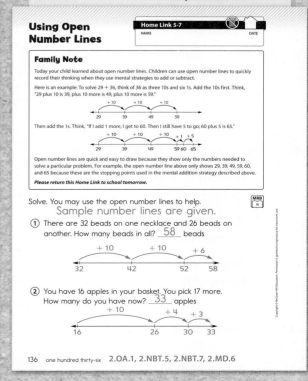

Lesson 5-7 **485**

Change-to-More Number Stories

Overview Children solve change-to-more number stories.

▶ **Before You Begin**
For Part 2 you will need a shopping bag and a copy of *My Reference Book*. You will also need to display a change diagram.

▶ **Vocabulary**
change-to-more number story • change diagram

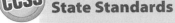 **Common Core State Standards**

Focus Clusters
- Represent and solve problems involving addition and subtraction.
- Use place value understanding and properties of operations to add and subtract.

 **Warm Up** 15–20 min

	Materials	
Mental Math and Fluency Children solve facts number stories.	slate	2.OA.1
Daily Routines Children complete daily routines.	See pages 4–43.	See pages xiv–xvii.

 **Focus** 30–40 min

Math Message Children determine whether a bag weighs more before or after a book is put in it.		2.OA.1
Discussing a Change-to-More Situation Children discuss how they know the weight of a bag changed to more.	*My Reference Book*, shopping bag	2.OA.1
Introducing the Change Diagram Children use a change diagram.	*Math Journal 2*, p. 120; *My Reference Book*, pp. 27–29	2.OA.1, 2.NBT.5, 2.NBT.7 SMP1, SMP2, SMP4
Solving Change-to-More Number Stories Children solve change-to-more number stories.	*Math Journal 2*, pp. 120–121; *Math Masters*, p. TA26 (optional)	2.OA.1, 2.NBT.5, 2.NBT.7
✓ **Assessment Check-In** See page 492.	*Math Journal 2*, p. 121	2.OA.1, 2.NBT.5, 2.NBT.7

CCSS 2.OA.1 Spiral Snapshot

GMC Use addition and subtraction to solve 1-step number stories.

5-1 Practice	5-3 Warm Up	5-7 Focus Practice	5-8 Warm Up Focus Practice	5-9 Focus Practice	5-10 Focus Practice	6-1 Warm Up Practice	6-2 Focus Practice

Spiral Tracker **Go Online** to see how mastery develops for all standards within the grade.

 Practice 15–20 min

Measuring Objects Children measure objects in centimeters and inches.	*Math Journal 2*, p. 122; centimeter ruler; inch ruler	2.MD.1, 2.MD.3
Math Boxes 5-8: Preview for Unit 6 Children practice and maintain skills.	*Math Journal 2*, p. 119	See page 493.
Home Link 5-8 **Homework** Children solve change-to-more number stories.	*Math Masters*, pp. 138–139	2.OA.1, 2.NBT.5, 2.NBT.7

---

connectED.mheducation.com

Plan your lessons online with these tools.

 ePresentations Student Learning Center Facts Workshop Game eToolkit Professional Development Home Connections Spiral Tracker Assessment and Reporting English Learners Support Differentiation Support

Differentiation Options

RtI

Readiness   5–15 min

WHOLE CLASS
SMALL GROUP
PARTNER
INDEPENDENT

Showing Change on Number Lines

Math Journal 2, inside back cover

For support working with change number stories, children use a number line to solve problems. Name a start number for children to locate on their number lines. Tell them that they are going to change this location on the number line by adding or subtracting another number. For example, suppose the start number is 14 and the change is "add 6." Children count up on the number line to find the end number, which is 20. Ask: *Is the end number more or less than the start number?* More Continue with similar problems.

Enrichment   5–15 min

WHOLE CLASS
SMALL GROUP
PARTNER
INDEPENDENT

Writing Change-to-More Stories

Activity 75; *Math Journal 2,* p. 120; *Math Masters,* p. TA7

To extend their understanding of change-to-more number stories, children use information from journal page 120 to write number stories on *Math Masters,* page TA7. They draw change diagrams and write number models for their stories. GMP4.1 Encourage children to write stories that involve either a missing start number or a missing change number.

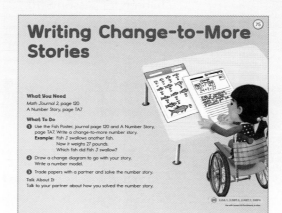

Extra Practice   5–15 min

WHOLE CLASS
SMALL GROUP
PARTNER
INDEPENDENT

Solving More "Fishy" Stories

Math Journal 2, p. 120; *Math Masters,* p. 137

For additional practice with change-to-more number stories, children complete *Math Masters,* page 137. They refer to the Fish Poster on journal page 120. Using a change diagram to help, children write a number model with an unknown for each problem. GMP4.1

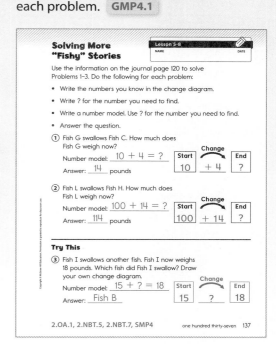

English Language Learners Support

Beginning ELL The word *change* may be confusing for English language learners. They have been making change in the context of money, but this lesson uses *change* to mean "to become different." Use concrete examples to introduce this meaning of *change.* For example, a balloon's appearance changes when you fill it with air, and a stack of books grows larger as you add more books. Show children pairs of objects in context, some identical and some reflecting a change, and use *yes* or *no* questions to prompt them to indicate when a change occurs.

Go Online ELL English Learners Support

Standards and Goals for
Mathematical Practice

SMP1 **Make sense of problems and persevere in solving them.**

  GMP1.1 Make sense of your problem.

  GMP1.6 Compare the strategies you and others use.

SMP2 **Reason abstractly and quantitatively.**

  GMP2.3 Make connections between representations.

SMP4 **Model with mathematics.**

  GMP4.1 Model real-world situations using graphs, drawings, tables, symbols, numbers, diagrams, and other representations.

  GMP4.2 Use mathematical models to solve problems and answer questions.

---

① Warm Up   15–20 min   Go Online   ePresentations   eToolkit

▶ Mental Math and Fluency

Pose number stories for children to solve on their slates. Have children share their strategies. *Leveled exercises:*

● ○ ○   Andrew had 12 carrot sticks in his lunch. He ate 7 of them. How many carrot sticks does he have left?   5 carrot sticks

● ● ○   Alexis has 9 toy animals in her room and some toy animals in her backpack. She has 14 toy animals all together. How many are in her backpack?   5 toy animals

● ● ●   Manny is making cards for his friends. He made 8 cards before lunch. He needs to have 17 cards in all. How many cards does he have left to make?   9 cards

▶ Daily Routines

Have children complete daily routines. See the Planning Ahead note in Lesson 5-1 for an adjustment to the Temperature Routine that will prepare children for Lesson 5-10.

---

② Focus   30–40 min   Go Online   ePresentations   eToolkit

▶ Math Message

Which bag weighs more—a bag that is empty or a bag with a book inside? Why?   It weighs more with the book inside. The book adds weight.

▶ Discussing a Change-to-More Situation

| WHOLE CLASS | SMALL GROUP | PARTNER | INDEPENDENT |

Math Message Follow-Up The bag's weight increases (or changes to more) when a book is placed inside. Demonstrate this by having a volunteer hold an empty shopping bag. Then place a copy of *My Reference Book* in the bag. Ask children to explain how they know whether the bag was heavier when it was empty or after the book was put in it. Sample answer: I could tell the bag was heavier after the book was put in it because it pulled my hand down more. Tell children that when the bag got heavier, its weight changed. Explain that children will solve number stories about situations where something changes.

▶ Introducing the Change Diagram

Math Journal 2, p. 120; My Reference Book, pp. 27–29

| WHOLE CLASS | SMALL GROUP | PARTNER | INDEPENDENT |

Change number stories begin with a starting quantity. This quantity is increased (or decreased), so the ending quantity is more (or less) than the starting quantity. This lesson focuses on **change-to-more number stories.** Change-to-less number stories will be the focus of Lesson 5-10.

Have children turn to the Fish Poster on journal page 120. If necessary, review the abbreviations *lb* and *in.* with children. Pose a change-to-more number story based on the poster. *For example:*

- Fish K weighs 35 pounds. It swallows Fish D, which weighs 5 pounds. How much does Fish K weigh now? 40 pounds

Draw a blank unit box on the board. Ask: *What label goes in the unit box?* Pounds

Unit
pounds

Display a **change diagram** and refer to it as you fill in the numbers. Model how it can be used as a helpful tool when solving number stories. Emphasize how easy it is to sketch a change diagram.

| Start | Change | End |

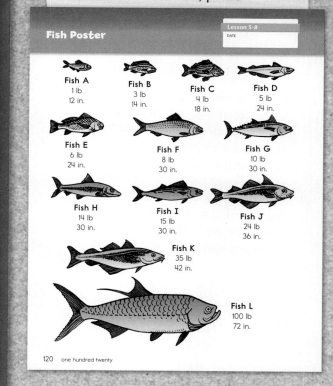

Math Journal 2, p. 120

Fish Poster

Lesson 5-8
DATE

Fish A
1 lb
12 in.

Fish B
3 lb
14 in.

Fish C
4 lb
18 in.

Fish D
5 lb
24 in.

Fish E
6 lb
24 in.

Fish F
8 lb
30 in.

Fish G
10 lb
30 in.

Fish H
14 lb
30 in.

Fish I
15 lb
30 in.

Fish J
24 lb
36 in.

Fish K
35 lb
42 in.

Fish L
100 lb
72 in.

120   one hundred twenty

Academic Language Development

To help children describe the various change contexts, provide them with a story frame:

- We **started** with 35.
- We **changed** it by adding 5.
- We **ended** with 40.

Ask questions such as the following:

- *What do we want to find out from the story?* How much Fish K weighs now GMP1.1
- *Do we know Fish K's weight before it swallowed Fish D? If so, what was it?* Yes. It weighed 35 pounds.

Put 35 in the Start box.

- *What change occurred?* Fish K swallowed Fish D. *Does Fish K weigh more or less than before it swallowed Fish D?* More *How much more?* 5 pounds GMP1.1

Put + 5 on the Change line.

Then put ? in the End box, reminding children that we don't yet know the result. Explain how we can use a question mark to represent what we want to find out (or what we don't know). Tell children that in this problem the question mark represents how much Fish K weighs after swallowing Fish D. GMP4.1

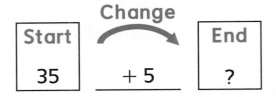

Explain to children that another way to represent a number story is by writing a number model. Write $35 + 5 = ?$ below the diagram. GMP4.1 Remind them that the question mark represents what they want to find out.

- *How do we find Fish K's weight after it has swallowed Fish D?* Add $35 + 5$.

Have children share their strategies for finding Fish K's total weight. Some possible strategies are as follows:

- I started with 35 and counted up by 1s.

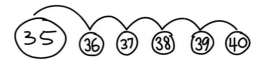

- I counted up by 5s. I know that 40 is 5 more than 35 because when I count by 5s, the next number after 35 is 40.

Ask: *What is Fish K's final weight?* 40 pounds Below the number model with the question mark, write a summary number model number with 40 substituted for the question mark: $35 + 5 = 40$. GMP4.2

Refer to the diagram and the number models. Ask children how each element in the diagram relates to the number models. Sample answer: The number in the Start box is 35, which is the first number in the number model. The Change line in the diagram says + 5, which relates to the second number in the number model. And the End box has ?, which represents what we do not know or what we need to find out: 35 + 5 = ? The second number model shows the answer to the number story in place of the question mark. **GMP2.3**

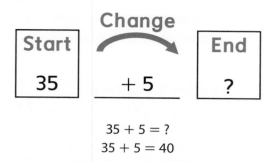

$$35 + 5 = ?$$
$$35 + 5 = 40$$

Help children make up other change-to-more number stories using the Fish Poster. Emphasize stories for which the start and change numbers are known and the end number is the unknown; also include some stories for which the start number or the change number is the unknown. *For example:*

- Fish J swallowed another fish. Fish J now weighs 29 pounds. How much did the fish weigh that Fish J swallowed? 24 + ? = 29; 5 pounds Which fish did Fish J swallow? Fish D

- A fish that swallowed Fish A now weighs 36 pounds. How much did the fish weigh before it swallowed Fish A? ? + 1 = 36; 35 pounds Which fish swallowed Fish A? Fish K

Model children's solutions on the board using these steps:

1. Fill in a change diagram for each problem. Write in the numbers that are known and write ? for the number that is unknown. **GMP4.1**

2. Write a number model for the problem, using ? to represent the unknown number. **GMP4.1**

3. Solve the problem. **GMP4.2**

Be sure to discuss the connection between each label in the diagram and the numbers or symbols in the number model. **GMP2.3**

With the class read more about change diagrams on *My Reference Book*, pages 27–29.

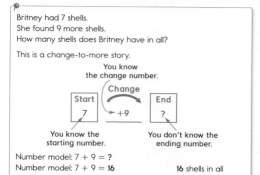

My Reference Book, p. 27

Operations and Algebraic Thinking

Some number stories are about a change. The number you start with changes to more or changes to less. You can use a **change diagram** to help you solve these stories.

Britney had 7 shells.
She found 9 more shells.
How many shells does Britney have in all?

This is a change-to-more story.

You know
the change number.

You know the
starting number.

You don't know the
ending number.

Number model: 7 + 9 = ?
Number model: 7 + 9 = 16   16 shells in all

MRB
twenty-seven   27

Math Journal 2, p. 121

Lesson 5-8

"Fishy" Stories   DATE

For each number story, do the following:
- Use the information on the Fish Poster to write the numbers you know in the change diagram.
- Write ? for the number you need to find or don't know.
- Write a number model with ? for the number you need to find.
- Solve the problem and write the answer.

① Fish J swallows Fish B. How much does Fish J weigh now?
Number model: $24 + 3 = ?$
Answer: 27 pounds

Change
Start 24 + 3 End ?

② Fish K swallows Fish G. How much does Fish K weigh now?
Number model: $35 + 10 = ?$
Answer: 45 pounds

Change
Start 35 + 10 End ?

Try This

③ Fish F swallows Fish B. How much does Fish F weigh now?
Draw your own change diagram.
Number model: $8 + 3 = ?$
Answer: 11 pounds

Change
Start 8 + 3 End ?

2.OA.1, 2.NBT.5, 2.NBT.7   one hundred twenty-one   121

Math Journal 2, p. 122

Measuring Objects

Lesson 5-8
DATE

Look around the room. Find some objects you think are about **1 centimeter** long. Use a ruler to check the lengths. List some 1-centimeter objects here.

Now find some objects you think are about **1 inch** long. Use a ruler to check the lengths. List some 1-inch objects here.

Now find some objects you think are about **10 centimeters** long. Use a ruler to check the lengths. List some 10-centimeter objects here.

Now find some objects you think are about **10 inches** long. Use a ruler to check the lengths. List some 10-inch objects here.

Math Journal 2, p. 119

Math Boxes
Preview for Unit 6

Lesson 5-8
DATE

① Write each number in expanded form.

34 _30 + 4_
66 _60 + 6_
52 _50 + 2_
81 _80 + 1_

MRB 73

② What number do the base-10 blocks show? _232_

Use base-10 shorthand to show the number another way.

Sample answer:

□□ ||| ..

MRB 71-72

③ Hannah scored 10 points. Jen scored 15 points. How many points did they score in all? _Sample_

Number model: _model:_
_10 + 15 = ?_

Answer: _25_ points

MRB 24-26

④
How many more books were read on Saturday than on Thursday? _5_ books

MRB 116

⑤ Chad has 12 pennies. Adam has 7 pennies. How many more pennies does Chad have than Adam? Draw a picture to show what you did to solve this problem. _Pictures vary._

Answer: _5_ pennies

MRB 24-26

① 2.NBT.1, 2.NBT.3 ② 2.NBT.1, 2.NBT.3, SMP2 ③ 2.OA.1, 2.NBT.7, 2.NBT.8 ④ 2.MD.10 ⑤ 2.OA.1, 2.NBT.5   one hundred nineteen 119

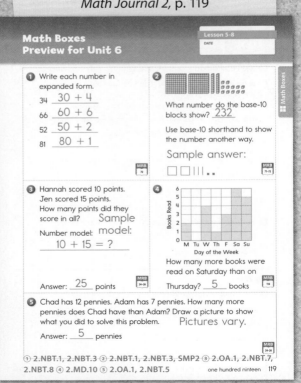

▶ Solving Change-to-More Number Stories

Math Journal 2, pp. 120–121

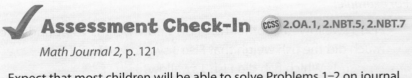

WHOLE CLASS   SMALL GROUP   **PARTNER**   INDEPENDENT

Partners use journal page 120 to complete the problems on journal page 121. Check that they record the known information in the correct places in the change diagrams, write ? for the unknown numbers, and write number models that represent the number stories. Encourage children to use a variety of strategies to solve the problems.

Differentiate   **Adjusting the Activity**

For children who have difficulty filling in the change diagram, use the following guiding questions. Ask: *Which fish did we start with? What change took place?* Tell children to write the names of the fish in the appropriate spaces in the diagram on *Math Masters*, page TA26. Then have them write the weights below each fish name.

Go Online ▶   Differentiation Support

✓ Assessment Check-In   CCSS 2.OA.1, 2.NBT.5, 2.NBT.7

Math Journal 2, p. 121

Expect that most children will be able to solve Problems 1–2 on journal page 121. Encourage those who have difficulty solving the number stories to use various tools, such as open number lines, drawings, or number grids. Some children may be able to draw the change diagram for Problem 3 and solve the number story. At this time, do not expect all children to write number models for these problems. Later lessons will offer opportunities for children to organize information in diagrams and write number models.

☑ Assessment and Reporting   **Go Online** ▶   to record student progress and to see trajectories toward mastery for these standards.

Summarize   Bring the class together and have partners share their solution strategies.   GMP1.6

 **Practice**   15–20 min

Go Online

ePresentations   eToolkit   Home Connections

▶ Measuring Objects

Math Journal 2, p. 122

| WHOLE CLASS | **SMALL GROUP** | **PARTNER** | INDEPENDENT |

To provide practice measuring objects, have children find objects in the classroom with specified lengths in both centimeters and inches. They record their findings on journal page 122.

▶ Math Boxes 5-8: Preview for Unit 6

Math Journal 2, p. 119

| WHOLE CLASS | **SMALL GROUP** | **PARTNER** | **INDEPENDENT** |

Mixed Practice   Math Boxes 5-8 are paired with Math Boxes 5-12. These problems focus on skills and understandings that are prerequisite for Unit 6. You may want to use information from these Math Boxes to plan instruction and groupings in Unit 6.

▶ Home Link 5-8

Math Masters, pp. 138–139

Homework   Children solve change-to-more number stories. They fill in a change diagram and write a number model for each problem.

Planning Ahead

For the Focus portion of Lesson 5-9, have children copy the information from *Math Journal 1,* pages 94–97 (Addition Facts Inventory Record, Parts 1 and 2) to *Math Journal 2,* pages 250–253. Or cut out the pages and staple them to the inside back cover of *Math Journal 2.* Children will continue to record the facts they know from memory and facts that need more practice. The Addition Facts Inventory Record will become a cumulative record of children's progress toward fact mastery.

Math Masters, p. 138

Change Number Stories

Home Link 5-8

NAME   DATE

Family Note

Your child has learned how to represent a problem by using a change diagram, which is shown in the example below. Using diagrams like this can help children organize the information in a problem. When the information is organized, it is easier to decide which operation (+, −, ×, ÷) to use to solve the problem. Change diagrams are used to represent problems in which a starting quantity is increased or decreased. For the number stories on this Home Link, the starting quantity is always increased.

Please return the second page of this Home Link to school tomorrow.

Do the following for each number story on the next page: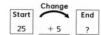

- Write the numbers you know in the change diagram.
- Write ? for the number you need to find.
- Write a number model. Use ? for the number you need to find.
- Answer the question.

Example: Twenty-five children are riding on a bus. At the next stop, 5 more children get on. How many children are on the bus now?

Start	Change	End
25	+ 5	?

The number of children on the bus has increased by 5.

Possible number model: 25 + 5 = ?

Answer: There are now 30 children on the bus.

138   one hundred thirty-eight

Math Masters, p. 139

Change Number Stories (continued)

Home Link 5-8

NAME   DATE

① Becky ate 11 grapes after lunch. She ate 7 more grapes after dinner. How many grapes did she eat in all?

Start	Change	End
11	+ 7	?

Number model:
<u>11 + 7 = ?</u>
<u>18</u> grapes

② Bob has 30 baseball cards. He buys 8 more. How many baseball cards does Bob have now?

Start	Change	End
30	+ 8	?

Number model:
<u>30 + 8 = ?</u>
<u>38</u> cards

③ A large fish weighs 42 pounds. A small fish weighs 10 pounds. The large fish swallows the small fish. How much does the large fish weigh now?

Draw your own change diagram.

Start	Change	End
42	+ 10	?

Number model: <u>42 + 10 = ?</u>
<u>52</u> pounds

2.OA.1, 2.NBT.5, 2.NBT.7   one hundred thirty-nine   139

Parts-and-Total Number Stories

Overview Children solve parts-and-total number stories.

▶ **Before You Begin**
Display a parts-and-total diagram. You can use *Math Masters,* page TA27. For Part 2 have children copy the information from *Math Journal 1,* pages 94–97 (Addition Facts Inventory Record, Parts 1 and 2) to *Math Journal 2,* pages 250–253 or cut out the pages and staple them to the inside back cover of *Math Journal 2.* Children will continue to record the facts they know from memory and facts that need more practice. The Addition Facts Inventory Record will become a cumulative record of children's progress toward fact mastery.

▶ **Vocabulary**
parts-and-total diagram • total • parts-and-total number story

Common Core State Standards

Focus Clusters
- Represent and solve problems involving addition and subtraction.
- Add and subtract within 20.
- Use place value understanding and properties of operations to add and subtract.

1 Warm Up   15–20 min

	Materials	
Mental Math and Fluency Children mentally add and subtract 10 and 100.	slate	2.NBT.8
Daily Routines Children complete daily routines.	See pages 4–43.	See pages xiv–xvii.

2 Focus   30–40 min

Math Message Children find the total number of dots on a domino.	slate	2.OA.2
Introducing the Parts-and-Total Diagram Children use the parts-and-total diagram.	*Math Masters,* p. 123 (optional); toolkit coins (optional); number grid (optional)	2.OA.1, 2.NBT.5, 2.NBT.7 SMP1, SMP2, SMP4
Solving Parts-and-Total Number Stories Children solve parts-and-total number stories.	*Math Journal 2,* p. 123	2.OA.1, 2.NBT.5, 2.NBT.7
✓ **Assessment Check-In** See page 499.	*Math Journal 2,* p. 123	2.OA.1, 2.NBT.5, 2.NBT.7

CCSS 2.NBT.7 **Spiral Snapshot**

GMC Add multidigit numbers using models or strategies.

5-4 Warm Up Focus Practice	5-6 through 5-8 Focus Practice	5-9 Focus Practice	5-10 Focus Practice	6-1 Practice	6-2 through 6-8 Focus Practice

Spiral Tracker   **Go Online** to see how mastery develops for all standards within the grade.

3 Practice   15–20 min

Practicing with Fact Triangles Children practice facts using Fact Triangles.	*Math Journal 2,* pp. 250–253; Fact Triangles	2.OA.2
Math Boxes 5-9 Children practice and maintain skills.	*Math Journal 2,* p. 124	See page 499.
Home Link 5-9 **Homework** Children solve parts-and-total number stories.	*Math Masters,* pp. 141–142	2.OA.1, 2.NBT.5, 2.NBT.7

connectED.mheducation.com

Plan your lessons online with these tools.

 ePresentations   Student Learning Center   Facts Workshop Game   eToolkit   Professional Development   Home Connections   Spiral Tracker   Assessment and Reporting   English Learners Support   Differentiation Support

Differentiation Options

 RtI

CCSS 2.OA.1, SMP4

Readiness — 5–15 min

Joining Objects

WHOLE CLASS
SMALL GROUP
PARTNER
INDEPENDENT

counters, slate, craft sticks

For experience with parts-and-total situations using a concrete model, children model addition number stories on their slates with counters. Show children how to divide their slates into two sections with a craft stick. Tell a parts-and-total number story. For example, say: *I had 5 white shells and 3 gray shells.* Children should place 5 counters on one side of the craft stick and 3 counters on the other. Ask: *How many shells do I have all together?* 8 They pick up the craft stick and use it to sweep the counters together. Then they count the total. **GMP4.1** Have children retell the number story and answer to a partner. Repeat with different number stories as needed.

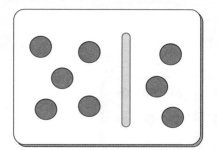

CCSS 2.OA.1, 2.NBT.5, 2.NBT.7

Enrichment — 5–15 min

Writing Missing-Part Number Stories

WHOLE CLASS
SMALL GROUP
PARTNER
INDEPENDENT

Activity Card 76;
Math Journal 2, p. 123;
Math Masters, p. TA7

To apply their understanding of parts-and-total situations, children make up their own number stories that involve a missing part.

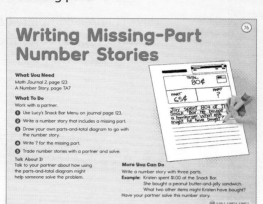

CCSS 2.OA.1, 2.NBT.5, 2.NBT.7

Extra Practice — 5–15 min

Solving More Parts-and-Total Stories

WHOLE CLASS
SMALL GROUP
PARTNER
INDEPENDENT

Math Masters, p. 140

For additional practice solving parts-and-total number stories, children complete *Math Masters,* page 140.

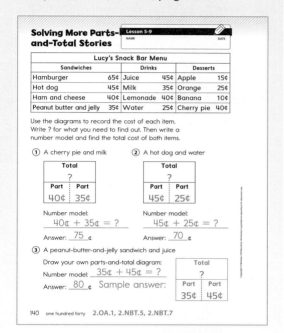

English Language Learners Support

Beginning ELL Use role play and teacher think-aloud statements to help children construct the meaning of the term *total* to mean "quantities combined to make a desired or targeted amount." Give children 5 or fewer of the same object, such as counters. Use statements such as these: *I want 6 _____. I want a total of 6 _____. Who can help me?* Encourage volunteers to count aloud to show that together they have the total requested. Model combining your objects with someone else's. Then have children complete sentence starters such as these: *I have _____. My friend has _____. Together we have a total of _____.* Repeat for totals through 10.

 Go Online   ELL English Learners Support

Standards and Goals for Mathematical Practice

SMP1 Make sense of problems and persevere in solving them.

  GMP1.1 Make sense of your problem.

SMP2 Reason abstractly and quantitatively.

  GMP2.1 Create mathematical representations using numbers, words, pictures, symbols, gestures, tables, graphs, and concrete objects.

  GMP2.3 Make connections between representations.

SMP4 Model with mathematics.

  GMP4.1 Model real-world situations using graphs, drawings, tables, symbols, numbers, diagrams, and other representations.

  GMP4.2 Use mathematical models to solve problems and answer questions.

① Warm Up 15–20 min ⟨Go Online⟩ ePresentations eToolkit

▶ Mental Math and Fluency

Dictate 2- and 3-digit numbers and have children mentally add and subtract 10 and 100, recording their answers on their slates.
Leveled exercises:

- ●○○ Add 10 to the following: 70, 122, and 140. 80; 132; 150
  Subtract 10 from the following: 40, 156, and 290. 30; 146; 280
- ●●○ Add 100 to the following: 66, 800, and 620. 166; 900; 720
  Subtract 100 from the following: 400, 212, and 707. 300; 112; 607
- ●●● Add 10 to the following: 193, 291, and 399. 203; 301; 409
  Subtract 10 from the following: 207, 406, and 102. 197; 396; 92

▶ Daily Routines

Have children complete daily routines. See the Planning Ahead note in Lesson 5-1 for an adjustment to the Temperature Routine that will prepare children for Lesson 5-10.

② Focus 30–40 min ⟨Go Online⟩ ePresentations eToolkit

▶ Math Message

What is the total number of dots? 17

► # Introducing the Parts-and-Total Diagram

WHOLE CLASS **SMALL GROUP** | PARTER | INDEPENDENT

Math Message Follow-Up Draw a unit box with the label *dots*. Display a **parts-and-total diagram.** (*See Before You Begin.*) Write 8 and 9 in the two boxes labeled Part. Write 17 in the box labeled Total.

Tell children that the diagram is a convenient way to represent the domino in the Math Message. The Part boxes show the number of dots on each side of the domino, and the Total box shows the **total** number of dots on the domino. **GMP2.1**

Academic Language Development Show children different examples of *diagrams,* such as simple Venn diagrams or data tables, and explain that diagrams are simple drawings that help to organize their thinking. **GMP1.1**

Pose an additional problem. Ask: *A hot dog costs 45¢. An orange costs 25¢. What is the total cost?*

Erase the label in the unit box and the numbers in the parts-and-total diagram. Write the label ¢ in the unit box. Discuss why the diagram is a good way to organize the information from the **parts-and-total number story. GMP4.1** Sample answer: The cost of a hot dog is one part of the total cost, and the cost of an orange is the other part. Write 45¢ and 25¢ in the two Part boxes. The total cost is unknown, so write ? in the Total box. **GMP1.1**

Below the diagram, write a number model that represents the problem, using a question mark for what children want to find out (or what they don't know): 45¢ + 25¢ = ? **GMP4.1** Ask: *How do we find the total cost?* Add 45 + 25. Ask children to share their strategies for finding the total cost. Sample strategies may include the following:

- Count up from the larger addend by using the values of dimes and nickels. 45¢, 55¢, 65¢, 70¢ Use an open number line to model counting up to solve the problem.

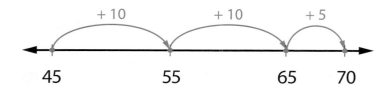

- Use toolkit coins. Model 45¢ with 4 dimes and 1 nickel. Think of 25¢ as 2 dimes and 1 nickel. Add the dimes and then add the nickels. Then find the total cost. 4 dimes + 2 dimes = 6 dimes; 1 nickel + 1 nickel = 2 nickels; 6 dimes + 2 nickels = 70¢

- Use the number grid. Start at 45. Go down two 10s (55, 65) and then go five 1s to the right. You land on 70.

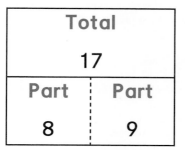

Unit
dots

Total
17

Part	Part
8	9

A parts-and-total diagram for the domino in the Math Message

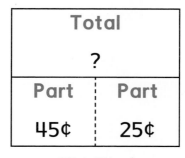

Unit
¢

Total
?

Part	Part
45¢	25¢

45¢ + 25¢ = ?
45¢ + 25¢ = 70¢

Below the number model with the question mark, write a summary number model for the problem with 70¢ substituted for the question mark: 45¢ + 25¢ = 70¢. **GMP4.2** Refer to the diagram and to the models. Ask children how each element in the diagram relates to the number models. Sample answer: The 45 and 25 in the Part boxes represent the two numbers we add in the number models. The ? in the Total box represents what we do not know or need to find out. It is the same as the ? in the first number model. The second number model shows the answer to the number story in place of the question mark. **GMP2.3**

▸ Solving Parts-and-Total Number Stories

Math Journal 2, p. 123

WHOLE CLASS | SMALL GROUP | PARTNER | INDEPENDENT

Working alone or as partners, children find the total cost of the items in Problem 1 on journal page 123. Go over children's answers, prompting them to share their solution strategies. The discussion will allow you to determine how much help children need to complete the rest of the journal page. If more practice is needed, make up and solve several number stories like Problem 1. Display and use parts-and-total diagrams.

Differentiate **Adjusting the Activity**

Challenge children by having them solve these number stories:

- Josh has $1.00. He buys a hot dog and milk. What can he buy for dessert? An apple or a banana
- What is the total cost of all the items on the snack bar menu? $4.20 Use a calculator to check the total.
- Choose a sandwich, a drink, and a dessert for yourself. How much will these items cost? Write a number model. Answers vary.

Go Online | Differentiation Support

Assessment Check-In CCSS 2.OA.1, 2.NBT.5, 2.NBT.7

Math Journal 2, p. 123

Expect most children to solve Problems 1–2 on journal page 123. If children have difficulty solving the number stories, encourage them to use various tools, such as an open number line, a drawing, or a number grid. Some children may be able to draw the parts-and-total diagram for Problem 3 and solve the number story. At this time, do not expect all children to write number models for these problems. Later lessons will provide opportunities for children to organize information in diagrams and write number models.

Assessment and Reporting | Go Online to record student progress and to see trajectories toward mastery for these standards.

Math Journal 2, p. 123

Parts-and-Total Number Stories

Lesson 5-9
DATE

Lucy's Snack Bar Menu

Sandwiches		Drinks		Desserts	
Hamburger	65¢	Juice	45¢	Apple	15¢
Hot dog	45¢	Milk	35¢	Orange	25¢
Ham and cheese	40¢	Lemonade	40¢	Banana	10¢
Peanut butter and jelly	35¢	Water	25¢	Cherry pie	40¢

For each problem, you are buying two items. Record the cost of each item in a diagram. Write ? for the number you need to find. Then write a number model and find the total cost of both items.

1. A lemonade and a banana

Total	
?	
Part	Part
40¢	10¢

Number model: 40¢ + 10¢ = ?

Answer: 50 ¢

2. A hot dog and an apple

Total	
?	
Part	Part
45¢	15¢

Number model: 45¢ + 15¢ = ?

Answer: 60 ¢

3. A hamburger and an orange

Draw your own parts-and-total diagram.

Sample answer:

Total	
?	
Part	Part
65¢	25¢

Number model: 65¢ + 25¢ = ?

Answer: 90 ¢

2.OA.1, 2.NBT.5, 2.NBT.7 — one hundred twenty-three 123

Math Journal 2, p. 124

Math Boxes

Lesson 5-9
DATE

1. You buy juice for 64¢. Use Ⓠ, Ⓓ, Ⓝ, and Ⓟ to show the coins you use to buy the juice.
Sample answer: ⓆⓆⒹⓅⓅⓅⓅ

2. Solve.
54 + 10 = 64
76 + 10 = 86
63 + 100 = 163
145 = 100 + 45

3. Put an X on the digit in the tens place.
4X6
3X9
5X0

4. Count up by 100s.
25, 125, 225, 325, 425, 525, 625

5. **Writing/Reasoning** What do you notice when you count up by 100 from 25 to 625 in Problem 4?
Sample answer: The digit in the hundreds place goes up by 1 with each count.

① 2.NBT.7, 2.MD.8 ② 2.NBT.5, 2.NBT.8 ③ 2.NBT.1 ④ 2.NBT.2, 2.NBT.8 ⑤ 2.NBT.2, SMP7

124 one hundred twenty-four

Summarize Have children share their solution strategies for the problems on journal page 123.

 Practice 15–20 min | Go Online | ePresentations | eToolkit | Home Connections

▶ Practicing with Fact Triangles

Math Journal 2, pp. 250–253

| WHOLE CLASS | SMALL GROUP | PARTNER | INDEPENDENT |

As children practice with Fact Triangles, have them take an inventory of the addition facts they know and facts that need more practice. They record their progress on the Addition Facts Inventory Record, Parts 1 and 2 on journal pages 250–253. See Lesson 3-3 for additional details.

▶ Math Boxes 5-9

Math Journal 2, p. 124

| WHOLE CLASS | SMALL GROUP | PARTNER | INDEPENDENT |

Mixed Practice Math Boxes 5-9 are paired with Math Boxes 5-11.

▶ Home Link 5-9

Math Masters, pp. 141–142

Homework Children solve parts-and-total number stories. They fill in a parts-and-total diagram and write a number model for each problem.

Math Masters, p. 141

Parts-and-Total Number Stories
Home Link 5-9
NAME | DATE

Family Note

Your child has learned how to represent and solve problems by using parts-and-total diagrams. Parts-and-total diagrams are used to organize the information in problems in which two or more quantities (parts) are combined to form a total quantity.

Please return the second page of this Home Link to school tomorrow.

Large suitcase 45 pounds | Small suitcase 30 pounds | Backpack 17 pounds | Package 15 pounds

Use the weights shown in the pictures above to do the following for each number story on the next page:

- Write the numbers you know in a parts-and-total diagram.
- Write ? for the number you need to find.
- Write a number model. Use ? for the number you need to find.
- Answer the question.

Example: You carry the small suitcase and the package. How many pounds do you carry in all?

The parts are known. The total is to be found.

Possible number model: $30 + 15 = ?$

Answer: 45 pounds

Total	
?	
Part	Part
30	15

one hundred forty-one 141

Math Masters, p. 142

Parts-and-Total Number Stories
Home Link 5-9
NAME | DATE
(continued)

① You wear the backpack and carry the small suitcase. How many pounds do you carry in all?

Total	
?	
Part	Part
17	30

Number model:
$17 + 30 = ?$

Answer: 47 pounds

② You carry the large suitcase and the small suitcase. How many pounds do you carry in all?

Total	
?	
Part	Part
45	30

Number model:
$45 + 30 = ?$

Answer: 75 pounds

③ You wear the backpack and carry the package. How many pounds do you carry in all?

Draw your own parts-and-total diagram:

Total	
?	
Part	Part
17	15

Number model: $17 + 15 = ?$

Answer: 32 pounds

142 one hundred forty-two 2.OA.1, 2.NBT.5, 2.NBT.7

Lesson 5-10

Change Number Stories

Overview Children solve change number stories involving temperature.

▶ **Vocabulary**

thermometer • degree Fahrenheit (°F) • change diagram • change-to-less number story

 **Common Core State Standards**

Focus Clusters
- Represent and solve problems involving addition and subtraction.
- Understand place value.
- Use place value understanding and properties of operations to add and subtract.

1 Warm Up 15–20 min	**Materials**	
Mental Math and Fluency Children count up and back by 2s and 10s.		**2.NBT.2**
Daily Routines Children complete daily routines.	See pages 4–43.	See pages xiv–xvii.

2 Focus 30–40 min		
Math Message Children make observations about a thermometer.	Class Thermometer Poster	
Discussing the Thermometer Children discuss thermometers and temperatures.	Class Thermometer Poster	**2.NBT.2**
Solving Change Number Stories Children solve change-to-more and change-to-less stories.	*My Reference Book,* pp. 27–29; *Math Masters,* p. TA26 (optional); slate	**2.OA.1, 2.NBT.5, 2.NBT.7** **SMP1, SMP4**
Using Change Diagrams Children solve temperature-change problems.	*Math Journal 2,* pp. 125–126	**2.OA.1, 2.NBT.5, 2.NBT.7**
✓ **Assessment Check-In** See page 505.	*Math Journal 2,* pp. 125–126	**2.OA.1, 2.NBT.5, 2.NBT.7**

CCSS 2.NBT.7  **Spiral Snapshot**

GMC Add multidigit numbers using models or strategies.

| 5-4
Warm Up
Focus
Practice | 5-6 through 5-9
Focus
Practice | 5-10
Focus
Practice | 6-1
Practice | 6-2 through 6-8
Warm Up
Focus
Practice | 7-1
Warm Up
Practice |

Spiral Tracker **Go Online** to see how mastery develops for all standards within the grade.

3 Practice 15–20 min		
Playing *Number Top-It* **Game** Children practice comparing 3-digit numbers.	*My Reference Book,* pp. 170–172; *Math Masters,* pp. G7–G8; 4 each of number cards 0–9; glue or tape; paper	**2.NBT.1, 2.NBT.3, 2.NBT.4** **SMP7**
Math Boxes 5-10 Children practice and maintain skills.	*Math Journal 2,* p. 127; centimeter ruler	See page 505.
Home Link 5-10 **Homework** Children solve change number stories.	*Math Masters,* pp. 146–147	**2.OA.1, 2.NBT.5, 2.NBT.7**

connectED.mheducation.com

Plan your lessons online with these tools.

 ePresentations   Student Learning Center   Facts Workshop Game   eToolkit   Professional Development   Home Connections   Spiral Tracker   Assessment and Reporting   English Learners Support   Differentiation Support

Differentiation Options

 RtI

Readiness
5–15 min

| WHOLE CLASS |
| SMALL GROUP |
| PARTNER |
| INDEPENDENT |

Showing Change on a Number Line

Math Journal 2, inside back cover

For support working with change-to-more or change-to-less situations, children use the number line on the inside back cover of their journals to solve problems. See the Readiness activity in Lesson 5-8.

Shown below are the answers to *Math Masters,* page 145 from the Extra Practice activity.

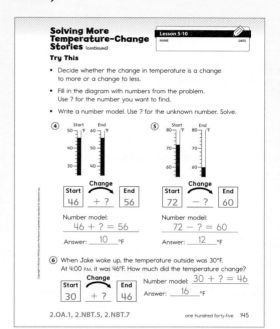

Enrichment
5–15 min

| WHOLE CLASS |
| SMALL GROUP |
| PARTNER |
| INDEPENDENT |

Finding Changes in Temperature

Math Masters, p. 143

To extend their understanding of change situations, children identify and use daily A.M. and P.M. temperatures in various cities or towns to solve change number stories. **GMP4.1**

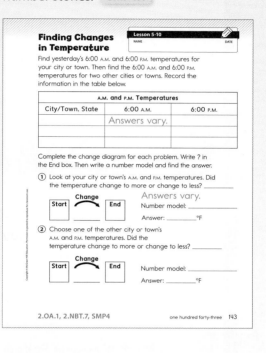

Extra Practice
5–15 min

| WHOLE CLASS |
| SMALL GROUP |
| PARTNER |
| INDEPENDENT |

Solving More Temperature-Change Stories

Math Masters, pp. 144–145

For additional practice solving temperature-change problems, children complete *Math Masters,* pages 144–145. See the Readiness column for answers to *Math Masters,* page 145.

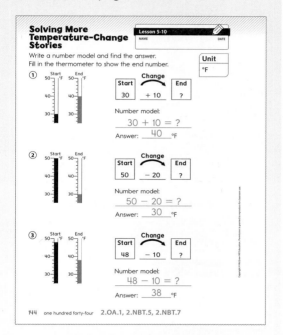

English Language Learners Support

Beginning ELL Use objects, pictures, and gestures to demonstrate the meanings of the words *hot* (for example, steam rising from a hot beverage in a cup), *cold* (an ice cube), *cool* (a fan), and *warm* (a blanket). Provide opportunities for children to have oral practice with the terms by showing pictures and using sentence starters such as the following: *This is an ice cube. It is _____. This is snow. It is _____. This is a sweater. It keeps me _____.*

 Go Online   ELL   English Learners Support

SMP1 Make sense of problems and persevere in solving them.

GMP1.1 Make sense of your problem.

SMP4 Model with mathematics.

GMP4.1 Model real-world situations using graphs, drawings, tables, symbols, numbers, diagrams, and other representations.

GMP4.2 Use mathematical models to solve problems and answer questions.

Class Thermometer Poster

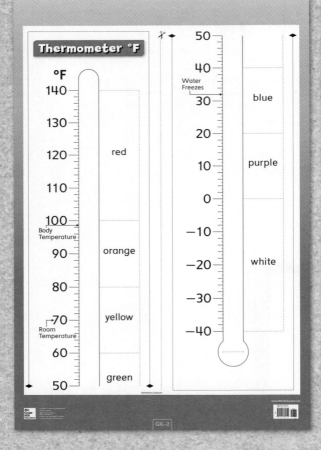

1 Warm Up 15–20 min   Go Online ePresentations eToolkit

▶ Mental Math and Fluency

Children practice counting up and back by 10s and 2s from a multiple of 10. Counting by 2s is a useful skill for reading temperatures on a thermometer with a scale marked at 2-degree intervals. Have the class count orally in unison. *Leveled exercises:*

- ●○○ Count up by 10s from 0 to 100. 0, 10, 20, 30, . . . , 100
  Count up by 2s from 0 to 20. 0, 2, 4, 6, . . . , 20

- ●●○ Count up by 2s from 60 to 70. 60, 62, 64, 66, 68, 70
  Count back by 10s from 100 to 0. 100, 90, 80, 70, . . . , 0

- ●●● Count back by 2s from 20 to 10. 20, 18, 16, 14, 12 , 10
  Count back by 2s from 60 to 50. 60, 58, 56, 54, 52, 50

▶ Daily Routines

Have children complete daily routines.

2 Focus 30–40 min   Go Online ePresentations eToolkit

▶ Math Message

Look at the Class Thermometer Poster. What do you notice about the thermometer?

▶ Discussing the Thermometer

| WHOLE CLASS | SMALL GROUP | PARTNER | INDEPENDENT |

Math Message Follow-Up Ask children to share their observations about the Class Thermometer Poster (°F). Point out that a **thermometer** measures temperature, or how hot or cold something is (relative to the number scale on the thermometer). Cover the following points:

- The narrow glass tube in a thermometer contains a liquid that expands when it gets warmer, causing the liquid to rise in the tube. The warmer the temperature, the higher the liquid rises.

- In the United States, everyday temperatures—such as those in weather reports and recipes—are usually given in **degrees Fahrenheit (°F).** In the sciences, temperatures are often given in degrees Celsius (°C), which is also becoming common in everyday life. Ask children to identify zones on the thermometer that might correspond to hot, cold, warm, and cool temperatures.

- Each multiple of 10 degrees is marked and labeled with a number on the thermometer's scale. Have a volunteer point to the thermometer as children skip count by 10s from 0 to 140.

- Between the multiples of 10 degrees, the degree marks are spaced at 2-degree intervals, such as at 72, 74, 76, and 78 degrees. Have a volunteer point to the marks on the thermometer as children skip count by 2s from 70 to 100. Point out that the scales on thermometers can vary.
- Ask: *At what temperature does water freeze?* 32°F

Remind children that in previous lessons they solved problems using parts-and-total diagrams and **change diagrams.** Today they will solve temperature-change problems with change diagrams.

Academic Language Development Show children different measuring tools, such as a ruler, a weighing scale, a measuring cup, and so on. Have them name items they can measure with each tool and specify each tool's unit(s) of measure. Talk about a thermometer as a tool for measuring temperature, with *degree* as the unit of measure.

▶ Solving Change Number Stories

My Reference Book, pp. 27–29

WHOLE CLASS	SMALL GROUP	PARTNER	INDEPENDENT

You may wish to review change diagrams by rereading *My Reference Book,* page 27 with your class. Then read *My Reference Book,* pages 28–29. Briefly discuss the similarities and differences between the diagrams. *My Reference Book,* page 29 provides an example using temperatures in which the change is unknown. Quickly draw or display a change diagram (*Math Masters,* page TA26) and a unit box labeled *degrees Fahrenheit* (°F).

Using temperatures that are multiples of 5 or 10, pose several temperature-change problems like the one below. Have children write the answers on their slates. Ask:

- *It was 50°F at 9:00 A.M. and 70°F at noon. Did it get warmer or cooler?* Warmer
- *What type of change happened?* Sample answers: It got warmer. The temperature went up.
- *Did it change to more or change to less?* Change to more **GMP1.1**

Then write the information in a change diagram. **GMP4.1** Write 50 in the Start box, 70 in the End box, and then + ? on the Change line.

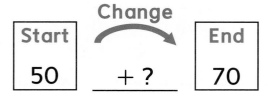

My Reference Book, p. 27

Operations and Algebraic Thinking

Some number stories are about a change. The number you start with changes to more or changes to less. You can use a **change diagram** to help you solve these stories.

> Britney had 7 shells.
> She found 9 more shells.
> How many shells does Britney have in all?
>
> This is a change-to-more story.
>
> You know the change number.
>
> Change
>
> | Start | | End |
> |---|---|---|
> | 7 | +9 | ? |
>
> You know the starting number.   You don't know the ending number.
>
> Number model: $7 + 9 = ?$
> Number model: $7 + 9 = 16$
>
> 16 shells in all

twenty-seven MRB 27

Math Journal 2, p. 125

Temperature Changes Lesson 5-10 DATE

Write ? in the End box. Then write a number model and find the answer. Fill in the thermometer to show the End number.

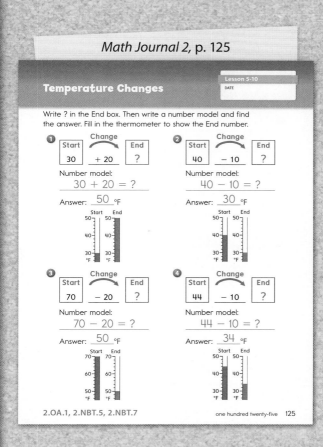

①
Start	Change	End
30	+ 20	?

Number model:
$30 + 20 = ?$

Answer: 50 °F

②
Start	Change	End
40	− 10	?

Number model:
$40 - 10 = ?$

Answer: 30 °F

③
Start	Change	End
70	− 20	?

Number model:
$70 - 20 = ?$

Answer: 50 °F

④
Start	Change	End
44	− 10	?

Number model:
$44 - 10 = ?$

Answer: 34 °F

2.OA.1, 2.NBT.5, 2.NBT.7   one hundred twenty-five 125

Math Journal 2, p. 126

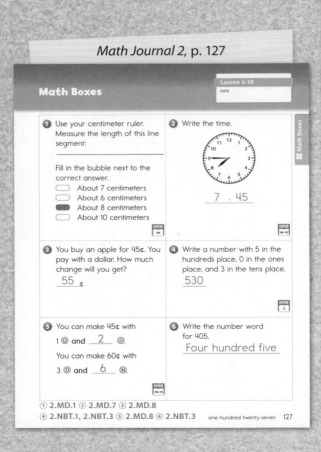

Math Journal 2, p. 126

Math Journal 2, p. 127

Math Journal 2, p. 127

Write the number model 50 + ? = 70 below the diagram. GMP4.1 Point to the relevant parts of the change diagram as you read the number model aloud. Ask: *How many degrees warmer did it get?* 20°F GMP4.2 Record the summary number model 50 + 20 = 70 and have children share their solution strategies. *For example:*

- Model the change on a number line. Sketch an open number line on the board to illustrate the problem.

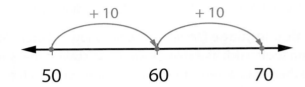

- Use a number grid. Start at 50. Count two 10s to 70.

Using temperatures that are multiples of 5 or 10, pose another problem like the following: *It was 40°F at six o'clock. By ten o'clock, the temperature had gone down 30°F. What was the temperature at ten o'clock?*

Fill in a change diagram and remind children that the minus sign in − 30 indicates that this is a **change-to-less number story.** The temperature goes down, and it gets colder.

Write the number model 40 − 30 = ? below the diagram. GMP4.1 Ask: *What was the temperature at ten o'clock?* 10°F Have children share their solution strategies. GMP4.2 Then record the summary number model 40 − 30 = 10.

Pose additional temperature problems. Include problems that use 2-digit temperatures and have differences that are multiples of 10. *For example:* If the temperature is 72°F and then goes down 20 degrees, what is the new temperature? 52°F

▶ Using Change Diagrams

Math Journal 2, pp. 125–126

| WHOLE CLASS | SMALL GROUP | **PARTNER** | INDEPENDENT |

Partners work through journal pages 125–126 and check each other's work.

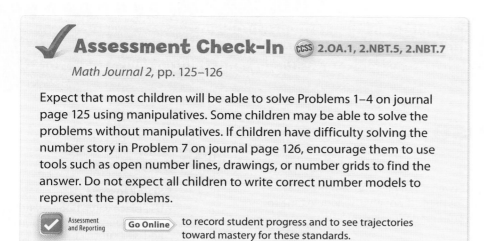

Assessment Check-In CCSS 2.OA.1, 2.NBT.5, 2.NBT.7

Math Journal 2, pp. 125–126

Expect that most children will be able to solve Problems 1–4 on journal page 125 using manipulatives. Some children may be able to solve the problems without manipulatives. If children have difficulty solving the number story in Problem 7 on journal page 126, encourage them to use tools such as open number lines, drawings, or number grids to find the answer. Do not expect all children to write correct number models to represent the problems.

Assessment and Reporting / Go Online to record student progress and to see trajectories toward mastery for these standards.

Summarize Discuss Problems 6–7 on journal page 126. Both problems give the start and end numbers but represent them in different ways.

3 Practice 15–20 min

Go Online — ePresentations, eToolkit, Home Connections

▶ Playing *Number Top-It*

My Reference Book, pp. 170–172; Math Masters, pp. G7–G8

WHOLE CLASS | **SMALL GROUP** | **PARTNER** | INDEPENDENT

Have children compare numbers by playing *Number Top-It.* See Lesson 4-5 or have children refer to *My Reference Book,* pages 170–172 for directions. **GMP7.2** Have children record their comparisons using <, >, or = on paper.

▶ Math Boxes 5-10

Math Journal 2, p. 127

WHOLE CLASS | **SMALL GROUP** | **PARTNER** | **INDEPENDENT**

Mixed Practice Math Boxes 5-10 are paired with Math Boxes 5-6.

▶ Home Link 5-10

Math Masters, pp. 146–147

Homework Children practice using change diagrams.

Math Masters, p. 146

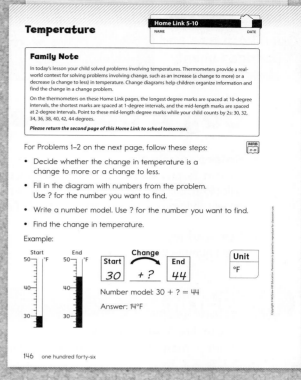

Math Masters, p. 147

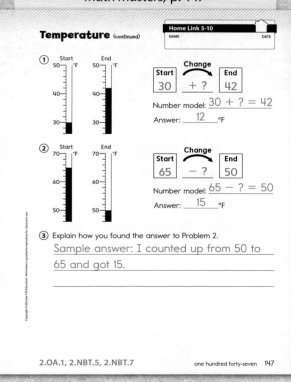

Lesson 5-10 **505**

Adding Multidigit Numbers

Overview **Day 1:** Children complete an open response problem by solving an addition problem using two different trategies.
Day 2: The class discusses selected strategies, and children revise their work.

Day 1: Open Response

▶ **Before You Begin**
For Part 2a, prepare packs of toolkit bills that include at least twelve $10 bills and seventeen $1 bills (*Math Masters*, pp. G12–G13) for children who want to use them. If possible, schedule time to review children's work and plan for Day 2 of the lesson with your grade-level team.

▶ **Vocabulary**
open number line

Common Core State Standards

Focus Clusters
• Use place value understanding and properties of operations to add and subtract.
• Work with time and money.

1 Warm Up	15–20 min	**Materials**	
Mental Math and Fluency Children mentally add and subtract 10 and 100.		slate	2.NBT.8
Daily Routines Children complete daily routines.		See pages 4–43.	See pages xiv–xvii.

2a Focus	45–55 min		
Math Message Children solve a subtraction number story and talk about their strategies with a partner.		*Math Journal 2,* p. 128; number grid	2.NBT.5, 2.NBT.9, 2.MD.8 SMP1, SMP4, SMP5
Buying a Clock Children discuss their subtraction strategies and the importance of attending to the units in the number story.		*Math Journal 2,* p. 128; number grid	2.NBT.5, 2.NBT.9, 2.MD.8 SMP1, SMP4, SMP5
Solving the Open Response Problem Children model a shopping problem and show two different strategies for finding the sum of two prices.		*Math Journal 2,* p. 129; *My Reference Book,* pp. 3–5 (optional); *Math Masters,* pp. 148–149; base-10 blocks; $10 and $1 toolkit bills (see Before You Begin); number grid; Standards for Mathematical Practice Poster	2.NBT.5, 2.NBT.9, 2.MD.8 SMP1, SMP4, SMP5

Getting Ready for Day 2 →

Review children's work and plan discussion for reengagement. *Math Masters,* p. TA5; children's work from Day 1

CCSS 2.NBT.5 Spiral Snapshot

GMC Add within 100 fluently.

| 5-3
Focus
Practice | 5-6 through 5-10
Focus
Practice | 5-11
Focus
Practice | 6-1
Warm Up
Practice | 6-2 through 6-8
Warm Up
Focus
Practice | 7-1
Warm Up
Focus
Practice |

Spiral Tracker   **Go Online**   to see how mastery develops for all standards within the grade.

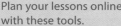

1 Warm Up 15–20 min Go Online ePresentations eToolkit

▶ Mental Math and Fluency

Dictate 2- and 3-digit numbers. Have children mentally add or subtract 10 or 100 and record their answers on their slates.

● ○ ○ Add 10 to the following: 62 72; 88 98; and 97 107
Subtract 10 from the following: 42 32; 78 68; and 110 100

● ● ○ Add 100 to the following: 121 221; 200 300; and 177 277
Subtract 100 from the following: 133 33; 220 120; and 700 600

● ● ● Add 10 to the following: 224 234; 298 308; and 409 419
Subtract 10 from the following: 600 590; 508 498; and 222 212

▶ Daily Routines

Have children complete daily routines.

2a Focus 45–55 min Go Online ePresentations eToolkit

▶ Math Message

Math Journal 2, p. 128

Turn to and complete journal page 128. **GMP1.5, GMP4.2, GMP5.2**

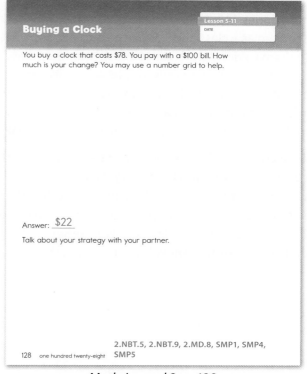

Buying a Clock Lesson 5-11 DATE

You buy a clock that costs $78. You pay with a $100 bill. How much is your change? You may use a number grid to help.

Answer: $22

Talk about your strategy with your partner.

128 one hundred twenty-eight 2.NBT.5, 2.NBT.9, 2.MD.8, SMP1, SMP4, SMP5

Math Journal 2, p. 128

Standards and Goals for Mathematical Practice

SMP1 Make sense of problems and persevere in solving them.
GMP1.5 Solve problems in more than one way.

SMP4 Model with mathematics.
GMP4.2 Use mathematical models to solve problems and answer questions.

SMP5 Use appropriate tools strategically.
GMP5.2 Use tools effectively and make sense of your results.

Professional Development

The focus of this lesson is **GMP1.5.** Children draw from a range of tools and strategies to calculate the cost of two items in two different ways. Using multiple strategies helps children deepen their understanding of the role of place value in addition. At this point, children are not expected to use a traditional paper-and-pencil algorithm for addition. This lesson allows children to experiment with tools and their own addition methods. Formal addition methods are addressed in Lessons 6-6, 6-7, and 6-8.

Go Online for information about **SMP1** in the *Implementation Guide*.

Math Masters, p. 148

Shopping at a Department Store   Lesson 5-11

NAME   DATE

Select two items to buy from the Shopping Poster. Write the names of the items and their prices.

Item 1: _____   Price of Item 1: _____

Item 2: _____   Price of Item 2: _____

The cash register at the store is broken. Find the total cost of your items <u>two different ways</u> to be sure you are correct.

① Show and explain your first strategy here.

Answers vary. See sample children's work on page 513 of the *Teacher's Lesson Guide*.

2.NBT.5, 2.NBT.9, 2.MD.8, SMP1, SMP4, SMP5

148   one hundred forty-eight

► Buying a Clock

Math Journal 2, p. 128

| WHOLE CLASS | SMALL GROUP | PARTNER | INDEPENDENT |

Math Message Follow-Up Ask children to share the strategies they used to solve the Math Message number story. Discuss the different tools children might use to solve the problem, such as number grids, **open number lines,** parts-and-total diagrams, or other drawings. GMP4.2, GMP5.2 Highlight different strategies children may use with the same tool, such as counting up and counting back. GMP1.5

> **Differentiate** **Common Misconception**
>
> Many activities involving money in Units 1–4 used cents as the unit, so children may report their answer to the Math Message problem in cents. Ask these children to complete a unit box for the problem and discuss the difference between 12 cents and 12 dollars. Emphasize that strategies for working with cents or other units also work for dollars. Have children discuss the similarities and differences in the symbols $ and ¢. For example, the dollar symbol is written before the number (as in $24) while the cents symbol is written after the number (as in 24¢).

Tell children that they are going to think more about how they can use different strategies to solve problems. GMP1.5

► Solving the Open Response Problem

Math Journal 2, p. 129; *Math Masters*, pp. 148–149

| WHOLE CLASS | SMALL GROUP | PARTNER | INDEPENDENT |

Distribute *Math Masters,* pages 148–149 and have children turn to journal page 129. Read the problem as a class and ask partners to discuss what the problem is asking them to do. Make base-10 blocks, number grids, and packs of toolkit bills ($1 and $10 bills only) available to children. Explain that an important part of the task is to show two different ways to find the total cost. Emphasize that solving a problem more than one way does not mean doing the problem a second time the same way. GMP1.5

As you circulate, look for children who use mental strategies to solve the problem but only explain their strategies by writing "I did it in my head" or "I added them." Ask children to explain orally exactly what they did, and then have them use words, numbers, or pictures to show what they did. Encourage children who use base-10 blocks, bills, or other manipulatives to include drawings that show how they used the tools. GMP4.2, GMP5.2

Differentiate Adjusting the Activity

If children have difficulty getting started, encourage them to model the cost of each item using base-10 blocks or toolkit bills (or drawings of these tools). Ask: *What do you need to do now to figure out the total cost of the two items?* GMP4.2, GMP5.2 Answers vary. For children who successfully show one strategy but are not sure how to begin a second, encourage them to try using a different tool. GMP1.5

Partners should work together to share ideas about the task, but children should select their own items and record their strategies.

Summarize You may wish to read with children pages 3–5 in *My Reference Book,* which describes children's thinking about solving problems and using more than one strategy. Ask: *Why do you think it is important to be able to solve a problem more than one way?* GMP1.5 Sample answers: to check your answer; to learn which strategies you like best Refer children to the Standards for Mathematical Practice Poster.

Collect children's work so that you can evaluate it and prepare for Day 2.

Getting Ready for Day 2

Math Masters, p. TA5

Planning a Follow-Up Discussion

Review children's work. Use the Reengagement Planning Form (*Math Masters,* page TA5) and the rubric on page 511 to plan ways to help children meet expectations for both the content and practice standards. Look for common misconceptions in addition strategies as well as use of a variety of tools and interesting strategies.

Organize the discussion in one of the ways below or in another way you choose. If children's work is unclear or if you prefer to show work anonymously, rewrite the work for display.

Go Online for sample children's work that you can use in your discussion.

Math Masters, p. 149

Shopping at a Department Store (continued) Lesson 5-11

NAME   DATE

② Show and explain your second strategy here.

Answers vary. See sample children's work on page 513 of the *Teacher's Lesson Guide.*

2.NBT.5, 2.NBT.9, 2.MD.8, SMP1, SMP4, SMP5

one hundred forty-nine 149

Math Journal 2, p. 129

Shopping Poster Lesson 5-11

DATE

Telephone $36	Camera $63	Basketball $25	Calculator $17
Toaster $39	Iron $32	Blank CDs $14	Chair $28

one hundred twenty-nine 129

Sample child's work, Child A

Item 1: _camera_ Price of Item 1: _63_

Item 2: _Basketball_ Price of Item 2: _25_

① Show and explain your **first** strategy here.

So I had 63 $ I added 25
and I got 88 as my total.

+10 +10 +5

63   88

Sample child's work, Child B

Item 1: _Camera_ Price of Item 1: _$63_

Item 2: _Blank CDs_ Price of Item 2: _$14_

① Show and explain your **first** strategy here.

Start at 63 then go up 7 then
go down that will be 80 so go
back 3 that will be 77.

63 + 7 = 70   70 + 10 = 80   80 − 3 = 77.

+7 +10 +3   77

63 70 80 77

Sample child's work, Child C

Item 1: _Phone_ Price of Item 1: _$36_

Item 2: _caclulater_ Price of Item 2: _$17_

② Show and explain your **second** strategy here.

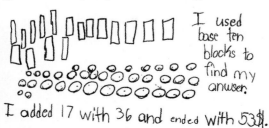

I used base ten blocks to find my answer.

I added 17 with 36 and ended with 53$.

1. Display responses that use the same tool but different strategies to find the total. See sample work for Child A and Child B. Have children interpret and compare the strategies. Ask:
   - *What tool and strategy do you think Child A used to find the total cost?* **GMP4.2, GMP5.2** Sample answer: The child used an open number line. The child started at $63, which is the camera price, and added 10, 10, and 5 because that makes $25.
   - *What could this child do to help us understand the strategy better?* **GMP5.2** Sample answer: The child could show what the number is after each jump on the number line.
   - *How are Child A's and Child B's work alike? Different?* **GMP1.5** Sample answer: Both used an open number line. The first child added 10s then 1s. The second made a 10, added 10, and subtracted 3.
   - *What could the second child do to help us understand the strategy better?* **GMP5.2** Sample answer: The child could show subtracting 3 on the number line by going back instead of going forward.

2. Display responses that show different tools but draw on similar strategies. For example, discuss a child's response that explains adding the 10s first, then the 1s, and then combining the 10s and 1s to find the total cost. Then discuss a child's response that shows combining base-10 blocks in a similar way. **GMP1.5, GMP5.2**

3. Display responses that show tools used in more and less efficient ways. For example, discuss a child's response that shows a strategy of counting by 1s on the number grid and have children compare it with a response that shows counting by 10s and 1s. **GMP1.5**

4. Display a response that shows an inappropriate use of a tool. See sample work for Child C. Ask:
   - *What tool and strategy did this child use to find the total cost?* **GMP4.2, GMP5.2** Sample answer: The child used base-10 blocks, but the longs and cubes are all used as 1s.
   - *How could this child use the base-10 blocks in a better way?* **GMP5.2** Sample answer: The child could use the longs as 10s and the cubes as 1s.

Planning for Revisions

Have copies of *Math Masters,* pages 148–149 or extra paper available for children to use in revisions. You might want to ask children to use colored pencils so you can see what they revised.

Adding Multidigit Numbers

Overview   Day 2: The class discusses selected strategies, and children revise their work.

Day 2: Reengagement

▶ **Before You Begin**
Have extra copies available of *Math Masters,* pages 148–149 for children to revise their work.

Common Core State Standards

Focus Clusters
- Use place value understanding and properties of operations to add and subtract.
- Work with time and money.

2b Focus   50–55 min

Materials

Setting Expectations
Children review the open response problem and discuss using pictures, numbers, and words to show their strategies. They also review how to respectfully discuss their own and others' work.

Guidelines for Discussions Poster

Reengaging in the Problem
Children analyze and compare strategies they used to find the sum of the two prices.

selected samples of children's work

2.NBT.5, 2.NBT.9, 2.MD.8
SMP1, SMP4, SMP5

Revising Work
Children improve the clarity and completeness of their drawings and explanations.

Math Masters, pp. 148–149 (optional); children's work from Day 1; colored pencils (optional); base-10 blocks; $10 and $1 toolkit bills (see Before You Begin from Day 1); number grid

2.NBT.5, 2.NBT.9, 2.MD.8
SMP1, SMP4, SMP5

✓ **Assessment Check-In**   See page 513 and the rubric below.

2.NBT.5
SMP1

Goal for Mathematical Practice	Not Meeting Expectations	Partially Meeting Expectations	Meeting Expectations	Exceeding Expectations
GMP1.5 Solve problems in more than one way.	Does not provide an explanation, using words, drawings, or number models, of an appropriate addition strategy.	Provides an explanation, using words, drawings, or number models, of one appropriate addition strategy and may attempt a second strategy.	Provides an explanation, using words, drawings, or number models, of two appropriate addition strategies.	Meets expectations and provides explanations for a strategy using two or three forms (words, drawings, or number models), each of which is an adequate explanation on its own.

3 Practice   10–15 min

Math Boxes 5-11
Children practice and maintain skills.

Math Journal 2, p. 130

See page 512.

Home Link 5-11
Homework Children solve an addition number story using two different strategies.

Math Masters, p. 150

2.NBT.5, 2.NBT.9, 2.MD.8

2b Focus

50–55 min | Go Online

ePresentations eToolkit

► Setting Expectations

| **WHOLE CLASS** | SMALL GROUP | PARTNER | INDEPENDENT |

Review the open response problem from Day 1. Ask: *What do you think a complete answer to this problem needs to include?* Sample answer: It should show two different ways of figuring out the total cost of the two items. *How can you show your tools and strategies?* Sample answer: using drawings, words, or numbers

You may wish to review the guidelines for discussion you began in Units 1 and 2. Tell children that they are going to look at other children's work and think about different strategies they used to solve the problem.

► Reengaging in the Problem

| **WHOLE CLASS** | SMALL GROUP | **PARTNER** | INDEPENDENT |

Children reengage with the problem by analyzing and critiquing other children's work in pairs and in a whole-group discussion. Have children discuss with partners before sharing with the whole group. Guide this discussion based on the decisions you made in Getting Ready for Day 2.
GMP1.5, GMP4.2, GMP5.2

► Revising Work

| WHOLE CLASS | SMALL GROUP | **PARTNER** | **INDEPENDENT** |

Pass back children's work from Day 1. Before children revise anything, have them examine their responses. Ask the questions below one at a time. Have partners discuss their responses and give a thumbs-up or thumbs-down based on their own work.

- *Did you show and explain how you found the total cost two different ways?* GMP1.5
- *If you used a tool, did you show how you used it?* GMP5.2
- *Are your drawing and explanation for your first strategy clear enough that someone else could understand them?*
- *Are your drawing and explanation for your second strategy clear enough that someone else could understand them?* GMP4.2

Tell children they now have a chance to revise their work. Tell them to add to their earlier work using colored pencils or to use another sheet of paper, instead of erasing their original work.

Differentiate Adjusting the Activity

For children who wrote a strong response on Day 1, provide a clean sheet of paper and ask them to show and explain how much change they would receive if they paid for their items with a $100 bill.

Math Journal 2, p. 130

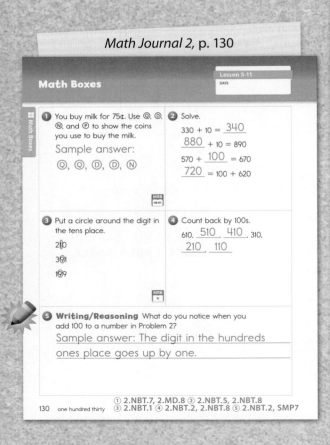

Summarize Ask children to reflect on their work and revisions. Ask: *What did you do to improve your work?* Answers vary.

 Assessment Check-In (CCSS) **2.NBT.5**

Collect and review children's revised work. Expect children to improve their work based on the class discussion. For the content standard, expect most children to correctly find the total cost for their items. You can use the rubric on page 511 to evaluate children's revised work for **GMP1.5**.

☑ Assessment and Reporting ▸ Go Online ⟩ to record student progress and to see trajectories toward mastery for these standards.

⟨ Go Online ⟩ for optional generic rubrics in the *Assessment Handbook* that can be used to assess any additional GMPs addressed in the lesson.

Sample Children's Work—Evaluated

See the sample in the margin. This work meets expectations for the content standard by showing that the sum of $63 and $17 is $80. The work meets expectations for the mathematical practice by showing and explaining two appropriate addition strategies—counting tallies and using base-10 blocks to add the 10s and then the 1s. GMP1.5

⟨ Go Online ⟩ for other samples of evaluated children's work.

③ Practice 10–15 min ⟨ Go Online ⟩ [ePresentations] [eToolkit] [Home Connections]

▶ Math Boxes 5-11 ✎

Math Journal 2, p. 130

| WHOLE CLASS | SMALL GROUP | PARTNER | INDEPENDENT |

Mixed Practice Math Boxes 5-11 are paired with Math Boxes 5-9.

▶ Home Link 5-11

Math Masters, p. 150

Homework Children solve an addition number story using two different strategies.

Sample child's work, "Meeting Expectations"

Item 1: ____Camera____ Price of Item 1: $63

Item 2: ___calculator___ Price of Item 2: $17

① Show and explain your **first** strategy here.

I yoused talees I [tally marks] started at $63 I added $17 I ended at $80.

② Show and explain your **second** strategy here.

$$\stackrel{6}{|}|||(... \times |^{17}_{...} = \stackrel{80}{|||||}$$

I started at $63 I added 17. I yoused Base ten block. I added plus 6 tens eaquls 7.7 plus 3 eaquls 10. Hen 7 eaquls 8.8 in the tens amb 0 in ther ones eaquls 80.

Math Masters, p. 150

Addition Strategies   Home Link 5-11
NAME   DATE

Family Note

In this lesson we added multidigit numbers. Your child solved an addition number story using two different strategies. Being able to solve problems more than one way and with different tools can help children confirm their answers and choose methods that work well in certain situations. Adding multidigit numbers will be revisited throughout the year.

Please return this Home Link to school tomorrow.

Uma bought a telephone for $36 and blank CDs for $14. What was her total cost? [MRB 76, 78]

① Show how to solve this problem using base-10 blocks.

Strategies vary.

Answer: $50

② Show how to solve this problem using an open number line.

Strategies vary.

Answer: $50

150 one hundred fifty 2.NBT.5, 2.NBT.9, 2.MD.8

Unit 5 Progress Check

Overview Day 1: Administer the Unit Assessments.
Day 2: Administer the Open Response Assessment.

2-Day Lesson

 Student Learning Center
Students may take
assessments digitally.

 Assessment and Reporting
Record results and track
progress toward mastery.

Day 1: Unit Assessments

1 Warm Up   5–10 min

Self Assessment
Children complete the Self Assessment.

Materials

Assessment Handbook, p. 31

2a Assess   35–50 min

⭐ **Unit 5 Assessment**
These items reflect mastery expectations to this point.

Assessment Handbook, pp. 32–33

Unit 5 Challenge (Optional)
Children may demonstrate progress beyond expectations.

Assessment Handbook, p. 34

CCSS Common Core State Standards	Goals for Mathematical Content (GMC)	Lessons	Self Assessment	Unit 5 Assessment	Unit 5 Challenge
2.OA.1	Model 1-step problems involving addition and subtraction.	5-7 to 5-10	3	6–9	
	Use addition and subtraction to solve 1-step number stories.	5-7 to 5-10	3	6–9	
2.OA.2	Subtract within 20 fluently.*			1b	
	Add within 20 fluently.*	5-1, 5-9		1a, 1c	
2.NBT.5	Add within 100 fluently.	5-3, 5-6 to 5-11		5a, 5c, 7, 8	1
	Subtract within 100 fluently.	5-6, 5-10, 5-11		5b, 9	1
2.NBT.7	Subtract multidigit numbers using models or strategies.	5-3, 5-6, 5-10		9	1
	Add multidigit numbers using models or strategies.	5-3, 5-4, 5-6 to 5-10		6–8	1, 2
2.NBT.8	Mentally add 10 to and subtract 10 from a given number.*	5-6, 5-7	4	5a–5e, 5h	2
	Mentally add 100 to and subtract 100 from a given number.*	5-6	5	5f, 5g	
2.MD.6	Represent sums and differences on a number-line diagram.*	5-7	6	6	
2.MD.8	Solve problems involving coins and bills.*	5-2 to 5-4, 5-11	1, 2	2–4	1
	Goals for Mathematical Practice (GMP)				
SMP3	Make sense of others' mathematical thinking. GMP3.2				1, 2

*Instruction and most practice on this content is complete.

/// **Spiral Tracker**   **Go Online** ⟩ to see how mastery develops for all standards within the grade.

1 Warm Up 5–10 min

▶ Self Assessment

Assessment Handbook, p. 31

| WHOLE CLASS | SMALL GROUP | PARTNER | **INDEPENDENT** |

Children complete the Self Assessment to reflect on their progress in Unit 5.

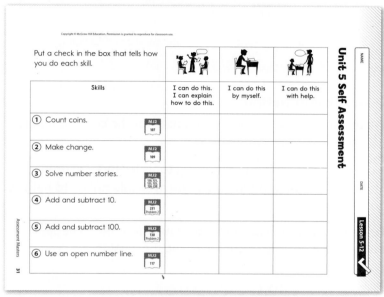

Assessment Handbook, p. 31

2a Assess 35–50 min   Go Online ✓ 👥
Assessment and Reporting   Differentiation Support

▶ Unit 5 Assessment

Assessment Handbook, pp. 32–33

| WHOLE CLASS | SMALL GROUP | PARTNER | **INDEPENDENT** |

Children complete the Unit 5 Assessment to demonstrate their progress on the Common Core State Standards covered in this unit.

Go Online for generic rubrics in the *Assessment Handbook* that can be used to evaluate children's progress on the Mathematical Practices.

Assessment Handbook, p. 32

NAME   DATE   **Lesson 5-12** ✓

Unit 5 Assessment

① Write + or − to make the number sentence true.

   a. $10 = 6 \underline{\;+\;} 4$

   b. $7 \underline{\;-\;} 5 = 2$

   c. $11 = 8 \underline{\;+\;} 3$

In Problems 2–4, use Ⓟ, Ⓝ, Ⓓ, and Ⓠ to show your answer.

② You buy a green pepper for 27¢. Show the coins you could use to pay the exact amount.

   Sample answer: Ⓓ Ⓓ Ⓝ Ⓟ Ⓟ

③ You buy some lettuce for 45¢. Show the coins you could use to pay the exact amount.

   Sample answer: Ⓠ Ⓓ Ⓓ

④ You buy yogurt for 70¢. Show the coins you could use to pay the exact amount.

   Sample answer: Ⓠ Ⓠ Ⓓ Ⓓ

⑤ Solve.

   a. $57 + 10 = \underline{67}$ b. $94 - 10 = \underline{84}$

   c. $98 + 10 = \underline{108}$ d. $\underline{90} = 100 - 10$

   e. $120 + 10 = \underline{130}$ f. $\underline{500} = 400 + 100$

   g. $200 - 100 = \underline{100}$ h. $\underline{120} = 130 - 10$

32   Assessment Handbook

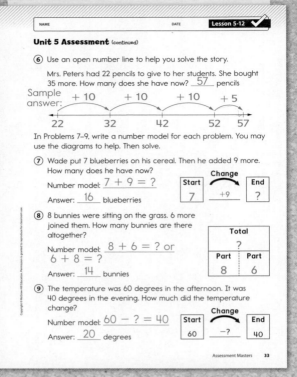

NAME   DATE   Lesson 5-12 ✓

Unit 5 Assessment (continued)

⑥ Use an open number line to help you solve the story.

Mrs. Peters had 22 pencils to give to her students. She bought 35 more. How many does she have now? __57__ pencils

Sample answer:

$+ 10$ $+ 10$ $+ 10$ $+ 5$

22   32   42   52   57

In Problems 7–9, write a number model for each problem. You may use the diagrams to help. Then solve.

⑦ Wade put 7 blueberries on his cereal. Then he added 9 more. How many does he have now?

Number model: $7 + 9 = ?$

Answer: __16__ blueberries

Start	Change $+9$	End
7		?

⑧ 8 bunnies were sitting on the grass. 6 more joined them. How many bunnies are there altogether?

Number model: $8 + 6 = ?$ or $6 + 8 = ?$

Answer: __14__ bunnies

Total ?	
Part	Part
8	6

⑨ The temperature was 60 degrees in the afternoon. It was 40 degrees in the evening. How much did the temperature change?

Number model: $60 - ? = 40$

Answer: __20__ degrees

Start	Change $-?$	End
60		40

Assessment Masters   **33**

Item(s)	Adjustments
1a–1c	To extend item 1, have children explain how they knew which symbol to use for each problem.
2–4	To scaffold items 2–4, provide coins.
5	To extend item 5, have children explain how they used mental strategies to solve Problems 5a–5h.
6	To scaffold item 6, provide access to the Class Number Line and Class Number Grid or number grid.
7–9	To scaffold items 7–9, provide counters and a number grid.

Advice for Differentiation

All instruction and most practice is complete for the content that is marked with an asterisk (*) on page 514.

Use the online assessment and reporting tools to track children's performance. Differentiation materials are available online to help you address children's needs.

> **NOTE** See the Unit Organizer on pages 438–439 or the online Spiral Tracker for details on Unit 5 focus topics and the spiral.

▶ Unit 5 Challenge (Optional)

Assessment Handbook, p. 34

| WHOLE CLASS | SMALL GROUP | PARTNER | **INDEPENDENT** |

Children can complete the Unit 5 Challenge after they complete the Unit 5 Assessment.

NAME   DATE   Lesson 5-12 ✓

Unit 5 Challenge

① Carlos buys a hamburger for 65¢ and gives the clerk $1. The clerk gives him a quarter for change. Carlos says the amount of change is incorrect. Do you agree with Carlos? Explain your answer.

Sample answer: I agree with Carlos. He started at 65¢ and added 10¢ to count up to 75¢ and then added 25¢ to count up to $1. He should get 35¢ back.

② Dana said that 192 + 10 = 102. Explain Dana's mistake.

Answers vary but should reflect an understanding of place value in addition. Sample answer: Dana should have said 192 + 10 = 202. When you have 9 tens and add another ten, the hundreds digit goes up 1 more.

34   Assessment Handbook

---

516   Unit 5 | Addition and Subtraction

Unit 5 Progress Check ✔

Day 2: Open Response Assessment

▶ **Before You Begin**
Have children's toolkit coins available.

2b Assess   50–55 min

Materials

Solving the Open Response Problem
Children find possible coin combinations for 75¢ and explain how they know one of the combinations equals 75¢.

Assessment Handbook, pp. 35–36; toolkit coins

Discussing the Problem
After completing the problem, children discuss solutions and explanations.

Assessment Handbook, pp. 35–36

CCSS Common Core State Standards

Goals for Mathematical Content (GMC)	Lessons
2.MD.8 Solve problems involving coins and bills.	5-2, 5-3, 5-4
Read and write monetary amounts.	5-2, 5-3, 5-4

Goal for Mathematical Practice (GMP)	
SMP3 Make mathematical conjectures and arguments. GMP3.1	5-2

/// **Spiral Tracker**   ⟨ Go Online ⟩ to see how mastery develops for all standards within the grade.

▶ **Evaluating Children's Responses**
Evaluate children's abilities to solve problems involving combinations of coins. Use the rubric below to evaluate their work based on **GMP3.1**.

Goal for Mathematical Practice GMP3.1 Make mathematical conjectures and arguments.	Not Meeting Expectations	Partially Meeting Expectations	Meeting Expectations	Exceeding Expectations
	Does not provide an argument that the coin combination has a value of 75 cents.	Provides an incomplete argument that the coin combination has a value of 75 cents by not referring to the value of the coins or by not providing evidence of adding the values of the coins.	Provides an argument using words, drawings, or number models that the coin combination has a value of 75 cents, referring to the value of the coins and providing evidence of adding the values of the coins.	Meets expectations and gives an argument in two or three forms (words, drawings, or number models), each of which would represent an adequate argument.

3 Look Ahead   10–15 min

Materials

Math Boxes 5-12
Children preview skills and concepts for Unit 6.

Math Journal 2, p. 131

Home Link 5-12
Children take home the Family Letter that introduces Unit 6.

Math Masters, pp. 151–154

Unit 5 Open Response Assessment
Buying from a Vending Machine

Carlos wants to buy chocolate milk from the vending machine. The milk costs 75¢. Carlos has 2 quarters, 5 dimes, and 5 nickels.

① Show at least four possible coin combinations Carlos could use to pay for the milk. Use Ⓝ, Ⓓ, and Ⓠ to record your answers.

Answers vary. See sample children's work on page 518 of the Teacher's Lesson Guide.

Sample child's work, "Meeting Expectations"

① Show at least four possible coin combinations
Carlos could use to pay for the milk.
Use Ⓝ, Ⓓ, and Ⓠ to record your answers.

① Q Q DD N   ② DDDDD NNNNN
③ Q Q NNNNN   ④ Q DDDDD

② Pick one of your coin combinations and show
or explain how you know it totals exactly 75¢.

I know two quarters makes
50¢ and I added two dimes
and that made 70¢ and then
I added a Nickele and that
made 75¢.

2b Assess 50–55 min [Go Online]
Assessment and Reporting

▶ Solving the Open Response Problem

Assessment Handbook, pp. 35–36

| WHOLE CLASS | SMALL GROUP | **PARTNER** | **INDEPENDENT** |

This open response problem requires children to apply skills and concepts from Unit 5 to determine coin combinations with a value of 75 cents. The focus of this task is **GMP3.1:** Make mathematical conjectures and arguments. Tell children that today they will figure out different coin combinations that they can use to buy chocolate milk.

Distribute *Assessment Handbook,* pages 35 and 36. Read the directions aloud. Remind children that they should find and show at least four different coin combinations that total 75 cents. Suggest that children divide the first page into four sections to separate their combinations. Make toolkit coins available. Emphasize that children will pick one of their coin combinations and show or explain how they know it totals 75 cents. GMP3.1

> **Differentiate**   **Adjusting the Assessment**
>
> If children have difficulty, work with them to create a chart that shows equivalencies for a dime (2 nickels) and a quarter (2 dimes and 1 nickel).

▶ Discussing the Problem

Assessment Handbook, pp. 35–36

| WHOLE CLASS | SMALL GROUP | PARTNER | INDEPENDENT |

After they complete their work, invite a few children to share a coin combination they found and how they know it totals exactly 75 cents.

Evaluating Children's Responses CCSS 2.MD.8

Collect children's work. For the content standard, expect most children to identify four coin combinations with a value of 75 cents and use ¢ symbols appropriately. You can use the rubric on page 517 to evaluate children's work for GMP3.1.

See the sample in the margin. This work meets expectations for the content standard because this child showed four coin combinations that have a value of 75 cents. The work meets expectations for the mathematical practice because the child provides an argument in words that the Ⓠ, Ⓠ, Ⓓ, Ⓓ, Ⓝ combination has a value of 75 cents, refers to the value of each coin, and explains how the values were added. GMP3.1

[Go Online] Assessment and Reporting

3 Look Ahead 10–15 min

Go Online | Home Connections

▶ **Math Boxes 5-12:** Preview for Unit 6

Math Journal 2, p. 131

| WHOLE CLASS | SMALL GROUP | PARTNER | INDEPENDENT |

Mixed Practice Math Boxes 5-12 are paired with Math Boxes 5-8. These problems focus on skills and understandings that are prerequisite for Unit 6. You may want to use information from these Math Boxes to plan instruction and grouping in Unit 6.

▶ **Home Link 5-12:** Unit 6 Family Letter

Math Masters, pp. 151–154

Home Connection The Unit 6 Family Letter provides parents and guardians with information and activities related to Unit 6 content.

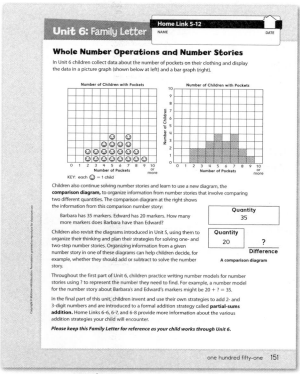

Math Masters, pp. 151–154

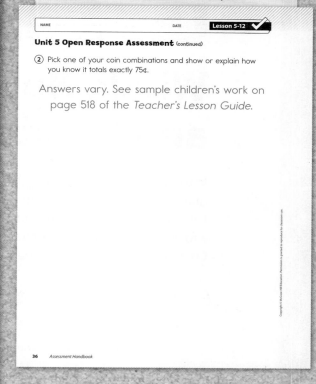

Math Journal 2, p. 131

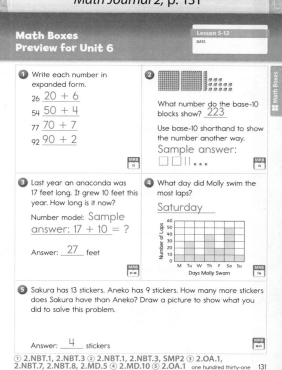

Unit 6 Organizer
Whole Number Operations and Number Stories

In this unit, children collect and display data about pockets on two different types of graphs. They are introduced to comparison number stories and two-step number stories. Later in the unit, they share and record their own invented strategies for addition and learn a formal addition strategy. Children's learning will focus on three clusters of the Common Core's content standards, as well as in-depth work on two of the Mathematical Practices.

CCSS Standards for Mathematical Content

Domain	Cluster
Operations and Algebraic Thinking	Represent and solve problems involving addition and subtraction.
Number and Operations in Base Ten	Use place value understanding and properties of operations to add and subtract.
Measurement and Data	Relate addition and subtraction to length.

Because the standards within each domain can be broad, *Everyday Mathematics* has unpacked each standard into Goals for Mathematical Content GMC. For a complete list of Standards and Goals, see page EM1.

For an overview of the CCSS domains, standards, and mastery expectations in this unit, see the **Spiral Trace** on pages 526–527. See the **Mathematical Background** (pages 528–530) for a discussion of the following key topics:

- Data Displays
- Number Stories
- Strategies for Addition

CCSS Standards for Mathematical Practice

SMP1 Make sense of problems and persevere in solving them.

SMP5 Use appropriate tools strategically.

For a discussion about how *Everyday Mathematics* develops these practices and a list of Goals for Mathematical Practice GMP, see page 531.

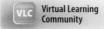 **VLC** Virtual Learning Community **Go Online** to **vlc.cemseprojects.org** to search for video clips on each practice.

©Mazer Creative Services

Go Digital with these tools at **connectED.mheducation.com**

 ePresentations
 Student Learning Center
 Facts Workshop Game
 eToolkit
 Professional Development
 Home Connections
 Spiral Tracker
 Assessment and Reporting
 English Learners Support
 Differentiation Support

Contents

*The standards listed here are addressed in the **Focus** of each lesson. For all the standards in a lesson, see the Lesson Opener.

Unit 6 Materials Virtual Learning Community

See how *Everyday Mathematics* teachers organize materials.
Search "Classroom Tours" at **vlc.cemseprojects.org**.

Lesson	Math Masters	Activity Cards	Manipulative Kit	Other Materials
6-1	pp. 155 (1 copy per partnership); 156–160; TA7		base-10 blocks (optional); per partnership: one 6-sided die	Class Data Pad (optional); number grid or number line (optional); slate
6-2	pp. 161–165; TA28		Quick Look Cards 89, 96, and 102	number grid or number line (optional); slate; Fact Triangles; 20 pennies or counters
6-3	pp. 166–167; TA29		per partnership: 2 tape measures; centimeter ruler	slate; number line, number grid, or other manipulatives
6-4	pp. 168–169; TA7	77	base-10 blocks (optional)	slate; Number-Grid Poster (optional)
6-5	pp. 170–171; TA7	78		slate; 20-foot length of masking tape; 21 index cards labeled 0–20
6-6	pp. 172–173; G21	79–80	base-10 blocks; per partnership: one 6-sided die; Class Number Line; 4 each of number cards 1–9	slate; number grid; number line (optional); poster paper; Pattern-Block Template
6-7	pp. 174–175; TA22	81	Quick Look Cards 81, 86, and 109; base-10 blocks; 4 each of number cards 1–9	slate; demonstration base-10 blocks; Class Data Pad
6-8	pp. 176; TA22	81–82	base-10 blocks; 4 each of number cards 0–10	slate
6-9	pp. 177–178; TA5		per partnership: base-10 blocks (at least 6 longs and 15 cubes)	slate; number grid; Number-Grid Poster; Standards for Mathematical Practice Poster; children's work from Day 1; Guidelines for Discussions Poster; selected samples of children's work; colored pencils (optional)
6-10	pp. 179–182; TA9 or TA10; *Assessment Handbook*, pp. 98–99	83–85	Quick Look Cards 98, 103, and 118; geoboard; rubber band; tape measure or yardstick; calculator; per group: 4 each of number cards 0–9; tape measure	slate; scissors; stick-on note; eraser; glue or tape; per group: large paper triangle (optional); 8 items of various lengths; index cards labeled *shorter than, longer than,* and *about the same as;* various lengths of string taped to labeled index cards
6-11	pp. 183–186; *Assessment Handbook*, pp. 37–45			

Literature Link   6-4 *Actual Size* (optional)   6-5 *Where the Sidewalk Ends* (optional)

Go Online for a complete literature list for Grade 2 and to download all Quick Look Cards.

Problem Solving  Professional Development

Everyday Mathematics emphasizes equally all three of the Common Core's dimensions of **rigor:**

- conceptual understanding
- procedural skill and fluency
- applications

Math Messages, other daily work, Explorations, and Open Response tasks provide many opportunities for children to apply what they know to solve problems.

▶ Math Message

Math Messages require children to solve a problem they have not previously been shown how to solve. Math Messages provide almost daily opportunities for problem solving.

▶ Daily Work

Journal pages, Home Links, Writing/Reasoning prompts, and Differentiation Options often require children to solve problems in mathematical contexts and real-life situations. **Minute Math+** offers Number Stories for transition times and spare moments throughout the day. See Routine 6, pages 38–43.

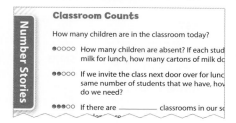

▶ Explorations

In Exploration C, children use rectangles and triangles to create and name other shapes.

▶ Open Response and Reengagement

In Lesson 6-9, the Open Response lesson in Unit 6, children use base-10 blocks to solve a subtraction problem that requires making a trade. They use words and pictures to show what they did. The reengagement discussion on Day 2 might focus on the different ways in which base-10 blocks can be used to subtract, or how to relate clearly in words and pictures how a tool was used. Using tools effectively and making sense of the results is the focus mathematical practice for the lesson.
GMP5.2 As children gain experience with different tools, they add new problem-solving techniques to their repertoires.

 Virtual Learning Community Go Online ▷ to watch an Open Response and Reengagement lesson in action. Search "Open Response" at **vlc.cemseprojects.org**.

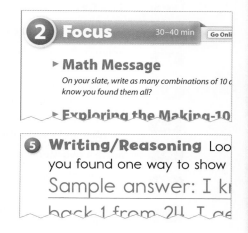

Look for GMP1.1–1.6 markers, which indicate opportunities for children to engage in SMP1: "Make sense of problems and persevere in solving them." Children also become better problem solvers as they engage in all of the CCSS Mathematical Practices. The yellow GMP markers throughout the lessons indicate places where you can emphasize the Mathematical Practices and develop children's problem-solving skills.

Assessment and Differentiation

See pages xxii–xxv to learn about a comprehensive online system for recording, monitoring, and reporting children's progress using core program assessments.

 **Go Online** to **vlc.cemseprojects.org** for tools and ideas related to assessment and differentiation from *Everyday Mathematics* teachers.

✔ Ongoing Assessment

In addition to frequent informal opportunities for "kid watching," every lesson (except Explorations) offers an **Assessment Check-In** to gauge children's performance on one or more of the standards addressed in that lesson.

Lesson	Task Description	CCSS Common Core State Standards
6-1	Use the data on a picture graph to complete a bar graph.	2.NBT.2, 2.MD.10
6-2	Solve comparison number stories using a number grid, a number line, drawings, or manipulatives.	2.OA.1, 2.NBT.5, 2.NBT.7, 2.MD.5
6-3	Write a number model to represent a number story and solve the story with or without diagrams or manipulatives.	2.OA.1, 2.NBT.5, 2.NBT.7
6-4	Use information from the Animal Heights and Lengths Poster to write at least one number story and solve it with or without tools.	2.OA.1, 2.NBT.5, 2.NBT.7, 2.MD.5
6-5	Solve a two-step number story with the help of drawings or diagrams.	2.OA.1, 2.NBT.5, 2.NBT.7, SMP1
6-6	Take an informal look at inventing strategies for solving 2-digit addition problems and making ballpark estimates.	2.NBT.5
6-7	Represent 2- and 3-digit addends with base-10 blocks and combine the blocks to find the partial sums.	2.NBT.1, 2.NBT.5, 2.NBT.7, SMP5
6-8	Solve 2- and 3-digit addition problems with or without base-10 blocks.	2.NBT.5, 2.NBT.7
6-9	Use strategies based on place value to accurately subtract.	2.NBT.7, SMP5

▶ Periodic Assessment

Unit 6 Progress Check This assessment focuses on the CCSS domains of *Operations and Algebraic Thinking, Number and Operations in Base Ten,* and *Measurement and Data.* It also contains a Cumulative Assessment to help monitor children's learning and retention of content that was the focus of Units 1–5.

NOTE Odd-numbered units include an **Open Response Assessment.** Even-numbered units include a **Cumulative Assessment.**

► # Unit 6 Differentiation Activities Differentiation Support English Learners Support

Differentiation Options Every regular lesson provides Readiness, Enrichment, **Extra Practice,** and **English Language Learners Support** activities that address the Focus standards of that lesson.

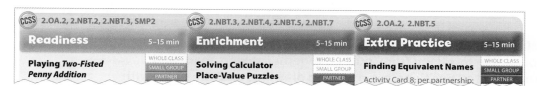

CCSS 2.OA.2, 2.NBT.2, 2.NBT.3, SMP2	CCSS 2.NBT.3, 2.NBT.4, 2.NBT.5, 2.NBT.7	CCSS 2.OA.2, 2.NBT.5
Readiness 5–15 min	**Enrichment** 5–15 min	**Extra Practice** 5–15 min
Playing *Two-Fisted* *Penny Addition*	**Solving Calculator Place-Value Puzzles**	**Finding Equivalent Names**
WHOLE CLASS / SMALL GROUP / PARTNER	WHOLE CLASS / SMALL GROUP / PARTNER	WHOLE CLASS / SMALL GROUP / PARTNER
		Activity Card 8; per partnership:

Activity Cards These activities, written to the children, enable you to differentiate Part 2 of the lesson through small-group work.

English Language Learners Activities and point-of-use support help children at different levels of English language proficiency succeed.

Differentiation Support Two online pages for most lessons provide suggestions for game modifications, ways to scaffold lessons for children who need additional support, and language development suggestions for Beginning, Intermediate, and Advanced English language learners.

Activity Card 83

Differentiation Support online pages

For **ongoing distributed practice,** see these activities:
• Mental Math and Fluency
• Differentiation Options: Extra Practice
• Part 3: Journal pages, Math Boxes, *Math Masters,* Home Links
• Print and online games

Ongoing Practice Differentiation Support

► # Games

Games in *Everyday Mathematics* are an essential tool for practicing skills and developing strategic thinking.

Lesson	Game	Skills and Concepts	CCSS Common Core State Standards
6-6	*The Exchange Game*	Making exchanges with base-10 blocks	2.NBT.1, 2.NBT.1a, 2.NBT.1b
6-8	*Salute!*	Practicing addition facts and finding missing addends	2.OA.2, SMP6
6-10	*Beat the Calculator*	Practicing addition facts	2.OA.2

VLC Virtual Learning Community | Go Online ⟩ to look for examples of *Everyday Mathematics* games at **vlc.cemseprojects.org**.

$\overset{\text{CCSS}}{\bigcirc}$ Spiral Trace: Skills, Concepts, and Applications

⭐ **Mastery Expectations** This Spiral Trace outlines instructional trajectories for key standards in Unit 6. For each standard, it highlights opportunities for Focus instruction, Warm Up and Practice activities, as well as formative and summative assessment. It describes the **degree of mastery**— as measured against the entire standard—expected at this point in the year.

Operations and Algebraic Thinking

2.OA.1 Use addition and subtraction with 100 to solve one- and two-step word problems involving situations of adding to, taking from, putting together, taking apart, and comparing, with unknowns in all positions, e.g., by using drawings and equations with a symbol for the unknown number to represent the problem.

⭐ By the end of Unit 6, expect children to **add and subtract within 100 to solve one-step word problems involving situations of adding to, taking from, putting together, and taking apart, e.g., by using drawings or equations to represent the problem.**

Number and Operations in Base Ten

2.NBT.5 Fluently add and subtract within 100 using strategies based on place value, properties of operations, and/or the relationship between addition and subtraction.

⭐ By the end of Unit 6, expect children to **add and subtract within 100 using strategies based on place value, properties of operations, and/or the relationship between addition and subtraction, with or without tools.**

2.NBT.7 Add and subtract within 1000, using concrete models or drawings and strategies based on place value, properties of operations, and/or the relationship between addition and subtraction; relate the strategy to a written method. Understand that in adding or subtracting three-digit numbers, one adds or subtracts hundreds and hundreds, tens and tens, ones and ones; and sometimes it is necessary to compose or decompose tens or hundreds.

⭐ By the end of Unit 6, expect children to **add and subtract numbers at least within 100 using concrete models or drawings and strategies based on place value, properties of operations, and/or the relationship between addition and subtraction. Expect children to understand that in adding 3-digit numbers, one adds hundreds and hundreds, tens and tens, ones and ones; and sometimes it is necessary to compose or decompose tens or hundreds.**

Spiral Tracker

Go to **connectED.mcgraw-hill.com** for comprehensive trajectories that show how in-depth mastery develops across the grade.

Measurement and Data

2.MD.5   Use addition and subtraction within 100 to solve word problems involving lengths that are given in the same units, e.g., by using drawings (such as drawings of rulers) and equations with a symbol for the unknown number to represent the problem.

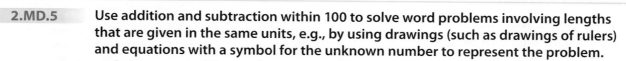

By the end of Unit 6, expect children to **use addition and subtraction within 100 to solve word problems involving lengths that are given in the same units by using drawings.**

2.MD.6   Represent whole numbers as lengths from 0 on a number line diagram with equally spaced points corresponding to the numbers 0, 1, 2, . . . , and represent whole-number sums and differences within 100 on a number line diagram.

By the end of Unit 6, expect children to **represent whole-number lengths from 0 on a number line diagram with equally spaced points corresponding to the numbers 0, 1, 2, . . . and sums within 100 on a number-line diagram.**

2.MD.10   Draw a picture graph and a bar graph (with single-unit scale) to represent a data set with up to four categories. Solve simple put-together, take-apart, and compare problems using information present in a bar graph.

By the end of Unit 6, expect children to **draw a picture graph using a tally chart.**

Key    = Assessment Check-In    = Progress Check Lesson   = Current Unit    = Previous or Upcoming Lessons

Mathematical Background: Content

 This discussion highlights the major content areas and the Common Core State Standards addressed in Unit 6. See the online Spiral Tracker for complete information about the learning trajectories for all standards.

▶ Data Displays (Lesson 6-1)

In Lesson 6-1 children collect data about the number of pockets on their clothing and display the data on a picture graph and a bar graph. **2.MD.10** Both data displays show the *counts,* or frequencies, of each category of data. A *picture graph* uses collections of simple pictures to represent the counts and is read by counting the pictures or symbols above a particular category. Picture graphs include a *key* explaining what each symbol in the graph represents. For example, the key on the picture graph below shows that each book symbol represents 1 book. The graph shows that Britney read 4 books because there are 4 book symbols above Britney's name. It is important for children to pay attention to the key of a picture graph to make sure that they interpret the graph correctly.

Books Read in Ms. Cook's Class

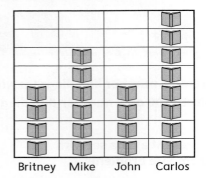

A *bar graph* uses bar heights to represent counts in each category of data and is read by comparing the height of each bar to the vertical scale. The bar graph at the right shows that Britney read 4 books because the top of the bar above Britney's name aligns with 4 on the vertical scale.

Both picture graphs and bar graphs are useful for comparing counts. For example, it is easy to see from these that Carlos read the most books, John and Britney read the same number of books, and Mike read 2 more books than the number read by either Britney or John.

Books Read in Ms. Cook's Class

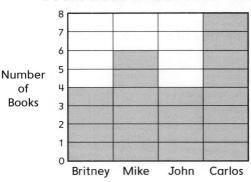

ballpark estimate
bar graph
comparison diagram
comparison number story
data

difference
geoboard
graph key
partial sums
partial-sums addition

picture graph
quantity
rectangular array
tally chart
two-step number story

▶ Number Stories (Lessons 6-2 through 6-5)

In Lesson 6-2 children solve number stories that describe comparison situations. **2.OA.1** Comparison situations involve two separate quantities and the *difference* between them. They can be about things that are counted or things that are measured. The following is a *comparison number story* involving counts: *Beth scored 14 points. Ivy scored 8 points. How many more points did Beth score than Ivy?* The following is a comparison number story involving measures: *Sam's pencil is 8 inches long. Mark's pencil is 2 inches shorter than Sam's pencil. How long is Mark's pencil?*

In *Everyday Mathematics* children use *comparison diagrams* to organize information in comparison number stories. They fill in boxes in the diagram with numbers that they know, using a question mark to represent the number they need to find. The margin contains comparison diagrams filled in for the number stories above.

Number stories such as these provide a context for problem solving. This unit extends work that children have been doing with problem solving in several ways. In past units (and also in Lesson 6-2), children were asked only to solve one type of number story at a time and use only one type of diagram (parts-and-total, change, or comparison). In Lesson 6-3 children may choose their own diagram. It is important for them to understand that there is no one correct choice of diagram. Many number stories can be interpreted in more than one way, and a child's choice of diagram should depend on the way he or she thinks about the problem. (See Lesson 6-3 for more information about this topic.) Children continue to practice interpreting number stories in Lesson 6-4, when they solve problems about lengths and heights of animals. **2.MD.5**

In Lesson 6-5 children begin solving *two-step number stories*. **2.OA.1** Children use manipulatives, draw pictures, or fill in situation diagrams to organize the information in the problems and decide how to solve them.

In Lessons 6-2 through 6-5, children continue to write number models with unknowns to represent number stories. Often more than one number model can fit a number story. For example, Sam and Mark's pencil story above could be represented by $? + 2 = 8$ or by $8 - 2 = ?$ In the case of two-step number stories, children represent the story by writing either one or two number models. Consider the following two-step number story: *Allen had 25 raspberries. He ate 10 of them. Then he picked 6 more. How many raspberries does Allen have now?* This story can be represented by two number models: $25 - 10 = ?$ for the first step and $15 + 6 = ?$ for the second step. The story can also be represented by one number model: $25 - 10 + 6 = ?$ to show both steps.

Quantity
Beth: 14

Quantity	
Ivy: 8	?

Difference

Quantity
Sam: 8 in.

Quantity	
Mark: ? in.	2 in.

Difference

Go Online to the *Implementation Guide* for more information about situation diagrams.

▶ Strategies for Addition (Lessons 6-6 through 6-8)

In Lesson 6-6 children share their own invented strategies for adding 2- and 3-digit numbers. **2.NBT.5, 2.NBT.7**   The addition strategies children invent are typically one of three types:

Counting Up   Children start with one addend and then count up the number of times specified by the other addend. To solve 24 + 45, a child might start at 24 and count up 45, often in jumps, or *multiples,* of 10 and 5. Counting-up strategies are based on several properties of addition. For example, when children count up, they often start with the larger addend, even when the larger addend is presented second in the problem. They might solve 24 + 45 by starting at 45 and counting up 24. This is an application of the *turn-around rule,* or the Commutative Property of Addition. **2.NBT.5, 2.NBT.7**

Combining Place-Value Groups   Children break numbers into tens and ones and add the parts separately. To solve 62 + 39, a child might think the following: *60 + 30 is 90, and 2 + 9 is 11. So the total is 90 + 11, or 101.* This strategy is based on place value, and it extends easily to 3-digit numbers, which require breaking the numbers into hundreds, tens, and ones. **2.NBT.5, 2.NBT.7**

Adjusting and Compensating   Children adjust the numbers in a problem to make them "friendly," or easier to add. To solve 49 + 17, a child might add 1 to the 49 to make it the friendly number 50. After adjusting 49 to 50 by adding the 1, however, it is necessary to compensate by subtracting 1 from the other number. Children can compensate either before or after they add the numbers. (*See margin.*) Adjusting and compensating strategies are based on the inverse relationship between addition and subtraction, which means you have to add to make up for subtracting, and vice versa. **2.NBT.5, 2.NBT.7**

In addition to sharing invented strategies in Lesson 6-6, children make written records of their thinking, which helps them understand and explain why the strategies work. **2.NBT.9**

In Lesson 6-8 children are introduced to a formal addition strategy called *partial-sums addition,* which is a formalization of the strategy of combining place-value groups described above. In Lesson 6-7 children build readiness for partial-sums addition by using base-10 blocks to represent addends. They combine blocks to find *partial sums* and then the final total. In Lesson 6-8 they represent addends using *expanded form,* which connects the concrete strategy of using the blocks to a written version of the same strategy. **2.NBT.3, 2.NBT.7**

Partial-sums addition is an example of an *algorithm,* or a step-by-step set of instructions for completing a task. Children in *Everyday Mathematics* learn several addition algorithms over the course of Grades 2–4. The curriculum introduces partial-sums addition first because it is closely related to methods that children invent, and it develops good number sense in children by emphasizing the values of the digits. Working with partial-sums addition lays the foundation for children to understand the standard algorithm for addition starting in Grade 4.

I can take 1 from the 17 and add it to the 49 to get 50 + 16 = 66.

A child compensates before adding 49 + 17.

I can add 1 to 49 to get 50. Then I add 50 to 17 to get 67. But I still have to take away the 1 I added, so the answer is 66.

A child compensates after adding 49 + 17.

Mathematical Background: Practices

 In Everyday Mathematics, *children learn the **content** of mathematics as they engage in the **practices** of mathematics. As such, the Standards for Mathematical Practice are embedded in children's everyday work, including hands-on activities, problem-solving tasks, discussions, and written work. Read here to see how Mathematical Practices 1 and 5 are emphasized in this unit.*

▶ Standard for Mathematical Practice 1

According to Mathematical Practice 1, "Mathematically proficient students start by explaining to themselves the meaning of a problem." In Lessons 6-2 through 6-5, children solve one- and two-step number stories. They use situation diagrams to help organize the information, identify what they know and what they need to find out, and decide on a solution strategy. In other words, they begin by "making sense of [their] problem." GMP1.1

Mathematical Practice 1 also states that mathematically proficient students "monitor and evaluate their progress" and "continually ask themselves, 'Does this make sense?'" In Lesson 6-6 children share their own invented addition strategies. They make written records of their thinking. Recording the steps of addition strategies helps children to "reflect on [their] thinking as [they] solve . . . problems" and keep track of what they have done and what they still have left to do. GMP1.2 In this unit children also use *ballpark estimates* to "check whether [their] answers make sense." GMP1.4

▶ Standard for Mathematical Practice 5

In the words of Mathematical Practice 5, "mathematically proficient students consider the available tools when solving a mathematical problem" and "make sound decisions about when each of these tools might be helpful." In Lessons 6-2 through 6-5, children solve a series of number stories, making use of tools to help them solve the problems. Because no particular tool is specified, children are free to "choose appropriate tools." GMP5.1 Recognizing the need to add 10s to solve a number story, children might use a number grid. Or they might use base-10 blocks if they sense a visual representation of the numbers might be helpful. It is important for children to use a variety of tools, rather than relying on just one.

This unit also encourages children to consider whether they need to use a tool at all. Lesson 6-4 raises the question of why children should use a number grid to add when they can do the computation mentally. Deciding whether a tool is needed or helpful is part of learning to "use tools effectively." GMP5.2 As children become more fluent with computation, they should become less reliant on tools and more inclined to use mental strategies. Finally, this unit also challenges students to use familiar tools in new ways. In Lesson 6-9 children use base-10 blocks to represent a subtraction problem that requires trading 1 ten for 10 ones. They discuss how to use base-10 blocks effectively for subtraction and make sense of their results, which will enable them to use base-10 blocks more flexibly in the future. GMP5.2

 Standards and Goals for
Mathematical Practice

SMP1 Make sense of problems and persevere in solving them.

  GMP1.1 Make sense of your problem.

  GMP1.2 Reflect on your thinking as you solve your problem.

  GMP1.3 Keep trying when your problem is hard.

  GMP1.4 Check whether your answer makes sense.

  GMP1.5 Solve problems in more than one way.

  GMP1.6 Compare the strategies you and others use.

SMP5 Use appropriate tools strategically.

  GMP5.1 Choose appropriate tools.

  GMP5.2 Use tools effectively and make sense of your results.

Go Online to the *Implementation Guide* for more information about the Mathematical Practices.

For children's information on the Mathematical Practices, see *My Reference Book,* pages 1–22.

Representing Data: Pockets

Overview Children draw picture graphs and bar graphs to represent a data set.

▶ **Before You Begin**
For Part 2 make one copy of *Math Masters,* page 157 for every two children. Cut the pages apart and place them near the Math Message. If your school requires a uniform, modify the Part 2 activities, from counting pockets to counting pencils, pens, or other objects children can tally.

▶ **Vocabulary**
data • tally chart • picture graph • graph key • bar graph

Common Core State Standards

Focus Clusters
• Understand place value.
• Relate addition and subtraction to length.
• Represent and interpret data.

1 Warm Up   15–20 min

	Materials	
Mental Math and Fluency Children solve comparison number stories.	slate	2.OA.1, 2.NBT.5
Daily Routines Children complete daily routines.	See pages 4–43.	See pages xiv–xvii.

2 Focus   30–40 min

Math Message Children count the number of pockets on their clothes.	*Math Masters,* p. 157	2.NBT.2
Tallying Pockets Data Children represent the class pockets data in tally charts.	*Math Journal 2,* p. 133; *Math Masters,* pp. 157–158; Class Data Pad (optional)	2.NBT.2 SMP4
Drawing a Picture Graph Children draw picture graphs to represent the pocket data.	*Math Journal 2,* p. 134; *My Reference Book,* p. 115; *Math Masters,* p. 159	2.NBT.2, 2.MD.10 SMP3, SMP4
Drawing a Bar Graph Children draw bar graphs to represent the pocket data.	*Math Journal 2,* p. 135; *My Reference Book,* p. 116 (optional)	2.NBT.2, 2.MD.6, 2.MD.10 SMP2, SMP4
✓ **Assessment Check-In** See page 536.	*Math Journal 2,* p. 135	2.NBT.2, 2.MD.10

CCSS 2.MD.10 **Spiral Snapshot**

GMC Organize and represent data on bar and picture graphs.

6-1 Focus Practice	6-4 Practice	6-5 Practice	6-10 Practice	7-3 Practice	7-9 Focus Practice

/// Spiral Tracker   Go Online   to see how mastery develops for all standards within the grade.

3 Practice   15–20 min

Solving Addition Problems Children add 2-digit numbers.	*Math Journal 2,* p. 136; number grid or number line (optional); base-10 blocks (optional)	2.NBT.5, 2.NBT.6, 2.NBT.7
Math Boxes 6-1 Children practice and maintain skills.	*Math Journal 2,* p. 132	See page 537.
Home Link 6-1 **Homework** Children draw a bar graph to represent data.	*Math Masters,* p. 160	2.NBT.2, 2.MD.6, 2.MD.10

connectED.mheducation.com

Plan your lessons online with these tools.

 ePresentations   Student Learning Center   Facts Workshop Game   eToolkit   Professional Development   Home Connections   Spiral Tracker   Assessment and Reporting   English Learners Support   Differentiation Support

Differentiation Options

RtI

2.NBT.2

Readiness — 5–15 min

Recording Tally Marks

Math Masters, p. 155
(1 copy per partnership); per
partnership: 1 die

| WHOLE CLASS |
| SMALL GROUP |
| PARTNER |
| INDEPENDENT |

For experience recording tally marks, partners take turns rolling a die and using a tally mark to record the number rolled on *Math Masters*, page 155. Play continues until one child has 5 or more tally marks next to each number. Then have children record the total number of tally marks in each row.

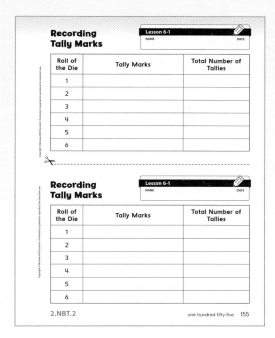

CCSS **2.OA.1, 2.MD.6, 2.MD.10**

Enrichment — 5–15 min

Generating Number Stories

Math Journal 2, p. 135;
Math Masters, p. TA7

| WHOLE CLASS |
| SMALL GROUP |
| PARTNER |
| INDEPENDENT |

To apply their understanding of bar graphs, children use information from the bar graph on journal page 135 to generate number stories involving comparison, putting-together, and taking-apart situations. Children write their number stories on *Math Masters,* page TA7. Have partners trade stories and solve.

CCSS **2.MD.6, 2.MD.10, SMP4**

Extra Practice — 5–15 min

Making a Bar Graph

Math Masters, p. 156

| WHOLE CLASS |
| SMALL GROUP |
| PARTNER |
| INDEPENDENT |

For additional practice representing data, children gather data and make a bar graph on *Math Masters,* page 156. **GMP4.1**

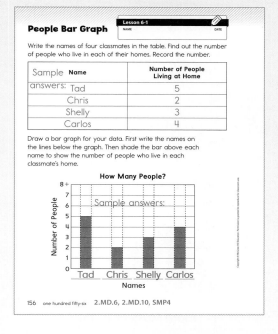

English Language Learners Support

Beginning ELL Scaffold with visuals and labels to help English language learners understand terms associated with bar graphs and picture graphs. Prepare separate displays of both kinds of graphs and add labels while introducing the following terms: *bar graph, bar, label, title, picture graph,* and *graph key.* Ask *yes* or *no* questions such as these: *Is this the title? Is this the graph key?* In addition, have children respond to Total Physical Response prompts by either pointing to or showing the different elements as you name them or by naming the elements as you point to them and use the terms.

 Go Online ELL English Learners Support

Standards and Goals for
Mathematical Practice

SMP2 Reason abstractly and quantitatively.
　GMP2.3 Make connections
　between representations.

SMP3 Construct viable arguments and critique the reasoning of others.
　GMP3.1 Make mathematical conjectures
　and arguments.

SMP4 Model with mathematics.
　GMP4.1 Model real-world situations using
　graphs, drawings, tables, symbols, numbers,
　diagrams, and other representations.

　GMP4.2 Use mathematical models to solve
　problems and answer questions.

Math Masters, p. 157

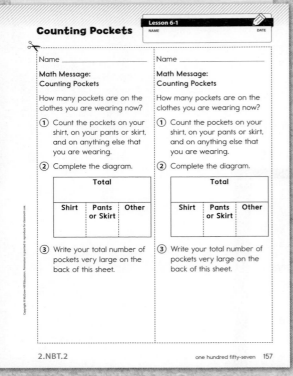

1 Warm Up　15–20 min　[Go Online]　ePresentations　eToolkit

▶ Mental Math and Fluency

Pose comparison number stories. You may want to use the names of children in your class. Have children solve the problems on their slates and share their strategies. *Leveled exercises:*

- ●○○ Ann's jump rope is 8 feet long. Melissa's jump rope is 6 feet long. How much longer is Ann's jump rope than Melissa's? 2 feet
- ●●○ Maurice is 48 inches tall. Lisa is 59 inches tall. How much shorter is Maurice than Lisa? 11 inches
- ●●● Mike's pet snake is 24 inches long. Marti's pet snake is 22 inches longer than Mike's pet snake. How long is Marti's snake? 46 inches

▶ Daily Routines

Have children complete daily routines.

2 Focus　30–40 min　[Go Online]　ePresentations　eToolkit

▶ Math Message

Math Masters, p. 157

Take a Counting Pockets *page. Follow the directions.*

▶ Tallying Pockets Data

Math Journal 2, p. 133; *Math Masters,* pp. 157–158

WHOLE CLASS | SMALL GROUP | PARTNER | INDEPENDENT

Math Message Follow-Up As children share the total number of pockets they have on their clothes, record the tally marks on *Math Masters,* page 158 or on a table drawn on the Class Data Pad. Have children record the tally marks on journal page 133. GMP4.1 Explain that the tally marks represent **data,** or information that is gathered by counting, measuring, questioning, or observing. Tell children that we counted pockets to gather data about pockets.

Count the tallies and have children complete the Number column. Then spend a few minutes talking about the table. Ask questions such as the following:

- *How many children have 5 pockets?* (Repeat for other numbers.)
  GMP4.2 Answers vary.
- *What is the most common number of pockets?* GMP4.2 Answers vary.
- Point to a number in the Number column. *What does this number mean?* GMP4.2 Sample answer: It means 5 children are wearing clothes with 3 pockets.

Explain that a **tally chart,** like the one children just made, is one way to display the pocket data. Tell children that they will also display the data in a picture graph and a bar graph.

Differentiate | **Adjusting the Activity**

To extend the activity, have children do one of the following:

• Determine the total number of pockets for the whole class.

• Find how many children have more than 3 pockets.

Chidren can use calculators if necessary.

Go Online ▸ Differentiation Support

▸ # Drawing a Picture Graph

Math Journal 2, p. 134; *My Reference Book,* p. 115; *Math Masters,* p. 159

WHOLE CLASS | **SMALL GROUP** | PARTNER | INDEPENDENT

Display *Math Masters,* page 159 or have children look at journal page 134. Explain that a **picture graph** is another way to diplay data. Instead of tally marks, picture graphs use pictures or symbols to represent numbers. Point out the title of the graph and the label for the horizontal axis. Ask children to explain what they think the smiley face in the **graph key** means. A smiley face stands for 1 child.

Academic Language Development Some children may be confused by the term *graph key.* Help them make the connection between a graph key and a key that opens a lock. Ask children how having a key could help them find out what is behind a closed door. Then ask them to think about how the graph key helps to "unlock" the meaning of symbols in a picture graph.

Demonstrate how to represent the class pockets data on a picture graph on *Math Masters,* page 159 as children follow along on journal page 134. When children seem comfortable, have them each complete the picture graph on their own. GMP4.1

> **NOTE** The last category for the Pockets data is "10 or more." If this category has many tallies, you may want to discuss the different numbers of pockets that this category includes.

When most of the class is finished, discuss the graph. Ask:

• *What does the picture graph show?* The number of children who have each number of pockets

• *How many children have 5 pockets?* GMP4.2 Answers vary.

• *How many children have 6 or more pockets?* GMP4.2 Answers vary.

• *Do more children have 3 pockets or 4 pockets?* GMP4.2 Answers vary.

• *Do you think the graph would look different if we were dressed in bathing suits? Explain.* GMP3.1 Yes. Bathing suits usually don't have any pockets.

With the class, read about picture graphs on *My Reference Book,* page 115.

Math Journal 2, p. 133

Making a Tally Chart | Lesson 6-1 DATE

Use tallies to show how many children in your class have each number of pockets.

Pockets	Children	
	Tallies	Number
0	Sample answer:	0
1		0
2	//	2
3	//	2
4	///	3
5	////	4
6	///	3
7	////	4
8	//	2
9	/	1
10 or more		0

2.NBT.2, SMP4   one hundred thirty-three 133

Math Journal 2, p. 134

Drawing a Picture Graph | Lesson 6-1 DATE

Draw a picture graph of the pockets data.

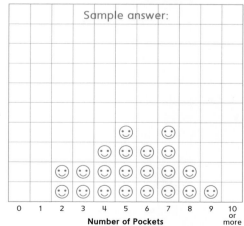

Number of Children with Pockets

Sample answer:

Number of Pockets

KEY: Each ☺ = 1 child

134 one hundred thirty-four 2.NBT.2, 2.MD.10, SMP4

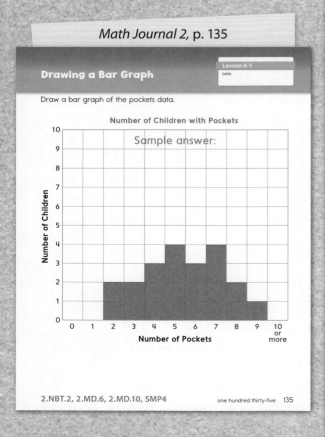

Measurement and Data

Graphs

Read It Together

A **picture graph** uses a picture or a symbol to show data.

Books Read in Ms. Cook's Class

Britney   Mike   John   Carlos

KEY: Each 📖 = 1 Book

There are 4 📖 above Britney.
This means that Britney read 4 books.

There are 6 📖 above Mike.
This means that Mike read 6 books.
Mike read 2 more books than Britney.

Try It Together

How many more books did Carlos read than John?

one hundred fifteen   **115** MRB

Math Journal 2, p. 135

Lesson 6-1

Drawing a Bar Graph

DATE

Draw a bar graph of the pockets data.

Number of Children with Pockets

Sample answer:

Number of Children

Number of Pockets

2.NBT.2, 2.MD.6, 2.MD.10, SMP4   one hundred thirty-five **135**

► # Drawing a Bar Graph

Math Journal 2, p. 135

WHOLE CLASS | SMALL GROUP | **PARTNER** | **INDEPENDENT**

Tell children that they will use the pocket data shown on their picture graph to make a **bar graph.** Refer children to *My Reference Book,* page 115 for a brief review of bar graphs, if needed. Point out that the height of the bars shows the number of children. Then have children complete journal page 135 independently or with a partner. GMP4.1

Differentiate   **Adjusting the Activity**

Point out the tally chart of pocket data shown on either journal page 133, a display of *Math Masters,* page 158, or the table on the Class Data Pad. Have children verbalize the number of classmates who have 0 pockets and fill in the bar above 0 on the bar graph accordingly. Continue with each number of pockets until their bar graphs are complete.

Go Online ▷   Differentiation Support

✓ # Assessment Check-In   CCSS 2.NBT.2, 2.MD.10

Math Journal 2, p. 135

Expect that most children will be able to use the data shown on the picture graph on journal page 134 to complete the bar graph on journal page 135. For those who struggle, consider the suggestion in the Adjusting the Activity note above. Additional practice drawing picture graphs and bar graphs to represent data will be provided throughout the year.

✓ Assessment and Reporting   Go Online ▷ to record student progress and to see trajectories toward mastery for these standards.

Summarize Have children compare the picture graph and the bar graph on journal pages 134–135. Ask: *How are the picture graph and the bar graph similar?* GMP2.3 Sample answers: Both graphs show the number of children who have various numbers of pockets. Both graphs have the same shape. *How are they different?* GMP2.3 Sample answers: The picture graph uses pictures and a key to show the number of children. The bar graph uses bars and numbers on the side to show the number of children.

3 Practice 15–20 min

Go Online

ePresentations eToolkit Home Connections

▶ Solving Addition Problems

Math Journal 2, p. 136

| WHOLE CLASS | **SMALL GROUP** | **PARTNER** | **INDEPENDENT** |

Children add 2-digit numbers. Encourage them to use a number grid, a number line, an open number line, or base-10 blocks as needed.

▶ Math Boxes 6-1

Math Journal 2, p. 132

| WHOLE CLASS | **SMALL GROUP** | **PARTNER** | **INDEPENDENT** |

Mixed Practice Math Boxes 6-1 are paired with Math Boxes 6-3.

▶ Home Link 6-1

Math Masters, p. 160

Homework Children count the pockets of four people at home and represent their data in a bar graph.

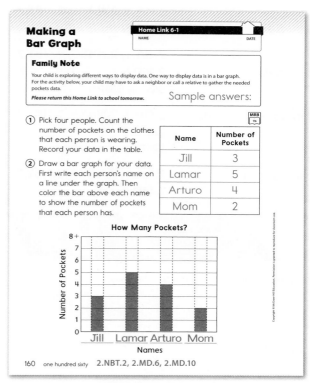

Math Masters, p. 160

Math Journal 2, p. 136

Solving Addition Problems

Lesson 6-1
DATE

Fill in the unit box. Then solve each problem. You can use base-10 blocks, an open number line, a number line, or a number grid to help you. Show your work.

Unit

① 20
+63
83

② 42
+39
81

③ 54
+38
92

Try This

④ 58
+67
125

136 one hundred thirty-six 2.NBT.5, 2.NBT.6, 2.NBT.7

Math Journal 2, p. 132

Math Boxes

Lesson 6-1
DATE

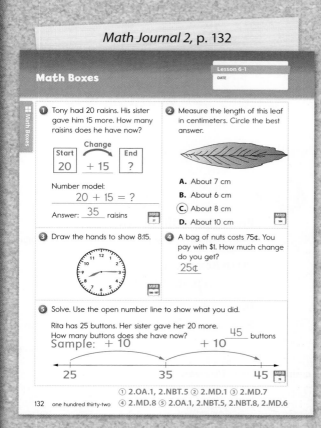

① Tony had 20 raisins. His sister gave him 15 more. How many raisins does he have now?

Start 20 Change + 15 End ?

Number model:
20 + 15 = ?

Answer: 35 raisins

② Measure the length of this leaf in centimeters. Circle the best answer.

A. About 7 cm
B. About 6 cm
C. About 8 cm
D. About 10 cm

③ Draw the hands to show 8:15.

④ A bag of nuts costs 75¢. You pay with $1. How much change do you get?
25¢

⑤ Solve. Use the open number line to show what you did.

Rita has 25 buttons. Her sister gave her 20 more. How many buttons does she have now? 45 buttons

Sample: + 10   + 10

25   35   45

① 2.OA.1, 2.NBT.5 ② 2.MD.1 ③ 2.MD.7
132 one hundred thirty-two ④ 2.MD.8 ⑤ 2.OA.1, 2.NBT.5, 2.NBT.8, 2.MD.6

Comparison Number Stories

Overview Children solve comparison number stories.

▶ **Before You Begin**
For Part 1 select and sequence Quick Look Cards 102, 89, and 96. For Part 2 decide how you will display a comparison diagram.

▶ **Vocabulary**
comparison number story • comparison diagram • quantity • difference

 Common Core State Standards

Focus Clusters
- Represent and solve problems involving addition and subtraction.
- Use place value understanding and properties of operations to add and subtract.
- Relate addition and subtraction to length.

1 Warm Up 15–20 min

Materials

Mental Math and Fluency Children view Quick Look Cards.	Quick Look Cards 102, 89, and 96; slate	2.OA.2
Daily Routines Children complete daily routines.	See pages 4–43.	See pages xiv–xvii.

2 Focus 20–30 min

Math Message Children solve a comparison number story.		2.OA.1, 2.NBT.5, 2.NBT.7, 2.MD.5
Solving Comparison Number Stories Children discuss how to use a comparison diagram to organize information in a number story.	*My Reference Book,* pp. 30–31 (optional); *Math Masters,* p. TA28	2.OA.1, 2.NBT.5, 2.NBT.7, 2.MD.5 SMP4, SMP6
Practicing with Comparison Number Stories Children solve comparison number stories.	*Math Journal 2,* pp. 137–138; number grid or number line (optional)	2.OA.1, 2.NBT.5, 2.NBT.7, 2.MD.5 SMP5
✓ **Assessment Check-In** See page 542.	*Math Journal 2,* pp. 137–138	2.OA.1, 2.NBT.5, 2.NBT.7, 2.MD.5

CCSS 2.NBT.5 **Spiral Snapshot**

GMC Subtract within 100 fluently.

| 5-10 Focus Practice | 5-11 Focus | 6-1 Warm Up | 6-2 Focus Practice | 6-3 through 6-5 Warm Up Focus Practice | 6-6 Warm Up | 6-9 Focus Practice |

Spiral Tracker | **Go Online** | to see how mastery develops for all standards within the grade.

3 Practice 15–20 min

Practicing with Fact Triangles Children practice facts using Fact Triangles.	*Math Journal 2,* pp. 250–253; Fact Triangles	2.OA.2
Math Boxes 6-2 Children practice and maintain skills.	*Math Journal 2,* p. 139	See page 543.
Home Link 6-2 **Homework** Children solve comparison number stories.	*Math Masters,* pp. 164–165	2.OA.1, 2.NBT.5, 2.NBT.7, 2.MD.5

connectED.mheducation.com

Plan your lessons online with these tools.

 ePresentations Student Learning Center Facts Workshop Game eToolkit Professional Development Home Connections Spiral Tracker Assessment and Reporting English Learners Support Differentiation Support

Differentiation Options RtI

Readiness   5–15 min

Comparing Penny Amounts

| WHOLE CLASS |
| SMALL GROUP |
| **PARTNER** |
| INDEPENDENT |

20 pennies or counters, paper

For experience comparing numbers using a concrete model, children complete the following activity. Partners place a pile of pennies (or other counters) between them. Each one grabs a handful, counts them, and records the number. Then partners compare numbers and find the difference by lining up the pennies as shown below. *For example:* Partner A grabs 9 pennies, and Partner B grabs 6 pennies. They line up their pennies to show that the difference is 3 pennies.

Partners work together to record number models representing the lined-up pennies. *For example:* $9 - 6 = 3$.

GMP2.2 Encourage children to describe comparisons using language such as "I have 3 more than you" and "You have 3 less than I have."

Enrichment   5–15 min

Comparing Number Stories

| WHOLE CLASS |
| SMALL GROUP |
| PARTNER |
| **INDEPENDENT** |

Math Masters, p. 161

To further explore comparison number stories, children complete *Math Masters,* page 161. First they solve comparison number stories and write number models. Then they compare the number stories and explain how to use one number story to help solve another. **GMP1.6**

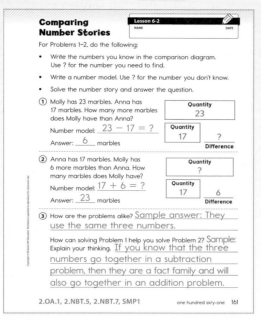

Extra Practice   5–15 min

More Comparison Stories

| WHOLE CLASS |
| SMALL GROUP |
| **PARTNER** |
| **INDEPENDENT** |

Math Masters, pp. 162–163

For more practice, partners complete *Math Masters,* pages 162–163.

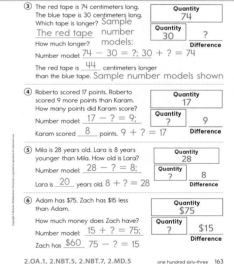

English Language Learners Support

Beginning ELL Use the diagrams that appear in the lesson as visual models to help English learners understand the actions in number stories. Using the diagrams, role-play number stories with relatively easy numbers. Use teacher think-alouds with each move in the diagram. Have children model the solution with counters. Reverse roles so that children use the diagrams as you retell the number story and use counters.

 Go Online   ELL   English Learners Support

Standards and Goals for
Mathematical Practice

SMP4 **Model with mathematics.**
   GMP4.1 Model real-world situations using
   graphs, drawings, tables, symbols, numbers,
   diagrams, and other representations.

SMP5 **Use appropriate tools strategically.**
   GMP5.1 Choose appropriate tools.

SMP6 **Attend to precision.**
   GMP6.1 Explain your mathematical thinking
   clearly and precisely.

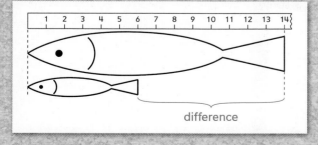

difference

1 Warm Up   15–20 min   Go Online   ePresentations   eToolkit

▶ Mental Math and Fluency

Flash the following Quick Look Cards in sequence: 102, 89, and 96.
Always allow for a second look. Follow up by asking children to write
number sentences to show how they found the total number of dots.

Previously children have been verbalizing how they see the ten frames.
Beginning in this unit, children are asked to write number sentences
to show what they see. Expect that most children will write the fact
represented by the double ten frame. For Quick Look Card 96, for
example, they will likely write $5 + 6 = 11$. Some children will write
a number sentence that represents their strategy for determining the
number of dots. For example, they may write $5 + 5 + 1 = 11$. Both
number sentences, as well as others, are appropriate. *Leveled exercises:*

●○○ Quick Look Card 102 Sample answer: $6 + 6 = 12$
●●○ Quick Look Card 89 Sample answer: $7 + 3 = 10$
●●● Quick Look Card 96 Sample answers: $5 + 6 = 11$; $5 + 5 + 1 = 11$

▶ Daily Routines

Have children complete daily routines.

2 Focus   20–30 min   Go Online   ePresentations   eToolkit

▶ Math Message

Fish A is 14 inches long. Fish B is 6 inches long. How many inches longer is
Fish A than Fish B? 8 inches

▶ Solving Comparison Number Stories

Math Masters, p. TA28

| WHOLE CLASS | SMALL GROUP | PARTNER | INDEPENDENT |

Math Message Follow-Up Use the Math Message problem to begin
a discussion of **comparison number stories.** Comparison number stories
involve differences between two quantities.

Draw a picture showing how children can solve the Math Message
problem by lining up the two fish (the two quantities) against a ruler.
(*See margin.*) Sketch a **comparison diagram** or display *Math Masters*,
page TA28. Write 14, 6, and ? in the diagram as shown in the margin on
the next page. **GMP4.1** Discuss the meanings of the words *quantity* and
difference as they appear in the diagram. Explain that **quantity** describes
an amount or a number of things, and the **difference** is the amount
unmatched between two quantities. In the Math Message problem,
the difference tells how many inches longer Fish A is than Fish B.

Point out that the comparison diagram offers a convenient way to organize information in the Math Message problem. The longer Quantity box shows the larger quantity—the length of Fish A. The shorter Quantity box shows the smaller quantity—the length of Fish B. The Difference line shows how many more inches Fish A measures than Fish B. Point out that the Quantity box at the top is as long as both the bottom Quantity box and the Difference line. This provides a good visual for children.

Differentiate **Adjusting the Activity**

Whenever they are working with a comparison diagram, children should write the known and missing information (shown with the symbol ?) in the diagram. Have children write words or short phrases in the diagram as a reminder of what the numbers mean. For the Math Message problem, the words *Fish A* and *Fish B* might be written as reminders.

Go Online | Differentiation Support

Ask children to represent the number story using a number model with a question mark for the unknown. **GMP4.1** Record it below the diagram. Ask: *Can you write different number models to represent this number story?* Yes. There are four possible number models: $14 - 6 = ?$; $14 - ? = 6$; $? + 6 = 14$; and $6 + ? = 14$. Ask children to explain how their number models fit the situation in the number story. After children solve the number story, write a summary number model. (*See margin.*) **GMP4.1**

You may wish to read about comparison diagrams as a class in *My Reference Book,* pages 30–31. Tell children that today they will work with number stories that compare two quantities. They will use a comparison diagram to organize the information from the story to decide whether to add or subtract.

> **NOTE** *Everyday Mathematics* does not recommend teaching a keyword approach to solving number stories. It is problematic, for example, to tell children that the keyword *more* indicates that they should add because this is not always true. In Example 1, the word *more* appears, but adding the numbers given in the problem does not produce the correct answer. Instead of teaching keywords, encourage children to think through the relationships between the given numbers before choosing an operation.

Work with the children to solve the following comparison stories.

Example 1

Joey scored 30 points. Max scored 10 points. How many more points did Joey score than Max? 20 points

With the class, fill in the comparison diagram as shown in the margin. **GMP4.1** Write ? for the difference. Ask children to generate number models to match the number story and explain how their number models fit the situation. Record children's number models under the diagram.

Invite children to share strategies for finding the difference between 30 and 10. **GMP6.1** *Sample strategies:*

• Think: "What must I add to 10 to get 30?"

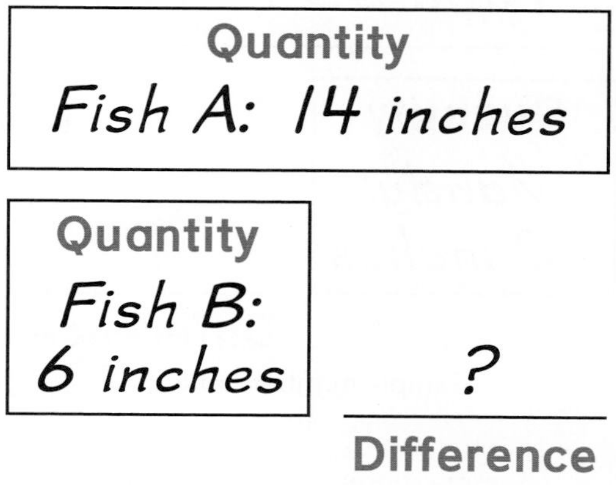

Quantity
Fish A: 14 inches

Quantity
Fish B: 6 inches   **?**
_____
                Difference

Possible number models:
$14 - 6 = ?$
$14 - ? = 6$
$? + 6 = 14$
$6 + ? = 14$

Sample summary number model:
$14 - 6 = 8$

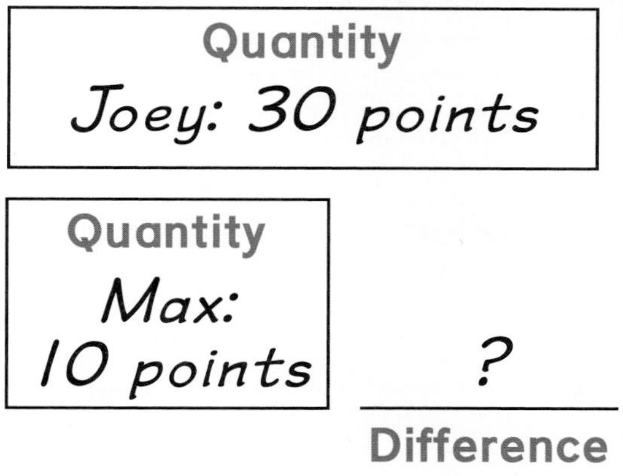

Quantity
Joey: 30 points

Quantity
Max: 10 points   **?**
_____
                Difference

Sample number model:
$30 - 10 = ?$

Sample summary number model:
$30 - 10 = 20$

Quantity
Nan: 55 inches

Quantity
Mandy: 40 inches

?
_____
Difference

Sample number model:
55 − 40 = ?

Sample summary number model:
55 − 40 = 15

Quantity
$47 radio

Quantity
?

$12
_____
Difference

Sample number model:
47 − ? = 35

Sample summary number model:
47 − 12 = 35

Math Journal 2, p. 137

Comparison Number Stories

Lesson 6-2
DATE

For each number story, follow these steps:

- Write the numbers you know in the comparison diagram. Use ? for the number you need to find.
- Write a number model. Use ? for the number you don't know.
- Solve the problem and answer the question.

1. Barb's ribbon is 27 inches long. Cindy's ribbon is 10 inches long. Which ribbon is longer?

   Quantity
   27

   <u>Barb's ribbon</u>

   Quantity
   10   ?
   Difference

   How much longer? Sample answer:

   Number model:
   27 − 10 = ?

   Barb's ribbon is <u>17</u> inches longer than Cindy's.

2. Frisky lives on the 16th floor. Fido lives on the 7th floor. Who lives on the higher floor?

   Quantity
   16

   <u>Frisky</u>

   Quantity
   7   ?
   Difference

   How many floors higher? Sample answer:

   Number model:
   7 + ? = 16

   Frisky lives <u>9</u> floors higher than Fido.

2.OA.1, 2.NBT.5, 2.NBT.7, 2.MD.5, SMP5   one hundred thirty-seven   137

- Think of the comparison diagram as a Fact Triangle. Think, "30 − 10 is the difference I want."

After children share their strategies and solutions, write a summary number model with 20 substituted for the question mark. (*See margin on page 541.*) **GMP4.1**

Example 2

Nan is 55 inches tall. Mandy is 40 inches tall. Who is shorter? How much shorter? Mandy; 15 inches

Fill in the comparison diagram as a class and write possible number models with ? for the unknown number. Have children explain how the number model fits the story; then have children solve the problem and write a summary number model with 15 substituted for the question mark. (*See margin.*) **GMP4.1** Invite children to share their strategies.

Example 3

A radio costs $47. A calculator costs $12 less than the radio. How much does the calculator cost? $35

Repeat the process followed for the other examples, making sure children understand that this example is different because the difference is known and the smaller quantity is unknown. (*See margin.*) One strategy for solving this problem could be to start at 47 and count back one 10 and two 1s. Another would begin by thinking "What plus 12 is 47?"

▶ Practicing with Comparison Number Stories

Math Journal 2, pp. 137–138

| WHOLE CLASS | SMALL GROUP | PARTNER | INDEPENDENT |

Partners complete the problems on journal pages 137–138. Check that children record the known information in the comparison diagrams and use ? to represent unknown numbers. For Problems 1–4, the difference will be the unknown number. For Problem 5, the difference will be known and one of the two quantities will be unknown. Children may draw pictures and use number lines, number grids, or any other tool to help them solve the problems. **GMP5.1**

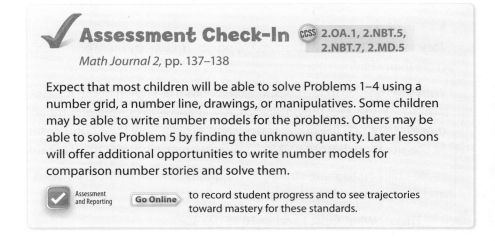

✓ Assessment Check-In CCSS 2.OA.1, 2.NBT.5, 2.NBT.7, 2.MD.5

Math Journal 2, pp. 137–138

Expect that most children will be able to solve Problems 1–4 using a number grid, a number line, drawings, or manipulatives. Some children may be able to write number models for the problems. Others may be able to solve Problem 5 by finding the unknown quantity. Later lessons will offer additional opportunities to write number models for comparison number stories and solve them.

☑ Assessment and Reporting   (Go Online)   to record student progress and to see trajectories toward mastery for these standards.

Summarize Have volunteers share with the class their solution strategies from journal page 137. GMP6.1

3 Practice 15–20 min

Go Online | ePresentations | eToolkit | Home Connections

▶ Practicing with Fact Triangles

Math Journal 2, pp. 250–253

WHOLE CLASS | SMALL GROUP | PARTNER | INDEPENDENT

As children practice with Fact Triangles, have them use the Addition Facts Inventory Record, Parts 1 and 2 (journal pages 250–253) to take inventory of the addition facts they know and facts that need more practice. See Lesson 3-3 for additional details.

▶ Math Boxes 6-2

Math Journal 2, p. 139

WHOLE CLASS | SMALL GROUP | PARTNER | INDEPENDENT

Mixed Practice Math Boxes 6-2 are paired with Math Boxes 6-4.

▶ Home Link 6-2

Math Masters, pp. 164–165

Homework Children solve comparison stories. They fill in a comparison diagram and write a number model for each problem.

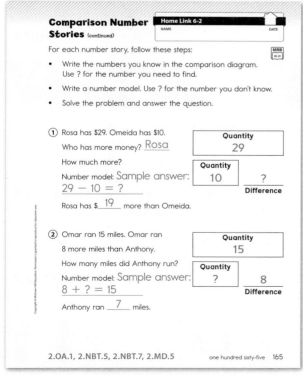

Math Masters, p. 165

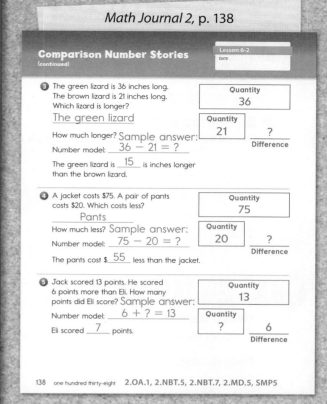

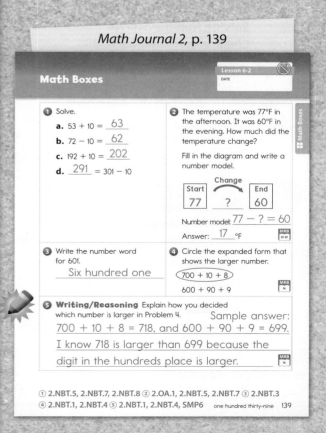

Lesson 6-3

Interpreting Number Stories

Overview Children choose diagrams to use for solving number stories.

▶ **Before You Begin**
Decide how you will display the diagrams for Part 2 of the lesson.

 Common Core State Standards

Focus Clusters
- Represent and solve problems involving addition and subtraction.
- Use place value understanding and properties of operations to add and subtract.
- Relate addition and subtraction to length.

1 Warm Up 15–20 min

	Materials	
Mental Math and Fluency Children mentally add and subtract 10 and 100.	slate	**2.NBT.8**
Daily Routines Children complete daily routines.	See pages 4–43.	See pages xiv–xvii.

2 Focus 30–40 min

Math Message Children solve a number story.	*Math Journal 2*, p. 140	**2.OA.1, 2.NBT.5, 2.NBT.7**
Sharing Strategies Children share strategies for solving a number story.	*Math Journal 2*, p. 140	**2.OA.1, 2.NBT.5, 2.NBT.7** **SMP4**
Selecting Diagrams Children select situation diagrams to help organize information from number stories.	*Math Journal 2*, p. 140	**2.OA.1, 2.NBT.5, 2.NBT.7,** **2.MD.5** **SMP1, SMP4**
Solving Number Stories Children solve number stories.	*Math Journal 2*, p. 141	**2.OA.1, 2.NBT.5, 2.NBT.7** **SMP4**
✓ **Assessment Check-In** See page 548.	*Math Journal 2*, p. 141	**2.OA.1, 2.NBT.5, 2.NBT.7**

CCSS 2.OA.1 **Spiral Snapshot**

GMC Model 1-step problems involving addition and subtraction.

Routine 6	5-8 through 5-10 Focus Practice	6-2 Focus Practice	6-3 Focus Practice	6-4 Focus Practice	6-5 Focus Practice	6-7 Practice

III **Spiral Tracker**   **Go Online** to see how mastery develops for all standards within the grade.

3 Practice 15–20 min

Comparing Lengths Children solve comparison number stories involving lengths.	*Math Journal 2*, pp. 142–143; per partnership: 2 tape measures	**2.OA.1, 2.NBT.5, 2.NBT.7,** **2.MD.5**
Math Boxes 6-3 Children practice and maintain skills.	*Math Journal 2*, p. 144; centimeter ruler	See page 549.
Home Link 6-3 **Homework** Children use diagrams to solve number stories.	*Math Masters*, p. 167	**2.OA.1, 2.NBT.5, 2.NBT.7**

 **connectED.mheducation.com**

Plan your lessons online with these tools.

 ePresentations   Student Learning Center   Facts Workshop Game   eToolkit   Professional Development   Home Connections   Spiral Tracker   Assessment and Reporting   ELL English Learners Support   Differentiation Support

Differentiation Options

RtI

Readiness

5–15 min

WHOLE CLASS	
SMALL GROUP	
PARTNER	
INDEPENDENT	

Organizing Information

Math Masters, p. TA29

For experience organizing information from a number story, children answer questions about a number story. For example, display the following problem: Robert has 6 cents, and Colleen has 7 cents. How much money do they have all together?

Ask: *What do I know? What do I need to find out?* Select and display an appropriate diagram, such as a parts-and-total diagram. Ask children to share their answers as you fill out the diagram. **GMP4.2** Repeat with other number stories as needed and encourage children to make up their own stories.

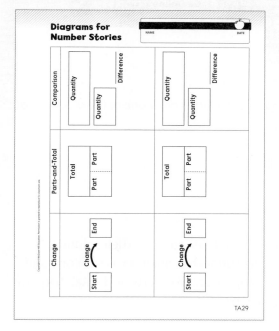

Enrichment

5–15 min

WHOLE CLASS	
SMALL GROUP	
PARTNER	
INDEPENDENT	

Writing Number Stories to Match Number Models

Math Masters, p. 166

To apply their understanding of number models, children make up their own number stories to fit given number models and situation diagrams.

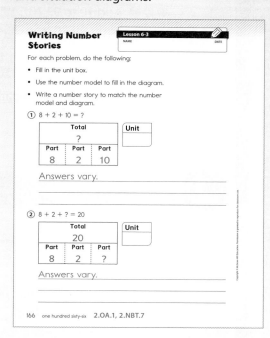

Extra Practice

5–15 min

WHOLE CLASS	
SMALL GROUP	
PARTNER	
INDEPENDENT	

Solving a Partner's Number Story

Math Masters, p. TA29; slate; number line, number grid, or other manipulatives

For practice with number stories, partners make up and solve number stories. One partner tells a number story and then partners work together to organize the information from the story in one of the diagrams on *Math Masters,* page TA29. Children record a number model on their slates and use ? for the unknown number. They solve the problem, using number lines, number grids, or other manipulatives if necessary. Then the other partner tells a number story and the process is repeated.

For example, the first partner tells this story: "Julia had 48 cents, and Marcus had 35 cents. If Marcus gives his money to Julia, how much money will she have?" After filling in the diagram and writing the number model $48 + 35 = ?$ on their slates, children might use a number grid to figure out the total. The answer is 83 cents.

English Language Learners Support

Beginning ELL To introduce the terms *select* and *choose*, role-play choosing between concrete objects. Use think-alouds such as the following: *I like this one. I will choose this one.* Use the terms *select* and *choose* interchangeably. For example, say: *I will choose this one. Yes, I will select this one.* Use Total Physical Response prompts to direct children to select or choose among groups of objects, such as crayons, markers, and pencils. Provide practice using different groups of items.

Go Online   ELL English Learners Support

Standards and Goals for
Mathematical Practice

SMP1 Make sense of problems and persevere in solving them.
GMP1.1 Make sense of your problem.

SMP4 Model with mathematics.
GMP4.1 Model real-world situations using graphs, drawings, tables, symbols, numbers, diagrams, and other representations.

GMP4.2 Use mathematical models to solve problems and answer questions.

1 Warm Up 15–20 min Go Online

ePresentations eToolkit

▶ Mental Math and Fluency

Dictate 2- and 3-digit numbers. Have children mentally add 10 or 100 to them, or subtract 10 or 100 from them, and record their answers on slates. *Leveled exercises:*

●○○ Add 10 to 40 50, 64 74, and 150. 160
Subtract 10 from 70 60, 37 27, and 340. 330

●●○ Add 100 to 44 144, 700 800, and 510. 610
Subtract 100 from 700 600, 318 218, and 809. 709

●●● Add 10 to 97 107, 193 203, and 398. 408
Subtract 10 from 304 294, 508 498, and 109. 99

▶ Daily Routines

Have children complete daily routines.

2 Focus 30–40 min Go Online

ePresentations eToolkit

▶ Math Message

Math Journal 2, p. 140

Solve Problem 1 on journal page 140. You may draw one of the diagrams at the top of the page to help.

▶ Sharing Strategies

Math Journal 2, p. 140

| WHOLE CLASS | SMALL GROUP | PARTNER | INDEPENDENT |

Math Message Follow-Up Ask children to share their solution strategies for Problem 1. If no one suggests a diagram, sketch one and model how to use it. Change, comparison, and parts-and-total diagrams all work for Problem 1. GMP4.1, GMP4.2

Professional Development

Until now, lessons have focused on one type of number story at a time. For example, all the problems in Lesson 6-2 were comparison stories, and the comparison diagram was the only diagram used. In this lesson, children are asked to categorize addition and subtraction number stories and then solve them.

Do not force any number story into a particular mold. There may be several ways to interpret a problem.

 Go Online Professional Development

▶ Selecting Diagrams

Math Journal 2, p. 140

| WHOLE CLASS | SMALL GROUP | PARTNER | INDEPENDENT |

Show children how different diagrams can be used to organize the information in Problem 1, which can be interpreted as a change situation, a comparison situation, and a parts-and-total situation. *For example:*

- **Change Situation** Last year there was some number of water slides at Rushing Waters (the Start number). Nine new slides were added (the Change number). Now there are 26 slides (the End number). I want to find the Start number. `GMP1.1`

- **Comparison Situation** I'm comparing the number of slides this year (the larger Quantity: 26) to the number of slides last year (the smaller Quantity). I know that there are more slides this year (the Difference: 9). I want to find the smaller Quantity. `GMP1.1`

- **Parts-and-Total Situation** I know there are 26 slides in all (the Total). Of these, some slides (the first Part) were there last year, and 9 new slides (the second Part) were added this year. I want to find the first Part. `GMP1.1`

NOTE Different children might think about a number story in different ways and will choose diagrams that match their thinking.

For Problem 2 on journal page 140, children do the following:

1. Write a number model with ? representing the unknown number. `GMP4.1` They may draw a situation diagram to help organize the information.

2. Calculate the sum or the difference to solve the problem. `GMP4.2`

3. Write the answer.

Encourage children to share strategies, making sure to demonstrate how to organize the information in a diagram.

| Differentiate | **Adjusting the Activity** |

Watch for children who have difficulty writing a number model with ? for the unknown. Ask them first to explain how they view the problem. Then direct them toward the diagram that best matches their way of thinking. Alternatively, pick a diagram and ask children to explain how to put the numbers from the problem into that diagram. Finally, have children use the completed diagram to help them write the number model.

Go Online ▶ Differentiation Support

Addition and Subtraction Number Stories — Lesson 6-3 — DATE

Do the following for each number story:

- Write a number model. Use ? to show what you need to find. To help, you may draw a
- Solve the problem and write the answer.

1 Rushing Waters now has 26 water slides. That is 9 more than last year. How many water slides were there last year?

Number model: Sample answers: $26 - 9 = ?$; $9 + ? = 26$

There were ___17___ water slides last year.

2 The Loop Slide is 65 feet high. The Tower Slide is 45 feet high. How much shorter is the Tower Slide?

Number model: Sample answers: $65 - 45 = ?$; $45 + ? = 65$

The Tower Slide is ___20___ feet shorter.

140 one hundred forty   2.OA.1, 2.NBT.5, 2.NBT.7, 2.MD.5, SMP1, SMP4

Addition and Subtraction Number Stories *(continued)* — Lesson 6-3 — DATE

List two things you and your partner like to collect:

___Answers vary.___   _____

Then do the following for each number story:

- Select an item from your collections list.
- Write a number model. Use ? to show the number you need to find. To help, you may draw a
- Solve the problem and write the answer.

3 Colin has 20 _____. Fiona has 30 _____.
How many _____ do they have in all?
Number model: Sample answer: $20 + 30 = ?$

Colin and Fiona have ___50___ _____ in all.
(answer)   (unit)

4 Alexi had 34 _____. He gave 12 _____ to Theo.
How many _____ does Alexi have now?
Number model: Sample answer: $34 - 12 = ?$

Alexi has ___22___ _____ now.
(answer)   (unit)

2.OA.1, 2.NBT.5, 2.NBT.7, SMP4   one hundred forty-one 141

Lesson 6-3   **547**

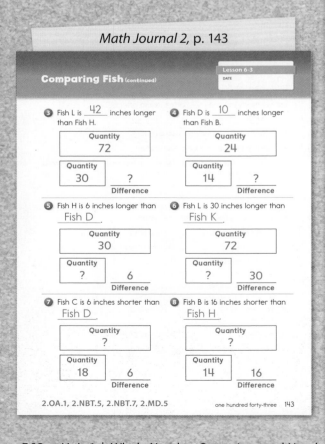

Math Journal 2, p. 142

Math Journal 2, p. 143

► Solving Number Stories

Math Journal 2, p. 141

WHOLE CLASS	SMALL GROUP	PARTNER	INDEPENDENT

Ask partners to discuss things they like to collect and have them record their two favorite items at the top of journal page 141. Invite them to share their lists with the class. Sample answers: Coins, shells, trading cards, dolls, stuffed animals Explain to children that they will use one of these items as their unit for Problem 3 and another for Problem 4.

Have children solve Problems 3–4 in small groups. Ask a volunteer to model their group's solution. GMP4.1, GMP4.2 When the problem has been solved one way, ask if any group solved it a different way.

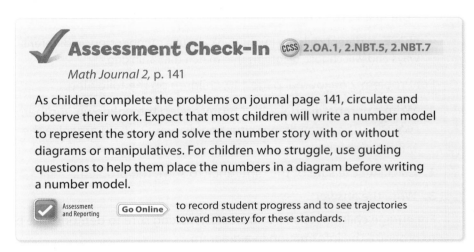

✓ **Assessment Check-In** (CCSS) **2.OA.1, 2.NBT.5, 2.NBT.7**

Math Journal 2, p. 141

As children complete the problems on journal page 141, circulate and observe their work. Expect that most children will write a number model to represent the story and solve the number story with or without diagrams or manipulatives. For children who struggle, use guiding questions to help them place the numbers in a diagram before writing a number model.

✓ Assessment and Reporting ⟨Go Online⟩ to record student progress and to see trajectories toward mastery for these standards.

Summarize Have children discuss how using diagrams can help them solve number stories. GMP4.2 Sample answer: A diagram helps me write a number model and decide whether to add or subtract.

3 Practice 15–20 min Go Online

ePresentations eToolkit Home Connections

▶ Comparing Lengths

Math Journal 2, pp. 142–143

| WHOLE CLASS | SMALL GROUP | **PARTNER** | INDEPENDENT |

Work through Problem 1 with the class. For larger numbers and measurements, have children use the numbers on two tape measures to represent the quantities.

Length of Fish B

Length of Fish C

These tape measures are not the actual size.

Have children work as partners to complete the journal pages.

▶ Math Boxes 6-3

Math Journal 2, p. 144

| WHOLE CLASS | **SMALL GROUP** | **PARTNER** | **INDEPENDENT** |

Mixed Practice Math Boxes 6-3 are paired with Math Boxes 6-1.

▶ Home Link 6-3

Math Masters, p. 167

Homework Children choose diagrams to organize information from addition and subtraction number stories and then solve the problems.

Math Journal 2, p. 144

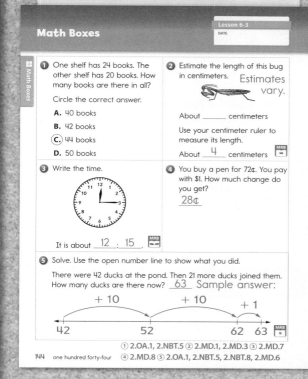

Math Boxes
Lesson 6-3
DATE

① One shelf has 24 books. The other shelf has 20 books. How many books are there in all?

Circle the correct answer.

A. 40 books
B. 42 books
C. 44 books
D. 50 books

② Estimate the length of this bug in centimeters. Estimates vary.

About _____ centimeters

Use your centimeter ruler to measure its length.

About __4__ centimeters

③ Write the time.

It is about __12__ : __15__

④ You buy a pen for 72¢. You pay with $1. How much change do you get?

__28¢__

⑤ Solve. Use the open number line to show what you did.

There were 42 ducks at the pond. Then 21 more ducks joined them. How many ducks are there now? __63__ Sample answer:

$+10$ $+10$ $+1$

42   52   62   63

144 one hundred forty-four

① 2.OA.1, 2.NBT.5 ② 2.MD.1, 2.MD.3 ③ 2.MD.7
④ 2.MD.8 ⑤ 2.OA.1, 2.NBT.5, 2.NBT.8, 2.MD.6

Math Masters, p. 167

Addition and Subtraction Number Stories

Home Link 6-3
NAME   DATE

Family Note

In today's lesson your child used diagrams to help solve addition and subtraction number stories. Diagrams help children organize the information from number stories, identify the missing information, and decide whether to add or subtract to solve the problem. Organizing information in a diagram also helps children write a number model using ? to represent what they don't know. Encourage your child to choose a diagram that best matches the way he or she sees the problem. There's no right or wrong diagram for a problem. What matters is that it matches the child's thinking.

Please return this Home Link to school tomorrow.

Do the following for each number story:

• Write a number model. Use ? to show what you need to find. To help, you may draw a

• Solve the problem and write the answer.

① It snowed 16 inches in Chicago on Friday night. It snowed 7 inches on Saturday night. How much snow did Chicago receive in all?

Number model: Sample answer: 16 + 7 = ?

Answer: __23__ inches

② Evelyn has 30 blocks. She used 24 blocks to build a tower. How many blocks are not used for the tower?

Number model: Sample answer: 24 + ? = 30

Answer: __6__ blocks

2.OA.1, 2.NBT.5, 2.NBT.7   one hundred sixty-seven 167

Lesson 6-3   **549**

Animal Number Stories

Overview **Children solve animal number stories.**

► **Before You Begin**
For the optional Extra Practice activity, obtain the book *Actual Size* by Steve Jenkins (Houghton Mifflin Harcourt, 2011).

 Common Core State Standards

Focus Clusters
• Represent and solve problems involving addition and subtraction.
• Use place value understanding and properties of operations to add and subtract.
• Relate addition and subtraction to length.

1 Warm Up 15–20 min

Materials

Mental Math and Fluency Children identify the values of digits.	slate	2.NBT.1, 2.NBT.3
Daily Routines Children complete daily routines.	See pages 4–43.	See pages xiv–xvii.

2 Focus 30–40 min

Math Message Children identify the longest and shortest animals.	*Math Journal 2*, p. 146	2.NBT.3, 2.NBT.4
Solving Silly Animal Stories Children solve number stories involving heights and lengths.	*Math Journal 2*, p. 146	2.OA.1, 2.NBT.5, 2.NBT.7, 2.MD.2, 2.MD.5, SMP1, SMP6
Writing Silly Animal Stories Children write and solve their own number stories.	*Math Journal 2*, pp. 146–147; base-10 blocks (optional)	2.OA.1, 2.NBT.5, 2.NBT.7, 2.MD.5 SMP5
✓ **Assessment Check-In** See page 554.	*Math Journal 2*, p. 147	2.OA.1, 2.NBT.5, 2.NBT.7, 2.MD.5

CCSS 2.MD.5 Spiral Snapshot

GMC Solve number stories involving length by adding or subtracting.

6-2 Focus Practice 6-3 Focus Practice 6-4 Focus Practice 7-1 Practice 7-9 Practice 9-9 Practice 9-11 Practice

▦ Spiral Tracker ⟨ Go Online ⟩ to see how mastery develops for all standards within the grade.

3 Practice 15–20 min

Drawing a Bar Graph Children draw a bar graph and use it to solve problems.	*Math Journal 2*, pp. 148–149	2.MD.6, 2.MD.10
Math Boxes 6-4 Children practice and maintain skills.	*Math Journal 2*, p. 145	See page 555.
Home Link 6-4 **Homework** Children solve number stories involving length.	*Math Masters*, p. 169	2.OA.1, 2.NBT.5, 2.NBT.7, 2.MD.5

⟨ connectED.mheducation.com ⟩

Plan your lessons online with these tools.

 ePresentations   Student Learning Center   Facts Workshop Game   eToolkit   Professional Development   Home Connections   Spiral Tracker   Assessment and Reporting   English Learners Support   Differentiation Support

Differentiation Options

RtI

CCSS 2.NBT.5

Readiness 5–15 min

Using Mental Strategies

WHOLE CLASS
SMALL GROUP
PARTNER
INDEPENDENT

Number-Grid Poster (optional)

For support using strategies based on place value, children mentally add numbers by visualizing a number grid. Say a number, such as 21. Tell children to close their eyes and picture 21 on a number grid. Ask: *What number is 10 more than 21?* 31 If needed, they can take a peek at the Number-Grid Poster. Pose similar problems involving multiples of 10. For example, ask: *What number is 30 more than 17?* 47

After children seem comfortable using a mental image of a number grid, have them add 2-digit numbers where the number being added on is not a multiple of 10. For example, ask them to find 13 more than 20. Encourage children to think about 13 as 10 and 3. First add 10 to 20. 30 Then add 3 more. 33

CCSS 2.OA.1, 2.NBT.5, 2.NBT.7, 2.MD.5

Enrichment 5–15 min

Matching Number Models

WHOLE CLASS
SMALL GROUP
PARTNER
INDEPENDENT

Math Masters, p. 168

To apply their understanding of how number models represent addition and subtraction situations, children match number models to number stories on *Math Masters,* page 168.

Matching Number Models   Lesson 6-4
NAME   DATE

Read each animal story. The height and length of each each animal is on journal page 146. Circle the number models that match each story.

Hint: There may be more than one correct number model.

① If the ostrich stood on top of the polar bear, how many inches tall would they be together?   **Unit** inches

$108 = 132 + ?$ ⟨$132 + 108 = ?$⟩   ⟨$108 + 132 = ?$⟩

② If the crocodile and the anaconda were lying nose to nose, what would the total length in inches be?

$276 + ? = 312$ ⟨$? = 276 + 312$⟩   ⟨$312 + 276 = ?$⟩

③ How much taller in inches is the giraffe than the polar bear?

⟨$228 - 132 = ?$⟩   ⟨$132 + ? = 228$⟩ $? = 228 + 132$

④ How much shorter in inches is the crocodile than the anaconda?

⟨$312 - 276 = ?$⟩ $312 + 276 = ?$ ⟨$276 + ? = 312$⟩

Try This

⑤ If the ostrich stood on top of the giraffe, how much taller would they be in feet than the polar bear?   **Unit** feet

⟨$9 + 19 - 11 = ?$⟩ $19 + 11 + 9 = ?$ ⟨$? = 19 + 9 - 11$⟩

168   one hundred sixty-eight   2.OA.1, 2.NBT.5, 2.NBT.7, 2.MD.5

CCSS 2.OA.1, 2.NBT.5, 2.NBT.7, 2.MD.5

Extra Practice 5–15 min

More Animal Stories

WHOLE CLASS
SMALL GROUP
PARTNER
INDEPENDENT

Activity Card 77;
Math Masters, p. TA7; *Actual Size*

Literature Link For additional experience with number stories, children use information from the book **Actual Size** by Steve Jenkins (Houghton Mifflin Harcourt, 2011) to write number stories. In this book the author illustrates animals both large and small at actual size by giving examples of the sizes of various animals and parts of animals. Remind children to use the same unit of measure when they compare the lengths or the heights of animals.

More Animal Stories   77

What You Need
A Number Story, page TA7
Actual Size by Steve Jenkins

What To Do
❶ Read *Actual Size.*
❷ Write a number story about the heights or lengths of two animals in the book. The units you use for your story should be feet or inches.
❸ Draw a picture to go with your number story.
❹ Trade papers with a partner. Solve your partner's number story.

Talk About It
Talk to your partner about how you solved the number story.

English Language Learners Support

Beginning ELL Help children understand that the words *long* and *length* can be used to convey the same meaning; the words *high* and *height* can also be used for the same purposes. The differences in the sounds of the words in each pair prevent their related meanings from being obvious to English language learners. Use think-aloud paired questions, such as these: *How long is this string? What is the length of this string? How high is this bookstand? What is the height of this bookstand?* Have children repeat your answers using sentence frames: "This piece of string is _____ long. The length of this piece of string is _____."

〉 **Go Online**   **ELL** English Learners Support

1 Warm Up 15–20 min

Go Online
ePresentations eToolkit

▶ Mental Math and Fluency

Have children do place-value exercises on slates. *Leveled exercises:*

● ○ ○ Write 473. Put an X on the digit in the ones place. 3
      Circle the digit in the tens place. 7
      Underline the digit in the hundreds place. 4

● ● ○ Write 825. Circle the digit in the tens place. 2
      Put an X on the digit in the ones place. 5
      What is the value of the digit that is not marked? 800

● ● ● Write 2,347. Put an X on the digit in the ones place. 7
      What is the value of the digit in the tens place? 40
      In what place is the 3? Hundreds
      How much is the 3 worth? 300

▶ Daily Routines

Have children complete daily routines.

2 Focus 30–40 min

Go Online
ePresentations eToolkit

▶ Math Message

Math Journal 2, p. 146

Turn to journal page 146. Talk to a partner about the lengths and the heights of the animals. Which animal is longest? Which animal is shortest?

▶ Solving Silly Animal Stories

Math Journal 2, p. 146

| WHOLE CLASS | SMALL GROUP | PARTNER | INDEPENDENT |

Math Message Follow-Up Have children share the names and the lengths of the longest animal and the shortest animal. The blue whale is the longest at 1,176 inches, or 98 feet; the ostrich is the shortest at 108 inches, or 9 feet.

Discuss why the height of an ostrich can be given as both 108 inches tall and 9 feet tall. Remind children that because an inch is shorter than a foot, there are more inches than feet in measurements of the same animal.

Math Journal 2, p. 146

Animal Heights and Lengths Poster

Lesson 6-4
DATE

Saltwater Crocodile
276 inches
23 feet

Blue Whale
1,176 inches
98 feet

White Rhinoceros
168 inches
14 feet

Green Anaconda
312 inches
26 feet

Giraffe
228 inches
19 feet

Ostrich
108 inches
9 feet

Polar Bear
132 inches
11 feet

Giant Squid
660 inches
55 feet

146 one hundred forty-six

Academic Language Development Discuss the meaning of the words *length* and *height*. Length is the distance across something from one end to the other (such as the length of a boat from end to end). Height refers to how tall someone or something is (such as the height of a young child from top to bottom). Ask: *Which word would you use to describe the size of the giraffe?* Height, because the giraffe can stand up on its feet *For what other animals would you use the word height?* Sample answers: Polar bear, ostrich

Tell the class they will use the data from the Animal Heights and Lengths Poster on journal page 146 to make up and solve their own number stories. Today they should use only the measures in feet; in later lessons they will use the measures in inches.

Have children imagine the crocodile and the giant squid lying nose to nose. Tell children that we would like to find the total length in feet. To help find the total length in feet, display a unit box with the label "feet" as well as a parts-and-total diagram. Remind children that diagrams are tools to help organize their thinking about number stories.

Ask: *What do you know about the animals in the story?* The crocodile is 23 feet long, and the giant squid is 55 feet long. Write those numbers in the diagram. Ask: *What do you need to find out?* The total length of the two animals GMP1.1 Write ? for the unknown number and have a volunteer write a number model. (*See margin.*)

Have children share their solution strategies. Some children may have added mentally by starting at 55 and first adding the tens from the second addend ($55 + 20 = 75$) and then adding the ones ($75 + 3 = 78$). The total length is 78 feet. Emphasize the importance of including the unit in the answer. GMP6.3

Have a volunteer demonstrate this strategy on an open number line. The child starts at 55, hops 20, and lands at 75. Then he or she hops 3 more and lands at 78. So, $55 + 20 + 3 = 78$. The total length is 78 feet.

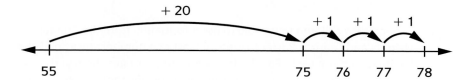

NOTE Children should learn to be flexible and efficient in their solution strategies. For example, there is no need for them to turn to the number grid or any other tool if they are able to find the correct answer mentally. Other times, representing a problem visually with pictures, counters, or base-10 blocks will result in a better understanding of how to solve it. If some children rely solely on the number grid to find answers, encourage them to try different strategies and then use the number grid to check their answers.

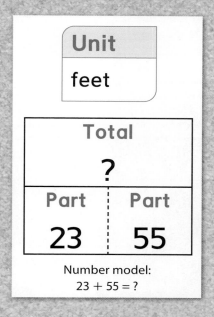

Unit
feet

Total
?

Part	Part
23	55

Number model:
$23 + 55 = ?$

Silly Animal Stories

Lesson 6-4
DATE

Write and solve two number stories. Use information from the Animal Heights and Lengths Poster.

Example:

How much longer is the giant squid than the giraffe?

Quantity **55**   Unit **feet**

Quantity **19**   ? Difference

Number model: 55 − 19 = ?

Answer: The giant squid is 36 feet longer.

1 Answers vary.   Unit

Number model: _____

Answer: _____

2 Answers vary.   Unit

Number model: _____

Answer: _____

2.OA.1, 2.NBT.5, 2.NBT.7, 2.MD.5, SMP5   one hundred forty-seven   147

Drawing a Bar Graph

Lesson 6-4
DATE

In one month a zoo recorded the number of eggs their birds laid. The number of eggs a bird lays at one time is called a "clutch." Use the data in the following table to draw a bar graph.

Type of Bird	American Robin	Canada Goose	Mallard Duck	Toucan
Number of Eggs in a Clutch	6	10	12	4

Bird Eggs Laid in a Month

Number of Eggs in a Clutch (vertical axis, 0–12)

American Robin   Canada Goose   Mallard Duck   Toucan

As needed, pose (or have children pose) additional number stories about comparing or adding the lengths of two animals. You might solve the problems as a class or have children solve them in partnerships and then share their solution strategies. *For example:*

- How much longer is the blue whale than the white rhinoceros? 84 feet

- How many feet would the giant squid have to grow to be as long as the blue whale? 43 feet

▶ Writing Silly Animal Stories

Math Journal 2, pp. 146–147

| WHOLE CLASS | SMALL GROUP | PARTNER | INDEPENDENT |

On journal page 147, children write two number stories. In each story they compare or add the lengths in feet of two animals from journal page 146. They also write a number model to represent each story, using ? for the unknown number. Encourage children to use mental strategies to solve the problems, although they may also use tools such as open number lines or base-10 blocks. They may also use paper and pencil to draw pictures or diagrams to help them think through the problems. **GMP5.1**

 Differentiate **Adjusting the Activity**

Whenever they are working with a diagram, children should write the known and missing information (shown with ?) on the diagram. Have children write the animal names or short phrases as a reminder of what the numbers mean.

Go Online ▶ Differentiation Support

✔ **Assessment Check-In** CCSS 2.OA.1, 2.NBT.5, 2.NBT.7, 2.MD.5

Math Journal 2, p. 147

Expect most children to be able to use information from the Animal Heights and Lengths Poster to write at least one number story and solve it with or without tools. For children who struggle, have them verbalize a story before they write it or recommend a specific tool to help them. Some children may draw a diagram to help them write a number model with ? for the unknown numbers. Later lessons will bring additional opportunities to solve number stories and write number models.

 Assessment and Reporting   **Go Online** ▶ to record student progress and to see trajectories toward mastery for these standards.

Summarize As time permits, have volunteers share their number stories and solution strategies with the class.

3 Practice 15–20 min

Go Online

ePresentations eToolkit Home Connections

▶ Drawing a Bar Graph

Math Journal 2, pp. 148–149

| WHOLE CLASS | SMALL GROUP | PARTNER | INDEPENDENT |

Children draw a bar graph to represent a data set and solve problems using information from the graph.

▶ Math Boxes 6-4

Math Journal 2, p. 145

| WHOLE CLASS | SMALL GROUP | PARTNER | INDEPENDENT |

Mixed Practice Math Boxes 6-4 are paired with Math Boxes 6-2.

▶ Home Link 6-4

Math Masters, p. 169

Homework Children solve number stories involving animal lengths.

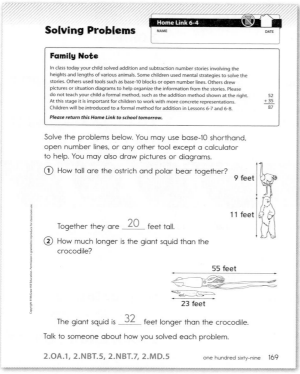

Math Masters, p. 169

Math Journal 2, p. 149

Drawing a Bar Graph (continued) — Lesson 6-4

Use the Bird Eggs Laid in a Month bar graph to solve the number stories.

1 The American robin and the toucan are tree birds. What is the total number of eggs laid by the tree birds?

Number model: Sample answers: 6 + 4 = ?;
Answer: 10 eggs
4 + 6 = 10

2 The Canada goose and the mallard duck are water birds. What is the total number of eggs laid by the water birds?

Number model: Sample answers: 10 + 12 = ?;
Answer: 22 eggs
10 + 12 = 22

3 How many more eggs did the mallard duck lay than the American robin?

Number model: Sample answers: 6 + ? = 12;
Answer: 6 eggs
12 − 6 = ?; 12 − 6 = 6

4 Use the data in the bar graph to write your own number story.

Answers vary.

Number model: _____

Answer: _____

2.MD.10   one hundred forty-nine 149

Math Journal 2, p. 145

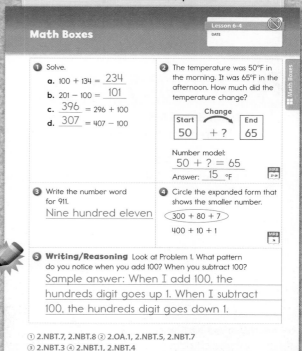

Math Boxes — Lesson 6-4

1 Solve.
   a. 100 + 134 = 234
   b. 201 − 100 = 101
   c. 396 = 296 + 100
   d. 307 = 407 − 100

2 The temperature was 50°F in the morning. It was 65°F in the afternoon. How much did the temperature change?

| Start | Change | End |
| 50 | + ? | 65 |

Number model:
50 + ? = 65
Answer: 15 °F

3 Write the number word for 911.
Nine hundred eleven

4 Circle the expanded form that shows the smaller number.
(300 + 80 + 7)
400 + 10 + 1

5 Writing/Reasoning Look at Problem 1. What pattern do you notice when you add 100? When you subtract 100?
Sample answer: When I add 100, the hundreds digit goes up 1. When I subtract 100, the hundreds digit goes down 1.

① 2.NBT.7, 2.NBT.8 ② 2.OA.1, 2.NBT.5, 2.NBT.7
③ 2.NBT.3 ④ 2.NBT.1, 2.NBT.4
⑤ 2.NBT.7, 2.NBT.8, SMP7   one hundred forty-five 145

Two-Step Number Stories

Overview Children solve two-step number stories.

▶ **Before You Begin**

For the optional Extra Practice activity, obtain the book *Where the Sidewalk Ends* by Shel Silverstein (HarperCollins Publishers, 2004). For Part 2 decide how you will display sample diagrams and number models for the problems.

▶ **Vocabulary**

two-step number story

Common Core State Standards

Focus Clusters
- Represent and solve problems involving addition and subtraction.
- Use place value understanding and properties of operations to add and subtract.

1 Warm Up   15–20 min

	Materials	
Mental Math and Fluency Children solve number stories.	slate	2.OA.1, 2.NBT.5
Daily Routines Children complete daily routines.	See pages 4–43.	See pages xiv–xvii.

2 Focus   30–40 min

Math Message Children write a number story to match a number model.	slate	2.OA.1
Sharing Number Stories Children share the number stories they wrote.	slate	2.OA.1
Solving Two-Step Number Stories Children solve two-step number stories.	*Math Journal 2*, pp. 150–151; *My Reference Book*, pp. 34–35	2.OA.1, 2.NBT.5, 2.NBT.7 SMP1, SMP4
✓ **Assessment Check-In** See page 561.	*Math Journal 2*, p. 151	2.OA.1, 2.NBT.5, 2.NBT.7, SMP1

CCSS 2.OA.1 **Spiral Snapshot**

GMC Use addition and subtraction to solve 2-step number stories.

| Routines
2 and 6 | 6-5
Focus
Practice | 6-7
Practice | 7-1
Practice | 7-2
Focus | 7-3
Practice | 7-5
Practice | 7-9
Practice |

/// **Spiral Tracker**   (Go Online) to see how mastery develops for all standards within the grade.

3 Practice   10–15 min

Math Boxes 6-5 Children practice and maintain skills.	*Math Journal 2*, p. 152	See page 561.
Home Link 6-5 **Homework** Children solve a two-step number story.	*Math Masters*, p. 171	2.OA.1, 2.NBT.5, 2.NBT.7

---

connectED.mheducation.com

Plan your lessons online
with these tools.

ePresentations Student Learning Facts Workshop eToolkit   Professional   Home   Spiral Tracker   Assessment   English Learners Differentiation
Center Game   Development Connections   and Reporting Support Support

Differentiation Options RtI

Readiness — 5–15 min

	WHOLE CLASS
	SMALL GROUP
	PARTNER
	INDEPENDENT

Acting Out Two-Step Problems

20-foot length of masking tape, 21 index cards labeled 0–20

Introduce solving two-step problems by having children hop on a number line to act out problems. Use masking tape to make a 20-foot number line on the floor, marked with 21 evenly spaced tick marks. Use index cards to label the tick marks 0–20. Give directions such as the following: *Begin at 2. Take 3 hops forward and then hop back 2. Where did you land?* 3 Have children write a number model or models to represent their hops. **GMP2.1** Sample answers: $2 + 3 - 2 = 3$; $2 + 3 = 5$ and $5 - 2 = 3$ Repeat the activity with similar problems.

Enrichment — 5–15 min

	WHOLE CLASS
	SMALL GROUP
	PARTNER
	INDEPENDENT

Writing a Two-Step Number Story

Activity Card 78;
Math Masters, p. TA7

To further explore two-step number stories, children write their own two-step number story using a given number model. They represent their story with a drawing or a diagram. **GMP2.1, GMP4.1**

Writing a Two-Step Number Story 78

What You Need
A Number Story, page TA7

What To Do
Work with a partner.
❶ Write a number story to match this number model: $13 + 7 - 8 = ?$
For example:
Mallory read for 13 minutes at lunchtime and 7 minutes at bedtime on Monday. Her brother Mike read 8 minutes less than her on Monday. How many minutes did Mike read?
❷ Draw pictures or diagrams to go with the number story.

Talk About It
Compare your number story and drawings with your partner.
How are they alike? How are they different?

More You Can Do
Write your own number model and story and have your partner solve.

Extra Practice — 5–15 min

	WHOLE CLASS
	SMALL GROUP
	PARTNER
	INDEPENDENT

Solving More Two-Step Problems

Math Masters, p. 170;
Where the Sidewalk Ends

Literature Link For additional practice with two-step problems, children read "Band-Aids," a poem from ***Where the Sidewalk Ends*** by Shel Silverstein (HarperCollins Publishers, 2004) and solve the two-step problems posed on *Math Masters,* page 170.

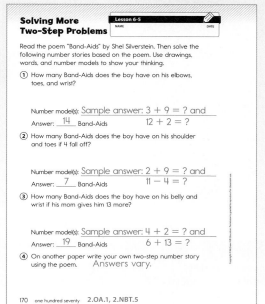

Solving More Two-Step Problems — Lesson 6-5

NAME   DATE

Read the poem "Band-Aids" by Shel Silverstein. Then solve the following number stories based on the poem. Use drawings, words, and number models to show your thinking.

① How many Band-Aids does the boy have on his elbows, toes, and wrist?

Number model(s): Sample answer: $3 + 9 = ?$ and $12 + 2 = ?$
Answer: __14__ Band-Aids

② How many Band-Aids does the boy have on his shoulder and toes if 4 fall off?

Number model(s): Sample answer: $2 + 9 = ?$ and $11 - 4 = ?$
Answer: __7__ Band-Aids

③ How many Band-Aids does the boy have on his belly and wrist if his mom gives him 13 more?

Number model(s): Sample answer: $4 + 2 = ?$ and $6 + 13 = ?$
Answer: __19__ Band-Aids

④ On another paper write your own two-step number story using the poem. Answers vary.

170   one hundred seventy   2.OA.1, 2.NBT.5

English Language Learners Support

Beginning ELL To help English language learners organize their solutions in preparation for sharing their strategies, provide a template illustrating what it means to use *drawings,* *words,* and *number models.* Doing so also provides children with a nonverbal way to show their thinking. Encourage beginning English learners to use their native language for the *words* section as needed.

Go Online | **ELL** English Learners Support

Standards and Goals for Mathematical Practice

SMP1 **Make sense of problems and persevere in solving them.**

  GMP1.1 Make sense of your problem.

SMP4 **Model with mathematics.**

  GMP4.1 Model real-world situations using graphs, drawings, tables, symbols, numbers, diagrams, and other representations.

  GMP4.2 Use mathematical models to solve problems and answer questions.

Math Journal 2, p. 150

Solving Two-Step Number Stories

Lesson 6-5
DATE

Do the following for each number story:

• Write a number model or number models. Use ? to show the number you need to find. To help, you may draw a

[diagrams]

• Solve the problem and write the answer.

❶ On Monday, Annabelle had 9 shells. On Tuesday, she found 12 more. On Wednesday, she gave 4 to her aunt. How many shells does Annabelle have now?

Number model(s):
Sample answer: 9 + 12 = ? and 21 − 4 = ?

Annabelle now has __17__ shells.

❷ Yvette bought 25 red balloons and 15 white balloons for a party. During the party 8 balloons popped. How many balloons did she have left when the party ended?

Number model(s):
Sample answer: 25 + 15 = ? and 40 − 8 = ?

Yvette had __32__ balloons left.

150 one hundred fifty   2.OA.1, 2.NBT.5, 2.NBT.7, SMP1, SMP4

1 **Warm Up** 15–20 min Go Online ePresentations eToolkit

▶ Mental Math and Fluency

Pose number stories. Have children solve them on their slates and then share their strategies. *Leveled exercises:*

● ○ ○ Jane has 8 toy cars. Theo has 7 toy cars. How many toy cars do they have in all? 15 toy cars

● ● ○ Shantell brought 14 thank-you notes to school. She gave out 8 during lunch. How many does she still have to give out? 6 notes

● ● ● Maya went down the slide 17 times. That was 9 more times than Jason. How many times did Jason go down the slide? 8 times

▶ Daily Routines

Have children complete daily routines.

2 **Focus** 30–40 min Go Online ePresentations eToolkit

▶ Math Message

Write a number story to match this number model: 15 − ? = 7.

▶ Sharing Number Stories

| WHOLE CLASS | SMALL GROUP | PARTNER | INDEPENDENT |

Math Message Follow-Up Invite a few children to share their number stories. Point out that each of their number stories is a one-step number story, or a number story that can be solved in only one step.

Tell children that today they will solve number stories that involve more than one step. They will see how using drawings and diagrams can help.

▶ Solving Two-Step Number Stories

Math Journal 2, pp. 150–151; My Reference Book, pp. 34–35

| WHOLE CLASS | SMALL GROUP | PARTNER | INDEPENDENT |

Have a volunteer read Problem 1 on journal page 150 aloud. Explain that you can break this number story into two steps to solve, so it is a **two-step number story.** Ask questions like the following:

- *What do we want to find out?* The number of shells Annabelle has now **GMP1.1**
- *Do we know how many shells Annabelle had to begin with?* Yes; 9 shells
- *What changes occurred?* Annabelle found 12 more shells; then she gave 4 shells away.
- *Two things happened. First Annabelle found 12 shells. Is this a change-to-more situation or a change-to-less situation?* Change-to-more *Then Annabelle gave away 4 shells. Is this a change-to-more situation or a change-to-less situation?* Change-to-less **GMP1.1**

Display and fill in a change diagram to model the first step. Write 9 in the Start box, + 12 on the Change line, and ? in the End box. (*See margin.*) Tell children that the ? represents the number of shells Annabelle has after adding 12 more to her collection. **GMP4.1** Have children suggest a number model with ? to represent the first step. Sample answer: $9 + 12 = ?$ Write the number model below the change diagram. (*See margin.*)

- *How can we find the number of shells Annabelle has at this point?* Add $9 + 12$. *How many shells does she have?* 21

Below the open number sentence, write the number model for the solution with 21 substituted for the question mark. (*See margin.*) Remind children that this is just the first of two steps needed to solve this number story. They now have the information they need to find the number of shells Annabelle has after giving 4 to her aunt.

Display a new change diagram. Write 21 in the Start box, − 4 on the Change line, and ? in the End box. (*See margin.*) Ask:

- *What does the ? represent?* **GMP4.1** The number of shells Annabelle has after giving 4 to her aunt
- *What is a number model for the second step?* Sample answer: $21 - 4 = ?$ Write the number model below the change diagram.
- *How can we find the number of shells Annabelle now has?* Subtract $21 - 4$.
- *How many shells does she have?* 17 shells

Below the open number sentence, write the number model for the solution with 17 substituted for the question mark. (*See margin.*) **GMP4.2**

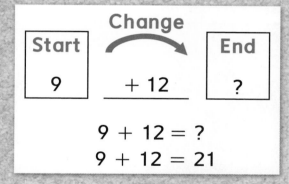

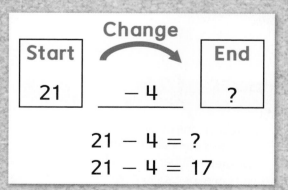

Solving Two-Step Number Stories (continued)

Lesson 6-5
DATE

3. Tommy is playing a board game with his sister. On his first turn, he earned 15 points. On his second turn, he lost 7 points. On his third turn, he earned 12 points. How many points does he have now?

Number model(s): Sample answer: $15 - 7 + 12 = ?$

Tommy has __20__ points.

4. Carrie had 23 markers. Luis gave her 7 more. She now has 15 more markers than Owen has. How many markers does Owen have?

Number model(s):
Sample answer: $23 + 7 = ?$ and $30 - 15 = ?$

Owen has __15__ markers.

Try This

5. On Monday Ellie had 18 gold stars. On Tuesday she got some more gold stars. On Wednesday she got 4 more gold stars. She now has 28 gold stars. How many gold stars did she get on Tuesday?

Number model(s):
Sample answers: $18 + 4 = ?$ and $28 - 22 = ?$;

Ellie got __6__ gold stars on Tuesday. $18 + ? + 4 = 28$

2.OA.1, 2.NBT.5, 2.NBT.7, SMP1, SMP4   one hundred fifty-one   151

Point out that it is also possible to write just one number model, $9 + 12 - 4 = ?$, for the problem. Although not all children will record the number model in this manner, it is important to show it as a possibility.

Have children work in small groups to solve Problem 2 on journal page 150. As you circulate, ask guiding questions such as the following:

- *What do we want to find out?* The number of balloons Yvette has now

  GMP1.1

- *What changes occurred?* She had 25 red balloons and 15 white balloons and then 8 popped.

- *What is the first step in this problem? The second step?* Sample answer: The first step is to find the total number of balloons she started with. The second step is to find how many she had after 8 popped.

- *How many balloons did Yvette have at the end?* 32 balloons

NOTE Some children may solve two-step number stories using different steps or different diagrams. Do not force a particular strategy for solving a two-step number story and don't suggest that there is a best set of diagrams for a two-step problem. Some children may be able to solve two-step problems without using diagrams.

Bring the class together to discuss their solutions. Share a variety of strategies. Have partners complete the problems on journal page 151. Emphasize to children the importance of recording their thinking for each problem using number models and drawings, diagrams, or words.

> **Differentiate**   **Common Misconception**
>
> Watch for children who mistakenly add all three quantities in the problem or add the first two numbers but do not know what to do with the third. Encourage them to break the number story into two parts. Ask: *What happened first? What happened next?*   Go Online   Differentiation Support

---

Professional Development

More than one number model can often fit a number story. In the case of two-step number stories, children can write one- or two-step number models to represent each story.

Go Online Professional Development

✓ Assessment Check-In 〈CCSS〉 2.OA.1, 2.NBT.5, 2.NBT.7

Math Journal 2, p. 151

As children complete Problems 3–5 on journal page 151, circulate and observe their work. Expect most children to be able to solve Problem 3 with the help of drawings or diagrams. Do not expect all children to solve Problems 4–5 because of the situation type (Problem 4 is a comparison problem) or the position of the unknown (Problem 5 involves an unknown change). Assist children as needed by helping them use drawings or situation diagrams to make sense of the problems. **GMP1.1**

☑ Assessment and Reporting | **Go Online** ▸ to record student progress and to see trajectories toward mastery for these standards.

Summarize Read *My Reference Book,* pages 34–35 with children and discuss.

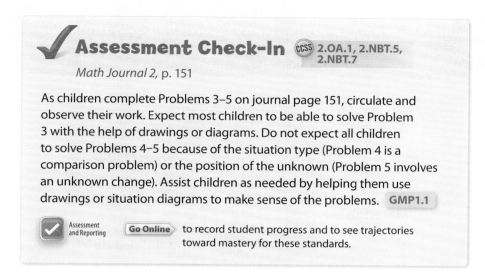

Math Journal 2, p. 152

---

③ Practice 10–15 min   **Go Online** ▸ ePresentations eToolkit Home Connections

▸ Math Boxes 6-5

Math Journal 2, p. 152

Mixed Practice Math Boxes 6-5 are paired with Math Boxes 6-8 and 6-10.

▸ Home Link 6-5

Math Masters, p. 171

Homework Children solve a two-step number story.

Math Masters, p. 171

Two-Step Number Stories Home Link 6-5

Family Note

In today's lesson your child solved two-step number stories, which can be broken into two parts and then solved in two steps. *For example:* Jonathan had 6 tickets for rides at the fair. His mother gave him 9 more. Then he gave 5 tickets to his friend. How many tickets does he have now?

To break this story into two parts, ask: What do you know from the story? (Jonathan had 6 tickets.) What happened first? (He received 9 more tickets.) What happened next? (He gave away 5 tickets.) What do you need to find out? (The number of tickets Jonathan has now.)

The first step is to figure out how many tickets Jonathan had after receiving some from his mother. The second step is to figure out how many tickets he had after giving some to his friend. Children are encouraged to solve two-step number stories using a variety of tools: drawings, open number lines, number grids, manipulatives, and diagrams.

They also learned to record either one or two number models for each number story—one for each part of the story or one number model to represent the whole story. *For example:* Use one number model, such as 6 + 9 − 5 = ?, for both parts. Or, use two number models, such as 6 + 9 = ? and 15 − 5 = ?, one for the first part and one for the second part. Answer: Jonathan now has 10 tickets.

Ask your child to explain the steps he or she takes to solve the problem below. Discuss how his or her number model(s) relates to the number story.

Please return this Home Link to school tomorrow.

• Write a number model or number models. Use ? to show the number you need to find. To help, you may draw a [diagram] or [diagram]

• Solve the problem and write the answer.

① At the beach, 11 children were playing in the sand. Then 6 more children joined them. Then 8 decided to go swimming. How many children were still playing in the sand?

Unit children

Number model(s): Sample answers: 11 + 6 − 8 = ?;
Answer: 9 children 11 + 6 = ? and 17 − 8 = ?

2.OA.1, 2.NBT.5, 2.NBT.7 one hundred seventy-one 171

Recording Addition Strategies

Overview Children make ballpark estimates and invent and record their own strategies for solving addition problems.

▶ **Vocabulary**
ballpark estimate

 Common Core State Standards

Focus Cluster
Use place value understanding and properties of operations to add and subtract.

1 Warm Up 15–20 min	Materials	
Mental Math and Fluency Children find differences between pairs of numbers.	number grid or number line (*Math Journal 2*, inside back cover; optional)	2.NBT.5, 2.NBT.7
Daily Routines Children complete daily routines.	See pages 4–43.	See pages xiv–xvii.

2 Focus 20–30 min		
Math Message Children solve an addition problem mentally and show why their strategies worked.	slate	2.NBT.5, 2.NBT.9
Sharing and Recording Strategies Children share addition strategies and practice recording their thinking.		2.NBT.5, 2.NBT.7, 2.NBT.9 SMP1, SMP3
Making Ballpark Estimates Children use ballpark estimates to check the reasonableness of their answers.	*Math Journal 2*, p. 153; number grid or number line (*Math Journal 2*, inside back cover; optional); base-10 blocks (optional)	2.NBT.5, 2.NBT.7 SMP1
✓ **Assessment Check-In** See page 566.	*Math Journal 2*, p. 153	2.NBT.5

CCSS 2.NBT.5 **Spiral Snapshot**

GMC Add within 100 fluently.

6-1 Warm Up Practice	6-2 through 6-5 Warm Up Focus Practice	6-6 Warm Up Focus Practice	6-7 Focus Practice	6-8 Focus Practice	7-1 Warm Up Focus Practice	7-3 Warm Up Focus Practice

/// Spiral Tracker   Go Online   to see how mastery develops for all standards within the grade.

3 Practice 15–20 min		
Playing *The Exchange Game* **Game** Children make exchanges with base-10 blocks.	*My Reference Book*, pp. 146–148; *Math Masters*, p. G21; base-10 blocks; per partnership: 1 die	2.NBT.1, 2.NBT.1a, 2.NBT.1b
Math Boxes 6-6 Children practice and maintain skills.	*Math Journal 2*, p. 154; Pattern-Block Template	See page 567.
Home Link 6-6 **Homework** Children make estimates and practice adding 2-digit numbers.	*Math Masters*, pp. 172–173	2.NBT.3, 2.NBT.5, 2.NBT.9

connectED.mheducation.com

Plan your lessons online with these tools.

 ePresentations

 Student Learning Center

 Facts Workshop Game

 eToolkit

 Professional Development

 Home Connections

 Spiral Tracker

 Assessment and Reporting

 English Learners Support

Differentiation Support

Differentiation Options RtI

CCSS 2.MD.6	**CCSS** 2.NBT.5, 2.NBT.7, SMP1	**CCSS** 2.NBT.5, SMP1

Readiness   5–15 min

WHOLE CLASS
SMALL GROUP
PARTNER
INDEPENDENT

Identifying Friendly Numbers

Class Number Line

For experience identifying friendly numbers using a visual model, children find friendly numbers on a number line. Tell children to point to the number 22 on the Class Number Line. Ask: *What friendly number is near 22?* Sample answers: 20; 25 *Why are these numbers friendly?* Sample answers: The number 20 is two 10s, and it is easy to add 10s. The number 25 is like a quarter. I know how to count by 25s. Repeat with other 2-digit numbers.

Enrichment   15–30 min

WHOLE CLASS
SMALL GROUP
PARTNER
INDEPENDENT

Addition Strategy Posters

Activity Card 79, base-10 blocks, number grid, 4 each of number cards 1–9, poster paper

To further explore adding 2-digit numbers, children use at least three different strategies to solve an addition problem. **GMP1.5** Children make posters comparing the strategies and share them with the class. **GMP1.6**

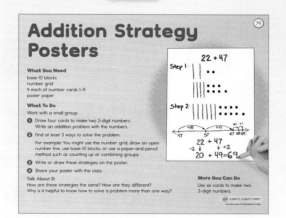

Extra Practice   5–15 min

WHOLE CLASS
SMALL GROUP
PARTNER
INDEPENDENT

Adding 2-Digit Numbers

Activity Card 80, 4 each of number cards 1–9, paper

For practice adding 2-digit numbers, children solve addition problems using paper-and-pencil strategies. Then they compare strategies with a partner. **GMP 1.6**

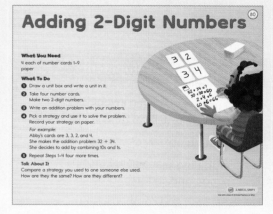

English Language Learners Support

Beginning ELL The expression *ballpark estimate* is a a familiar usage in American English. Provide context for this term for English language learners by displaying visuals of a ballpark. Introduce the expressions *in the ballpark* and *out of the ballpark*. Gesture to one of the visuals to demonstrate what happens when a home run is hit out of the ballpark. Connect that to an estimate that is far away from the actual answer, or *out of the ballpark*. Estimates close to the actual answer are *in the ballpark* (such as a ground ball hit in the infield) and are therefore called *ballpark estimates*.

Go Online > **ELL** English Learners Support

Standards and Goals for
Mathematical Practice

SMP1 **Make sense of problems and persevere in solving them.**
    GMP1.2 Reflect on your thinking as you solve your problem.
    GMP1.4 Check whether your answer makes sense.

SMP3 **Construct viable arguments and critique the reasoning of others.**
    GMP3.2 Make sense of others' mathematical thinking.

Professional Development

The strategies children invent for addition are usually of three major types:

- counting up
- combining place-value groups (1s, 10s, and so forth)
- adjusting and compensating

Examples of each type of strategy are given in this lesson, although adjusting and compensating is called by the more child-friendly name "making friendly numbers." For more information about these strategies, see the Mathematical Background section of the Unit 6 Organizer.

Go Online Professional Development

---

1 Warm Up   15–20 min   Go Online   ePresentations   eToolkit

▶ Mental Math and Fluency

Have children find the difference between pairs of numbers. They may use the number lines or number grids on the inside back covers of their journals. *Leveled exercises:*

- ●○○ 12 and 20 8
  -       33 and 63 30
- ●●○ 7 and 18 11
  -       45 and 87 42
- ●●● 3 and 21 18
  -       29 and 66 37

▶ Daily Routines

Have children complete daily routines.

2 Focus   20–30 min   Go Online   ePresentations   eToolkit

▶ Math Message

Use a mental strategy to solve 26 + 74. On your slate use words, numbers, or pictures to show why your strategy works.

▶ Sharing and Recording Strategies

| WHOLE CLASS | SMALL GROUP | PARTNER | INDEPENDENT |

Math Message Follow-Up Ask volunteers to share their strategies. As children share, model how to record their strategies by writing the key words and number sentences they used. (*See margin on next page.*)

Sample strategies for 26 + 74:

- **Counting Up** I used an open number line. I started at 74 because that was the bigger number. I jumped up 6 to get to an easy number, 80. I know 26 is the same as $20 + 6$, so I had 20 more to go. I made two more jumps of 10 and landed at 100.

- **Combining 10s and 1s** I added the 10s together: $20 + 70 = 90$. Then I added the 1s together: $6 + 4 = 10$. When I put 90 and 10 together, I got 100.

- **Making Friendly Numbers** I know 26 is 1 more than 25 and 74 is 1 less than 75, so I moved 1 from 26 to 74 to make the problem $25 + 75$. That's like counting quarters, and I know $25 + 75 = 100$.

Explain to children that making written records of their strategies can help them keep track of their own thinking. **GMP1.2** Written records can also help children understand one another's strategies. Ask volunteers to explain the written records of children's strategies you produced. **GMP3.2**

Tell children that they will practice adding numbers and recording their strategies. Display the problems shown below. Have small groups or partners solve the problems and make written records of their strategies.

$$76 + 23 = ?\quad 99 \qquad 52 + 29 = ?\quad 81 \qquad 129 + 237 = ?\quad 366$$

> **Differentiate** **Adjusting the Activity**
>
> If children struggle solving the problems mentally, they may use a number grid, a number line, or base-10 blocks to assist them.
>
> **Go Online** Differentiation Support

Circulate and observe as children are working. Encourage children to "talk out" and record what they are doing to help them think logically about what to do next. **GMP1.2** Ask guiding questions like the following:

- *How many have you counted up? How much do you have left to go?*
- *Did you combine all the 100s? All the 10s? All the 1s? What will you do next?*
- *How did you make a friendly number? Did you adjust both numbers so that you still get the same total?*

After groups have finished working, select children to share their strategies. Have children share both correct and incorrect strategies and prompt the class to discuss and resolve any issues. **GMP3.2** Highlight a strategy from each of the three major types: counting up, combining place-value groups, and making friendly numbers.

As children share their strategies, have them also share their written records. Sample written records are shown in the margin on this page.

- **Counting up (on an open number line):**

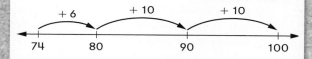

- **Combining 10s and 1s:**
  $$20 + 70 = 90$$
  $$6 + 4 = 10$$
  $$90 + 10 = 100$$

- **Making friendly numbers:**

$$26 + 74$$
$$-1 \downarrow \qquad \downarrow +1$$
$$25 + 75 = 100$$

Sample records of strategies

- **Counting up to solve 76 + 23:**
  $$76 + 20 = 96$$
  $$96 + 3 = 99$$
  $$76 + 23 = 99$$

- **Making friendly numbers to solve 52 + 29:**
  $$52 + 29 \leftarrow \text{Add 1 to 29.}$$
  $$52 + 30 = 82 \leftarrow \text{Subtract 1 from 82.}$$
  $$52 + 29 = 81$$

- **Combining 100s, 10s, and 1s to solve 129 + 237:**
  $$100 + 200 = 300$$
  $$20 + 30 = 50$$
  $$9 + 7 = 16$$
  $$300 + 50 + 16 = 366$$

Sample records of strategies

Math Journal 2, p. 153

Estimating and Adding

Lesson 6-6
DATE

Fill in the unit box. For each problem: *Sample estimates* Unit
- Make a ballpark estimate. *and strategies are given.*
- Solve the problem using any strategy you choose. Use words, numbers, or drawings to show your thinking.
- Explain how your estimate shows whether your answer makes sense.

1 32 + 26 = ?   Ballpark estimate: _30 + 30 = 60_

Strategy:

    + 10   + 10   + 6
  32   42   52   58

32 + 26 = _58_

Does your answer make sense? How do you know?
Sample answer: 58 is only 2 away from 60, so my answer makes sense.

2 18 + 44 = ?   Ballpark estimate: _20 + 40 = 60_

Strategy: 10 + 40 = 50, 8 + 4 = 12, and 50 + 12 = 62

18 + 44 = _62_

Does your answer make sense? How do you know?
Sample answer: 60 and 62 are close, so my answer makes sense.

2.NBT.5, 2.NBT.7, SMP1   one hundred fifty-three   153

▶ Making Ballpark Estimates

Math Journal 2, p. 153

WHOLE CLASS **SMALL GROUP** **PARTNER** INDEPENDENT

Tell children that now that they are adding larger numbers, it is important to check the *reasonableness* of their answers, or whether their answers make sense. One way to do this is to make a **ballpark estimate** and compare the estimate to the exact answer. **GMP1.4**

Academic Language Development Explain that estimates can vary but still be near the exact answer. Therefore, they are *in the ballpark* and are called *ballpark estimates*. Draw on children's background knowledge of baseball parks and contrast paired expressions, such as *ballpark estimate* and *exact answer* and *in the ballpark* and *out of the ballpark*.

Explain that one way to estimate a sum is to change the addends to close-but-easier numbers and then add those numbers. Refer to 52 + 29 from the previous activity and ask: *What friendly number is close to 52?* 50 *To 29?* 30 *What do I get when I add those numbers?* 50 + 30 = 80 Have children compare the ballpark estimate 80 to the exact answer 81. Ask: *Is our exact answer close to the ballpark estimate?* Yes. *Does the answer make sense? Does it seem reasonable? Why or why not?* Yes. I know from my estimate that the answer should be close to 80. **GMP1.4**

Tell children that they can make an estimate either before or after solving a problem. Display the problem 61 + 27 and ask children to make a ballpark estimate for it. Point out that there is more than one reasonable estimate. Sample answers: 60 + 30 = 90; 60 + 25 = 85 Have children find the exact answer and use the estimate to check whether the answer seems reasonable. Sample answer: The exact answer is 88. Because 88 is close to my estimate of 90, it seems right. Repeat the process for a problem with 3-digit addends, such as 203 + 118. Sample answer: One estimate is 200 + 120 = 320. My exact answer is 321. My answer seems reasonable because 321 is close to 320. **GMP1.4**

Have partners work together to complete journal page 153. They may use number lines, number grids, or base-10 blocks as needed.

✓ Assessment Check-In   CCSS 2.NBT.5

This is an early exposure to both inventing strategies for solving 2-digit addition problems and making ballpark estimates. Some children might be able to solve the problems on journal page 153 mentally, while others might use number lines, number grids, or base-10 blocks to solve the problems. A formal addition strategy is introduced in Lessons 6-7 and 6-8. Additional practice with invented and formal addition strategies, as well as with using estimates to judge the reasonableness of answers, will be provided throughout the rest of the year.

 Assessment and Reporting   **Go Online** to record student progress and to see trajectories toward mastery for these standards.

Summarize Have partners discuss ways they can check their thinking when they solve addition problems. Sample answers: I can review my written work to see if I made a mistake. I can use my estimate to see if my answer makes sense.

3 Practice 15–20 min

Go Online

ePresentations eToolkit Home Connections

▶ Playing *The Exchange Game*

My Reference Book, pp. 146–148; *Math Masters*, p. G21

| WHOLE CLASS | SMALL GROUP | PARTNER | INDEPENDENT |

Children play *The Exchange Game* with base-10 blocks. If necessary, review the game rules on *My Reference Book*, pages 146–148 with children.

Observe
- Which children know when to make exchanges?
- Which children can name the number represented by their blocks?

Discuss
- *How did you know when to make an exchange?*
- *How did you check your partner's trades?*

▶ Math Boxes 6-6

Math Journal 2, p. 154

| WHOLE CLASS | SMALL GROUP | PARTNER | INDEPENDENT |

Mixed Practice Math Boxes 6-6 are paired with Math Boxes 6-9.

▶ Home Link 6-6

Math Masters, pp. 172–173

Homework Children practice adding 2-digit numbers and making ballpark estimates. The Family Note provides information to parents about invented addition strategies.

Math Journal 2, p. 154

Math Boxes

Lesson 6-6
DATE

① Use your Pattern-Block Template to draw a shape that has three sides.

② Solve.
 a. $217 + \underline{10} = 227$
 b. $10 + 483 = \underline{493}$
 c. $507 = 497 + \underline{10}$
 d. $409 = 10 + \underline{399}$

③ Write each number in expanded form.
 69 $\underline{60 + 9}$
 24 $\underline{20 + 4}$
 345 $\underline{300 + 40 + 5}$
 180 Possible answers:
 $100 + 80;$
 $100 + 80 + 0$

④ What number do the base-10 blocks show? $\underline{124}$
 Use base-10 shorthand to show the number another way.
 Sample answer:

⑤ **Writing/Reasoning** Explain how you figured out the number shown by the base-10 blocks in Problem 4.
 Sample answer: 1 flat shows 100, 1 long shows 10, and 14 cubes show 14. So, $100 + 10 + 14 = 124.$

① 2.G.1 ② 2.NBT.8 ③ 2.NBT.1, 2.NBT.3 ④ 2.NBT.1, 2.NBT.1a,
154 one hundred fifty-four   2.NBT.3 ⑤ 2.NBT.1, 2.NBT.3, SMP2

Math Masters, p. 173

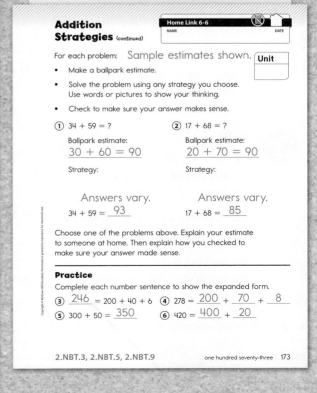

Addition Strategies (continued)

Home Link 6-6
NAME
DATE

For each problem: Sample estimates shown. Unit
- Make a ballpark estimate.
- Solve the problem using any strategy you choose. Use words or pictures to show your thinking.
- Check to make sure your answer makes sense.

① $34 + 59 = ?$
 Ballpark estimate:
 $\underline{30 + 60 = 90}$
 Strategy:

 Answers vary.
 $34 + 59 = \underline{93}$

② $17 + 68 = ?$
 Ballpark estimate:
 $\underline{20 + 70 = 90}$
 Strategy:

 Answers vary.
 $17 + 68 = \underline{85}$

Choose one of the problems above. Explain your estimate to someone at home. Then explain how you checked to make sure your answer made sense.

Practice
Complete each number sentence to show the expanded form.
③ $\underline{246} = 200 + 40 + 6$ ④ $278 = \underline{200} + \underline{70} + \underline{8}$
⑤ $300 + 50 = \underline{350}$ ⑥ $420 = \underline{400} + \underline{20}$

2.NBT.3, 2.NBT.5, 2.NBT.9   one hundred seventy-three 173

Partial-Sums Addition, Part 1

Overview Children use base-10 blocks to find partial sums and build readiness for partial-sums addition.

▶ **Before You Begin**
For Part 1, select and sequence Quick Look Cards 81, 86, and 109.

▶ **Vocabulary**
partial sums

 Common Core State Standards

Focus Clusters
- Understand that the three digits of a three digit number represents amounts of hundreds, tens, and ones; e.g., 706 equals 7 hundreds, 0 tens, and 6 ones.
- Use place value understanding and properties of operations to add and subtract.

---

1 Warm Up 15–20 min

Materials

Mental Math and Fluency Children view Quick Look Cards.	Quick Look Cards 81, 86, and 109; slate	**2.OA.2**
Daily Routines Children complete daily routines.	See pages 4–43.	See pages xiv–xvii.

---

2 Focus 20–30 min

Math Message Children solve a 2-digit addition problem.		**2.NBT.5, 2.NBT.9**
Sharing Strategies Children share their strategies for solving the Math Message.		**2.NBT.5, 2.NBT.9 SMP3**
Finding Partial Sums with Base-10 Blocks Children represent addends with base-10 blocks and combine the blocks to find the sum.	base-10 blocks; demonstration base-10 blocks; Class Data Pad	**2.NBT.1, 2.NBT.5, 2.NBT.7, 2.NBT.9 SMP1, SMP5**
Practicing Addition with Base-10 Blocks Children practice finding partial sums with base-10 blocks.	*Math Journal 2*, p. 155; base-10 blocks	**2.NBT.1, 2.NBT.5, 2.NBT.7, 2.NBT.9, SMP5**
✔ **Assessment Check-In** See page 572.	*Math Journal 2*, p. 155	**2.NBT.1, 2.NBT.5, 2.NBT.7, SMP5**

 2.NBT.7 Spiral Snapshot

GMC Add multidigit numbers using models or strategies.

6-1 Practice	6-2 through 6-6 Warm Up Focus Practice	6-7 Focus Practice	6-8 Focus Practice	7-1 Warm Up Practice	7-3 Warm Up	7-9 Practice

III Spiral Tracker   Go Online   to see how mastery develops for all standards within the grade.

---

3 Practice 15–20 min

Solving Number Stories Children solve number stories.	*Math Journal 2*, p. 156	**2.OA.1, 2.NBT.5, 2.NBT.6 SMP4**
Math Boxes 6-7: Preview for Unit 7 Children practice and maintain skills.	*Math Journal 2*, p. 157	See page 573.
Home Link 6-7 **Homework** Children practice addition with base-10 blocks.	*Math Masters*, p. 175	**2.NBT.1, 2.NBT.5, 2.NBT.7**

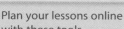 **connectED.mheducation.com**

Plan your lessons online with these tools.

 ePresentations   Student Learning Center   Facts Workshop Game   eToolkit   Professional Development   Home Connections   Spiral Tracker   Assessment and Reporting   English Learners Support   Differentiation Support

Differentiation Options

RtI

Readiness   15–30 min

WHOLE CLASS
SMALL GROUP
PARTNER
INDEPENDENT

Reviewing Place Value

Math Masters, p. TA22; paper;
base-10 blocks (longs and
cubes); per partnership:
4 each of number cards 1–9

Children review place value by
representing 2-digit numbers with as
few base-10 blocks as possible. **GMP2.1**
Partners share a Place-Value Mat
(*Math Masters*, page TA22). Each
partnership shuffles the number cards,
places the deck facedown, and draws
two cards. The first card goes in the ones
column on the Place-Value Mat, and the
second card in the tens column. Children
record on paper the number shown and
how many 10s (longs) and 1s (cubes) they
used. They switch the two cards on the
mat and repeat the procedure.

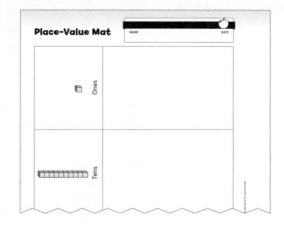

Enrichment   5–15 min

WHOLE CLASS
SMALL GROUP
PARTNER
INDEPENDENT

Comparing Addition Strategies

Math Masters, p. 174

To extend their work with multidigit
addition, children examine two addition
strategies on *Math Masters*, page 174.
GMP1.6, GMP3.2 Children write about
which strategy they thought was easier
to use and why. Then they use one of
the strategies to solve new problems.

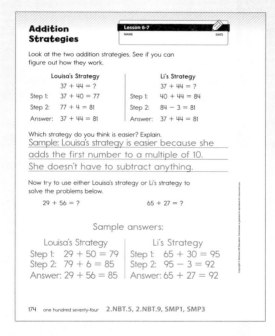

Extra Practice   5–15 min

WHOLE CLASS
SMALL GROUP
PARTNER
INDEPENDENT

Finding Partial Sums with Base-10 Blocks

Activity Card 81,
4 each of number cards 1–9,
base-10 blocks, slate

For practice finding partial sums, children
use base-10 blocks to solve addition
problems. **GMP5.2**

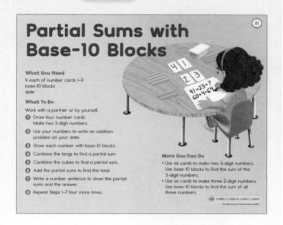

English Language Learners Support

Beginning ELL To build background knowledge to help children understand the
meaning of the term *partial*, demonstrate the meanings of *part* and *partial* using simple
jigsaw puzzles. Contrast the meanings of *parts*, *partial*, and *whole*, using the terms with
teacher think-aloud statements like these: *This is a part of the puzzle. This is a partial section
of the puzzle. I have made the whole puzzle.* Provide for student practice with the terms,
using Total Physical Response prompts like these: *Show me one part. Count the parts. Use
the parts to make the whole. Put the partial sections together. Make the whole puzzle with
the partial sections.*

Go Online   ELL   English Learners Support

Professional Development

As children represent numbers with base-10 blocks and use the blocks to find partial sums, they are building readiness for partial-sums addition, which is formally introduced in the next lesson. Partial-sums addition is an example of an algorithm, or a step-by-step procedure for completing a task. For more information about algorithms in *Everyday Mathematics* and beyond, see the Mathematical Background section of the Unit 6 Organizer.

Go Online > Professional Development

---

1 Warm Up   15–20 min

Go Online > ePresentations eToolkit

▶ Mental Math and Fluency

Flash the following sequence of Quick Look Cards: 81, 86, and 109. Always allow children a second look and follow up by asking them to write a sentence on their slates that describes how they found the total number of dots. *Leveled exercises:*

● ○ ○ Quick Look Card 81   Sample answer: I saw 4 and 4 and 1 more, so that makes 9.

● ● ○ Quick Look Card 86   Sample answer: I moved all 4 over from one ten frame to fill the other frame, so that's 10.

● ● ● Quick Look Card 109   Sample answer: I moved 2 over to fill one ten frame, and that left 3, so there are 13 in all.

▶ Daily Routines

Have children complete daily routines.

---

2 Focus   20–30 min

Go Online > ePresentations eToolkit

▶ Math Message

Solve 37 + 52. Explain your strategy to a partner.

▶ Sharing Strategies

WHOLE CLASS | SMALL GROUP | PARTNER | INDEPENDENT

Math Message Follow-Up   Circulate and observe as children solve the Math Message problem and explain their strategies to their partners.

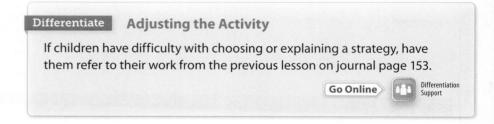

Differentiate   **Adjusting the Activity**

If children have difficulty with choosing or explaining a strategy, have them refer to their work from the previous lesson on journal page 153.

Go Online > Differentiation Support

Encourage partners to try using each other's strategies to solve the problem. GMP3.2 Have volunteers share a variety of strategies. Make sure the following strategy is included:

- **Combining 10s and 1s** I added the 10s to get 30 + 50 = 80. Then I added the 1s to get 7 + 2 = 9. When I put 80 and 9 together, I got 89.

Record this strategy with children's help. (*See margin.*) Tell children that today they will practice this strategy, using base-10 blocks to help them keep track of and show their thinking.

$$30 + 50 = 80$$
$$7 + 2 = 9$$
$$80 + 9 = 89$$

Sample record of the combining-10s-and-1s strategy to find 37 + 52

▶ Finding Partial Sums with Base-10 Blocks

| WHOLE CLASS | SMALL GROUP | PARTNER | INDEPENDENT |

Display the problem 45 + 26 in vertical form. Distribute base-10 blocks and instruct children to represent 45 and 26 using as few blocks as possible. Display an image of the base-10 blocks arranged to resemble the vertical addition problem. (*See margin.*) Have children follow along with their blocks as you model using them to solve 45 + 26.

Review the combining-10s-and-1s strategy discussed during the Math Message Follow-Up. Explain that children can follow the same steps with their base-10 blocks: add the 10s, add the 1s, and then add the two parts together. Ask:

- *Which blocks show the 10s?* GMP5.2 The longs Collect all the longs into one group.
- *Count the 10s. How many are there?* Six 10s, or 60
- *Which blocks show the 1s?* GMP5.2 The cubes Collect all the cubes into another group.
- *Count the 1s. How many are there?* Eleven 1s, or 11
- *What do we still need to do to find the answer?* GMP1.2 Add the 10s and 1s together: 60 + 11 = 71

Some children may notice that they have more than 10 cubes after they combine them, so they exchange 10 cubes for 1 long. Making the exchange is fine if they recognize the possibility on their own, but it is not necessary at this time.

Point out that children used the blocks to find parts of the sum, or **partial sums,** and then added the partial sums together to find the total. Have children explain in their own words how to use base-10 blocks to find partial sums. Record an explanation on the Class Data Pad for children to use for reference as needed. Sample answer: Show the numbers with base-10 blocks. Put the 10s together and the 1s together. Then add the partial sums together.

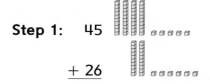

Step 1: 45
+ 26

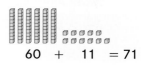

Step 2: Combine 10s and 1s.

60 + 11 = 71

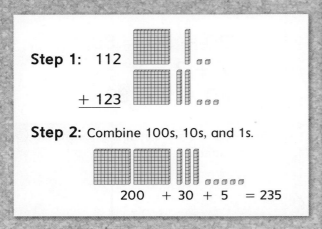

Step 1: 112

+ 123

Step 2: Combine 100s, 10s, and 1s.

200 + 30 + 5 = 235

Math Journal 2, p. 155

Addition with Base-10 Blocks

Lesson 6-7
DATE

Fill in the unit box.
Solve each problem using base-10 blocks.
Use base-10 shorthand to show what you did.
On the lines, record the partial sums and the answer.

Unit

Block	Flat	Long	Cube
Base-10 Shorthand	□			.

Example: 23
+ 46

Answer: 60 + 9 = 69

① 41
+ 35

Answer: 70 + 6 = 76

② 67
+ 38

Answer: 90 + 15 = 105

③ 123
+ 128

Answer: 200 + 40 + 11 = 251

2.NBT.5, 2.NBT.7, SMP5   one hundred fifty-five 155

Academic Language Development Given the pronunciation of the word *partial,* children will not be able to hear the word *part.* Display the word *partial* and highlight or underline the word *part.* Have them practice saying the word *partial.*

Display the problem 112 + 123 in vertical form and have children discuss how they might use base-10 blocks to solve it. Repeat the process described on the previous page, modeling how to find the partial sums using base-10 blocks as children follow along with their own blocks: represent the numbers, combine each type of block, and count the number of each type of block to find the total. 235 (*See margin.*)

Add an explanation to the Class Data Pad on how to use base-10 blocks to add 3-digit numbers.

Next have children work in partnerships to solve the following problems:

68 + 24 = ? 92   243 + 155 = ? 398

When children are finished, have volunteers demonstrate and explain to the class how they solved the problems using base-10 blocks.

Refer children to 68 + 24 again and ask: *What happens if I add the 1s first?* Sample answer: You would get the same partial sums and the same answer. Help children recognize that partial sums may be calculated in any order, but children must be careful to not miss any partial sums as they complete their calculations.

▶ Practicing Addition with Base-10 Blocks

Math Journal 2, p. 155

WHOLE CLASS	SMALL GROUP	PARTNER	INDEPENDENT

Children work in partnerships to complete journal page 155. They solve the problems using base-10 blocks and record their work on the journal page. **GMP5.2** Remind children to use base-10 shorthand to record their work.

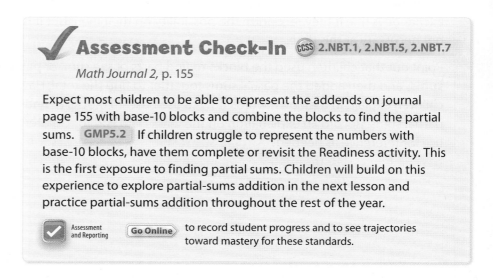

✓ **Assessment Check-In** (CCSS) 2.NBT.1, 2.NBT.5, 2.NBT.7

Math Journal 2, p. 155

Expect most children to be able to represent the addends on journal page 155 with base-10 blocks and combine the blocks to find the partial sums. **GMP5.2** If children struggle to represent the numbers with base-10 blocks, have them complete or revisit the Readiness activity. This is the first exposure to finding partial sums. Children will build on this experience to explore partial-sums addition in the next lesson and practice partial-sums addition throughout the rest of the year.

Assessment and Reporting   Go Online   to record student progress and to see trajectories toward mastery for these standards.

Summarize Have partners talk about why base-10 blocks are helpful for finding partial sums. Sample answer: They help me see the 100s, 10s, and 1s.

(3) Practice 15–20 min Go Online ePresentations eToolkit Home Connections

▶ Solving Number Stories

Math Journal 2, p. 156

| WHOLE CLASS | SMALL GROUP | **PARTNER** | **INDEPENDENT** |

Children solve number stories. They practice writing number models using ? for the unknown. **GMP4.1, GMP4.2**

▶ Math Boxes 6-7: Preview for Unit 7

Math Journal 2, p. 157

| WHOLE CLASS | **SMALL GROUP** | PARTNER | INDEPENDENT |

Mixed Practice Math Boxes 6-7 are paired with Math Boxes 6-11. These problems focus on skills and understandings that are prerequisite for Unit 7. You may want to use information from these Math Boxes to plan instruction and grouping in Unit 7.

▶ Home Link 6-7

Math Masters, p. 175

Homework Children practice addition with base-10 blocks.

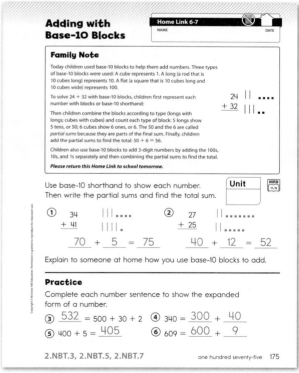

Math Masters, p. 175

Math Journal 2, p. 156

Number Stories Lesson 6-7 DATE

For each number story: Sample number models shown.

- Write a number model. Use ? to show the number you need to find. To help, you may draw a

- Solve the problem and write the answer. Include the unit.

(1) Jack drove 40 miles to a gas station. Then he drove 30 miles from the gas station to his friend's house. How many miles did Jack drive in all?

Number model: __40 + 30 = ?__

Answer: __70 miles__

(2) Emma found two leaves. One leaf was 9 centimeters longer than the other. The longer leaf was 20 centimeters long. How long was the shorter leaf?

Number model: __? + 9 = 20__

Answer: __11 centimeters__

Try This

(3) A fish weighs 35 pounds. An octopus weighs 20 pounds. A crab weighs 2 pounds. How much do all three weigh together?

Number model: __35 + 20 + 2 = ?__

Answer: __57 pounds__

156 one hundred fifty-six   2.OA.1, 2.NBT.5, 2.NBT.6, SMP4

Math Journal 2, p. 157

Math Boxes
Preview for Unit 7 Lesson 6-7 DATE

(1) Solve.   Unit: pencils

$14 + \underline{6} = 20$

$20 = 12 + \underline{8}$

$11 + \underline{9} = 20$

$\underline{7} + 13 = 20$

(2) Meg has ☆☆☆.
Dan has ☆☆☆☆☆☆☆.
Jen has ☆☆.

How many stars are there in all? __12__ stars

(3) Name something that is about 1 foot long.

__Answers vary.__

(4) Erin can jump about 1 foot high. John can jump about 2 feet high. How much higher can John jump?

About __1__ ft

(5) Books Read

Who read the most books? __Grace__

Who read the fewest books? __Lilly__

(6) Write a number story to match the number model.

$4 + 6 - 5 = ?$

__Sample answer: Tina had 4 red pens and 6 blue pens. She lost 5 pens. How many pens does Tina have left?__

① 2.OA.2, 2.NBT.5 ② 2.OA.1, 2.NBT.5, 2.NBT.6 ③ 2.MD.3
④ 2.MD.5 ⑤ 2.MD.10 ⑥ 2.NBT.5, SMP2   one hundred fifty-seven 157

Partial-Sums Addition, Part 2

Overview Children are introduced to partial-sums addition.

▶ **Vocabulary**
partial-sums addition

 Common Core State Standards

Focus Clusters
- Understand place value.
- Use place value understanding and properties of operations to add and subtract.

1 Warm Up 15–20 min	**Materials**	
Mental Math and Fluency Children compare 3-digit numbers.	slate	2.NBT.4
Daily Routines Children complete daily routines.	See pages 4–43.	See pages xiv–xvii.

2 Focus 30–40 min		
Math Message Children represent 2-digit numbers with base-10 blocks and expanded form.	base-10 blocks	2.NBT.1, 2.NBT.3 SMP2
Using Expanded Form to Find Partial Sums Children use expanded form to find partial sums and are introduced to partial-sums addition.	base-10 blocks	2.NBT.3, 2.NBT.5, 2.NBT.7, 2.NBT.9 SMP1, SMP2, SMP6
Estimating and Adding with Partial Sums Children practice using partial-sums addition and use ballpark estimates to check that their answers make sense.	*Math Journal 2*, p. 158; base-10 blocks (optional)	2.NBT.3, 2.NBT.5, 2.NBT.7, 2.NBT.9 SMP1
✓ **Assessment Check-In** See page 579.	*Math Journal 2*, p. 158	2.NBT.5, 2.NBT.7

CCSS 2.NBT.7 **Spiral Snapshot**

GMC Add multidigit numbers using models or strategies.

| 5-10
Focus
Practice | 6-1
Practice | 6-2 through 6-7
Warm Up
Focus
Practice | 6-8
Focus
Practice | 7-1
Warm Up
Practice | 7-3
Warm Up | 7-9
Practice |

Spiral Tracker **Go Online** to see how mastery develops for all standards within the grade.

3 Practice 15–20 min		
Playing *Salute!* **Game** Children find missing addends.	*My Reference Book*, pp. 162–163; per group: 4 each of number cards 0–10	2.OA.2 SMP6
Math Boxes 6-8 Children practice and maintain skills.	*Math Journal 2*, p. 159	See page 579.
Home Link 6-8 **Homework** Children practice partial-sums addition.	*Math Masters*, p. 176	2.NBT.5, 2.NBT.7

connectED.mheducation.com

Plan your lessons online with these tools.

 ePresentations   Student Learning Center   Facts Workshop Game   eToolkit   Professional Development   Home Connections   Spiral Tracker ✓ Assessment and Reporting   ELL English Learners Support   Differentiation Support

Differentiation Options

 RtI

CCSS 2.NBT.1, 2.NBT.3, SMP2

Readiness
5–15 min

Reviewing Place Value

WHOLE CLASS
SMALL GROUP
PARTNER
INDEPENDENT

Math Masters, p. TA22;
paper; base-10 blocks;
per partnership: 4 each of
number cards 1–9

Children review place value by
representing numbers using as few
base-10 blocks as possible. **GMP2.1** See
the Readiness activity in Lesson 6-7. If
children seem ready, you may wish to
adjust the directions so that children draw
three cards to create a 3-digit number. To
extend the activity, have children record
each number in expanded form after they
build it with base-10 blocks.

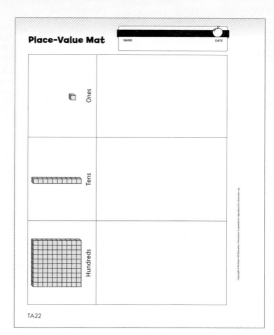

Place-Value Mat

Ones

Tens

Hundreds

TA22

CCSS 2.NBT.5, 2.NBT.7, SMP1

Enrichment
5–15 min

Comparing Addition Strategies

WHOLE CLASS
SMALL GROUP
PARTNER
INDEPENDENT

Activity Card 82,
4 each of number cards 1–9,
paper

To extend their work with multidigit
addition, children add numbers using
partial-sums addition and one other
strategy. Partners discuss which strategy
was easier to use. **GMP1.6**

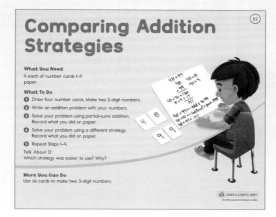

Comparing Addition Strategies

What You Need
4 each of number cards 1–9
paper

What To Do
1. Draw four number cards. Make two 2-digit numbers.
2. Write an addition problem with your numbers.
3. Solve your problem using partial-sums addition. Record what you did on paper.
4. Solve your problem using a different strategy. Record what you did on paper.
5. Repeat Steps 1–4.

Talk About It
Which strategy was easier to use? Why?

More You Can Do
Use six cards to make two 3-digit numbers.

CCSS 2.NBT.5, 2.NBT.6, 2.NBT.7, SMP5

Extra Practice
5–15 min

Finding Partial Sums with Base-10 Blocks

WHOLE CLASS
SMALL GROUP
PARTNER
INDEPENDENT

Activity Card 81,
4 each of number cards 1–9,
base-10 blocks, slate

To practice finding partial sums, children
solve addition problems using base-10
blocks. **GMP5.2** See the Extra Practice
activity in Lesson 6-7 for details. Some
children may be ready to find partial sums
without base-10 blocks. Encourage these
children to record their work on slates
using expanded form.

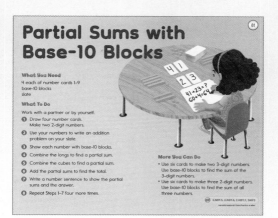

Partial Sums with Base-10 Blocks

What You Need
4 each of number cards 1–9
base-10 blocks
slate

What To Do
Work with a partner or by yourself.
1. Draw four number cards. Make two 2-digit numbers.
2. Use your numbers to write an addition problem on your slate.
3. Show each number with base-10 blocks.
4. Combine the longs to find a partial sum.
5. Combine the cubes to find a partial sum.
6. Add the partial sums to find the total.
7. Write a number sentence to show the partial sums and the answer.
8. Repeat Steps 1–7 four more times.

More You Can Do
• Use six cards to make two 3-digit numbers. Use base-10 blocks to find the sum of the 3-digit numbers.
• Use six cards to make three 2-digit numbers. Use base-10 blocks to find the sum of all three numbers.

English Language Learners Support

Beginning ELL To help English language learners understand the meaning of *expand*,
demonstrate an action such as playing an accordion or blowing up a balloon. For example,
show that a balloon is still the same balloon when you blow it up; it is just stretched out.
Using both numbers and base-10 blocks, show children how a number such as 352 gets
"stretched out," or expanded, into 300 + 50 + 2.

 Go Online ELL English Learners Support

1 Warm Up 15–20 min 〈Go Online〉 ePresentations eToolkit

▶ Mental Math and Fluency

Dictate pairs of 3-digit numbers. Have children compare the numbers and use <, >, or = to record the comparisons on their slates. *Leveled exercises:*

● ○ ○ 157 > 152
       349 < 649
       780 > 770

● ● ○ 878 < 882
       762 > 678
       450 < 540

● ● ● 987 > 978
       606 < 660
       461 > 451

▶ Daily Routines

Have children complete daily routines.

2 Focus 30–40 min 〈Go Online〉 ePresentations eToolkit

▶ Math Message

Show 53 and 44 with the fewest possible base-10 blocks. 5 longs and 3 cubes; 4 longs and 4 cubes

Then write both numbers in expanded form. 50 + 3; 40 + 4

Talk about these questions with a partner: What is the same about showing numbers with base-10 blocks and writing them in expanded form? What is different? GMP2.3

▶ Using Expanded Form to Find Partial Sums

| WHOLE CLASS | SMALL GROUP | PARTNER | INDEPENDENT |

Math Message Follow-Up Ask children to share their ideas about how representations using base-10 blocks and expanded form are similar and different. GMP2.3 Sample answers: Expanded form and base-10 blocks both show how much each digit of a number is worth. The base-10 blocks are objects, but I can write the expanded form on paper. Remind children that in the previous lesson they used base-10 blocks to help them find partial sums and add numbers. Tell them that today they will use expanded form to help them find partial sums.

Academic Language Development Use examples to contrast the terms *standard form* and *expanded form*. Encourage children to use the terms as they describe their strategies.

Have a volunteer demonstrate or explain how to use base-10 blocks to solve 53 + 44. 97 Refer as needed to the description of how to find partial sums on the Class Data Pad from Lesson 6-7. (*See margin.*)

Display the expanded form for 53 and 44 as shown below. Ask: *How could this expanded form help us do the same thing we just did with the base-10 blocks?* GMP2.3 Sample answer: We can add 50 and 40 to find the 10s and add 3 and 4 to find the 1s. Think aloud as you point to the relevant parts of the expanded form and record the partial sums: *First I add the 10s: 50 plus 40 is 90. Then I add the 1s: 3 plus 4 is 7. What is 90 + 7?* 97 Record the answer as shown below.

$$53 = 50 + 3$$
$$44 = 40 + 4$$
$$\overline{90 + 7 = 97}$$

Pose another 2-digit addition problem, such as 36 + 75. 111 Write the problem vertically. Ask children to tell you how to write each number in expanded form, and record the expanded form off to the side. You may want to use a second color. (*See margin.*) Challenge children to work with a partner to use expanded form to help them solve the problem without base-10 blocks. Encourage them to think about adding the 10s, adding the 1s, and then adding the partial sums.

> **Differentiate** **Adjusting the Activity**
>
> If children struggle writing the expanded form for each number, have them model the numbers with base-10 blocks or sketch base-10 shorthand and record their work on paper.
>
> Go Online Differentiation Support

After children have had time to work, ask them to share their thinking. GMP6.1 Point to the relevant parts of the expanded form as you use children's descriptions of their steps to complete the record shown in the margin. Tell children that this method is called **partial-sums addition.** When using partial-sums addition, children first find the partial sums, then add the partial sums together to find the total.

Remind children that one way to check whether their answers are reasonable is to make a ballpark estimate. Ask: *How could we make a ballpark estimate for this problem?* Sample answer: Use close-but-easier numbers: 40 + 70 = 110 *Does our answer, 111, seem reasonable?* GMP1.4 Sample answer: Yes. 111 is close to 110. Encourage children to use ballpark estimates to check whether their answers make sense whenever they add or subtract multidigit numbers, especially when they do not use tools. They can make their estimates before or after they find the exact answer.

Step 1: 53
$$+ \ 44$$

Step 2: Combine 10s and 1s.

$$90 \ + \ 7 \ = 97$$

$$
\begin{array}{rr}
 & 36 \quad 30 + 6 \\
 & + \ 75 \quad 70 + 5 \\
30 + 70 = & 100 \\
6 + 5 = & 11 \\
\hline
 & 111
\end{array}
$$

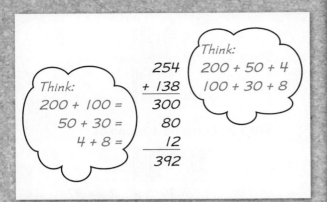

Repeat the activity with an addition problem with 3-digit addends, such as 254 + 138. **392** Explain to children that they should continue writing the expanded form for the addends if it helps them think about the partial sums, but if they are able to find the partial sums mentally, they do not have to write the expanded form every time. Ask them to tell you the expanded form for each addend. Record it and draw a thought bubble around it to emphasize that it is ok if they do not write it out. (*See margin.*) Then have children use partial-sums addition to solve the problem and ballpark estimates to check whether their answers make sense. Sample answer: 300 + 100 = 400; My answer makes sense because 392 is close to 400.

Ask children to share their answers and explain what they did in each step. **GMP6.1** As you make the record shown in the margin, think aloud: *Add the 100s: 200 plus 100 is 300. Add the 10s: 50 plus 30 is 80. Add the 1s: 4 plus 8 is 12. Find the total: 300 plus 80 plus 12 is 392.* Explain that children do not have to record the steps shown in blue if they are able to find the sums mentally, but they can write the steps if it helps them think through the problem.

Pose additional 2- and 3-digit addition problems as needed. Children should solve the problems using partial-sums addition, explain their thinking to partners, and use ballpark estimates to check the reasonableness of their answers. **GMP6.1, GMP1.4** *Suggestions:*

- 87 + 26 = ? 113
- 246 + 132 = ? 378
- 125 + 63 = ? 188

Differentiate | **Common Misconception**

Some children may struggle when using partial-sums addition to add a 3-digit number to a 2-digit number. For example, when adding 125 and 63, they may write 100 + 600 = 700 as the first partial sum. Encourage these children to write the addends in expanded form and use a zero to represent the hundreds in 63: 63 = 0 + 60 + 3. **Go Online** Differentiation Support

---

► # Estimating and Adding with Partial Sums

Math Journal 2, p. 158

| WHOLE CLASS | SMALL GROUP | **PARTNER** | INDEPENDENT |

Children work with partners to complete journal page 158. They should make ballpark estimates and use partial-sums addition to solve the problems. Explain to children that they can write out the parts shown in the think bubbles if it is helpful, but they do not have to do so for every problem.

Math Journal 2, p. 158

Partial-Sums Addition

Lesson 6-8
DATE

For Problems 1–3, make a ballpark estimate. Then solve the problem using partial-sums addition. Show your work. Use your estimate to check that your answer makes sense.

Unit

Example: 59 + 26 = ?
Ballpark estimate:
60 + 30 = 90

Think: 50 + 20 = / 9 + 6 =
59 + 26
Think: 50 + 9 / 20 + 6
70
15
85

① Ballpark estimate: Sample:
35 + 70 = 105

34
+ 71
105

② Ballpark estimate: Sample:
140 + 160 = 300

136
+ 157
293

③ Ballpark estimate: Sample:
120 + 50 = 170

122
+ 53
175

④ Solve one of the problems a different way. Explain your strategy.
Sample: For Problem 1, I changed 34 + 71 into 35 + 70 since it's easier to add: 35 + 70 = 105.

158   one hundred fifty-eight   2.NBT.3, 2.NBT.5, 2.NBT.7, 2.NBT.9, SMP1

✓ Assessment Check-In CCSS 2.NBT.5, 2.NBT.7

Math Journal 2, p. 158

Expect that most children will be able to correctly solve the problems on journal page 158. Some may need to use base-10 blocks to help them write the expanded form before adding. Because this is the first exposure to partial-sums addition, do not expect all children to accurately record all of their steps or clearly describe what they did. Practice with partial-sums addition will continue throughout the year.

☑ Assessment and Reporting Go Online to record student progress and to see trajectories toward mastery for these standards.

Summarize Partners explain to each other how they used their estimates to check that their answers on journal page 158 made sense. **GMP1.4** Sample answer: If my estimate was close to my answer, I knew the answer made sense. If it wasn't, I tried again to find the answer.

③ Practice 15–20 min

Go Online ePresentations eToolkit Home Connections

▶ Playing *Salute!*

My Reference Book, pp. 162–163

| WHOLE CLASS | **SMALL GROUP** | PARTNER | INDEPENDENT |

Have children play *Salute!* See Lesson 3-4 for detailed directions.

Observe
- What strategies are children using to find the missing addends?
- Which children understand the relationships between the numbers?

Discuss
- *How did you figure out the number on your card?*
- *Which numbers were easy to figure out? Which numbers were hard to figure out?* **GMP6.4**

▶ Math Boxes 6-8

Math Journal 2, p. 159

| WHOLE CLASS | SMALL GROUP | PARTNER | **INDEPENDENT** |

Mixed Practice Math Boxes 6-8 are paired with Math Boxes 6-5 and 6-10.

▶ Home Link 6-8

Math Masters, p. 176

Homework Children practice partial-sums addition.

Math Journal 2, p. 159

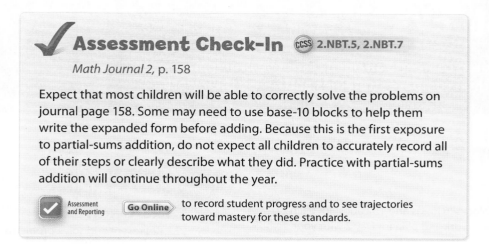

Math Masters, p. 176

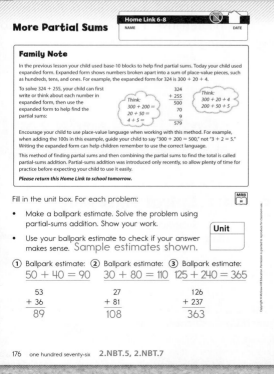

Subtracting with Base-10 Blocks

2-Day Lesson

Overview **Day 1:** Children complete an open response problem.
Day 2: Children compare strategies and revise their work.

Day 1: Open Response

▶ **Before You Begin**
Make the Number-Grid Poster visible. Prepare base-10 blocks (6 longs and 15 cubes) for each partnership. If possible, schedule time to review children's work and plan for Day 2 with your grade-level team.

Common Core State Standards

Focus Clusters
• Represent and solve problems involving addition and subtraction.
• Use place value understanding and properties of operations to add and subtract.

1 Warm Up   15–20 min

	Materials	
Mental Math and Fluency Children determine numbers shown in base-10 drawings.	slate	2.NBT.1, 2.NBT.1a
Daily Routines Children complete daily routines.	See pages 4–43.	See pages xiv–xvii.

2 Focus   45–55 min

Math Message Children solve a subtraction number story.	*Math Journal 2,* p. 160; number grid; per partnership: base-10 blocks (at least 6 longs and 15 cubes)	2.OA.1, 2.NBT.5, 2.NBT.7
Sharing Strategies Children discuss strategies they can use to solve a subtraction number story.	*Math Journal 2,* p. 160; Standards for Mathematical Practice Poster; per partnership: base-10 blocks (at least 6 longs and 15 cubes)	2.OA.1, 2.NBT.5, 2.NBT.7 SMP2, SMP5
Solving the Open Response Problem Children invent strategies to solve a subtraction problem using base-10 blocks.	*Math Masters,* p. 177; per partnership: base-10 blocks (at least 6 longs and 15 cubes); *My Reference Book,* pp. 16–17	2.OA.1, 2.NBT.5, 2.NBT.7 SMP1, SMP2, SMP5

Getting Ready for Day 2 →
Review children's work and plan discussion for reengagement. *Math Masters,* p. TA5; children's work from Day 1

CCSS 2.NBT.7 **Spiral Snapshot**
GMC Subtract multidigit numbers using models or strategies.

6-2 through 6-5 Focus Practice	6-6 Warm Up	6-9 Focus Practice	7-1 Warm Up Practice	7-3 Warm Up	7-7 Practice	7-9 Practice

Spiral Tracker   **Go Online** to see how mastery develops for all standards within the grade.

connectED.mheducation.com
Plan your lessons online with these tools.

 ePresentations
Student Learning Center
 Facts Workshop Game
 eToolkit
 Professional Development
Home Connections
 Spiral Tracker
 Assessment and Reporting
 ELL English Learners Support
 Differentiation Support

1 Warm Up 15–20 min Go Online

ePresentations eToolkit

▶ Mental Math and Fluency

Tell the numbers shown in the base-10 drawings.

●○○ | 15

●●○ | :::::. 21

●●● ||| ::::: 45

▶ Daily Routines

Have children complete daily routines.

2a Focus 45–55 min Go Online

ePresentations eToolkit

▶ Math Message

Math Journal 2, p. 160

Complete the problem on journal page 160.

▶ Sharing Strategies

Math Journal 2, p. 160

| WHOLE CLASS | SMALL GROUP | PARTNER | INDEPENDENT |

Math Message Follow-Up Give partners time to compare and discuss strategies before having volunteers share with the class. Children may describe a strategy of subtracting by counting back 24 from 56 on the number grid or number line. Children may also count up from 24 to 56 by tens and ones using a number grid or number line. GMP2.2

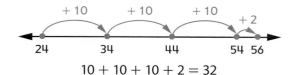

$$10 + 10 + 10 + 2 = 32$$

Counting up on an open number line: $56 - 24 = 32$

CCSS Standards and Goals for
Mathematical Practice

SMP1 Make sense of problems and persevere in solving them.
  GMP1.3 Keep trying when your problem is hard.

SMP2 Reason abstractly and quantitatively.
  GMP2.2 Make sense of the representations you and others use.

SMP5 Use appropriate tools strategically.
  GMP5.2 Use tools effectively and make sense of your results.

Professional Development

The focus of this lesson is GMP5.2. In the Math Message, children discuss how to use base-10 blocks to represent subtraction by simply removing blocks. The open response problem challenges children to invent ways to use base-10 blocks to represent subtraction when it is not possible to directly remove blocks. Children will need to use other strategies, such as covering up, crossing out, or making a trade.

Go Online for information about SMP5 in the *Implementation Guide*.

Math Journal 1, p. 160

A Subtraction Number Story Lesson 6-9
DATE

Bakers make 56 loaves of bread. They sell 24 loaves in the morning. How many loaves are left for sale in the afternoon?

Solve the problem using any strategy. Be ready to explain how you found the answer.

Number model: __56 − 24 = 32__ loaves

160 one hundred sixty 2.OA.1, 2.NBT.5, 2.NBT.7

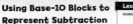

Using Base-10 Blocks to Represent Subtraction

One day the bakers make 53 dinner rolls. They sell 38 of the rolls that day. How many rolls do they have left over?

Solve the problem using base-10 blocks. Draw a picture to show how you used the blocks and write an explanation of how you solved the problem.

Answers vary. See sample children's work on page 589 of the *Teacher's Lesson Guide*.

__15__ rolls

2.OA.1, 2.NBT.5, 2.NBT.7,
SMP1, SMP2, SMP5

one hundred seventy-seven   177

ELL Support

Prior to the lesson, use visuals to review the vocabulary from the lesson: baker, loaf (loaves), dinner rolls. Then use simpler problems so children can practice communicating their strategies using base-10 blocks, such as: *The bakers make 11 loaves of bread. They sell 6 in the morning. How many are left for sale in the afternoon?*

NOTE In *Everyday Mathematics*, children solve subtraction problems using strategies before introduction of a standard algorithm. When children first use place value and number properties to invent their own strategies, they develop the conceptual understanding necessary for work with subtraction algorithms later. If children suggest a standard paper-and-pencil algorithm now, record it on the board, but do not take the time to teach it.

−9	−8	−7	−6	−5	−4	−3	−2	−1	0
1	2	3	4	5	6	7	8	9	10
11	12	13	14	15	16	17	18	19	20
21	22	23	24	25	26	27	28	29	30
31	(32)	33	34	35	36	37	38	39	40
41	42	43	44	45	46	47	48	49	50
51	52	53	54	55	[56]	57	58	59	60
61	62	63	64	65	66	67	68	69	70

Counting back on a number grid: $56 - 24 = 32$

Differentiate   Common Misconception

Look for children who include the starting number in their counts when they count up or back by ones. Consider demonstrating two methods for solving the problem using the number grid, one that counts the starting number (resulting in an answer of 33) and one that leads to the correct answer (32). Ask: *I get either 32 or 33 for my answer. Which method is correct? How do you know?* Sample answer: 32 is correct. To subtract 24 from 56, you count back 2 tens from 56 like this: 46, 36. You land on 36, but you don't say the 36 twice. Count back 4 ones: 35, 34, 33, 32. You land on 32.

Distribute base-10 blocks so each partnership has at least 6 longs and 15 cubes. Ask partners to discuss how to model the Math Message problem using the blocks. **GMP2.2, GMP5.2** Invite children to demonstrate and explain what they did. An appropriate use of the blocks will include the following steps:

1. Model the 56 loaves of bread with 5 longs and 6 cubes.
2. Remove 2 longs and 4 cubes to represent the 24 loaves that were sold.
3. Count the remaining blocks and record the answer. 32

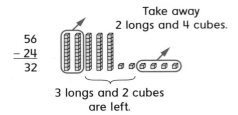

$$\begin{array}{r} 56 \\ -\ 24 \\ \hline 32 \end{array}$$

Take away 2 longs and 4 cubes.

3 longs and 2 cubes are left.

Refer children to **GMP5.2** on the Standards for Mathematical Practice Poster. Tell children that they are going to do more subtraction using base-10 blocks as a tool. **GMP5.2**

► Solving the Open Response Problem

Math Masters, p. 177; *My Reference Book.* pp. 16–17

| WHOLE CLASS | SMALL GROUP | **PARTNER** | **INDEPENDENT** |

Distribute *Math Masters,* page 177. Read the problem as a class. Tell children to solve the problem using base-10 blocks. Point out that they should expect to take more time to solve this problem than the Math Message problem. GMP1.3 Tell children that drawing blocks may help them explain their strategy. Remind them that even though they are focusing on one tool (base-10 blocks), there are different ways they can use the tool to solve the problem.

Circulate and observe. Expect to see children represent 53 with 5 longs and 3 cubes and to start showing the subtraction by removing 3 longs. Let them begin this way and try to make sense of what to do next. If you see children struggle at this point, ask: *What have you done so far?* Sample answer: I showed 53 and took away 3 longs, or 3 tens. *What do you still need to do?* Sample answer: I need to take away 8 cubes, or 8 ones. *How do you think you can do that?* Answers vary. Do not tell children to make a trade. Instead, have them develop their own ideas. Ask: *Are you saying you don't have the right blocks? What can you do about that?* Answers vary.

Watch for children who start by representing both the numbers in the problem (53 and 38) with blocks, as if they were adding. It is possible to solve the problem this way by comparing the numbers using the blocks and finding how much larger 53 is than 38. However, this is often harder for children to visualize and carry out than representing 53 and finding a way to take away 38. If children have trouble using this method, remind them of the context of the problem and ask them to use the blocks to represent the 53 dinner rolls. Then ask: *How can you represent selling 38 of the rolls?* GMP2.2, GMP5.2 Sample answer: I can take away 38 of the blocks.

> **Differentiate** **Adjusting the Activity**
>
> If children have difficulty, ask them to show you how to represent subtraction in the Math Message problem by taking away the correct number of base-10 blocks. When they can do this successfully, bring their attention back to the open response problem. This may help them verbalize that they need to find a way to take away 3 tens and 8 ones.

It can be helpful for partners to work together as they explore the base-10 blocks and share ideas about the task, but children should complete their own explanations and drawings of how they used blocks to represent subtraction.

NOTE Some children may represent the subtraction by covering up part of one of their longs using their hand or by drawing a long and crossing out individual cubes. In these cases, base-10 shorthand drawings may not be adequate because it is not possible to see the individual cubes. If you see children using base-10 shorthand in these or similar ways, ask them to clarify how many cubes they are covering or crossing out in their drawing.

My Reference Book, p. 16

Standards for Mathematical Practice

Choose Tools to Solve Problems

17 + 20 = ?

I started by putting out 17 counters. But using all those counters would take too long, and it's easy to make a mistake. So, I will try base-10 blocks instead.

Emma

Choose appropriate tools.
Emma decides that base-10 blocks are better than counters for this problem.

Emma shows 17 and 20 with base-10 blocks.

1 ten + 2 tens = 3 tens.

There are 3 tens and 7 ones in all. The sum is 37.

 MRB 16 | sixteen

Standards for Mathematical Practice

To check her answer, Emma uses a number grid to add 17 + 20.

11	12	13	14	15	16	[17]	18	19	20
21	22	23	24	25	26	27	28	29	30
31	32	33	34	35	36	(37)	38	39	40

Emma starts at 17 and moves down 2 rows to add 2 tens. She lands on 37, the same answer she got with base-10 blocks.

Use tools efficiently and make sense of your results. *Emma uses base-10 blocks to find the answer. She uses the number grid to check her answer.*

Mathematical Practice 5: Use appropriate tools strategically.

Try It Together

What tool would you use to solve the problem? Explain how you would use it.

seventeen | MRB 17

Summarize Have children read pages 16–17 in *My Reference Book*. Ask: *What tools did Emma use to solve the addition problem?* Sample answer: She used base-10 blocks and a number grid. *What tools did we use to solve the subtraction problems we did today?* Answers vary.

Collect children's work so you can evaluate it and prepare for Day 2. You may also want to make notes about children's strategies.

Getting Ready for Day 2

Math Masters, p. TA5

Planning a Follow-Up Discussion

Review children's work. Use the Reengagement Planning Form (*Math Masters*, page TA5) and the rubric on page 586 to plan ways to help children meet expectations for both the content and practice standards. Look for common misconceptions in children's use of base-10 blocks as well as a variety of successful subtraction strategies.

Organize the discussion in one of the ways described below or in another way you choose. If children's work is unclear or if you prefer to show work anonymously, rewrite the work for display.

(Go Online) for sample children's work that you can use in your discussion.

1. Display work with a viable strategy using base-10 blocks, but an error that resulted in an incorrect answer. See Child A's work. Ask:
   - *Do you agree or disagree with this child's answer?* Answers vary.
   - *Look carefully at the drawing and explanation. Is it clear enough so you understand what the child did?* Yes. *What base-10 blocks did the child start with?* 5 longs *What do they represent?* Sample answers: 50, 5 tens, or 50 rolls *How do you know?* GMP2.2, GMP5.2 Sample answer: There are 4 lines for the longs and 10 dots for the other ten. The child made a long out of cubes.
   - *Is there a problem with this step?* Sample answer: Yes. You need to start with 53 for the 53 rolls, not just 50.
   - *What did the child do next?* Sample answer: The child took away 3 longs and 8 cubes. *How do you know?* Sample answer: You can see 3 lines and 8 dots crossed out. *Why do you think this child did that?* GMP2.2, GMP5.2 Sample answer: The child wanted to subtract 38. *Was that the correct number to subtract?* Yes.
   - *What was this child's answer?* 12 *Do you think it is correct?* Sample answer: No. I think it should be 15.
   - *What would you say to help this child get the right answer? What did this child do right? How could this child use the blocks to correct the answer?* GMP1.3, GMP5.2 Sample answers: I would tell this child to count the number of blocks that the child started with to check that there are 53. The problem was to subtract 38 from 53 rolls. The child subtracted the right number, but started at 50.
   - *Let's work together to get the correct answer.*

Sample child's work, Child A

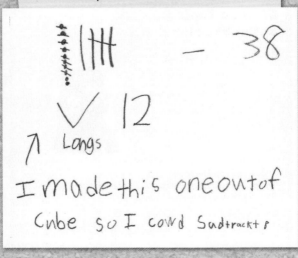

2. Display a response with the correct answer, but with an explanation or drawing that is unclear or doesn't match the answer. See Child B's work. Ask:

- *Do you agree or disagree with this child's answer?* Agree.
- *How do you think this child started? Do you see 53 shown with base-10 blocks?* **GMP2.2, GMP5.2** Sample answer: Yes. There are 5 longs, 2 here (to the left) and 3 here (to the right) and 3 cubes.
- *Did the child show or tell what was taken away?* **GMP5.2** Sample answer: Yes, the child wrote, "take 3 longs and take 8 cubes." That's 38.
- *Does the drawing clearly show that 3 longs and 8 cubes were taken away?* **GMP2.2** Sample answers: The drawing shows 3 longs taken away over here (to the right), but it is hard to see how many cubes were taken away. The drawing shows 5 crosses on the long, but it is hard to see exactly how many.
- *How can this child clearly show what was taken away?* Sample answer: It would be better to show each cube in the long, instead of a single stick, and you need to cross out 8 cubes, not 5.
- *If we show that clearly, what will be left?* 15

3. Display a response that clearly shows a correct solution, such as Child C's. Ask:

- *Look carefully at this work. Can you tell the steps this child took to solve the problem?* Yes. *How did this child start?* Sample answer: This child started with 5 longs and 3 cubes. That's 53.
- *What did the child do next? How can you tell?* Sample answer: This child took away 3 longs. You can see them crossed out. *What do the 3 longs stand for?* 3 tens, or 30
- *How many rolls were sold? How many do we need to take away?* 38 rolls *How many more does the child need to take away?* 8 *How did this child show that?* Sample answer: The child showed all 10 cubes on one long and crossed out 8 of the them.
- *Show us where the blocks are that are left.* Point to the 1 long and 5 cubes that are not crossed out. *How many is that?* 15

Planning for Revisions

Have base-10 blocks and copies of *Math Masters,* page 177 or extra paper available for children to use in revisions. You might want to ask children to use colored pencils so you can see what they revised.

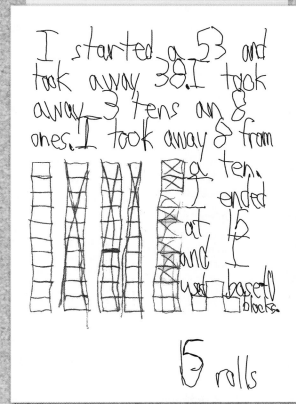

Sample child's work, Child B

Sample child's work, Child C

Subtracting With Base-10 Blocks

Overview   Day 2: Children compare strategies and revise their work.

Day 2: Reengagement

▶ **Before You Begin**
Have extra copies available of *Math Masters*, page 177 for children to revise their work.

Common Core State Standards

Focus Clusters
• Represent and solve problems involving addition and subtraction.
• Use place value understanding and properties of operations to add and subtract.

2b Focus 50–55 min	**Materials**	
Setting Expectations Children review the open response problem and discuss how to clearly represent their solutions with base-10 blocks.	Guidelines for Discussions Poster	
Reengaging in the Problem Children discuss different ways of representing subtraction using base-10 blocks.	selected samples of children's work	2.OA.1, 2.NBT.5, 2.NBT.7 SMP1, SMP2, SMP5
Revising Work Children improve their drawings and explanations.	*Math Masters,* p. 177 (optional); children's work from Day 1; colored pencils (optional); per partnership: base-10 blocks (at least 6 longs and 15 cubes)	2.OA.1, 2.NBT.5, 2.NBT.7 SMP1, SMP2, SMP5
✓ **Assessment Check-In** See page 588 and the rubric below.		2.NBT.7 SMP5

Goal for Mathematical Practice	Not Meeting Expectations	Partially Meeting Expectations	Meeting Expectations	Exceeding Expectations
GMP5.2 Use tools effectively and make sense of your results.	Does not provide an explanation in words or in a drawing of a strategy for representing the subtraction problem using base-10 blocks.	Provides an incomplete explanation of a strategy for representing the subtraction problem using base-10 blocks that may include words, a drawing, or both. **OR** Provides an explanation that includes contradictions between the words and drawing.	Provides a complete explanation of a strategy for representing the subtraction problem using base-10 blocks that includes either words or a drawing, or possibly a combination of the two. (Note: If the strategies are different between the two representations, but at least one is complete and does not contradict the other, the work meets expectations.)	Meets expectations and provides a complete explanation both in words and in a drawing, each showing a strategy for representing the subtraction problem using base-10 blocks. (Note: If the strategies are different between the two representations, and each one is complete and does not contradict the other, then the work exceeds expectations.)

3 Practice   10–15 min

Math Boxes 6-9 Children practice and maintain skills.	*Math Journal 2,* p. 161	See page 588.
Home Link 6-9 **Homework** Children solve number stories.	*Math Masters,* p. 178	2.OA.1, 2.NBT.5, 2.NBT.7

2b Focus   50–55 min   ⟨Go Online⟩

▶ Setting Expectations

| **WHOLE CLASS** | SMALL GROUP | PARTNER | INDEPENDENT |

Briefly review the open response problem from Day 1. Ask: *What did you need to think about as you were using base-10 blocks to solve this problem?* Sample answer: We could not just take away 38 blocks. We had to figure out how to take away 8 cubes when there were only 3 cubes when we first made 53.

Tell children they are going to look at other children's work and think about their drawings and explanations. Remind them that it is OK to make mistakes, and that they should help each other learn from their mistakes. Refer to your list of discussion guidelines and encourage children to use sentence frames such as the following:

- I like how you _____.
- Why did you _____?

▶ Reengaging in the Problem

| **WHOLE CLASS** | SMALL GROUP | **PARTNER** | INDEPENDENT |

Children reengage in the problem by analyzing and critiquing other children's work in pairs and in a whole-group discussion. Have children discuss with partners before sharing with the whole group. Guide this discussion based on the decisions you made in Getting Ready for Day 2.
GMP1.3, GMP2.2, GMP5.2

NOTE These Day 2 activities will ideally take place within a few days of Day 1. Prior to beginning Day 2, see Planning a Follow-Up Discussion from Day 1.

▶ Revising Work

| WHOLE CLASS | SMALL GROUP | **PARTNER** | **INDEPENDENT** |

Pass back children's work from Day 1. Before children revise anything, ask them to examine their drawings and explanations and decide how to improve them. Ask the questions below one at a time. Have partners discuss their responses and give a thumbs-up or thumbs-down based on their own work.

- *Were you able to solve the problem using base-10 blocks on Day 1? If not, you will spend more time doing this.* **GMP1.3**
- *Did you make a drawing showing how you used the blocks to represent subtraction?* **GMP2.2, GMP5.2**
- *Did you write an explanation of your strategy that is clear enough that someone else could use it? Does your partner understand it?*

Summarize Have children reflect on their work and revisions. Ask: *How did you show how to use base-10 pieces to subtract? Did you find a better way to show your thinking today?* Answers vary.

✓ Assessment Check-In   ⓒⓒⓢⓢ 2.NBT.7

Collect and review children's revised work. Expect children to improve their work based on the class discussion. For the content standard, expect most children to show that they can use strategies based on place value to accurately subtract. Expect that children will have attempted to improve or add to their drawings and explanations of their strategy for using base-10 blocks to represent the subtraction problem. You can use the rubric on page 586 to evaluate children's revised work for **GMP5.2**.

| ✓ | Assessment and Reporting | Go Online | to record student progress and to see trajectories toward mastery for these standards. |

Go Online for optional generic rubrics in the *Assessment Handbook* that can be used to assess any additional GMPs addressed in the lesson.

Math Journal 2, p. 161

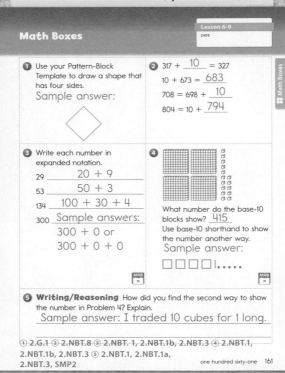

Math Boxes   Lesson 6-9

1. Use your Pattern-Block Template to draw a shape that has four sides.
Sample answer:

2. 317 + __10__ = 327
10 + 673 = __683__
708 = 698 + __10__
804 = 10 + __794__

3. Write each number in expanded notation.
29 __20 + 9__
53 __50 + 3__
134 __100 + 30 + 4__
300 Sample answers: 300 + 0 or 300 + 0 + 0

4. What number do the base-10 blocks show? __415__
Use base-10 shorthand to show the number another way.
Sample answer:

5. **Writing/Reasoning** How did you find the second way to show the number in Problem 4? Explain.
Sample answer: I traded 10 cubes for 1 long.

① 2.G.1 ② 2.NBT.8 ③ 2.NBT.1, 2.NBT.1b, 2.NBT.3 ④ 2.NBT.1, 2.NBT.1b, 2.NBT.3 ⑤ 2.NBT.1, 2.NBT.1a, 2.NBT.3, SMP2

one hundred sixty-one   161

Sample Children's Work—Evaluated

See the sample in the margin. This work meets expectations for the content standard because it shows a correct solution to the subtraction problem that uses place-value concepts. The work meets expectations for the mathematical practice because the child provides a complete explanation of a strategy using base-10 blocks in words and a drawing. This child showed 53 with 5 longs and 3 cubes, with 1 long made up of 10 cubes. Three longs and 8 cubes are crossed out. If the written explanation had included additional words about trading 1 long for 10 cubes, the work would exceed expectations. **GMP5.2**

Go Online for other samples of evaluated children's work.

Sample child's work, "Meeting Expectations"

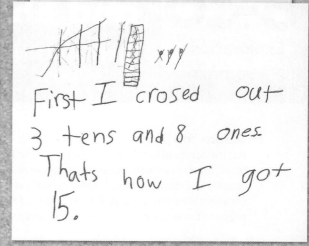

First I crosed out 3 tens and 8 ones Thats how I got 15.

3 Practice 10–15 min

Go Online ePresentations eToolkit Home Connections

▶ Math Boxes 6-9

Math Journal 2, p. 161

WHOLE CLASS | SMALL GROUP | PARTNER | INDEPENDENT

Mixed Practice Math Boxes 6-9 are paired with Math Boxes 6-6.

▶ Home Link 6-9

Math Masters, p. 178

Homework Children solve number stories.

Math Masters, p. 178

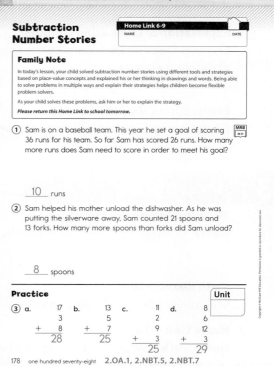

Subtraction Number Stories

Home Link 6-9

NAME | DATE

Family Note

In today's lesson, your child solved subtraction number stories using different tools and strategies based on place-value concepts and explained his or her thinking in drawings and words. Being able to solve problems in multiple ways and explain their strategies helps children become flexible problem solvers.

As your child solves these problems, ask him or her to explain the strategy.

Please return this Home Link to school tomorrow.

① Sam is on a baseball team. This year he set a goal of scoring 36 runs for his team. So far Sam has scored 26 runs. How many more runs does Sam need to score in order to meet his goal?

__10__ runs

② Sam helped his mother unload the dishwasher. As he was putting the silverware away, Sam counted 21 spoons and 13 forks. How many more spoons than forks did Sam unload?

__8__ spoons

Practice | Unit |

③ a. 17   b. 13   c. 11   d.   8
         3   5   2   6
     +   8   +   7   +   9   + 12
     —————   —————   ————— —————
        28   25   25   29

178   one hundred seventy-eight   2.OA.1, 2.NBT.5, 2.NBT.7

Lesson 6-10 Explorations

Exploring Arrays, Length, and Shapes

Overview Children build arrays on geoboards, measure and compare lengths, and create shapes.

▶ **Before You Begin**

For the optional Readiness activity, select at least eight classroom objects of various lengths and label three note cards *shorter than, longer than,* and *about the same as.* For the optional Extra Practice activity, cut string in various lengths, such as 12, 14, 18, and 21 inches. Tape each length to an index card and label the cards A, B, C, and so on. See the Exploration activities in Part 2 of the lesson for the preparation required. For Part 1 select and sequence Quick Look Cards 98, 103, and 118.

▶ **Vocabulary**

geoboard · rectangular array

Common Core State Standards

Focus Clusters
- Work with equal groups of objects to gain foundations for multiplication.
- Understand place value.
- Measure and estimate lengths in standard units.
- Reason with shapes and their attributes.

1 Warm Up 15–20 min

	Materials	
Mental Math and Fluency Children view Quick Look Cards.	Quick Look Cards 98, 103, and 118; slate	2.OA.2
Daily Routines Children complete daily routines.	See pages 4–43.	See pages xiv–xvii.

2 Focus 30–40 min

Math Message Children determine the number of objects in an array.		2.OA.4
Discussing Arrays Children make an array and write a related number model.	For demonstration: geoboard and rubber band	2.OA.4, 2.NBT.2 SMP2
Exploration A: Making Geoboard Arrays Children make arrays on geoboards, record these arrays on dot paper, and write number models for the arrays.	Activity Card 83; *Math Masters,* p. 180; *Math Masters,* p. TA9 or TA10; geoboard; rubber band; scissors	2.OA.4, 2.NBT.2 SMP2
Exploration B: Comparing Lengths Children measure and compare the lengths of different objects.	Activity Card 84; *Math Journal 2,* p. 162; *My Reference Book;* tape measure or yardstick; stick-on note; calculator; eraser	2.MD.1, 2.MD.4 SMP5
Exploration C: Making Shapes Children create shapes with triangles and rectangles.	Activity Card 85; *My Reference Book,* pp. 123–124 (optional); *Math Masters,* p. 181; scissors; glue or tape; paper	2.G.1

3 Practice 15–20 min

Playing *Beat the Calculator* **Game** Children practice addition facts.	*Assessment Handbook,* pp. 98–99; per group: 4 each of number cards 0–9, calculator, large paper triangle (optional)	2.OA.2
Math Boxes 6-10 Children practice and maintain skills.	*Math Journal 2,* p. 163	See page 595.
Home Link 6-10 **Homework** Children draw arrays and write number models.	*Math Masters,* p. 182	2.OA.4, 2.NBT.2

Differentiation Options

RtI

Readiness · 5–15 min

Comparing Objects by Length

| WHOLE CLASS |
| SMALL GROUP |
| PARTNER |
| INDEPENDENT |

8 items of various lengths; index cards labeled *shorter than, longer than,* and *about the same as* (see Before You Begin)

To explore linear measurement using a concrete model, children compare the lengths of objects. Display the objects and the labeled index cards faceup on a table. Ask a volunteer to pick one item. Then have another child select a different item, compare its length to the first item, and put it next to the card with the appropriate label. To compare, children can place the objects next to the other, aligning them at one end. Encourage children to verbalize their comparisons, for example, "The pencil is longer than the eraser."

Enrichment · 5–15 min

Comparing Lengths of Body Parts

| WHOLE CLASS |
| SMALL GROUP |
| PARTNER |
| INDEPENDENT |

Math Masters, p. 179; tape measure

To apply their understanding of comparing lengths, children estimate and then measure the lengths of different body parts. They record the data and compare the measurements on *Math Masters,* page 179.

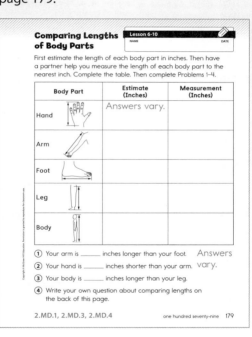

Extra Practice · 5–15 min

Comparing Lengths of String

| WHOLE CLASS |
| SMALL GROUP |
| PARTNER |
| INDEPENDENT |

various lengths of string taped to labeled index cards (see Before You Begin); per child: tape measure

For practice comparing lengths, children find the difference between the lengths of two pieces of string. As children choose two pieces of string and estimate their lengths, ask: *How long do you think each piece of string is?* Answers vary.

Then have children measure the lengths of the two strings. Ask: *Which string is longer? How much longer?* Sample answer: String A is longer than String B. String A is 20 inches long, and String B is 15 inches long, so String A is 5 inches longer. *How much longer does String B need to be so that it is the same length as string A? Explain how you know.* Sample answer: It would need to be 5 inches longer because 5 plus 15 is 20 inches.

English Language Learners Support

Beginning ELL To scaffold the Quick Looks activity for English language learners, provide them with simple posted sentence frames, such as the following:

• I moved _____.   • I filled in _____.   • I put together _____.

Have children use the sentence frames to respond to questions such as these: *What did you do? How did you see the dots?* For children not yet ready to communicate in English, provide a double ten frame (*Math Masters,* page TA6) and encourage them to show their responses using counters or drawings.

Go Online ELL English Learners Support

Standards and Goals for
Mathematical Practice

SMP2 Reason abstractly and quantitatively.

GMP2.1 Create mathematical representations using numbers, words, pictures, symbols, gestures, tables, graphs, and concrete objects.

SMP5 Use appropriate tools strategically.

GMP5.2 Use tools effectively and make sense of your results.

Math Masters, p. 180

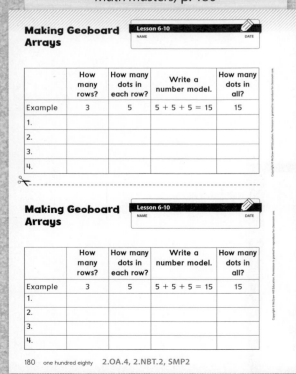

1 Warm Up   15–20 min   Go Online   ePresentations   eToolkit

▶ Mental Math and Fluency

Flash the following sequence of Quick Look Cards: 98, 103, and 118. Always allow children a second look and follow up by asking them to use their slates to record in words or number sentences how they found the total number of dots. *Leveled exercises:*

- ●○○ Quick Look Card 98 Sample answers: $9 + 2 = 11$; $10 + 1 = 11$
- ●●○ Quick Look Card 103 Sample answers: $5 + 7 = 12$; $5 + 5 + 2 = 12$
- ●●● Quick Look Card 118 Sample answers: $7 + 8 = 15$; $5 + 5 + 5 = 15$

▶ Daily Routines

Have children complete daily routines.

2 Focus   30–40 min   Go Online   ePresentations   eToolkit

▶ Math Message

How many muffins can fit in this pan?

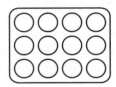

▶ Discussing Arrays

| WHOLE CLASS | SMALL GROUP | PARTNER | INDEPENDENT |

Math Message Follow-Up Ask:

- *How many rows are there in the pan?* 3
- *How many muffins can fit in each row?* 4
- *How many muffins can fit in the pan in all?* 12

Record the number model $4 + 4 + 4 = 12$ to represent the 3-by-4 array. **GMP2.1** Then show children how to use a rubber band to enclose a 3-by-4 array of pegs in a rectangle on a **geoboard.** Explain that in today's lesson children will use geoboards to make **rectangular arrays** and then write addition number models representing the arrays. **GMP2.1**

After explaining the Explorations activities, assign groups to each one. Plan to spend most of your time with children working on Exploration A.

▶ Exploration A: Making Geoboard Arrays

Activity Card 83; *Math Masters*, p. 180; p. TA9 or TA10

| WHOLE CLASS | SMALL GROUP | PARTNER | INDEPENDENT |

To prepare for this Exploration, check the size of your class's geoboards. If the geoboards are 5-by-5, use *Math Masters*, page TA9; if they are 7-by-7, use *Math Masters*, page TA10. Whichever page you use, you will need several copies for each child. You will need one copy of *Math Masters*, page 180 for every two children.

Children create rectangular arrays on geoboards and record them on dot paper. For each array, they count the number of enclosed pegs, including the ones touching the rubber bands. Children then write an addition number model to represent the total number of pegs in each array. GMP2.1

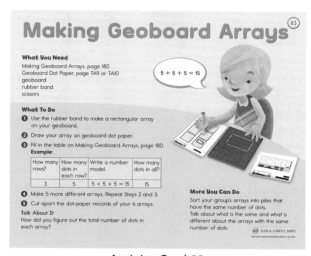

Activity Card 83

▶ Exploration B: Comparing Lengths

Activity Card 84; *Math Journal 2*, p. 162

| WHOLE CLASS | SMALL GROUP | PARTNER | INDEPENDENT |

Tape a yardstick or a tape measure (with the inch side up) to a table. Children measure four different objects and compare their lengths. GMP5.2 They find the difference in length between pairs of objects.

Differentiate **Adjusting the Activity**

For children who struggle to find the difference in length between two objects, suggest that they align the two objects with 0 on the yardstick or the tape measure, with one object on either side of the measuring tool. Children then measure the section of the longer object that extends beyond the end of the shorter object.

Go Online ▸ Differentiation Support

Math Journal, p. 162

Comparing Lengths   Lesson 6-10   DATE

Get an eraser, a calculator, a stick-on note, and *My Reference Book*. Measure each object's length to the nearest inch. Record the measures.

Eraser: about _____ inches   Answers vary.

Calculator: about _____ inches   Answers vary.

My Reference Book: about _____ inches   Sample answers: 9; 10

Stick-on note: about _____ inches   Answers vary.

Circle the name of the longer object.	Find the difference in length between the two objects.
Eraser Calculator Answers vary.	About _____ inches
Eraser Stick-on note	About _____ inches
Eraser *My Reference Book*	About _____ inches
My Reference Book Calculator	About _____ inches
My Reference Book Stick-on note	About _____ inches
Stick-on note Calculator	About _____ inches

Which object is the longest?   Answers vary.

Which object is the shortest?   Answers vary.

Explain how you found the differences in length between pairs of objects.

Answers vary.

162   one hundred sixty-two   2.MD.1, 2.MD.4, SMP5

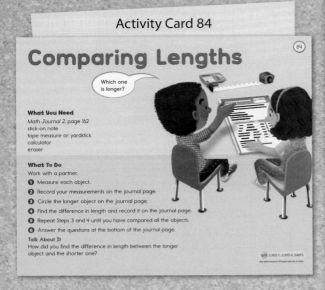

Comparing Lengths

Which one is longer?

What You Need
Math Journal 2, page 162
stick-on note
tape measure or yardstick
calculator
eraser

What To Do
Work with a partner.
1. Measure each object.
2. Record your measurements on the journal page.
3. Circle the longer object on the journal page.
4. Find the difference in length and record it on the journal page.
5. Repeat Steps 3 and 4 until you have compared all the objects.
6. Answer the questions at the bottom of the journal page.

Talk About It
How did you find the difference in length between the longer object and the shorter one?

▶ Exploration C: Making Shapes

Activity Card 85; *Math Masters*, p. 181

| WHOLE CLASS | SMALL GROUP | PARTNER | INDEPENDENT |

Children cut out the triangles and the rectangles on *Math Masters*, page 181. They put them together to form various shapes, which they paste or tape onto sheets of paper. Have additional copies of the page ready for children who need more triangles and rectangles.

Below are some possible constructions. Children may refer to *My Reference Book*, pages 123–124 for pictures of the shapes. Encourage children who enjoy this activity to attempt more complex constructions.

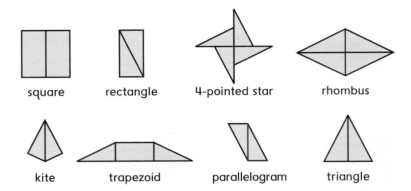

square   rectangle   4-pointed star   rhombus

kite   trapezoid   parallelogram   triangle

Summarize Children share a strategy they used to find the difference in length between the pairs of objects in Exploration B. Sample answer: I lined up the objects and measured the part that did not overlap.

③ Practice   15–20 min   Go Online   ePresentations   eToolkit   Home Connections

▶ Playing *Beat the Calculator*

Assessment Handbook, pp. 98–99

| WHOLE CLASS | SMALL GROUP | PARTNER | INDEPENDENT |

Have small groups play the game as introduced in Lesson 5-1. As you circulate and observe, monitor children's progress with addition facts using *Assessment Handbook*, pages 98–99. By the end of Grade 2, children are expected to know from memory all sums of two 1-digit numbers.

Observe
• Which facts do children know from memory?
• Which children need additional support to play the game?

Discuss
• *What strategies did you use to solve the facts you did not know?*
• *Why is it helpful to know addition facts?*

Math Masters, p. 181

Making Shapes

| Lesson 6-10 |
| NAME   DATE |

one hundred eighty-one   181

► # Math Boxes 6-10

Math Journal 2, p. 163

| WHOLE CLASS | SMALL GROUP | PARTNER | INDEPENDENT |

Mixed Practice Math Boxes 6-10 are paired with Math Boxes 6-5 and 6-8.

► # Home Link 6-10

Math Masters, p. 182

Homework Children draw arrays and write addition number models to show the total number of Xs.

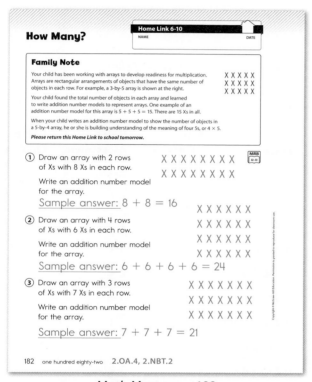

Math Masters, p. 182

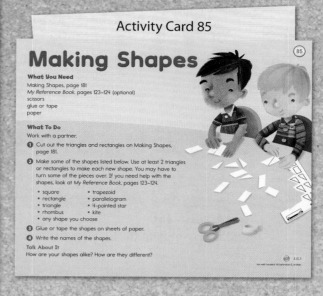

Activity Card 85

Making Shapes

What You Need
Making Shapes, page 181
My Reference Book, pages 123–124 (optional)
scissors
glue or tape
paper

What To Do
Work with a partner.
1. Cut out the triangles and rectangles on Making Shapes, page 181.
2. Make some of the shapes listed below. Use at least 2 triangles or rectangles to make each new shape. You may have to turn some of the pieces over. If you need help with the shapes, look at My Reference Book, pages 123–124.
   - square
   - rectangle
   - triangle
   - rhombus
   - any shape you choose
   - trapezoid
   - parallelogram
   - 4-pointed star
   - kite
3. Glue or tape the shapes on sheets of paper.
4. Write the names of the shapes.

Talk About It
How are your shapes alike? How are they different?

Math Journal 2, p. 163

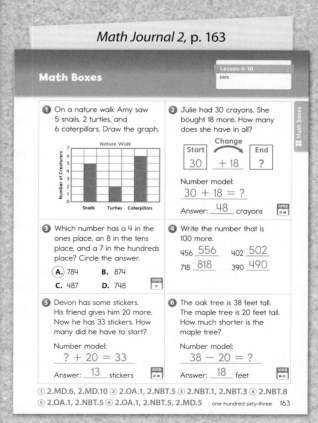

Unit 6 Progress Check

Overview Day 1: Administer the Unit Assessments.
Day 2: Administer the Cumulative Assessment.

2-Day Lesson

 Student Learning Center
Students may take assessments digitally.

 Assessment and Reporting
Record results and track progress toward mastery.

Day 1: Unit Assessments

1 Warm Up 5–10 min

Materials

Self Assessment
Children complete the Self Assessment.

Assessment Handbook, p. 37

2a Assess 35–50 min

Unit 6 Assessment
These items reflect mastery expectations to this point.

Assessment Handbook, pp. 38–41

Unit 6 Challenge (Optional)
Children may demonstrate progress beyond expectations.

Assessment Handbook, p. 42

CCSS Common Core State Standards	Goals for Mathematical Content (GMC)	Lessons	Self Assessment	Unit 6 Assessment	Unit 6 Challenge
2.OA.1	Use addition and subtraction to solve 1-step number stories.	6-2 to 6-5, 6-9	2	2–5	
	Use addition and subtraction to solve 2-step number stories.	6-5	3	6	1
	Model 1-step problems involving addition and subtraction.	6-2 to 6-5, 6-9		2–5	
2.NBT.5	Add within 100 fluently.	6-3 to 6-8	5, 6	3–8	
	Subtract within 100 fluently.	6-2 to 6-5, 6-9		2–6	
2.NBT.7	Add multidigit numbers using models or strategies.	6-3 to 6-8	5, 6	3–8	
	Subtract multidigit numbers using models or strategies.	6-2 to 6-5, 6-9		2–6	
2.NBT.9	Explain why addition and subtraction strategies work.	6-6 to 6-8			2
2.MD.5	Solve number stories involving length by adding or subtracting.	6-2 to 6-4	4	2, 4–6	
	Model number stories involving length.	6-2 to 6-4			
2.MD.10	Answer questions using information in graphs.	6-1	1	1	
	Goals for Mathematical Practice (GMP)				
SMP1	Make sense of your problem. **GMP1.1**	6-3 to 6-9		2–6	
SMP2	Create mathematical representations using numbers, words, pictures, symbols, gestures, tables, graphs, and concrete objects. **GMP2.1**	6-1, 6-5, 6-8 to 6-10		2–6	
SMP3	Make sense of others' mathematical thinking. **GMP3.2**	6-1, 6-6, 6-7			1, 2
SMP4	Model real-world situations using graphs, drawings, tables, symbols, numbers, diagrams, and other representations. **GMP4.1**	6-1 to 6-3, 6-5		2–6	
	Use mathematical models to solve problems and answer questions. **GMP4.2**	6-1, 6-3, 6-5		2–6	

Spiral Tracker **Go Online** to see how mastery develops for all standards within the grade.

1 Warm Up   5–10 min

▶ Self Assessment

Assessment Handbook, p. 37

| WHOLE CLASS | SMALL GROUP | PARTNER | **INDEPENDENT** |

Children complete the Self Assessment to reflect on their progress in Unit 6.

2a Assess   35–50 min

Go Online ✓ Assessment and Reporting   Differentiation Support

▶ Unit 6 Assessment

Assessment Handbook, pp. 38-41

| WHOLE CLASS | SMALL GROUP | PARTNER | **INDEPENDENT** |

Children complete the Unit 6 Assessment to demonstrate their progress on the Common Core State Standards covered in this unit.

Go Online for generic rubrics in the *Assessment Handbook* that can be used to evaluate children's progress on the Mathematical Practices.

Assessment Handbook, p. 39

NAME   DATE   Lesson 6-11 ✓

Unit 6 Assessment (continued)

For Problems 2–5:

- Write a number model with a ? to show what you need to find. You may draw a [Total / Part Part], [Start ⌢ End], or [Quantity / Quantity / Difference] to help.
- Solve the problem.
- Write the answer.

② Fish E is 40 inches long. Fish F is 30 inches long.

How much longer is Fish E than Fish F?

Number model: Sample answers: 40 − 30 = ?
or 30 + ? = 40

Fish E is ___10___ inches longer than Fish F.

③ Connor needed 20 juice boxes for her class party. She bought 22. How many extra juice boxes did she have?

Number model: Sample answers: 20 + ? = 22
or 22 − 20 = ?

Answer: ___2___ juice boxes

Assessment Masters   **39**

NAME   DATE   Lesson 6-11 ✓

Unit 6 Assessment

① Use the picture graph below to answer the questions.

How Many Pencils?

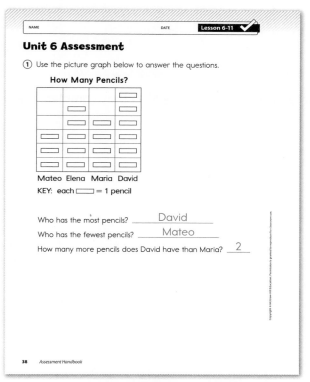

Mateo Elena Maria David
KEY: each ☐ = 1 pencil

Who has the most pencils? ___David___
Who has the fewest pencils? ___Mateo___
How many more pencils does David have than Maria? ___2___

38   Assessment Handbook

Assessment Handbook, p. 38

Assessment Handbook, p. 40

NAME   DATE   Lesson 6-11 ✓

Unit 6 Assessment (continued)

④ The giraffe is 13 feet tall when standing. Its legs are 6 feet long.

How tall is the giraffe when it is lying down?

Number model: Sample answers: 13 − 6 = ?, or
6 + ? = 13

Answer: ___7___ feet

⑤ The green ribbon is 15 feet long. The white ribbon is 10 feet long. Which is longer, the green ribbon or the white ribbon?

Green ribbon

How much longer?

Number model: Sample answers: 15 − 10 = ? or
10 + ? = 15

Answer: ___5___ feet

40   Assessment Handbook

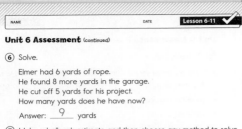

NAME   DATE   Lesson 6-11 ✓

Unit 6 Assessment (continued)

⑥ Solve.

Elmer had 6 yards of rope.
He found 8 more yards in the garage.
He cut off 5 yards for his project.
How many yards does he have now?

Answer: ___9___ yards

⑦ Make a ballpark estimate and then choose any method to solve.
Use your estimate to check if your answer makes sense.

Show your work.

39 + 46

Ballpark estimate: Sample answer: 40 + 50 = 90

Strategies vary.

39 + 46 = ___85___

⑧ Solve using partial-sums addition. Show your work.
You may use base-10 blocks to help. If you use blocks,
record your work in base-10 shorthand.

```
   26            194
 + 42          + 235
 ----          -----
   68            429
```

Assessment Masters   **41**

NAME   DATE   Lesson 6-11 ✓

Unit 6 Challenge

① Addie gathered 5 eggs in the hen house. 3 eggs fell out of her
basket and broke. She then found 12 more eggs. How many
eggs does Addie have now?

Jim wrote these number models to help him solve this problem.

5 + 3 = ?
8 + 12 = ?

Is he correct? Explain.

No, Addie lost 3 eggs so the first number
model should be 5 − 3 = ? and the second
number model should be 2 + 12 = ?.

② Melissa and Ebe both solved the same problem using partial
sums. Their work is shown below.

```
  Melissa          Ebe
    156            156
  + 338          + 338
  -----          -----
    400             14
     80             80
     14            400
  -----          -----
    494            494
```

Explain why both strategies work.

Sample answer: They both found the
partial sums, but in a different order.
They still added the same sums to get 494.

42   Assessment Handbook

Differentiate   **Adjusting the Assessment**

Item	Adjustments
1	To extend item 1, have children make up questions to ask about the data.
2–6	To scaffold items 2–5, have children draw a picture prior to writing a number model with a ? for the number they need to find.
7	To scaffold item 7, ask children guiding questions to help them determine what they know and what they need to find out.
8	To scaffold item 8, provide children with base-10 blocks.
9	To extend item 9, have children explain why the partial-sums method works.

Advice for Differentiation

All of the content included on the Unit 6 Assessment was recently
introduced and will be revisited in subsequent units.

Use the online assessment and reporting tools to track children's
performance. Differentiation materials are available online to help you
address children's needs.

> **NOTE** See the Unit Organizer on pages 526–527 or the online Spiral
> Tracker for details on Unit 6 focus topics and the spiral.

▶ Unit 6 Challenge (Optional)

Assessment Handbook, p. 42

| WHOLE CLASS | SMALL GROUP | PARTNER | **INDEPENDENT** |

Children can complete the Unit 6 Challenge after they complete the
Unit 6 Assessment.

Unit 6 Progress Check

Day 2: Cumulative Assessment

2b Assess   35–55 min

Materials

⭐ **Cumulative Assessment**
These items reflect mastery expectations to this point.

Assessment Handbook, pp. 43–45

CCSS

Common Core State Standards	Goals for Mathematical Content (GMC)	Cumulative Assessment
2.NBT.1	Understand 3-digit place value.*	2a–2c
2.NBT.1a	Understand exchanging tens and hundreds.*	5
2.NBT.1b	Understand 100, 200…900 as some hundreds and no tens and no ones.*	2a–2c, 10a–10b
2.NBT.3	Read and write numbers in expanded form.*	1a–1d, 10a–10b
	Read and write numbers.*	2a–2c
	Read and write number names.*	8, 9
2.NBT.4	Compare and order numbers.*	4a–4d
	Record comparisons using >, =, or <.*	4a–4d
2.NBT.8	Mentally add 10 to and subtract 10 from a given number.*	7a, 7b, 7e, 7g, 7j
	Mentally add 100 to and subtract 100 from a given number.*	7c, 7d, 7f, 7h, 7i
2.MD.1	Measure the length of an object.*	6a, 6b
2.MD.2	Measure an object using two different units of length.*	6a, 6b
	Describe how length measurements relate to the size of the unit.*	6c
2.MD.8	Read and write monetary amounts.	3a, 3b, 11a, 11b
	Solve problems involving coins and bills.	3a, 3b, 11a, 11b

	Goals for Mathematical Practice (GMP)	
SMP1	Solve problems in more than one way. **GMP1.5**	11a, 11b
SMP2	Make connections between representations. **GMP2.3**	11a, 11b
SMP3	Make mathematical conjectures and arguments. **GMP3.1**	10a, 10b
	Make sense of others' mathematical thinking. **GMP3.2**	10a, 10b
SMP4	Model real-world situations using graphs, drawings, tables, symbols, numbers, diagrams, and other representations. **GMP4.1**	11a, 11b
	Use mathematical models to solve problems and answer questions. **GMP4.2**	11a, 11b
SMP6	Explain you mathematical thinking clearly and precisely. **GMP6.1**	10a, 10b, 11b

*Instruction and most practice on this content is complete.

 Spiral Tracker   **Go Online** ▷ to see how mastery develops for all standards within the grade.

3 Look Ahead   10–15 min

Materials

Math Boxes 6-11
Children preview skills and concepts for Unit 7.

Math Journal 2, p. 164

Home Link 6-11
Children take home the Family Letter introducing Unit 7.

Math Masters, pp. 183–186

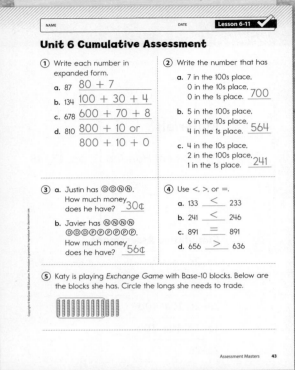

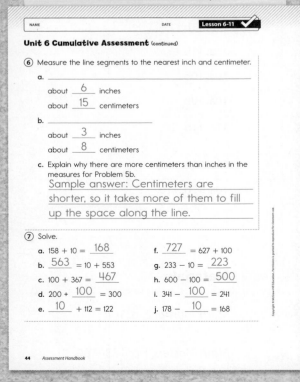

 **Assess**   35–55 min

Go Online

Assessment and Reporting   Differentiation Support

▶ Cumulative Assessment

Assessment Handbook, pp. 43–45

| WHOLE CLASS | SMALL GROUP | PARTNER | **INDEPENDENT** |

Children complete the Cumulative Assessment. The problems in the Cumulative Assessment address content from Units 1–5.

Monitor children's progress on the Common Core State Standards using the online assessment and reporting tools.

Go Online for generic rubrics in the *Assessment Handbook* that can be used to evaluate children's progress on the Mathematical Practices.

Differentiate   Adjusting the Assessment

Item(s)	Adjustments
1, 2, 4, 5	To scaffold items 1, 2, 4, and 5, provide children with base-10 blocks.
3	To scaffold item 3, have children use real coins to show the amounts.
6	To scaffold item 6, have children use the square pattern blocks and centimeter cubes alongside the ruler to measure the line segments.
7	To extend item 7, provide children with problems involving transition numbers, such as 298 + 10 or 990 to 1000.
8, 9	To scaffold items 8 and 9, display a poster with the number words for 1–9 and the decade numbers.

Advice for Differentiation

All instruction and most practice is complete for the content that is marked with an asterisk (*) on page 599.

Use the online assessment and reporting tools to track children's performance. Differentiation materials are available online to help you address children's needs.

3 Look Ahead 10–15 min Go Online Home Connections

▶ Math Boxes 6-11: Preview for Unit 7

Math Journal 2, p. 164

WHOLE CLASS | SMALL GROUP | PARTNER | INDEPENDENT

Mixed Practice Math Boxes 6-11 are paired with Math Boxes 6-7. These problems focus on skills and understandings that are prerequisite for Unit 7. You may want to use information from these Math Boxes to plan instruction and grouping in Unit 7.

▶ Home Link 6-11: Unit 7 Family Letter

Math Masters, pp. 183–186

Home Connection The Unit 7 Family Letter provides information and activities related to Unit 7 content.

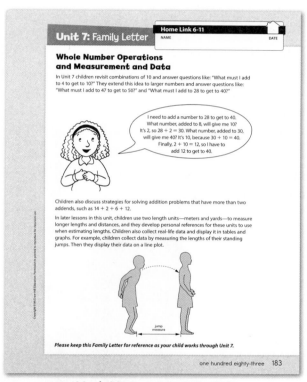

Math Masters, pp. 183–186

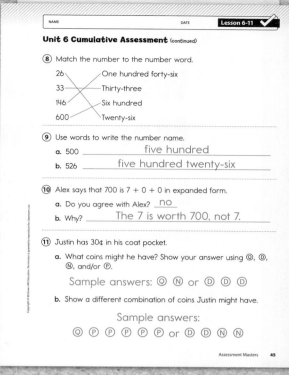

Assessment Handbook, p. 45

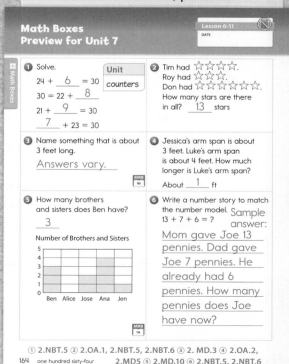

Math Journal 2, p. 164

Unit 7 Organizer
Whole Number Operations and Measurement and Data

In this unit, children further explore addition and subtraction strategies and use them to add three or more numbers. They use units of yards and meters to measure distances. At the end of the unit, they collect data and display it in a frequency table and a line plot. Children's learning will focus on three clusters of the Common Core's content standards, as well as in-depth work on two of the Mathematical Practices.

CCSS Standards for Mathematical Content

Domain	Cluster
Number and Operations in Base Ten	Use place value understanding and properties of operations to add and subtract.
Measurement and Data	Measure and estimate lengths in standard units. Represent and interpret data.

Because the standards within each domain can be broad, *Everyday Mathematics* has unpacked each standard into Goals for Mathematical Content **GMC**. For a complete list of Standards and Goals, see page EM1.

For an overview of the CCSS domains, standards, and mastery expectations in this unit, see the **Spiral Trace** on pages 608–609. See the **Mathematical Background** (pages 610–612) for a discussion of the following key topics:

- Addition and Subtraction Strategies
- Length Measurement Units and Tools
- Data Displays

CCSS Standards for Mathematical Practice

SMP3	Construct viable arguments and critique the reasoning of others.
SMP4	Model with mathematics.

For a discussion about how *Everyday Mathematics* develops these practices and a list of Goals for Mathematical Practice **GMP**, see page 613.

 **VLC** Virtual Learning Community **Go Online** to **vlc.cemseprojects.org** to search for video clips on each practice.

Go Digital with these tools at connectED.mheducation.com

ePresentations · Student Learning Center · Facts Workshop Game · eToolkit · Professional Development · Home Connections · Spiral Tracker · Assessment and Reporting · English Learners Support · Differentiation Support

Contents

*The standards listed here are addressed in the **Focus** of each lesson. For all the standards in a lesson, see the Lesson Opener.

Unit 7 Materials

VLC Virtual Learning Community

See how *Everyday Mathematics* teachers organize materials. Search "Classroom Tours" at **vlc.cemseprojects.org**.

Lesson	Math Masters	Activity Cards	Manipulative Kit	Other Materials
7-1	pp. 187–188; TA3; TA11; G25	86	calculator; yardstick for demonstration (optional); counters; 4 each of number cards 1–10	slate; Number-Grid Poster; number grid
7-2	pp. 189–191; TA5		base-10 blocks	number grid; children's work from Day 1; Guidelines for Discussions Poster; selected samples of children's work; Standards for Mathematical Practice Poster; colored pencils (optional)
7-3	pp. 192–193; TA8; TA30; G24; G26–G27 (optional)	87–89	one 20-sided polyhedral die or three 6-sided dice; base-10 blocks; 1 each of number cards 1–20; 4 counters	slate; paper clip; egg carton
7-4	pp. 194; G26–G27	90	square pattern block; yardstick; per partnership: tape measure; per group: one 20-sided polyhedral die or three 6-sided dice	slate; Class Data Pad; 12-inch ruler; 10-centimeter ruler; path marked with masking tape
7-5	p. 195	91	meterstick; yardstick; tape measure; centimeter cube (optional)	slate; Class Data Pad; 10-centimeter ruler; Fact Triangles; Pattern-Block Template; path marked with masking tape
7-6	pp. 196–198; TA7–TA8		tape measure; per partnership: meterstick; 1-inch-square pattern blocks	slate; chalk, penny, or other marker; per partnership: 10-centimeter ruler; centimeter cubes
7-7	pp. 199–201	92–93	tape measure; base-10 blocks (optional); one 6-sided die	slate; stick-on notes; stick-on notes from Math Message; number grid or number line (optional); number line drawn by teacher; 12-inch ruler; objects from classroom varying in length from 1 to 12 inches
7-8	pp. 155 (1 copy per partnership); 202–206	94–95	tape measure; per partnership: one 6-sided die	slate; stick-on notes; stick-on notes from Math Message
7-9	pp. 207–209; G23; G24 (optional)	96–98	tape measure; yardstick; meterstick; per partnership: die (prepared in Lesson 5-6), calculator	slate; scissors; envelope; paper clip; Our Favorite Fruits data; *Math Journal 2*, Activity Sheets 7–12
7-10	pp. 210–213; *Assessment Handbook*, pp. 46–51		base-10 blocks	

Problem Solving Professional Development

Everyday Mathematics emphasizes equally all three of the Common Core's dimensions of **rigor:** conceptual understanding, procedural skill and fluency, and applications. Math Messages, other daily work, Explorations, and Open Response tasks provide many opportunities for children to apply what they know to solve problems.

▶ Math Message

Math Messages require children to solve a problem they have not previously been shown how to solve. Math Messages provide almost daily opportunities for problem solving.

▶ Daily Work

Journal pages, Home Links, Writing/Reasoning prompts, and Differentiation Options often require children to solve problems in mathematical contexts and real-life situations. **Minute Math+** offers Number Stories for transition times and spare moments throughout the day. See Routine 6, pages 38–43.

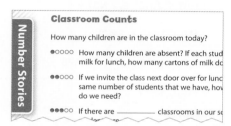

▶ Explorations

In Exploration A children find ways to sort shapes into different groups. In Exploration B children use data gathered about the class's favorite fruit to draw a picture graph.

▶ Open Response and Reengagement

In Lesson 7-2 children solve a number story involving adding four 2-digit numbers and explain their strategies. The reengagement discussion on Day 2 could focus on how children used place value or properties of addition to find the answer, or it could focus on comparing different strategies. Using structures to solve problems and answer questions is the focus practice for the lesson. Looking for and making use of structures such as the place-value structure of numbers can help children see similarities among problems and apply familiar strategies to new problems. GMP7.2

 **Virtual Learning Community** (Go Online) to watch an Open Response and Reengagement lesson in action. Search "Open Response" at **vlc.cemseprojects.org**.

▶ Open Response Assessment

Progress Check Open Response tasks offer opportunities to assess children's problem-solving abilities. In Progress Check Lesson 7-10, children decide if two different sets of base-10 blocks represent the same number and explain their thinking. GMP2.2

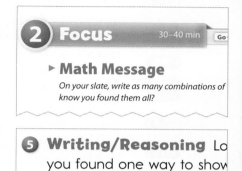

Assessment and Differentiation

 Assessment and Reporting

See pages xxii–xxv to learn about a comprehensive online system for recording, monitoring, and reporting children's progress using core program assessments.

 VLC Virtual Learning Community **Go Online** to **vlc.cemseprojects.org** for tools and ideas related to assessment and differentiation from *Everyday Mathematics* teachers.

✓ Ongoing Assessment

In addition to frequent informal opportunities for "kid watching," every lesson (except Explorations) offers an **Assessment Check-In** to gauge children's performance on one or more of the standards addressed in that lesson.

Lesson	Task Description	**CCSS** Common Core State Standards
7-1	Hit the target number in rounds of *Hit the Target* within three changes, with or without a number grid.	2.NBT.5
7-2	Find the sum of four numbers and attempt to improve the drawing and explanation of how to figure out whether the theater could hold all of the children.	2.NBT.6, SMP7
7-3	Find the sum of the four addends mentally, using a paper-and-pencil strategy or using tools such as a number grid or number line.	2.NBT.5, 2.NBT.6, 2.NBT.9
7-4	Use a yardstick to measure and record at least one distance to the nearest yard.	2.MD.1, 2.MD.3, SMP5
7-5	Select appropriate measuring tools and use them to correctly measure at least two lengths.	2.MD.1, 2.MD.3, SMP5, SMP6
7-6	Measure to both the nearest centimeter and inch with a tape measure.	2.MD.1, SMP5
7-7	Find the difference between the longest and shortest jumps without using a tool.	2.NBT.5
7-8	Accurately represent data in a line plot.	2.MD.9

▶ Periodic Assessment

Unit 7 Progress Check This assessment focuses on the CCSS domains of *Number and Operations in Base Ten* and *Measurement and Data*. It also contains an Open Response Assessment to test children's ability to decide if two different sets of base-10 blocks represent the same number and explain their thinking. GMP2.2

> **NOTE** Odd-numbered units include an **Open Response Assessment.** Even-numbered units include a **Cumulative Assessment.**

▶ Unit 7 Differentiation Activities

 Differentiation Support   English Learners Support

Differentiation Options Every regular lesson provides **Readiness, Enrichment, Extra Practice,** and **English Language Learners Support** activities that address the Focus standards of that lesson.

Activity Card 98

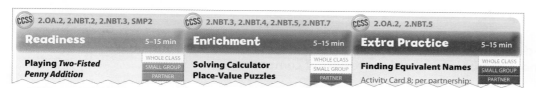

CCSS 2.OA.2, 2.NBT.2, 2.NBT.3, SMP2	CCSS 2.NBT.3, 2.NBT.4, 2.NBT.5, 2.NBT.7	CCSS 2.OA.2, 2.NBT.5
Readiness 5–15 min	**Enrichment** 5–15 min	**Extra Practice** 5–15 min
Playing *Two-Fisted Penny Addition* WHOLE CLASS SMALL GROUP PARTNER	**Solving Calculator Place-Value Puzzles** WHOLE CLASS SMALL GROUP PARTNER	**Finding Equivalent Names** WHOLE CLASS SMALL GROUP PARTNER Activity Card 8; per partnership:

Activity Cards These activities, written to the children, enable you to differentiate Part 2 of the lesson through small-group work.

English Language Learners Activities and point-of-use support help children at different levels of English language proficiency succeed.

Differentiation Support online pages

Differentiation Support Two online pages for most lessons provide suggestions for game modifications, ways to scaffold lessons for children who need additional support, and language development suggestions for Beginning, Intermediate, and Advanced English language learners.

Ongoing Practice Differentiation Support

▶ Games

Games in *Everyday Mathematics* are an essential tool for practicing skills and developing strategic thinking.

For **ongoing distributed practice,** see these activities:
- Mental Math and Fluency
- Differentiation Options: Extra Practice
- Part 3: Journal pages, Math Boxes, *Math Masters*, Home Links
- Print and online games

Lesson	Game	Skills and Concepts	CCSS Common Core State Standards
7-1	*Hit the Target*	Finding differences between 2-digit numbers and multiples of 10	2.NBT.5, 2.NBT.9, SMP1, SMP3
7-1	*Hit the Target* with Other Numbers	Finding differences between 2-digit numbers and 3-digit numbers	2.NBT.5, 2.NBT.7, SMP3
7-3 7-4	*Basketball Addition*	Adding three or more numbers	2.NBT.5, 2.NBT.6, 2.NBT.9, SMP1, SMP7
7-8	*Beat the Calculator*	Practicing addition facts	2.OA.2
7-9	*Addition/Subtraction Spin*	Adding and subtracting 10 and 100 mentally with 3-digit numbers	2.NBT.5, 2.NBT.7, 2.NBT.8, SMP8

VLC Virtual Learning Community **Go Online** to look for examples of *Everyday Mathematics* games at **vlc.cemseprojects.org**.

Spiral Trace: Skills, Concepts, and Applications

⭐ **Mastery Expectations** This Spiral Trace outlines instructional trajectories for key standards in Unit 7. For each standard, it highlights opportunities for Focus instruction, Warm Up and Practice activities, as well as formative and summative assessment. It describes the **degree of mastery**— as measured against the entire standard—expected at this point in the year.

Number and Operations in Base Ten

2.NBT.4 Compare two three-digit numbers based on meanings of the hundreds, tens, and ones digits, using >, =, and < symbols to record the results of comparisons.

| 4-12 Progress Check | 5-1 Warm Up Practice | 5-5 Practice | 5-10 Practice | 6-4 Practice | 6-8 Warm Up | 6-11 Progress Check | 7-6 Practice | 7-7 Warm Up | 7-8 Focus | 9-5 Focus Practice | 9-12 Progress Check |

⭐ By the end of Unit 7, expect children to **compare two 3-digit numbers based on meanings of the hundreds, tens, and ones digits, using <, >, and = symbols to record the results of comparisons.**

2.NBT.5 Fluently add and subtract within 100 using strategies based on place value, properties of operations, and/or the relationship between addition and subtraction.

| 6-2 through 6-9 Warm Up Focus Practice | 6-11 Progress Check | 7-1 Warm Up Focus Practice | 7-3 Warm Up Focus Practice | 7-4 Warm Up Practice | 7-6 Practice | 7-7 Focus Practice | 7-8 Focus Practice | 7-9 Warm Up Practice | 7-10 Progress Check | 8-1 Warm Up Practice |

⭐ By the end of Unit 7, expect children to **add and subtract within 100, using strategies based on place value, properties of operations, and/or the relationship between addition and subtraction.**

2.NBT.6 Add up to four two-digit numbers using strategies based on place value and properties of operations.

| 7-2 Focus Practice | 7-3 Focus Practice | 7-4 Practice | 7-5 Warm Up Practice | 7-9 Warm Up Practice | 7-10 Progress Check | 8-1 through 8-4 Practice | 8-9 Practice | 8-12 Progress Check | 9-2 Warm Up | 9-3 Warm Up |

⭐ By the end of Unit 7, expect children to **add three or more 2-digit numbers (up to 20) with the use of a paper-and-pencil strategy, a number grid, or a number line.**

Spiral Tracker

Go to **connectED.mheducation.com** for comprehensive trajectories that show how in-depth mastery develops across the grade.

Measurement and Data

2.MD.1 Measure the length of an object by selecting and using appropriate tools such as rulers, yardsticks, meter sticks, and measuring tapes.

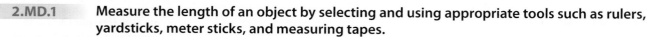

4-8 through 4-11 Focus Practice | 4-12 Progress Check | 5-8 Practice | 6-10 Focus | 6-11 Progress Check | 7-4 through 7-6 Focus Practice | 7-9 Focus | 7-10 Progress Check | 9-1 Practice | 9-4 Focus Practice

⭐ By the end of Unit 7, expect children to **select appropriate measuring tools and measure the length of an object or distance to the nearest inch, foot, or centimeter.**

2.MD.2 Measure the length of an object twice, using length units of different lengths for the two measurements; describe how the two measurements relate to the size of the unit chosen.

4-8 Focus | 4-10 Focus | 4-11 Focus | 5-3 Practice | 6-11 Progress Check | 7-4 Focus | 7-8 Focus | 7-9 Focus | 8-5 Practice

⭐ By the end of Unit 7, expect children to **measure the length of an object in both inches and centimeters (or in two different units of length) and describe how the two measurements relate to the size of the unit chosen.**

2.MD.10 Draw a picture graph and a bar graph (with single-unit scale) to represent a data set with up to four categories. Solve simple put-together, take-apart, and compare problems using information presented in a bar graph.

6-1 Focus Practice | 6-4 Practice | 6-5 Practice | 6-7 Practice | 6-8 Practice | 6-10 Practice | 6-11 Progress Check | 7-2 through 7-4 Practice | 7-9 Focus Practice | 8-12 Progress Check | 9-5 Practice

⭐ By the end of Unit 7, expect children to **answer simple questions about the data on a picture graph and bar graph.**

Geometry

2.G.1 Recognize and draw shapes having specified attributes, such as a given number of angles or a given number of equal faces. Identify triangles, quadrilaterals, pentagons, hexagons, and cubes.

5-5 Practice | 6-10 Focus | 7-6 Practice | 7-9 Focus | 7-10 Practice | 8-1 through 8-5 Focus Practice | 8-11 Focus Practice | 8-12 Progress Check | 9-5 Practice

⭐ By the end of Unit 7, expect children to **sort shapes and identify common attributes.**

Key  ✓ = Assessment Check-In   ✓ = Progress Check Lesson   ▨ = Current Unit   ▬ = Previous or Upcoming Lessons

Mathematical Background: Content

 This discussion highlights the major content areas and the Common Core State Standards addressed in Unit 7. See the online Spiral Tracker for complete information about the learning trajectories for all standards.

▶ Addition and Subtraction Strategies
(Lessons 7-1 through 7-3)

In Units 2 and 3, children used the number 10 as a stopping point when they applied the making-10 and going-through-10 strategies for addition and subtraction facts. **2.OA.2** In Lesson 7-1 children extend this idea to using *multiples of 10* as breaking points when mentally adding or subtracting 2-digit numbers. **2.NBT.5**

Children begin by solving problems—such as "32 plus what equals 40?"—and reason that because they know that it takes 8 to get from 2 to 10, they also know that it takes 8 to get from 32 to 40. They then extend the strategy to problems with differences that are larger than 10. For example, to solve "44 plus what equals 70?" a child might reason as follows: *It takes 6 to get from 44 to 50, and then it takes 20 more to get from 50 to 70. That's 6 + 20 = 26 altogether. So 44 + 26 = 70.* By this time in the year, children are familiar with combinations of 10 and adding 10s. Thus, this strategy takes advantage of the place-value structure of base-10 numbers to convert a challenging problem into two easier problems. Children are encouraged to use concrete models, such as a number grid, to solve the problems and explain why the strategy works. **2.NBT.7, 2.NBT.9**

In Lessons 7-2 and 7-3, children solve addition problems involving three or more addends. **2.NBT.6** They may apply two properties of addition while solving these problems. The Commutative Property of Addition says that two numbers can be added in either order without changing the sum. This means, for example, 3 + 4 = 4 + 3. In *Everyday Mathematics,* this property is called the *turn-around rule* and is often applied to fact work. Children may also apply this property when they add 3 or more numbers. For example, to solve 3 + 17 + 5, children may find it easier to start with the largest number and think of the problem as 17 + 3 + 5. When they switch the order of 17 and 3, children are applying the Commutative Property of Addition.

According to the Associative Property of Addition, when adding three numbers, the grouping of addends can be changed without changing the sum. When you group addends, you add them first. For example, when solving 18 + 16 + 4, you can group 18 and 16 or 16 and 4. If you group 18 and 16, add them first to get 34 and then add 4 more to get 38. Alternatively, if you group 16 and 4, add them first to get 20, and then add 18 more to get 38. When solving this problem, children may choose to add 16 and 4 first because they make 20. When they choose which numbers to add first, children are applying the Associative Property of Addition.

Children are not expected to name these properties or even be aware that they are applying them. Rather, they are expected to develop an understanding that reordering and regrouping addends can make addition problems easier to solve.

Standards and Goals for
 Mathematical Content

Because the standards within each domain can be broad, *Everyday Mathematics* **has unpacked each standard into Goals for Mathematical Content GMC. For a complete list of Standards and Goals, see page EM1.**

						0
4	5	6	7	8	9	10
14	15	16	17	18	19	20
24	25	26	27	28	29	30
34	35	36	37	38	39	40

It takes 6 units to go from 4 to 10, so it takes 6 units to go from 24 to 30.

NOTE The Associative Property of Addition is often illustrated with parentheses: (18 + 16) + 4 = 18 + (16 + 4). However, second graders are not expected to use parentheses.

Unit 7 Vocabulary

addend	line plot	partial-sums addition
arm span	meter (m); yard (yd)	personal reference
frequency table	multiple of 10	standard unit

▶ Length Measurement Units and Tools

(Lessons 7-4 and 7-5)

In Lessons 7-4 and 7-5, children use yardsticks and metersticks to measure lengths and distances in *yards* and *meters*. **2.MD.1** They also discuss how to use personal measurement references to estimate lengths. **2.MD.3** For example, children determine that their arm spans are a little more than a yard or a meter. They then can estimate a length in meters or yards by measuring how many times their arm span fits along the length. Making estimates and then checking them by measuring can help children develop good measurement sense.

By the end of this unit, children will have been introduced to many length measurement units (centimeters, meters, inches, feet, and yards) and many measurement tools (inch rulers, centimeter rulers, yardsticks, metersticks, and tape measures). When they encounter length measurement problems later in the year or in real life, they will be faced with two questions: *What unit should I use?* and *What tool should I use?*

When choosing a length unit, one must first consider the size of an object or the distance to be measured. Ideally, there will be rough proportionality between the length of an object or distance and the length unit used. Said another way, it is best to use shorter length units to measure shorter objects or distances, and longer length units to measure longer objects or distances. For example, it doesn't make sense to measure the length of a calculator in yards because a calculator is much shorter than a yard. Similarly, it doesn't make sense to measure the length of a football field in inches because it is difficult for most people to make sense of a measurement such as 3,600 inches. It is much easier to understand and visualize a field that is 100 yards long. The activities in these lessons provide practice with choosing a unit of the appropriate size.

When choosing a measurement tool, several things must be considered. For example, *Is my tool marked in the units I want to use?* It is extremely difficult to measure in inches when using a ruler marked in centimeters. *Can I line the tool up with the length I want to measure?* It is difficult to measure the distance around a head or a wrist using a ruler if the ruler will not bend. A tape measure may be a better choice.

The tool used may also affect the accuracy of a measure. For example, moving a foot ruler along a 25-foot length may produce a less accurate measurement than using a 30-foot tape measure because the ruler may not be accurately lined up each time it is moved. Children are not required to focus extensively on the possibility of measurement error, however. The measurement activities in Unit 7 focus on examining the more straightforward issues of choosing a tool that is marked in the correct units and is appropriate to use in a given situation. **2.MD.1**

Both centimeters and inches are appropriate units for measuring the length of a banana.

NOTE In the United States, U.S. customary units are used for most everyday purposes, whereas metric units are used in science and industry. Second graders in *Everyday Mathematics* are not expected to consider the appropriate measurement system for a given situation. However, they are expected to gain familiarity with both systems.

▶ Data Displays (Lessons 7-6 through 7-9)

In Lesson 7-6 children generate measurement data by measuring the lengths of arm spans and standing jumps to the nearest centimeter and inch. **2.MD.9** They note that the measures in centimeters are larger because centimeters are shorter units. **2.MD.2**

In Lesson 7-7 children organize the standing-jump data into a line plot. **2.MD.9** Like a picture graph and a bar graph, a line plot is useful for displaying frequencies. A *line plot* looks quite similar to a picture graph and is read in a similar way: the frequency of a particular number is determined by counting the number of Xs or other symbols above it. For example, the line plot below shows that four children have an arm span of 46 inches because there are four squares above 46. Line plots are also called *dot plots* or *sketch graphs* because they are quick and easy to draw and show a rough sketch of the data.

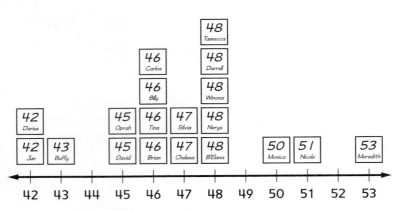

Arm Spans

Length of Arm Spans (inches)

One difference between how picture graphs and line plots are used is that picture graphs are more often used to display categorical data. For example, the labels across the bottom of a picture graph may be the categories *spring, summer, fall,* and *winter*. The data displayed may then be the number of children who chose each season as their favorite. For a line plot, however, the horizontal axis is usually a number line, so the data displayed would be numerical. Line plots are particularly useful for displaying measurement data.

In Lesson 7-7 children organize the standing-jump data into a line plot by placing stick-on notes above a number line. In Lesson 7-8 they organize the arm-span data by making a *frequency table,* or tally chart, of the data. They then use the tally chart to make a line plot by drawing Xs above a number line.

NOTE Graphs and tables are examples of mathematical models. **SMP4** For more information, see the Mathematical Background: Practices section on the next page.

Mathematical Background: Practices

 In Everyday Mathematics, *children learn the **content** of mathematics as they engage in the **practices** of mathematics. As such, the Standards for Mathematical Practice are embedded in children's everyday work, including hands-on activities, problem-solving tasks, discussions, and written work. Read here to see how Mathematical Practices 3 and 4 are emphasized in this unit.*

▶ Standard for Mathematical Practice 3

According to **SMP3**, mathematically proficient students "make conjectures and build a logical progression of statements to explore the truth of their conjectures." In Lesson 7-1 children enter a starting number into their calculators and work to change it into a given multiple of 10. For example, they enter 17 and are asked to change it to 40. When children state whether they need to add or subtract and what number they need to add or subtract, they are making *conjectures,* or educated guesses, based on some given information. When they explain how they determined what number to add or subtract, they are making *arguments,* or logical progressions of statements, that support their conjectures.

Children also test their conjectures by entering their changes into calculators to see if they made the correct change. This allows them to reexamine their thinking if necessary and make new, revised conjectures and arguments. Children continue to make mathematical conjectures and arguments as they play *Hit the Target.* **GMP3.1**

▶ Standard for Mathematical Practice 4

According to **SMP4**, mathematically proficient students "are able to identify important quantities in a practical situation and map their relationships using such tools as diagrams, two-way tables, [and] graphs." In Lesson 7-6 children collect data about the lengths of arm spans and standing jumps. In Lessons 7-7 and 7-8, they organize the data into frequency tables and line plots.

Frequency tables and line plots are mathematical models that help children identify important quantities and relationships in their data sets. For example, both models make it easy to identify the largest number in a data set, the smallest number in a data set, and the number that appears most often. Tables and line plots also help children see relationships among the numbers in a data set. For example, they can easily see that more children have a 45-inch arm span than a 50-inch arm span. In this way, children model real-world situations using tables and graphs repeatedly in Unit 7 and use their models to answer questions. **GMP4.1, GMP4.2**

 Standards and Goals for Mathematical Practice

SMP3 Construct viable arguments and critique the reasoning of others.

   GMP3.1 Make mathematical conjectures and arguments.

   GMP3.2 Make sense of others' mathematical thinking.

SMP4 Model with mathematics.

   GMP4.1 Model real-world situations using graphs, drawings, tables, symbols, numbers, diagrams, and other representations.

   GMP4.2 Use mathematical models to solve problems and answer questions.

Go Online to the *Implementation Guide* for more information about the Mathematical Practices.

For children's information on the Mathematical Practices, see *My Reference Book,* pages 1–22.

Playing *Hit the Target*

Overview Children practice finding differences between 2-digit numbers and multiples of 10.

▶ **Before You Begin**
Have copies of *Math Masters*, page G25 available for children to record extra rounds of *Hit the Target*.

▶ **Vocabulary**
multiple of 10

 Common Core State Standards

Focus Clusters
• Add and subtract within 20.
• Understand place value.
• Use place value understanding and properties of operations to add and subtract.

1 Warm Up 15–20 min

	Materials	
Mental Math and Fluency Children name the nearest multiple of 10 for each number.	slate	2.NBT.5, 2.NBT.7
Daily Routines Children complete daily routines.	See pages 4–43.	See pages xiv–xvii.

2 Focus 20–30 min

Math Message Children find missing addends in combinations of 10.		2.OA.2
Making Multiples of 10 Children use combinations of 10 to find missing addends.	Number-Grid Poster	2.NBT.5, 2.NBT.9 SMP7
Solving Calculator Change Puzzles Children add or subtract to change numbers to multiples of 10.	calculator	2.NBT.1, 2.NBT.1a, 2.NBT.5, 2.NBT.9 SMP3
Introducing and Playing *Hit the Target* **Game** Children find differences between 2-digit numbers and multiples of 10.	*Math Journal 2*, p. 165; *Math Masters*, p. G25 (optional); per partnership: calculator	2.NBT.5, 2.NBT.9 SMP1, SMP3
✓ **Assessment Check-In** See page 618.	*Math Journal 2*, p. 165	2.NBT.5

CCSS 2.NBT.5 Spiral Snapshot

GMC Subtract within 100 fluently.

| 6-5
Warm Up
Focus
Practice | 6-6
Warm Up | 6-9
Focus
Practice | 7-1
Warm Up
Focus
Practice | 7-3
Warm Up | 7-4
Warm Up | 7-6
Practice | 7-7
Focus
Practice |

Spiral Tracker **Go Online** to see how mastery develops for all standards within the grade.

3 Practice 10–15 min

Bamboo Plant Number Stories Children solve number stories about a bamboo plant.	*Math Journal 2*, pp. 166–167; yardstick for demonstration (optional)	2.OA.1, 2.NBT.5, 2.NBT.7, 2.MD.5
Math Boxes 7-1 Children practice and maintain skills.	*Math Journal 2*, p. 168	See page 619.
Home Link 7-1 **Homework** Children find missing addends.	*Math Masters*, p. 188	2.OA.2, 2.NBT.5

connectED.mheducation.com

Plan your lessons online with these tools.

 ePresentations   Student Learning Center   Facts Workshop Game   eToolkit   Professional Development   Home Connections   Spiral Tracker   Assessment and Reporting   English Learners Support   Differentiation Support

Differentiation Options

RtI

Readiness
5–15 min

| WHOLE CLASS |
| SMALL GROUP |
| PARTNER |
| INDEPENDENT |

Making Multiples of 10

Math Masters, pp. 187 and TA11; counters

To gain experience with multiples of 10, children find the number of missing dots on ten frames on *Math Masters*, page 187. Children may want to represent the problems using counters and the ten-frame card on *Math Masters*, page TA11. GMP2.1

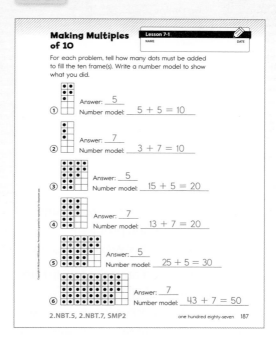

Making Multiples of 10   Lesson 7-1

For each problem, tell how many dots must be added to fill the ten frame(s). Write a number model to show what you did.

① Answer: 5   Number model: 5 + 5 = 10
② Answer: 7   Number model: 3 + 7 = 10
③ Answer: 5   Number model: 15 + 5 = 20
④ Answer: 7   Number model: 13 + 7 = 20
⑤ Answer: 5   Number model: 25 + 5 = 30
⑥ Answer: 7   Number model: 43 + 7 = 50

2.NBT.5, 2.NBT.7, SMP2   one hundred eighty-seven 187

Enrichment
5–15 min

| WHOLE CLASS |
| SMALL GROUP |
| PARTNER |
| INDEPENDENT |

Playing *Hit the Target* with Other Numbers

Math Masters, p. G25; per partnership: calculator

To apply their understanding of finding differences between multidigit numbers, children play *Hit the Target* with any 2-digit number (not necessarily a multiple of 10) as the target number. GMP3.1 Children may also choose 3-digit numbers as the start and target numbers. Children record their rounds on *Math Masters*, page G25.

Hit the Target Record Sheet

Round 1
Target number: _____

Starting Number	Change	Result	Change	Result	Change	Result

Round 2
Target number: _____

Starting Number	Change	Result	Change	Result	Change	Result

Round 3
Target number: _____

Starting Number	Change	Result	Change	Result	Change	Result

Round 4
Target number: _____

Starting Number	Change	Result	Change	Result	Change	Result

G25

Extra Practice
5–15 min

| WHOLE CLASS |
| SMALL GROUP |
| PARTNER |
| INDEPENDENT |

Finding Differences

Activity Card 86; number grid (*Math Journal 2*, inside back cover); *Math Masters*, p. TA3; 4 each of number cards 1–10

For more practice finding the difference between a 2-digit number and a multiple of 10, children pick two cards and create a 2-digit number. They locate that number on the number grid on the inside back cover of their journals, name a multiple of 10 that is larger than their 2-digit number, and find the difference between the two numbers.

Finding Differences   86

What You Need
Number Grid, page TA3
4 each of number cards 1–9

What To Do
Work with a partner.
1. Shuffle the cards and place them facedown on the table.
2. Pick two cards and use them to make a 2-digit number.
3. Point to your number on the number grid.
4. Your partner points to a multiple of 10 that is higher than your number.
5. Find the difference between your number and the multiple of 10.
6. Trade jobs. Repeat 5 times.

Talk About It
What helped you to find the difference between your number and the multiple of 10?

English Language Learners Support

Beginning ELL Build children's understanding of the term *change to* as meaning "to make something different or turn it into something else" by changing the color or the shape of concrete objects. For example, take a ball of clay and change it to look like a pancake. Use a think-aloud to accompany demonstrations. Say: *I have a ball of clay. I am going to make it into something different. The ball will change to a pancake.* Invite children to change balls of clay into other shapes. Ask: *What did the ball of clay change to? How is it different?* Accept and restate one-word answers into simple sentences, using a sentence frame such as the following: "The ball changed to a _____."

Go Online   ELL English Learners Support

Standards and Goals for
Mathematical Practice

SMP1 Make sense of problems and persevere in solving them.
GMP1.3 Keep trying when your problem is hard.

SMP3 Construct viable arguments and critique the reasoning of others.
GMP3.1 Make mathematical conjectures and arguments.

SMP7 Look for and make use of structure.
GMP7.1 Look for mathematical structures such as categories, patterns, and properties.

① Warm Up 15–20 min [Go Online]
ePresentations eToolkit

▶ Mental Math and Fluency

Pose numbers and have children write the nearest multiple of 10 on their slates. *Leveled exercises:*

●○○ 31 30, 42 40, 19 20
●●○ 57 60, 74 70, 65 60 or 70
●●● 98 100, 151 150, 227 230

▶ Daily Routines

Have children complete daily routines.

② Focus 20–30 min [Go Online]
ePresentations eToolkit

▶ Math Message

Find the missing numbers.

$5 + \underline{\ 5\ } = 10$ $\underline{\ 7\ } + 3 = 10$ $4 + \underline{\ 6\ } = 10$

▶ Making Multiples of 10

| WHOLE CLASS | SMALL GROUP | PARTNER | INDEPENDENT |

Math Message Follow-Up Have volunteers share how they found the missing numbers. Most children will recognize the problems as combinations of 10, but not everyone will find the answers in the same way.

Tell children that the numbers 10, 20, 30, and so on are called **multiples of 10.** Point out that the ones digit in a multiple of 10 is always 0, as in the number 10. Extend the Math Message by having children find the difference between a given 2-digit number and the next multiple of 10. Pose problems such as the following and, if needed, suggest that children use the Number-Grid Poster to help them solve the problems. *Suggestions:*

$25 + \underline{\ 5\ } = 30$ $\underline{\ 6\ } + 44 = 50$

Ask: *What patterns do you notice?* GMP7.1 Sample answer: The ones digits make a combination of 10. Guide children to see how to use their knowledge of combinations of 10 to solve problems like these. Then tell them they will add and subtract to change numbers into multiples of 10.

▶ Solving Calculator Change Puzzles

| WHOLE CLASS | SMALL GROUP | PARTNER | INDEPENDENT |

Have children try a few problems like these:

• Enter 45 into your calculator. Change it to 50. Did you add or subtract? Add What number did you add? 5

- Enter 33. Change it to 40. Did you add or subtract? Add What number did you add? 7
- Enter 60. Change it to 52. Did you add or subtract? Subtract What number did you subtract? 8

Extend the activity to problems with differences of 10 or more. For each problem, ask volunteers to explain how they knew what number to add or subtract. GMP3.1 Encourage children to first add or subtract to find the nearest multiple of 10. Then they can add or subtract 10s to get to the change-to number. *Suggestions:*

- Enter 32. Change it to 50. Did you add or subtract? Add What number did you add? 18 A child might reason as follows: "What number, added to 2, will give me 10? It's 8, so 32 + 8 = 40. Which number, added to 40, will give me 50? It's 10, because 40 + 10 = 50. Finally, I know that 8 + 10 = 18, so I have to add 18 in all."

- Enter 74. Change it to 100. Did you add or subtract? Add What number did you add? 26 A child might reason as follows: "I know that 74 + 6 = 80. Because 80 is eight 10s and 100 is ten 10s, I need to add two more 10s, or 20. So I have to add 6 + 20 = 26 in all."

- Enter 86. Change it to 40. Did you add or subtract? Subtract What number did you subtract? 46 One child might count up, reasoning as follows: "What number, added to 40, will give me 80? It's 40. What number, added to 80, will give me 86? It's 6. I have to subtract 46."

  Another child might count back, reasoning like this: "What number, subtracted from 86, will give me 80? The answer is 6. What number, subtracted from 80, will give me 40? The answer is 40. Finally, I know that 6 + 40 = 46, so I have to subtract 46 in all."

▶ Introducing and Playing *Hit the Target*

Math Journal 2, p. 165

| WHOLE CLASS | SMALL GROUP | PARTNER | INDEPENDENT |

Playing *Hit the Target* gives children practice mentally finding differences between multiples of 10 and smaller or larger 2-digit numbers. Players try to change a start number to a target number by adding and subtracting. They make conjectures, or educated guesses, about what number they need to add or subtract and check their guesses on a calculator. GMP3.1

Review the following directions with the class. Children play with a partner and share a calculator.

Directions

1. Players agree on a 2-digit multiple of 10 as a target number. Each player records the target number on the *Hit the Target* Record Sheet.

2. Player A chooses a 2-digit start number less than or more than the target number and records it on Player B's record sheet.

3. Player B enters the start number into a calculator and tries to change it into the target number by adding or subtracting on the calculator.

4. Player B continues adding or subtracting until the target number is reached. Player A records each change and its result on Player B's record sheet.

Math Journal 2, p. 165

Hit the Target Record Sheet Lesson 7-1

DATE

Record three rounds of *Hit the Target.*

Example Round
Target number: 40

Starting Number	Change	Result	Change	Result	Change	Result
12	+ 38	50	− 10	40		

Round 1
Target number: _____

Starting Number	Change	Result	Change	Result	Change	Result

Round 2
Target number: _____

Starting Number	Change	Result	Change	Result	Change	Result

Round 3
Target number: _____

Starting Number	Change	Result	Change	Result	Change	Result

2.NBT.5, 2.NBT.9, SMP1, SMP3   one hundred sixty-five 165

Lesson 7-1   **617**

Academic Language Development

Encourage children to use academic language in their discussion by providing sentence frames such as these:

- When I didn't hit the target number on the first try, I decided to add _____ because I knew that _____.
- When I didn't hit the number on the first try, I decided to subtract _____ because I figured that _____.

Provide these sentence frames in writing and encourage children to choose one for reporting out during class discussion, taking into account that these sentence frames are more appropriate for children with significant English language proficiency.

5. Players then switch roles. Player B selects a start number and records it on Player A's record sheet. Player B fills in Player A's record sheet as Player A adds and subtracts on the calculator.

6. The player who hits the target number in fewer tries wins the round.

7. Partners choose a new target number and begin the next round.

Talk through the example at the top of journal page 165 and play a few practice rounds with the class. Then have partners play the game. Children should record their first three rounds of play on journal page 165. Some children may want to record additional rounds on the *Hit the Target* Record Sheet on *Math Masters,* page G25.

Observe

- Which children seem to have a strategy for hitting the target number?
- Which children need additional support to understand and play the game?

Discuss

- *How did you decide what number to add or subtract?* GMP3.1
- *If you didn't hit the target number on your first try, how did you decide what to do next?* GMP1.3

Differentiate **Adjusting the Activity**

Suggest that children begin by locating the start number on a number grid and counting by 1s to get to the next multiple of 10. They record this as the first change. Children next count by 10s to get to the target number. They record this as the second change. Go Online Differentiation Support

---

Differentiate **Game Modifications** Go Online Differentiation Support

---

✓ Assessment Check-In CCSS 2.NBT.5

Math Journal 2, p. 165

Expect most children to successfully hit the target numbers in rounds of *Hit the Target* within three changes, with or without the number grid. If children struggle to hit the target numbers even while using the number grid, suggest that they first count up by 1s to the next multiple of 10 and then count by 10s to the target number, as suggested in the Adjusting the Activity note. Some children may be able to hit the target number without the number grid or with fewer than three changes.

 Assessment and Reporting Go Online to record student progress and to see trajectories toward mastery for these standards.

Summarize Have partners explain to each other how they decided what number to add or subtract in one of their *Hit the Target* rounds. GMP3.1

Math Journal 2, p. 166

Bamboo Plant Number Stories Lesson 7-1 DATE

Bamboo is one of the world's fastest-growing plants. Some types of bamboo grow more than 24 inches per day and reach heights close to 100 feet! For one week a growing bamboo plant was measured. The chart below shows its height at the beginning of each day.

Bamboo Plant Growth for One Week

Sun.	Mon.	Tues.	Wed.	Thurs.	Fri.	Sat.
12 in.	26 in.	40 in.	57 in.	63 in.	80 in.	99 in.

Use the information above to solve the following number stories.

❶ How many inches did the bamboo plant grow from Tuesday to Friday?

❷ How many inches did the bamboo plant grow from Thursday to Friday?

Sample number models:
Number model:
80 − 40 = ?; 40 + ? = 80

Answer: __40__ inches

Sample number models:
Number model:
63 + ? = 80; 80 − 63 = ?

Answer: __17__ inches

166 one hundred sixty-six   2.OA.1, 2.NBT.5, 2.NBT.7, 2.MD.5

3 Practice   10–15 min

Go Online
ePresentations   eToolkit   Home Connections

▶ Bamboo Plant Number Stories

Math Journal 2, pp. 166–167

| WHOLE CLASS | SMALL GROUP | PARTNER | **INDEPENDENT** |

Children solve number stories about the growth of a bamboo plant. Read the first paragraph on journal page 166 as a class and discuss the information in the chart. To help children understand how quickly bamboo grows, you might use a yardstick to demonstrate the difference between Sunday's and Monday's heights. Children complete the page independently. Encourage them to sketch situation diagrams as needed to help them organize information from the problems and write the number models.

▶ Math Boxes 7-1

Math Journal 2, p. 168

| WHOLE CLASS | **SMALL GROUP** | **PARTNER** | INDEPENDENT |

Mixed Practice   Math Boxes 7-1 are paired with Math Boxes 7-3.

▶ Home Link 7-1

Math Masters, p. 188

Homework   Children find missing addends to make multiples of 10.

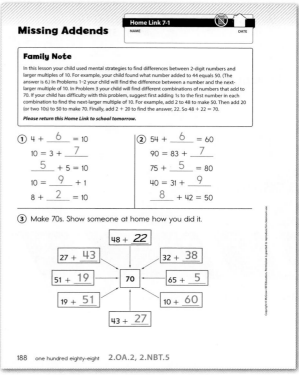

Math Masters, p. 188

Math Journal 2, p. 167

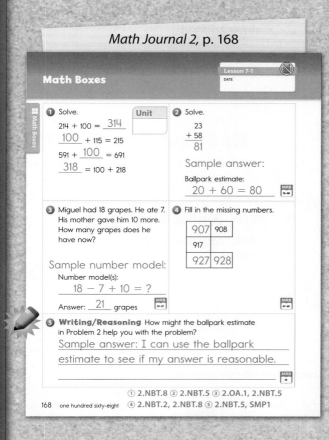

Math Journal 2, p. 168

Four or More Addends

2-Day Lesson

Overview **Day 1:** Children solve an open response problem by applying place-value concepts and addition properties.
Day 2: The class discusses selected solutions, and children revise their work.

Day 1: Open Response

▶ **Before You Begin**
If possible, schedule time to review children's work and plan for Day 2 of this lesson with your grade-level team.

▶ **Vocabulary**
addends • partial-sums addition

CCSS **Common Core State Standards**

Focus Clusters
• Represent and solve problems involving addition and subtraction.
• Add and subtract within 20.
• Use place value understanding and properties of operations to add and subtract.

1 Warm Up   15–20 min

Materials

Mental Math and Fluency
Children solve addition problems with three or more addends.

Daily Routines
Children complete daily routines.

See pages 4–43.

See pages xiv–xvii.

2a Focus   45–55 min

	Materials	
Math Message Children solve and discuss a number story with three addends.	*Math Journal 2,* p. 169; number grid	2.OA.1, 2.OA.2, 2.NBT.6
Counting Pencils Children discuss strategies for solving a number story with three addends.	*Math Journal 2,* p. 169; number grid	2.OA.1, 2.NBT.6 SMP1, SMP7
Solving the Open Response Problem Children solve a number story with four two-digit addends.	*Math Masters,* p. 189; number grid; base-10 blocks	2.OA.1, 2.NBT.6 SMP1, SMP4, SMP7

Getting Ready for Day 2 →

Review children's work and plan discussion for reengagement.   *Math Masters,* p. TA5; children's work from Day 1

CCSS 2.NBT.6 **Spiral Snapshot**

GMC Add up to four 2-digit numbers.

7-2 Warm Up Focus Practice	7-3 Focus Practice	7-4 Practice	7-5 Warm Up Practice	7-9 Warm Up Practice	8-1 through 8-4 Practice	8-9 Practice

 Spiral Tracker   **Go Online** to see how mastery develops for all standards within the grade.

connectED.mheducation.com

Plan your lessons online with these tools.

ePresentations   Student Learning Center   Facts Workshop Game   eToolkit   Professional Development   Home Connections   Spiral Tracker   Assessment and Reporting   English Learners Support   Differentiation Support

1 Warm Up   15–20 min
Go Online ePresentations eToolkit

▶ Mental Math and Fluency

Display problems one at a time. Encourage children to use mental strategies to solve.

● ○ ○ $7 + 7 + 3 = ?$ 17

$? = 4 + 6 + 5$ 15

● ● ○ $5 + 5 + 6 = ?$ 16

$? = 6 + 6 + 2$ 14

● ● ● $? = 5 + 6 + 1$ 12

$? = 1 + 1 + 2 + 2$ 6

▶ Daily Routines

Have children complete daily routines.

2a Focus   45–55 min
Go Online ePresentations eToolkit

▶ Math Message

Math Journal 2, p. 169

Complete journal page 169.

▶ Counting Pencils

Math Journal 2, p. 169

| WHOLE CLASS | SMALL GROUP | PARTNER | INDEPENDENT |

Math Message Follow-Up Invite children to share their strategies for solving the Math Message problem first with their partners and then with the class. Possible strategies include using mental arithmetic, counting up on the number grid, drawing tally marks, or using an open number line. **GMP1.6** Consider highlighting strategies from children who included a parts-and-total diagram to organize their thinking, or who created a unit box to keep track of the units.

CCSS Standards and Goals for Mathematical Practice

SMP1 Make sense of problems and persevere in solving them.
 GMP1.6 Compare the strategies you and others use.

SMP4 Model with mathematics.
 GMP4.1 Model real-world situations using graphs, drawings, tables, symbols, numbers, diagrams, and other representations.

SMP7 Look for and make use of structure.
 GMP7.2 Use structures to solve problems and answer questions.

Professional Development

The focus of this lesson is GMP7.2. Structures can include properties of operations and place value, both of which are prominent in this lesson. In the Math Message, children discuss useful ways to apply the Associative Property or Commutative Property of Addition or both. For more information about these properties and children's strategies, see the Mathematical Background section of the Unit 7 Organizer. Children discuss these structures without using the terms. To solve the open response problem, children can also use their understanding of place-value structures to add four addends.

Go Online for information about SMP7 in the *Implementation Guide.*

Collecting Pencils

Lesson 7-2
DATE

Lia has 3 pencils. Thomas has 6 pencils. Nate has 7 pencils. How many pencils do they have in all?

Solve the problem using any strategy. Share your strategy with your partner.

Number model: $\underline{3 + 6 + 7 = 16}$ pencils

2.OA.1, 2.OA.2, 2.NBT.6　　　　　　　　one hundred sixty-nine　169

Expect that most children will add the numbers in the order they are presented in the number story. Look for and highlight strategies in which children make a ten first (3 + 7) and then add the 6 to make the mental arithmetic easier. Have a child share this strategy.

Display the three number models shown below that re-group or re-order the addends. Show them first without the carrots or a second number model.

$$3 + 6 + 7 =$$
$$9 + 7 = 16$$

$$3 + 7 + 6 =$$

$$6 + 7 + 3 =$$

Ask:

- *How are these number models alike?* Sample answers: All of them have the same numbers; none of them has a total.
- *What is different about these number models?* Sample answer: The order of the numbers is different in each one.

Use "carrots" as shown to indicate the addition that happens first. Have children determine the sums and write a new number model below with two addends and a total as shown in the first example. After completing all three examples, ask:

- *Were any of the three ways easier for you to find the sum? Why?* GMP1.6 Sample answer: The second one was easier because I know that 3 + 7 makes 10, and 10 is easy to add.
- *Does it make a difference in what order the three numbers are added?* No. *Why?* GMP7.2 Sample answer: It's like the turn-around rule. You can mix up the numbers to add, and you get the same answer.

If no child mentions the turn-around rule, ask: *How do you think this is related to the turn-around rule for addition?* GMP7.2 Sample answer: The turn-around rule says we can switch the numbers we add in an addition fact and it's the same answer. Here we can switch the three numbers and the answer is the same.

Tell children that they are going to solve another problem involving more than two **addends** using different strategies.

ELL Support

English language learners may struggle to explain their strategies for solving the problems. Prior to the lesson and as needed, use simple problems with three or four addends—similar to the Math Message problem—to give children practice communicating their strategies. Encourage the use of diagrams, tools (e.g., base-10 blocks, number grids, number lines), and verbal explanations.

Use photographs, models, or drawings to preview some of the vocabulary for the open response problem: *puppet show, show* (as a noun), *seats, to send, enough*. Also review ordinal numbers used in the problem: *first, second, third, fourth*.

► Solving the Open Response Problem

Math Masters, p. 189

| WHOLE CLASS | SMALL GROUP | **PARTNER** | **INDEPENDENT** |

Distribute *Math Masters,* page 189. Make base-10 blocks and number grids available. Read the problem as a class and check that children understand the question. Tell children that they can use any strategy they wish to determine whether there are enough seats in the theater for all the children. **GMP4.1, GMP7.2**

Circulate and observe children as they work. Expect that only a few children will strategically order the addends to make their work easier (such as starting with 19 and 21 to make a ten, or starting with 42 and 21 to facilitate a mental strategy that does not require regrouping). If you see children counting up by ones on a number grid, allow this, but ask these children to try to find the sum using another strategy. You may wish to make notes about children's strategies.

Differentiate **Adjusting the Activity**

If children have difficulty, ask them how many children would be going to the puppet show if only the first two schools sent children. Have them represent two of the addends using base-10 blocks. After they model the total from two schools with base-10 blocks, encourage them to represent the blocks in a drawing and to think about how to find the total number of children from all four schools.

If children quickly solve the problem and write a complete explanation, ask them to check their work by using a second strategy to solve the problem.

Partners can work together to share ideas about the task, but children should complete their own explanations and drawings. Encourage children to use simple pictures or drawings of base-10 blocks along with numbers in their explanations. **GMP4.1, GMP7.2**

Summarize Ask: *How were the strategies you and your partner used similar or different?* **GMP1.6** Answers vary.

Collect children's work so you can evaluate it and prepare for Day 2 of the lesson.

Math Masters, p. 189

Going to the Puppet Show

Lesson 7-2

NAME   DATE

A theater has 100 seats. Four schools are sending children to a puppet show at the theater. The first school will send 21 children. The second will send 13. The third will send 42, and the fourth will send 19.

Are there enough seats in the theater for all the children? __Yes.__

Show and explain how you figured out your answer.

Answers vary. See sample children's work on page 628 of the *Teacher's Lesson Guide.*

2.OA.1, 2.NBT.6, SMP1, SMP4, SMP7   one hundred eighty-nine   189

Are there enough seats in the theater for all children?

Yes

Show and explain how you figured out your answer.

$$
\begin{array}{r}
+\ 21 \\
13 \\
+19 \\
42 \\
\hline
15 \\
80 \\
\hline
9 \ 5
\end{array}
$$

Getting Ready for Day 2

Math Masters, p. TA5

Planning a Follow-Up Discussion

Review children's work and your notes about their thinking. Use the Reengagement Planning Form (*Math Masters,* page TA5) and the rubric on page 626 to plan ways to help children meet expectations on both the content and practice standards. Look for common misconceptions or errors in the addition as well as strategies that applied place-value understanding or addition properties.

Reengagement Planning Form

Common Core State Standard (CCSS):
2.NBT.6: Add up to four two-digit numbers using strategies based on place value and properties of operation

Goal for Mathematical Practice (GMP): *GMP7.2 Use structures to solve problems and answer questions.*

Organize the discussion in one of the ways below or in another way you choose. If children's work is unclear or if you prefer to show work anonymously, rewrite the work for display.

(Go Online) for sample children's work that you can use in your discussion.

1. Display a response that shows a **partial-sums addition** strategy, as in Child A's work. Ask:

   - *What do these four numbers above this line represent?* GMP4.1 Sample answer: Those are the numbers of children from the four schools.
   - *How do you think this child came up with the numbers 15 and 80?* GMP7.2 Sample answer: The child added the ones and got 15, then added the tens and got 80.
   - *What is the name of this addition strategy?* GMP7.2 Sample answer: partial-sums addition
   - *How could this explanation be improved so someone else can understand how to use this strategy?* Sample answer: The child could explain where the 15 and 80 came from and say the units.

2. Display a response that shows a strategy other than partial-sums addition, but that still relies on place-value understanding. Possibilities include using base-10 blocks or a number grid, as long as the response shows that the child worked with tens and ones and not just with ones. See the sample work from Child B. Ask:

- *What do you think these numbers are across the top?* GMP4.1 Sample answer: They are the numbers of children from each school.
- *What is this child showing in the drawing?* Sample answer: The child is using base-10 blocks to show the numbers. *Did this child represent each of the numbers correctly?* Yes. *Explain how you know.* GMP4.1, GMP7.2 Sample answer: The child drew longs for the tens in each number and cubes for the ones.
- *What do you think this child did after showing the four numbers with the drawing?* Sample answer: The words say the child counted first by tens and then by ones. *Show how to count these base-10 blocks this way.* GMP7.2 Sample answer: 10, 20, 30, . . ., 80, 81, 82, 83, . . ., 95
- *How could this child improve the explanation?* Sample answers: The child could show the counts. The child could show units.

3. Display work from a child who did not use place-value understanding to solve the problem. See the sample work from Child C. Ask:

- *What do you think these numbers and arrows are showing?* GMP4.1 Sample answer: They are the numbers of children from each school. Maybe the arrows show the order the child added them.
- *What are all these tick marks that are half-erased? What do you think the child was doing with those?* GMP4.1 Sample answer: Maybe the child was showing each number with tick marks and then counted all the marks.
- *What do you think about the strategy of making a tick mark for every child from every school and counting them one by one?* GMP1.6 Sample answers: It works, but it can take a long time; it's easy to lose count; counting by fives with tallies or tens and ones with base-10 blocks can be easier.

4. Display work from a child who strategically ordered the addends to facilitate the addition by making a ten or in another way. Invite this child to explain the strategy to the others. GMP1.6, GMP7.2

Planning for Revisions

Have copies of *Math Masters,* page 189 or extra paper available for children to use in revisions. You might want to ask children to use colored pencils so you can see what they revised.

Sample child's work, Child B

21   13   42 + 19

II. I...   IIII... I...........

95

Ecould ten fist
I codding the one Next

Theaner is Yes

Sample child's work, Child C

Are there enough seats in the theater for all children? Show and explain how you figured out your answer.

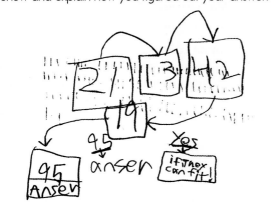

Four or More Addends

Overview **Day 2:** The class discusses selected solutions, and children revise their work.

Day 2: Reengagement

Common Core State Standards

▶ **Before You Begin**
Have extra copies available of *Math Masters,* page 189 for children to revise their work.

Focus Clusters
• Represent and solve problems involving addition and subtraction.
• Use place value understanding and properties of operations to add and subtract.

2b Focus   50–55 min

Materials

Setting Expectations Children review the open response problem and discuss what a complete explanation would include. They review how to respectfully discuss other's work.	Guidelines for Discussions Poster	SMP1
Reengaging in the Problem Children discuss and compare their strategies for adding four addends.	selected samples of children's work	2.OA.1, 2.NBT.6 SMP1, SMP4, SMP7
Revising Work Children revise their work from Day 1.	*Math Masters,* p. 189 (optional); *My Reference Book,* pp. 20–21 (optional); children's work from Day 1; Standards for Mathematical Practice Poster; colored pencils (optional); number grid; base-10 blocks	2.OA.1, 2.NBT.6 SMP7

✓ **Assessment Check-In**   See page 628 and the rubric below.

2.NBT.6
SMP7

Goal for Mathematical Practice **GMP7.2** Use structures to solve problems and answer questions.	Not Meeting Expectations	Partially Meeting Expectations	Meeting Expectations	Exceeding Expectations
	Does not provide evidence of an addition strategy that is based on place value.	Provides evidence of an addition strategy that is based on place value but includes significant errors, such as confusing tens and ones.	Provides evidence of an accurate addition strategy that is based on place value **or** that involves strategically ordering the addends.	Meets expectations and shows a second strategy for adding the numbers.

3 Practice   10–15 min

Math Boxes 7-2 Children practice and maintain skills.	*Math Journal 2,* p. 170	See page 629.
Home Link 7-2 **Homework** Children use strategies and partial-sums addition to find the sum of three addends.	*Math Masters,* pp. 190–191	2.NBT.6

2b Focus

50–55 min | Go Online

ePresentations   eToolkit

NOTE These Day 2 activities will ideally take place within a few days of Day 1. Prior to beginning Day 2, see Planning a Follow-Up Discussion from Day 1.

▶ Setting Expectations

| **WHOLE CLASS** | SMALL GROUP | PARTNER | INDEPENDENT |

Briefly review the open response problem from Day 1. Ask: *What did you do to decide whether there were enough seats in the theater?* Sample answer: I added the number of children from the four schools. *What would a complete response to this problem include?* Sample answer: It would say whether the theater has enough seats and explain how we figured it out.

Then tell children that they are going to look at other children's work and compare the strategies they used. GMP1.6

Point out that other children's work will likely be different from theirs. Remind them that they should listen closely to each other to see if they can learn a new strategy. Refer to your list of discussion guidelines and encourage children to use these sentence frames:

- I like how _____.
- I don't understand _____.

▶ Reengaging in the Problem

| **WHOLE CLASS** | SMALL GROUP | **PARTNER** | INDEPENDENT |

Children reengage in the problem by analyzing and critiquing other children's work in pairs and in a whole-group discussion. Have children discuss with partners before sharing with the whole group. Guide this discussion based on the decisions you made in Getting Ready for Day 2. GMP1.6, GMP4.1, GMP7.2

My Reference Book, pp. 20–21

Find and Use Patterns and Properties

Find each sum: 22 + 10   34 + 10   49 + 10

Mason counts by 1s on the number grid to solve these problems.

21	22	23	24	25	26	27	28	29	30
31	32	33	34	35	36	37	38	39	40
41	42	43	44	45	46	47	48	49	50
51	52	53	54	55	56	57	58	59	60

22 + 10 = 32
34 + 10 = 44
49 + 10 = 59

> There is a pattern. When I add 10, I can move down one row to find the answer.

Mason

Look for mathematical structures such as categories, patterns, and properties.

Mason sees a pattern for adding 10 to numbers on the number grid.

MRB
20 twenty

Are there enough seats in the theater for all children?

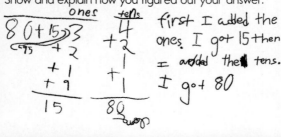

there would be anough seats for the teachers!

Show and explain how you figured out your answer.

first I added the ones I got 15 then I added the tens. I got 80

▶ Revising Work

WHOLE CLASS | SMALL GROUP | **PARTNER** | **INDEPENDENT**

Pass back children's work from Day 1. Make number grids and base-10 blocks available. Before children revise anything, ask them to examine their drawings and explanations and decide how to improve them. Ask the following questions one at a time. Have partners discuss their responses and give a thumbs-up or thumbs-down based on their own work.

- *Did you say whether the theater has enough seats for all the children?*
- *Did you explain how you figured out your answer? Is your explanation clear enough that someone else can use your strategy?*
- *Did you include any drawings that help to show how you used tools to figure out your answer?*

Tell children they now have a chance to revise their work. Remind children that the strategies discussed are not the only correct ones. Encourage children to add to their original work using colored pencils or to use another sheet of paper, instead of erasing their original work.

✓ Assessment Check-In   CCSS 2.NBT.6

Collect and review children's revised work. Expect children to improve their work based on the class discussion. For the content standard, expect most children to find that the sum of the four numbers is 95 using any strategy. You can use the rubric on page 626 to evaluate children's revised work for **GMP7.2**.

 Assessment and Reporting   Go Online to record student progress and to see trajectories toward mastery for these standards.

Go Online for optional generic rubrics in the *Assessment Handbook* that can be used to assess any additional GMPs addressed in the lesson.

Sample Children's Work—Evaluated

See the sample in the margin. This work meets expectations for the content standard by correctly showing the sum of the four numbers. The work meets expectations for the mathematical practice because it shows evidence of a partial-sums addition strategy. The child uses place value to add tens and ones separately before combining them.   **GMP7.2**

Go Online for other samples of evaluated children's work.

Summarize Ask: *What strategy or tools did you use to solve the problem?* Answers vary. You may wish to read with children *My Reference Book* pages 20–21. Ask: *What pattern or structure did Mason use to solve his problem?* He added tens by moving down a row on the number grid. *What property or structure did Emma use?* She used the turn-around rule. *What structures did we use to solve our problem?* Sample answers: We used tens and ones to add. Some children changed the order to make adding easier. Refer children to **GMP7.2** on the Standards for Mathematical Practice Poster.

3 Practice 10–15 min

Go Online ePresentations eToolkit Home Connections

▶ Math Boxes 7-2

Math Journal 2, p. 170

WHOLE CLASS | SMALL GROUP | PARTNER | INDEPENDENT

Mixed Practice Math Boxes 7-2 are paired with Math Boxes 7-4.

▶ Home Link 7-2

Math Masters, pp. 190–191

Homework Children change the order of addends and use partial sums to add three or more addends.

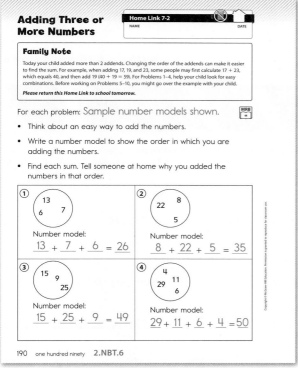

Math Masters, p. 190

Math Journal 2, p. 170

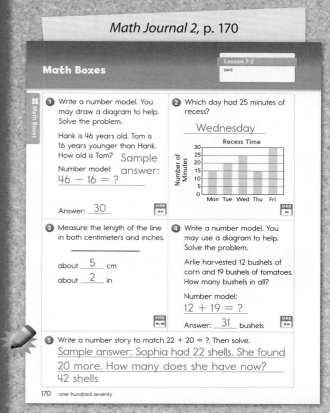

Math Masters, p. 191

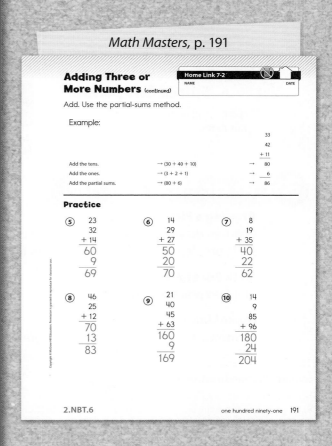

Playing *Basketball Addition*

Overview Children solve addition problems with three or more addends.

▶ Before You Begin

For Part 2 display *Math Masters,* page 192. Gather either 20-sided dice or three 6-sided dice to play *Basketball Addition.* You may wish to have copies of *Math Masters,* pages G26–G27 available. For the optional Enrichment activity, prepare one spinner for each child, partnership, or group by selecting and writing random numbers between 19 and 99 on a spinner from *Math Masters,* page G24. For the optional Extra Practice activity, prepare an egg carton for each child or partnership by writing a number between 1 and 19 in each cup.

Common Core State Standards

Focus Cluster
Use place value understanding and properties of operations to add and subtract.

1 Warm Up 15–20 min

Materials

Mental Math and Fluency Children add and subtract with multiples of 10.	slate	2.NBT.5, 2.NBT.7
Daily Routines Children complete daily routines.	See pages 4–43.	See pages xiv–xvii.

2 Focus 20–30 min

Math Message Children find sums of three or four numbers.		2.NBT.5, 2.NBT.6
Sharing Strategies Children share strategies for adding three or four numbers.		2.NBT.5, 2.NBT.6, 2.NBT.9 SMP7
Introducing *Basketball Addition* Children learn a game for adding three or more numbers.	*Math Masters,* p. 192; one 20-sided polyhedral die or three 6-sided dice	2.NBT.5, 2.NBT.6, 2.NBT.9 SMP1, SMP3, SMP7
Playing *Basketball Addition* **Game** Children practice adding three or more numbers.	*Math Journal 2,* p. 171; *Math Masters,* p. TA8; *Math Masters,* pp. G26–G27 (optional); per group: one 20-sided die or three 6-sided dice	2.NBT.5, 2.NBT.6, 2.NBT.9 SMP1, SMP7
✓ **Assessment Check-In** See page 634.	*Math Masters,* p. TA8	2.NBT.5, 2.NBT.6, 2.NBT.9

CCSS 2.NBT.6 Spiral Snapshot

GMC Add up to four 2-digit numbers.

7-2 Warm Up Focus Practice	7-3 Focus Practice	7-4 Practice	7-5 Warm Up Practice	7-9 Warm Up Practice	8-1 through 8-4 Practice	8-9 Practice

Spiral Tracker   **Go Online** to see how mastery develops for all standards within the grade.

---

3 Practice 10–15 min

Drawing a Picture Graph Children draw a picture graph to represent a data set and solve problems using information from the graph.	*Math Journal 2,* pp. 172–173	2.MD.10
Math Boxes 7-3 Children practice and maintain skills.	*Math Journal 2,* p. 174	See page 635.
Home Link 7-3 **Homework** Children add three or more numbers.	*Math Masters,* p. 193	2.NBT.4, 2.NBT.5, 2.NBT.6

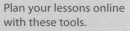
connectED.mheducation.com

Plan your lessons online with these tools.

 ePresentations   Student Learning Center   Facts Workshop Game   eToolkit   Professional Development   Home Connections   Spiral Tracker   Assessment and Reporting   English Learners Support   Differentiation Support

Differentiation Options RtI

 CCSS 2.NBT.5, 2.NBT.6, 2.NBT.7, SMP5

Readiness
5–15 min

WHOLE CLASS
SMALL GROUP
PARTNER
INDEPENDENT

Using Base-10 Blocks to Add

Activity Card 87;
Math Masters, p. TA30;
1 each of number cards 1–20;
base-10 blocks

To explore finding sums of three addends using a concrete model, children use base-10 blocks to solve addition problems. **GMP5.2** Children record their number models on *Math Masters,* page TA30.

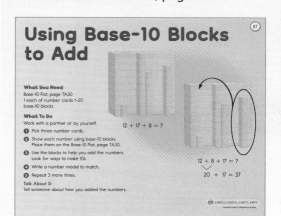

Base-10 Flat

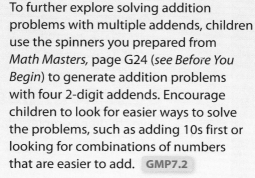 **CCSS** 2.NBT.5, 2.NBT.6, SMP7

Enrichment
5–15 min

WHOLE CLASS
SMALL GROUP
PARTNER
INDEPENDENT

Adding Four 2-Digit Numbers

Activity Card 88;
Math Masters, p. G24;
paper clip; pencil; paper

To further explore solving addition problems with multiple addends, children use the spinners you prepared from *Math Masters,* page G24 (*see Before You Begin*) to generate addition problems with four 2-digit addends. Encourage children to look for easier ways to solve the problems, such as adding 10s first or looking for combinations of numbers that are easier to add. **GMP7.2**

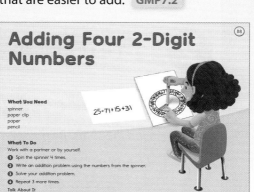
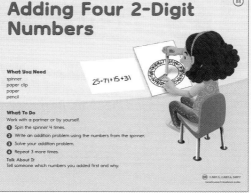

CCSS 2.NBT.5, 2.NBT.6, SMP7

Extra Practice
5–15 min

WHOLE CLASS
SMALL GROUP
PARTNER
INDEPENDENT

Adding Four Numbers

Activity Card 89, egg carton, 4 counters, paper

For more practice solving addition problems with multiple addends, children use a prepared egg carton (*see Before You Begin*) to generate addition problems with four addends between 1 and 19. Encourage children to look for easier ways to solve the problems, such as adding 10s first or looking for combinations of numbers that are easier to add. **GMP7.2**

English Language Learners Support

Beginning ELL To build or activate background knowledge about basketball, consider asking the physical education teacher to teach children how to play the game. Alternatively, show a video clip from a basketball game. Display labeled pictures of a basketball court, the scoreboard, and the clock. Point to how the numbers on the scoreboard change as players score points. Show how the clock counts down the time during each period. Demonstrate the meaning of the term *halftime* with a sketch of a circle divided in half with 20 minutes written on each side and the dividing line labeled *halftime*.

 Go Online **ELL** English Learners Support

Standards and Goals for
Mathematical Practice

SMP1 Make sense of problems and persevere in solving them.
   GMP1.5 Solve problems in more than one way.

SMP3 Construct viable arguments and critique the reasoning of others.
   GMP3.1 Make mathematical conjectures and arguments.

SMP7 Look for and make use of structure.
   GMP7.2 Use structures to solve problems and answer questions.

Professional Development

As children solve the Math Message problems and other problems in this lesson, they may apply the Commutative Property of Addition, the Associative Property of Addition, or both. For more information about these properties and how they relate to children's strategies, see the Mathematical Background section of the Unit 7 Organizer. You do not need to use the names of these properties with children. The goal is for children to begin to develop an understanding that addends can be reordered or grouped (or both) in ways that make mental addition easier.

 Go Online   Professional Development

---

① Warm Up   15–20 min   Go Online

ePresentations   eToolkit

▶ Mental Math and Fluency

Pose addition and subtraction problems involving multiples of 10. Children write the answers on their slates. Have volunteers share their strategies. *Leveled exercises:*

● ○ ○ $30 + ? = 80$ 50
        $60 - 40 = ?$ 20
        $20 + ? = 90$ 70

● ● ○ $27 + ? = 80$ 53
        $50 = ? + 22$ 28
        $70 - 23 = ?$ 47

● ● ● $140 - 18 = ?$ 122
        $24 + ? = 190$ 166
        $150 - 134 = ?$ 16

▶ Daily Routines

Have children complete daily routines.

---

② Focus   20–30 min   Go Online

ePresentations   eToolkit

▶ Math Message

Find the sums. Be prepared to share your strategy.

$12 + 17 + 8 =$ __37__

__40__ $= 4 + 9 + 16 + 11$

Unit
points

▶ Sharing Strategies

| WHOLE CLASS | SMALL GROUP | PARTNER | INDEPENDENT |

Math Message Follow-Up Have children share their solution strategies. Discuss which two numbers, when added first, make it easier to add all of the numbers. **GMP7.2** Ask:

- *For 12 + 17 + 8, which two numbers would you add first: 12 and 17, 12 and 8, or 17 and 8?* Sample answer: 12 and 8 *Why?* Sample answer: I know $2 + 8 = 10$, so $12 + 8 = 20$. Then it's easy to add $20 + 17 = 37$.

- *For 4 + 9 + 16 + 11, why might it be easier to start by adding 4 and 16 and then 9 and 11?* Both pairs of numbers add to 20, and adding $20 + 20 = 40$ is easy.

Tell children that they will play a game to practice adding three or more numbers.

▶ Introducing *Basketball Addition*

Math Masters, p. 192

| WHOLE CLASS | SMALL GROUP | PARTNER | INDEPENDENT |

Basketball Addition is played by two teams of 3 to 5 players each. The number of points scored by each player in each half is determined by rolling one 20-sided polyhedral die or by rolling three 6-sided dice and using their sum. The team that scores the most points wins the game.

Divide children into two groups and play a sample game as a class. Display the scoreboard on *Math Masters,* page 192 and use it to record information. Follow these directions for the first half of the game:

1. Choose teams made up of 3 to 5 children from each group. Players take turns rolling one 20-sided die or three 6-sided dice, with players from opposing teams alternating turns.

2. Select a volunteer from each group to record each player's roll on the scoreboard.

3. After all of the players have taken a turn, ask children to find their team's score for the first half by adding the scores of all their players. Allow them plenty of time. Remind those who finish early to check their work and try to solve the problem another way. **GMP1.5**

4. Ask children to share different ways to add their team members' points. Encourage them to look for ways to simplify the problem. For example, they might add all the 10s first or look for combinations of numbers that are easier to add. **GMP7.2** Sample strategies for adding 13, 7, 15, and 12 are shown in the margin.

5. Ask children to determine which team won the first half and by how many points. Allow children plenty of time to solve the problem.

Differentiate **Common Misconception**

Some children may think they have to add the numbers in the order they appear. Ask these children to recall the Math Message Follow-Up discussion and remind them that they can add any two numbers first.

Go Online Differentiation Support

At halftime, invite children to examine the scoreboard. Ask: *Is it possible for the team that is behind to win?* Yes. *Why or why not?* **GMP3.1** Sample answer: In the second half, the team that is behind could roll a lot of large numbers, and the team that is ahead could roll a lot of small numbers.

Academic Language Development Provide sentence frames for children to use in justifying their answers to the question above. *For example:* "It is possible for Team _____ to win because _____."

Play the second half of the game following the same directions. Choose different children to roll the dice and record the scores. After children have figured out the scores for the second half, ask them to find out which team won the game and by how many points.

Math Masters, p. 192

Basketball Addition

Lesson 7-3
NAME   DATE

	Points Scored			
	Team 1		Team 2	
	1st Half	2nd Half	1st Half	2nd Half
Player 1				
Player 2				
Player 3				
Player 4				
Player 5				
Team Score				

Point Totals	1st Half	2nd Half	Final
Team 1	_____	_____	_____
Team 2	_____	_____	_____

① Which team won the first half? _____
By how much? _____ points

② Which team won the second half? _____
By how much? _____ points

③ Which team won the game? _____
By how much? _____ points

192 one hundred ninety-two   2.NBT.5, 2.NBT.6

$$13 = 10 + 3$$
$$7 = 0 + 7$$
$$15 = 10 + 5$$
$$12 = \underline{10 + 2}$$
$$30 + 17 = 47$$

One child used expanded form to add the 10s and 1s separately.

$$13 + 7 + 15 + 12$$
$$20 \quad + 15 + 12$$
$$35 \quad + 12 = 47$$

Another child combined numbers that were easier to add.

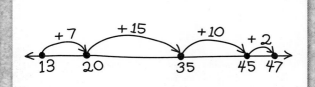

A third child drew an open number line.

Basketball Addition

Lesson 7-3

DATE

	Points Scored			
	Team 1		Team 2	
	1st Half	2nd Half	1st Half	2nd Half
Player 1				
Player 2				
Player 3				
Player 4				
Player 5				
Team Score				

Point Totals	1st Half	2nd Half	Final
Team 1	____	____	____
Team 2	____	____	____

❶ Which team won the first half? __Answers vary.__

By how much? _____ points

❷ Which team won the second half? _____

By how much? _____ points

❸ Which team won the game? _____

By how much? _____ points

2.NBT.5, 2.NBT.6, 2.NBT.9, SMP1, SMP7   one hundred seventy-one   171

Drawing a Picture Graph

Lesson 7-3

DATE

An *astronaut* is a person who is trained to fly into space. This table shows how many trips four American astronauts took into space. Use the information to draw a picture graph.

Astronaut Name	Curtis Brown	Bonnie Dunbar	Jerry Ross	James Wetherbee
Number of Trips into Space	6	5	7	6

Trips Astronauts Took into Space

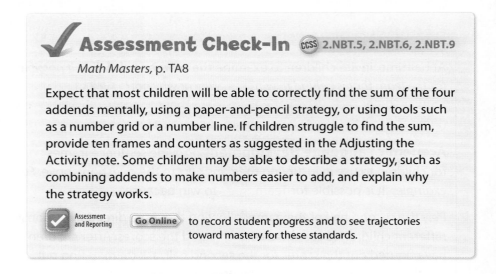

KEY: ◯ = 1 trip into space

172   one hundred seventy-two   2.MD.10

▶ Playing *Basketball Addition*

Math Journal 2, p. 171; *Math Masters*, p. TA8

| WHOLE CLASS | **SMALL GROUP** | PARTNER | INDEPENDENT |

Divide the class into groups so that each group has two teams of 3 to 5 players. Each group keeps score on the scoreboard on one player's journal page 171. Groups may play the game several times using a different player's journal page each time.

You may wish to distribute a copy of the directions for *Basketball Addition* on *Math Masters*, page G26 to each group for children to refer to as needed. If groups need additional scoreboards, provide copies of *Math Masters*, page G27.

Circulate and observe as children play the game. When they finish playing the first half, children should complete Problem 1 on the journal page. At the end of each game, they should complete Problems 2–3.

Differentiate   **Adjusting the Activity**

Provide counters and ten frames (*Math Masters*, page TA11) to help children identify the addends that combine to make multiples of 10.

Go Online ▶   Differentiation Support

Observe

- Which children successfully add the numbers to find the total score?
- Which children need support to understand and play the game?

Discuss

- *Which numbers did you choose to add first? Why?*   GMP7.2
- *Can you use another strategy to add the numbers?*   GMP1.5

After groups have played several games, distribute My Exit Slip (*Math Masters,* page TA8) to each child. Have children find the sum of $6 + 11 + 14 + 19$ and explain their strategy on My Exit Slip.   50

✓ **Assessment Check-In**   CCSS   2.NBT.5, 2.NBT.6, 2.NBT.9

Math Masters, p. TA8

Expect that most children will be able to correctly find the sum of the four addends mentally, using a paper-and-pencil strategy, or using tools such as a number grid or a number line. If children struggle to find the sum, provide ten frames and counters as suggested in the Adjusting the Activity note. Some children may be able to describe a strategy, such as combining addends to make numbers easier to add, and explain why the strategy works.

✓ Assessment and Reporting   **Go Online** ▶   to record student progress and to see trajectories toward mastery for these standards.

Summarize Children share the strategies they wrote on My Exit Slip, explaining their choices with the whole class or with partners.

3 Practice 10–15 min

Go Online ePresentations eToolkit Home Connections

▶ Drawing a Picture Graph

Math Journal 2, pp. 172–173

| WHOLE CLASS | SMALL GROUP | **PARTNER** | **INDEPENDENT** |

Children draw picture graphs to represent data and answer questions based on the graphs. Before children begin, point out the graph key on journal page 172. Tell children to use a circle to show one trip into space. Then have them complete the pages independently or in partnerships.

▶ Math Boxes 7-3

Math Journal 2, p. 174

| WHOLE CLASS | **SMALL GROUP** | PARTNER | INDEPENDENT |

Mixed Practice Math Boxes 7-3 are paired with Math Boxes 7-1.

▶ Home Link 7-3

Math Masters, p. 193

Homework Children practice adding three or more numbers.

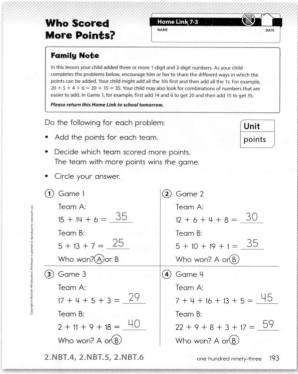

Math Masters, p. 193

Math Journal 2, p. 173

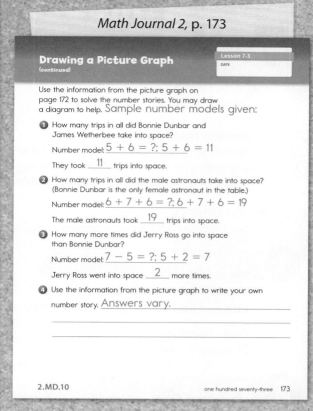

Math Journal 2, p. 174

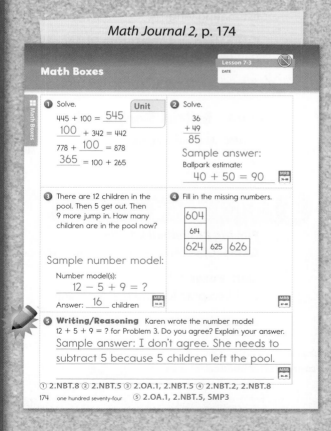

Measuring with Yards

Overview Children explore U.S. customary length units and measure to the nearest yard.

▶ **Before You Begin**
Identify your own personal measurement reference for 1 yard. For Part 2 each partnership will need a yardstick. For the optional Enrichment activity, mark a path from your classroom to another location on the same floor. Leave the path marked for use in the Enrichment activity in Lesson 7-5.

▶ **Vocabulary**
standard unit • yard (yd) • personal reference

**Common Core
State Standards**

Focus Cluster
Measure and estimate lengths in standard units.

	Materials	
① Warm Up 15–20 min		
Mental Math and Fluency
Children find differences between pairs of numbers.		2.NBT.5
Daily Routines
Children complete daily routines.	See pages 4–43.	See pages xiv–xvii.

② Focus 30–40 min		
Math Message
Children use a nonstandard unit to estimate length.	slate	SMP1
Measuring with a Nonstandard Unit
Children explore nonstandard and standard units.	Class Data Pad (optional)	2.MD.2
SMP1, SMP5, SMP6
Introducing the Yard
Children discuss the yard as a standard unit and compare a yardstick to a tape measure.	Class Data Pad; square pattern block, 12-inch ruler, yardstick (optional); per partnership: yardstick, tape measure	2.MD.1, 2.MD.4
SMP6
Finding Personal References for U.S. Customary Units
Children find personal references to use for estimating lengths.	*Math Journal 2*, p. 175; Class Data Pad; per partnership: yardstick, tape measure, 12-inch ruler	2.MD.1
SMP5
Estimating and Measuring Distances
Children estimate and measure distances to the nearest yard.	*Math Journal 2*, p. 176; Class Data Pad; per partnership: yardstick	2.MD.1, 2.MD.3
SMP5
✓ **Assessment Check-In** See page 641.	*Math Journal 2*, p. 176	2.MD.1, 2.MD.3, SMP5

CCSS 2.MD.3 Spiral Snapshot

GMC Estimate lengths.

| 4-8
Focus
Practice | 5-1 through 5-8
Practice | 7-4
Focus | 7-5
Focus | 7-6
Focus | 7-9
Focus |

Spiral Tracker **Go Online** to see how mastery develops for all standards within the grade.

③ Practice 10–15 min		
Playing Basketball Addition
Game Children practice adding three or more numbers.	*Math Masters*, pp. G26–G27; per group: three 6-sided dice	2.NBT.5, 2.NBT.6, 2.NBT.9
SMP1, SMP7
Math Boxes 7-4
Children practice and maintain skills.	*Math Journal 2*, p. 177; 12-inch ruler; 10-centimeter ruler	See page 641.
Home Link 7 Homework Children explore measurement.	*Math Masters*, p. 194	2.NBT.5, SMP5

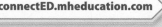

Plan your lessons online with these tools.

 ePresentations   Student Learning Center   Facts Workshop Game   eToolkit   Professional Development   Home Connections   Spiral Tracker   Assessment and Reporting   English Learners Support   Differentiation Support

Differentiation Options

RtI

CCSS 2.MD.1, 2.MD.2, SMP5

Readiness   5–15 min

| WHOLE CLASS |
| SMALL GROUP |
| **PARTNER** |
| INDEPENDENT |

Comparing Standard Units

12-inch ruler, yardstick

To explore relationships among different length units, children measure an object's length using three different units and then compare the measures. Choose an object longer than 1 yard, such as a table. Have children measure the length of the table to the nearest yard, foot, and inch.

Ask: *What do you notice when you use different units to measure?* GMP5.2 Sample answer: A yard is longer than a foot and an inch, so the number of yards in the measure is smaller than the number of feet or inches. *Why might you choose to use one unit instead of another?* Sample answers: Using yards is faster because I have to move the yardstick only twice, but I have to move the foot ruler many times. Using feet is better because the table is only a little more than 7 feet long, so a measurement of 7 feet is close, but a measurement of 2 yards is not as close.

CCSS 2.MD.1, 2.MD.3, SMP5

Enrichment   5–15 min

| WHOLE CLASS |
| SMALL GROUP |
| **PARTNER** |
| INDEPENDENT |

Using Yards to Measure a Path

yardstick or tape measure, paper, path marked with masking tape (see Before You Begin)

To apply their understanding of measuring distances, partners use a yardstick or a tape measure to measure the length of a crooked path to the nearest yard. GMP5.2 Before measuring, have children walk along the path and use a personal reference for 1 yard—such as the distance from their waist to the floor or the width of the classroom door—to help them estimate the length of the path to the nearest yard.

To help children keep track, suggest that they record an estimate for each section of the path and then add the estimates together to get an estimate for the total length. They should keep track of their actual measurements in the same way to find the total length of the path. As children report their estimates and measurements, remind them to use phrases, such as the following:

- about _____ yards
- between _____ and _____ yards
- a little more than _____ yards
- a little less than _____ yards

CCSS 2.MD.1, 2.MD.3, SMP5

Extra Practice   5–15 min

| WHOLE CLASS |
| SMALL GROUP |
| **PARTNER** |
| INDEPENDENT |

Estimating and Measuring with Yards

Activity Card 90, paper, yardstick

For practice measuring, children use personal references and yardsticks to estimate and then measure distances to the nearest yard. Children record their estimates and measurements. GMP5.2

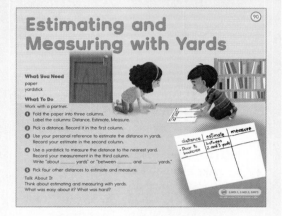

English Language Learners Support

Beginning ELL Display labeled pictures of various kinds of *yards,* such as a backyard, a front yard, and a schoolyard. In simple sentences, describe what might go on in such settings. For example, say: *At recess we play in the schoolyard. When it's warm, some people have barbecues in their backyard.* Show children both a toolkit tape measure marked at 1 yard and a yardstick. As you show how to use them, say: *This is a yard for measuring.* Show children that the word *yard* is always spelled the same way. As you call out the names of the various kinds of yards, ask children to point to the corresponding picture or tool.

 Go Online   ELL   English Learners Support

NOTE Some children are hesitant to make estimates and will avoid doing so by measuring first and then recording a number close to the actual measure as their "estimate." Others will estimate before measuring but then erase their first estimate and record a new "estimate" that is closer to the actual measure. Whenever there are estimation opportunities, use think-alouds to model the process for making estimates. For example, say: *I know that if I lie along the wall, I would stretch from the corner to that desk. I think 5 more of me could fit along the whole wall, so my estimate is that about 6 of me will fit.*

1 Warm Up   15–20 min   [Go Online]   ePresentations   eToolkit

▶ Mental Math and Fluency

Have children find the difference between pairs of numbers.
Leveled exercises:

● ○ ○ 15 and 25 10   ● ● ○ 31 and 19 12   ● ● ● 89 and 31 58
         35 and 20 15   57 and 92 35   74 and 47 27

▶ Daily Routines

Have children complete daily routines.

2 Focus   30–40 min   [Go Online]   ePresentations   eToolkit

▶ Math Message

About how many children in our class can lie head-to-foot along the longest wall of our classroom? Write your estimate on your slate.

▶ Measuring with a Nonstandard Unit

| **WHOLE CLASS** | SMALL GROUP | PARTNER | INDEPENDENT |

Math Message Follow-Up Expect children's estimates to vary widely. You might record some of their estimates on the Class Data Pad to show the range of children's estimates and reinforce the idea that any reasonable estimate is acceptable. Ask: *How did you make your estimate?* GMP1.1 Sample answer: I thought about how far I would stretch along the wall and how many times I would fit.

Ask volunteers to measure the length of the classroom by lying head-to-foot along the longest wall. The first child lies on the floor with his or her feet at one end of the room. The second child lies down head-to-head with the first child, the third child lies down feet-to-feet with the second child, and so on. The length of the classroom can be given as "_____ second graders long."

Ask: *Are all second graders the same height?* No. *Why is it important that we measure with units that are all the same length?* Sample answer: If we measure accurately with units that are all the same length, we all should get the same measurement. *What could we do to make sure our units are all the same length?* GMP5.2 Sample answers: We could use second graders who are all the same height. We could use just one second grader.

Select a volunteer. Measure the length of the classroom using this volunteer as the unit. First have the child lie on the floor with his or her feet at one end of the room. Mark the place at the top of the child's head before he or she gets up and moves. (*See margin on next page.*) Mark the second length and so on. The class keeps count as the child moves along the wall. Report the length of the classroom as, for example, "5 Shanes."

Ask: *How could we tell someone in another state how long our classroom is?* Discuss some of the difficulties of using people as units of measure:

- Unless we use the same person each time, we may get a different measurement each time we measure. **GMP6.4**
- Even if we try to use the same person each time, we can't take that person with us whenever we want to measure something. And as the person grows, the length of the unit will change. **GMP5.2**

Ask: *How can we make sure that we get the same measurement no matter who measures an object?* Sample answers: We can use the same unit and choose one that everyone knows about and that does not change. We can line up the units without gaps or overlaps. Review the meaning of **standard unit** as a unit of measure that is defined by a government or a standards organization. Standard units are used in many places and are familiar to many people.

Ask children to recall names of standard units they know. Sample answer: Inch Explain that they will investigate another standard unit called a yard.

The length of the classroom is measured using one second grader as the unit.

► Introducing the Yard

| WHOLE CLASS | SMALL GROUP | PARTNER | INDEPENDENT |

Children used inches, feet, and centimeters to measure short lengths in earlier lessons. For longer lengths, such as the length of a classroom, other standard units can be used. **GMP6.4** Display a yardstick and explain that a **yard (yd)** is a unit in the U.S. customary system that is 36 inches long. On the Class Data Pad write, "A yard is 36 inches long." Discuss the everyday and mathematical meanings of *yard*. People in the United States commonly use the yard to measure lengths longer than a few feet.

To illustrate the relative sizes of inch, foot, and yard, you may want to create a U.S. Customary Units poster. Tape a 1-inch-square pattern block, a 12-inch ruler, and a yardstick to poster paper and label their lengths.

Distribute a yardstick to each partnership. Have children compare the lengths of the yardsticks to their tape measures. Ask: *Which is longer?* Tape measure *How much longer?* 24 inches Some children might notice the star at the 36-inch mark on their tape measures. Explain that the distance from the beginning of the tape measure to the star is 1 yard, or 36 inches.

► Finding Personal References for U.S. Customary Units

Math Journal 2, p. 175

| WHOLE CLASS | SMALL GROUP | PARTNER | INDEPENDENT |

Explain that estimation is an important measurement skill because sometimes we need to measure when we don't have tools with us. Children might estimate the height of the slide on the playground. Adults might estimate the length of a rug to determine if it will fit in a room. The more we practice estimating, the more accurate our estimates get.

Math Journal 2, p. 175

Personal References Hunt   Lesson 7-4
DATE

1. Find things that are about 1 inch long, about 1 foot long, and about 1 yard long. Use a ruler, a tape measure, or a yardstick to help you. List them below. You can use these as your personal references to help you estimate lengths.

My Personal References		
About 1 in.	About 1 ft	About 1 yd
Answers vary.		

2. Find things that are about 1 centimeter long, about 10 centimeters long, and about 1 meter long. Use a ruler, a tape measure, or a meterstick to help you. List them below. You can use these as

My Personal References		
About 1 cm	About 10 cm	About 1 m
Answers vary.		

2.MD.1, SMP5   one hundred seventy-five 175

Math Journal 2, p. 176

Yards

Lesson 7-4

DATE

For Problems 1–3, do the following:

- Choose a distance.
- Estimate the distance in yards.
- Measure the distance to the nearest yard.
- Compare your measurement with your estimate.
  Talk to your partner about why they might be different.

Distance	My Estimate	My Yardstick Measurement
Example: *From the door to the window*	About __5__ yards or Between ____ and ____ yards	About ____ yards or Between __8__ and __9__ yards
❶ _____	About ____ yards or Between ____ and ____ yards	About ____ yards or Between ____ and ____ yards
❷ _____	About ____ yards or Between ____ and ____ yards	About ____ yards or Between ____ and ____ yards
❸ _____	About ____ yards or Between ____ and ____ yards	About ____ yards or Between ____ and ____ yards

176 one hundred seventy-six 2.MD.1, 2.MD.3, SMP5

One way to estimate an object's length is to compare it to a **personal reference,** which may be a body part or a familiar object. For example, a child's personal reference for an inch might be the length from the first joint to the tip of the thumb. To estimate an object's length in inches, children might imagine how many times this part of their thumb will fit along the object. Personal references for a foot and a yard might be a sheet of notebook paper and the width of a classroom door, respectively.

Children find personal references for U.S. customary units of length and record their work for Problem 1 on journal page 175. GMP5.1 In Lesson 7-5 they will complete Problem 2 by finding personal references for metric units of length.

After children have completed Problem 1, bring the class together. On the Class Data Pad, list some of the personal references children found.

Differentiate **Adjusting the Activity**

Have children measure and cut strips of paper or string that are 1 inch long, 1 foot long, and 1 yard long. Then can use these to search for personal references by comparing the lengths of these strips to parts of their bodies or objects in the classroom.

Go Online Differentiation Support

▶ Estimating and Measuring Distances

Math Journal 2, p. 176

| WHOLE CLASS | SMALL GROUP | PARTNER | INDEPENDENT |

Standing near the shortest wall of the classroom, show children your own personal reference for 1 yard. Ask children to imagine about how many times your personal reference will fit along the length of the wall. Record children's estimates on the Class Data Pad.

Next use a yardstick to model measuring the length of the same wall to the nearest yard. Emphasize how important it is to note where the end of the yardstick is before moving it to avoid gaps and overlaps as you measure. Either you or a child can show the class how to mark where the endpoint of the yardstick lands. Because the measurement is to the nearest yard, use language such as "about ____ yards" or "between ____ and ____ yards." Compare children's estimates to the measurement and briefly discuss why estimates and measurements might differ. Be sure to include the following ideas:

- A personal reference for a yard is probably not exactly 1 yard long.
- Because we want our estimates to be quick, we don't carefully lay out personal references. We try to imagine how many will fit. This can lead to some differences in estimates.

Have children measure several objects or distances to make sure they are measuring without gaps or overlaps. GMP5.2 Then have partners complete journal page 176. Children select distances to measure in or near the classroom, estimate the distances in yards using their personal references, and measure them to the nearest yard using a yardstick.

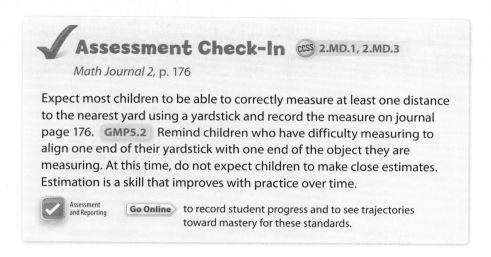

✔ Assessment Check-In CCSS 2.MD.1, 2.MD.3

Math Journal 2, p. 176

Expect most children to be able to correctly measure at least one distance to the nearest yard using a yardstick and record the measure on journal page 176. **GMP5.2** Remind children who have difficulty measuring to align one end of their yardstick with one end of the object they are measuring. At this time, do not expect children to make close estimates. Estimation is a skill that improves with practice over time.

✔ Assessment and Reporting **Go Online** to record student progress and to see trajectories toward mastery for these standards.

Summarize Children share their estimation strategies with the group. Point out that using their personal references to help estimate and then using a measuring tool to find the actual measurement will improve their estimation skills.

---

3 Practice 10–15 min

Go Online

ePresentations eToolkit Home Connections

▶ Playing *Basketball Addition*

Math Masters, pp. G26–G27

WHOLE CLASS | **SMALL GROUP** | PARTNER | INDEPENDENT

Children play *Basketball Addition* to practice adding three or more 1- and 2-digit numbers. See Lesson 7-3 for detailed directions.

Observe
- Which children can successfully add the numbers to find the total score?
- Which children need additional support to play the game?

Discuss
- *Which numbers did you choose to add first? Why?* **GMP7.2**
- *Can you use another strategy to add the numbers?* **GMP1.5**

▶ Math Boxes 7-4

Math Journal 2, p. 177

WHOLE CLASS | SMALL GROUP | **PARTNER** | **INDEPENDENT**

Mixed Practice Math Boxes 7-4 are paired with Math Boxes 7-2.

▶ Home Link 7-4

Math Masters, p. 194

Homework Children discuss how measurements are used at home, at work, or during other activities. They list tools that are used to measure and find pictures of measurements. **GMP5.1**

Math Journal 2, p. 177

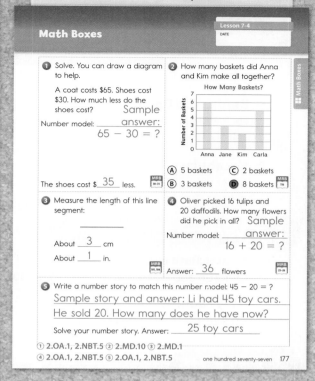

Math Masters, p. 194

Measuring with Meters

Overview Children find personal references for metric units of measure; they choose appropriate units and tools to estimate and measure lengths.

▶ **Before You Begin**
For Part 2 each pair of children will need a yardstick and a meterstick. For the optional Enrichment activity, mark a path from your classroom to another location on the same floor.

▶ **Vocabulary**
meter (m)

Common Core State Standards

Focus Cluster
Measure and estimate lengths in standard units.

1 Warm Up 15–20 min

	Materials	
Mental Math and Fluency Children add up to four 1- and 2-digit numbers.	slate	2.NBT.6
Daily Routines Children complete daily routines.	See pages 4–43.	See pages xiv–xvii.

2 Focus 30–40 min

Math Message Children list units and tools used to measure length.		2.MD.1
Introducing the Meter Children compare a meterstick to a tape measure and a yardstick.	Class Data Pad; centimeter cube (optional); per partnership: meterstick, yardstick, tape measure	2.MD.1, 2.MD.4
Finding Personal References for Metric Units Children find personal references for metric units of length.	Math Journal 2, p. 175; 10-centimeter ruler; meterstick	2.MD.1 SMP5
Estimating and Measuring Lengths Children use personal references to estimate lengths in metric units and use tools to measure the lengths.	Math Journal 2, p. 178; 10-centimeter ruler; tape measure or meterstick	2.MD.1, 2.MD.3 SMP5, SMP6
✓ **Assessment Check-In** See page 647.	Math Journal 2, p. 178	2.MD.1, 2.MD.3, SMP5, SMP6

CCSS 2.MD.1 **Spiral Snapshot**

GMC Measure the length of an object.

4-8 through 4-11 Focus Practice | 5-8 Practice | 6-10 Focus | 7-4 Focus Practice | **7-5 Focus Practice** | 7-6 Focus Practice | 7-9 Focus

/// Spiral Tracker   Go Online   to see how mastery develops for all standards within the grade.

3 Practice 10–15 min

Practicing with Fact Triangles Children practice addition and subtraction facts.	Math Journal 2, pp. 250–253; Fact Triangles	2.OA.2
Math Boxes 7-5 Children practice and maintain skills.	Math Journal 2, p. 179; 10-centimeter ruler	See page 647.
Home Link 7-5 **Homework** Children measure heights in metric units.	Math Masters, p. 195	2.NBT.5, 2.MD.1, 2.MD.2 SMP5

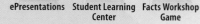
connectED.mheducation.com

Plan your lessons online with these tools.

 ePresentations Student Learning Center Facts Workshop Game eToolkit Professional Development Home Connections Spiral Tracker Assessment and Reporting English Learners Support Differentiation Support

Differentiation Options RtI

Readiness 5–15 min

| WHOLE CLASS |
| SMALL GROUP |
| **PARTNER** |
| INDEPENDENT |

Selecting Tools

yardstick, Pattern-Block Template, and tape measure (for demonstration)

For experience selecting tools for measurement, children discuss reasons for choosing a particular tool. Show the three tools and ask: *Would you use the tape measure, the yardstick, or the ruler on your Pattern-Block Template to measure the length of the door?* Sample answer: Tape measure *Why?* Sample answer: It is easier to measure longer distances with a longer tool. *Would you use the ruler or your Pattern-Block Template or a tape measure to measure around a trash can?* Tape measure *Why?* Sample answer: A tape measure can go around the can easier because it is flexible. **GMP5.1** Continue posing similar questions as needed.

Enrichment 5–15 min

| WHOLE CLASS |
| SMALL GROUP |
| **PARTNER** |
| INDEPENDENT |

Using Meters to Measure a Path

meterstick or tape measure path marked with masking tape (see Before You Begin)

To apply their understanding of measuring distances, partners use a meterstick or a tape measure to measure the length of a crooked path to the nearest meter. **GMP5.2** Before measuring, have children walk along the path and use a personal reference to help them estimate the length to the nearest meter.

To help children keep track, suggest that they record an estimate for each section of the path and then add the estimates together to estimate the total. They should keep track of their actual measurements in the same way to find the total length of the path. As children report their estimates and measurements, remind them to use phrases, such as the following:

- about _____ meters
- between _____ and _____ meters
- a little more than _____ meters
- a little less than _____ meters

Extra Practice 5–15 min

| WHOLE CLASS |
| SMALL GROUP |
| **PARTNER** |
| INDEPENDENT |

Estimating and Measuring with Meters

Activity Card 91, paper, meterstick

For practice measuring, children use personal references and metersticks to estimate and measure distances to the nearest meter. Children record their estimates and measurements. **GMP5.2**

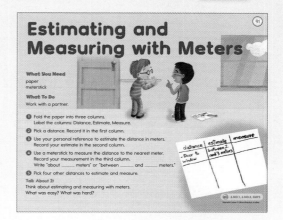

English Language Learners Support

Beginning ELL Display a labeled yardstick, meterstick, and tape measure. Use Total Physical Response (TPR) prompts to model naming the different tools. For example, say: *This is a yardstick.* Direct children to repeat the statement. Then ask: *What is this called?* Direct children to respond by saying, "This is a yardstick." Repeat several times. After all three items have been introduced this way, use TPR prompts such as the following: *Point to the meterstick. Put the yardstick on your desk. Bring me the tape measure.* Ask children to respond with the name of each tool as you point to it and ask: *What is this called?*

Go Online ELL English Learners Support

CCSS

Standards and Goals for Mathematical Practice

SMP5 **Use appropriate tools strategically.**

    GMP5.1 Choose appropriate tools.

    GMP5.2 Use tools effectively and make sense of your results.

SMP6 **Attend to precision.**

    GMP6.3 Use clear labels, units, and mathematical language.

① Warm Up   15–20 min   Go Online

ePresentations   eToolkit

▶ Mental Math and Fluency

Display problems one at a time. Children solve them and display the answers on their slates. Encourage children to use mental strategies. *Leveled exercises:*

●○○ $7 + 3 + 9 = ?$ 19

        $? = 8 + 6 + 14$ 28

●●○ $3 + 17 + 15 + 5 = ?$ 40

        $? = 12 + 26 + 14 + 8$ 60

●●● $? = 11 + 17 + 19 + 33$ 80

        $? = 45 + 12 + 15 + 28$ 100

▶ Daily Routines

Have children complete daily routines.

② Focus   30–40 min   Go Online

ePresentations   eToolkit

▶ Math Message

On paper write the names of units we use to measure length. Then write the names of tools we use to measure length.

▶ Introducing the Meter

| WHOLE CLASS | SMALL GROUP | PARTNER | INDEPENDENT |

Math Message Follow-Up Most children will probably name inches, feet, yards, and/or centimeters because those units have been used previously in second-grade activities. Tools mentioned will likely include an inch ruler, a centimeter ruler, a foot ruler, a tape measure, and a yardstick. Some children may mention personal references. Point out that personal references are helpful when estimating lengths, but they are not measuring tools.

Remind children that inches, feet, and yards are part of the U.S. customary system and centimeters are part of the metric system. In the metric system, the **meter** is another commonly used standard unit of length. Tell children that the abbreviation for meter is **m.** Show the class a meterstick. On the Class Data Pad, write "A meter is 100 centimeters long."

To draw attention to the relative sizes of the meter and the centimeter, you may want to create a Metric Units poster. Tape a meterstick and a centimeter cube to poster paper and label the length of each.

    NOTE In the metric system, all linear units are defined in terms of a meter.

Remind children that in the United States, the U.S. customary system is used for everyday purposes, whereas the metric system is used mostly for scientific purposes. Most labels on canned and packaged foods show both metric and U.S. customary units of measure. Most other countries in the world use only the metric system.

Distribute a meterstick and a yardstick to each partnership. Have children compare their metersticks to their toolkit tape measures and their yardsticks. After several minutes, bring the class back together to share their findings. Expect children's comments to include the following:

- The tape measure is the longest of the three.
- The yardstick is the shortest of the three.
- The meterstick is a little longer than the yardstick.
- The tape measure is easy to use when measuring around things or when measuring longer things.
- All three tools are easy to use when measuring things that are straight.

Ask: *About how much longer in centimeters is the meterstick than the yardstick?* Between 8 and 9 centimeters longer *About how much longer in centimeters is the tape measure than the meterstick?* About 50 centimeters longer

NOTE It is not important that children know the exact length difference between 1 yard and 1 meter. Knowing that 1 meter is a little longer than 1 yard is sufficient.

Differentiate　**Adjusting the Activity**

Display a meterstick and a pulled-out tape measure with the centimeter side facing up. Align the 0 marks to help children visualize the difference between their lengths. Ask: *If the endpoint of the meterstick is at 100 centimeters and the endpoint of the tape measure is at 150 centimeters, then how many centimeters longer is the tape measure?* 50 centimeters

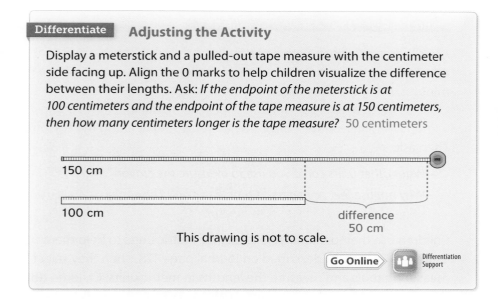

150 cm

100 cm

difference
50 cm

This drawing is not to scale.

Go Online ▸　Differentiation Support

▸ Finding Personal References for Metric Units

Math Journal 2, p. 175

| WHOLE CLASS | SMALL GROUP | PARTNER | INDEPENDENT |

Remind children that personal references are useful when estimating lengths and that in Lesson 7-4 they found personal references for U.S.

Math Journal 2, p. 175

Personal References Hunt　Lesson 7-4　DATE

① Find things that are about 1 inch long, about 1 foot long, and about 1 yard long. Use a ruler, a tape measure, or a yardstick to help you. List them below. You can use these as your personal references to help you estimate lengths.

My Personal References		
About 1 in.	About 1 ft	About 1 yd
Answers vary.		

② Find things that are about 1 centimeter long, about 10 centimeters long, and about 1 meter long. Use a ruler, a tape measure, or a meterstick to help you. List them below. You can use these as

My Personal References		
About 1 cm	About 10 cm	About 1 m
Answers vary.		

2.MD.1, SMP5　one hundred seventy-five　175

Math Journal 2, p. 178

Estimating and Measuring

Lesson 7-5
DATE

For each length:

- Choose one metric unit: centimeters (cm) or meters (m).
- Use your personal references to estimate the length. Record your estimate. Be sure to write the unit.
- Choose a measuring tool and measure the length. Record your measurement. Be sure to write the unit.
- Choose one U.S. customary unit: inches (in.), feet (ft), or yards (yd). Repeat the steps again using that unit.

Lengths	Metric Units		U.S. Customary Units	
	Estimate	Measurement	Estimate	Measurement
① Height of desk				
② Side of calculator		Answers vary.		
③ Width of door				
④ Side of classroom				

Choose your own lengths to estimate and measure.

Lengths	Metric Units		U.S. Customary Units	
	Estimate	Measurement	Estimate	Measurement
⑤				
⑥		Answers vary.		

178 one hundred seventy-eight   2.MD.1, 2.MD.3, SMP5, SMP6

customary units of length. Today children will work with a partner to find personal references for 1 centimeter, 10 centimeters, and 1 meter. For example, the width of a second grader's little finger might be about 1 centimeter. Because 1 yard and 1 meter are close in length, expect some children to select the same personal references for both lengths.

> **NOTE** A decimeter is a unit of length equivalent to 10 centimeters, or one-tenth of a meter. Although children will identify personal references for 10 centimeters, do not expect them to use the term *decimeter*.

Children find things that are personal references for metric units and record their work for Problem 2 on journal page 175. GMP5.1

> **Differentiate**   **Adjusting the Activity**
>
> Have children measure and cut strips of paper or string that are 1 centimeter long, 10 centimeters long, and 1 meter long. Then have them compare the lengths of these strips to objects in the classroom as they search for personal references.
>
> **Go Online** Differentiation Support

After children have completed Problem 2, bring the class together. On the Class Data Pad, list some of the personal references that children found.

▶ Estimating and Measuring Lengths

Math Journal 2, p. 178

WHOLE CLASS   **SMALL GROUP**   **PARTNER**   INDEPENDENT

Ask children to discuss with a partner how they might use a personal reference for 1 centimeter to estimate the length of a crayon. After a few minutes, bring the class together to share ideas. Sample answer: I know that the width of my little finger is about 1 centimeter. I have to think about how many widths of my little finger would fit along the crayon. Then ask:

- *What other units could you use to measure the crayon?* Inches
- *Why not use feet or meters?* Sample answer: The crayon is not very long, so we don't need long units to measure it.

Children use their personal references for metric units to help them estimate the lengths described on journal page 178. Then they select measuring tools and measure the lengths in metric units. Children do the same for U.S. customary units. GMP5.1, GMP5.2 Remind children of the importance of labeling their measures with units. GMP6.3 They record their work on the journal page.

Assessment Check-In CCSS **2.MD.1, 2.MD.3**

Math Journal 2, p. 178

Expect most children to be able to select appropriate measuring tools and use them to correctly measure at least two of the lengths on journal page 178. **GMP5.1, GMP5.2** They should also record and label their measurements with correct units. **GMP6.3** If children have difficulty measuring, remind them to align the 0 end of their tool with one end of the object they are measuring. At this time, do not expect children to make close estimates. Estimation is a skill that improves with practice over time. Later lessons will provide further opportunities for children to estimate and then measure lengths and distances.

☑ Assessment and Reporting   **Go Online** to record student progress and to see trajectories toward mastery for these standards.

Summarize Invite children to discuss the units and the tools they used to estimate and measure the lengths on journal page 178.

3 Practice 10–15 min **Go Online** ePresentations | eToolkit | Home Connections

▸ Practicing with Fact Triangles

Math Journal 2, pp. 250–253

| WHOLE CLASS | SMALL GROUP | PARTNER | INDEPENDENT |

As children practice with Fact Triangles, have them list the addition facts they know and the facts that need more practice on the Addition Facts Inventory Record, Parts 1 and 2 (journal pages 250–253). See Lesson 3-3 for additional details.

▸ Math Boxes 7-5

Math Journal 2, p. 179

| WHOLE CLASS | SMALL GROUP | PARTNER | INDEPENDENT |

Mixed Practice Math Boxes 7-5 are paired with Math Boxes 7-8.

▸ Home Link 7-5

Math Masters, p. 195

Homework Children use tools to measure two heights: the height of a table and the height of an adult. **GMP5.2**

Math Journal 2, p. 179

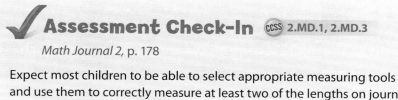

Math Boxes Lesson 7-5 DATE

① Solve. **Unit stickers**
$45 + \underline{5} = 50$
$70 = 61 + \underline{9}$
$87 + \underline{3} = 90$
$\underline{7} + 33 = 40$

② In my building there are 15 dogs, 13 cats, and 12 birds. How many pets are there in all?

Answer: $\underline{40}$ pets

③ Measure this line segment in centimeters.

About $\underline{9}$ cm

Draw a line segment that is 3 centimeters shorter than the one above.

What is the length of the line you drew?

About $\underline{6}$ cm

④ Solve. You can draw a diagram to help.

In the morning it was 60°F. At noon it was 75°F. How much did the temperature change?

Sample answers:
Number model:
$60 + ? = 75; 75 - 60 = ?$

Answer: $\underline{15}$ °F

⑤ **Writing/Reasoning** How did you add the numbers in Problem 2? Explain your strategy. Sample answer:
I added the 10s and got 30. Then I added the 1s and got 10. Then I added $30 + 10 = 40$.

① 2.NBT.5 ② 2.OA.1, 2.NBT.5, 6 ③ 2.MD.1, 4 ④ 2.OA.1, 2. NBT.5 ⑤ 2.NBT.5, 2.NBT.6, 2.NBT.9, SMP6   one hundred seventy-nine   179

Math Masters, p. 195

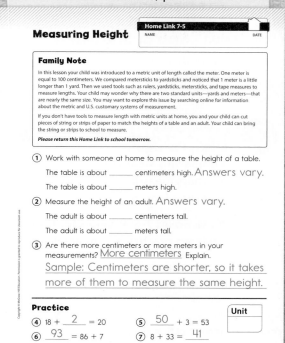

Measuring Height Home Link 7-5 NAME DATE

Family Note

In this lesson your child was introduced to a metric unit of length called the meter. One meter is equal to 100 centimeters. We compared metersticks to yardsticks and noticed that 1 meter is a little longer than 1 yard. Then we used tools such as rulers, yardsticks, metersticks, and tape measures to measure lengths. Your child may wonder why there are two standard units—yards and meters—that are nearly the same size. You may want to explore this issue by searching online for information about the metric and U.S. customary systems of measurement.

If you don't have tools to measure length with metric units at home, you and your child can cut pieces of string or strips of paper to match the heights of a table and an adult. Your child can bring the string or strips to school to measure.

Please return this Home Link to school tomorrow.

① Work with someone at home to measure the height of a table.

The table is about _____ centimeters high. Answers vary.

The table is about _____ meters high.

② Measure the height of an adult. Answers vary.

The adult is about _____ centimeters tall.

The adult is about _____ meters tall.

③ Are there more centimeters or more meters in your measurements? More centimeters Explain.
Sample: Centimeters are shorter, so it takes more of them to measure the same height.

Practice

④ $18 + \underline{2} = 20$ ⑤ $\underline{50} + 3 = 53$ **Unit**

⑥ $\underline{93} = 86 + 7$ ⑦ $8 + 33 = \underline{41}$

2.NBT.5, 2.MD.1, 2.MD.2, SMP5   one hundred ninety-five   195

Generating Data: Standing Jumps and Arm Spans

Overview Children measure lengths to the nearest centimeter and to the nearest inch.

▶ **Before You Begin**
For the Math Message, display the figure shown in the margin on page 650 and prepare one copy of *Math Masters,* page TA8 for every two children. For the standing jump activity in Part 2, make masking-tape lines on the floor in various spots in the classroom for children to use as starting lines.

▶ **Vocabulary**
arm span

Common Core State Standards

Focus Clusters
• Measure and estimate lengths in standard units.
• Represent and interpret data.

1 Warm Up   15–20 min

Materials

Mental Math and Fluency Children mentally add and subtract 100.	slate	2.NBT.8
Daily Routines Children complete daily routines.	See pages 4–43.	See pages xiv–xvii.

2 Focus   30–40 min

Math Message Children make predictions about measurements.	*Math Masters,* p. TA8	2.MD.2
Measuring Arm Spans Children learn how to measure arm spans.	*Math Masters,* p. TA8; tape measure (for demonstration)	2.MD.1, 2.MD.2, 2.MD.3 SMP5, SMP6
Collecting and Recording Arm Span Data Children measure arm spans in centimeters and in inches.	*Math Journal 2,* p. 180; per group: tape measure	2.MD.1, 2.MD.2, 2.MD.9 SMP5, SMP6
Collecting and Recording Standing Jump Data Children measure standing jumps in inches and centimeters.	*Math Journal 2,* p. 181; tape measure; chalk, penny, or other marker	2.MD.1, 2.MD.2, 2.MD.9 SMP5, SMP6
✓ **Assessment Check-In** See page 652.	*Math Journal 2,* pp. 180–181	2.MD.1, SMP5

CCSS 2.MD.1 **Spiral Snapshot**

GMC Measure the length of
an object.

4-8 through 4-11 Focus Practice	5-8 Practice	6-10 Focus	7-4 Focus Practice	7-5 Focus Practice	7-6 Focus Practice	7-9 Focus

/// **Spiral Tracker**   **Go Online**   to see how mastery develops for all standards within the grade.

3 Practice   10–15 min

Comparing Measurements Partners measure and compare height, head size, and shoe length.	*Math Journal 2,* p. 182; per partnership: meterstick, tape measure, 10-centimeter ruler	2.NBT.4, 2.NBT.5, 2.MD.1, 2.MD.4 SMP5
Math Boxes 7-6: Preview for Unit 8 Children practice and maintain skills.	*Math Journal 2,* p. 183	See page 653.
Home Link 7-6 **Homework** Children compare arm spans.	*Math Masters,* p. 198	2.NBT.5, 2.MD.1

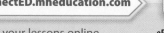
connectED.mheducation.com

Plan your lessons online
with these tools.

      ///   ELL

ePresentations   Student Learning   Facts Workshop   eToolkit   Professional   Home   Spiral Tracker   Assessment   English Learners   Differentiation
Center   Game   Development   Connections   and Reporting   Support   Support

Differentiation Options

RtI

Readiness
5–15 min

WHOLE CLASS
SMALL GROUP
PARTNER
INDEPENDENT

Comparing Units

1-inch-square pattern blocks, centimeter cubes

To explore how measurements relate to unit size, children compare the length of a line of 1-inch-square pattern blocks to a line of centimeter cubes. Have them place ten 1-inch-square pattern blocks in a straight line. Directly below, have children place ten centimeter cubes in a line, making sure that the first centimeter cube is aligned with the first 1-inch-square pattern block. Prompt children to verbalize their comparison of the two lines. *For example:* "Even though I used the same number of blocks and cubes, the line of centimeter cubes is shorter than the line of pattern blocks."

Enrichment
5–15 min

WHOLE CLASS
SMALL GROUP
PARTNER
INDEPENDENT

Making Up and Solving Number Stories

Math Masters, pp. 196 and TA7

To further explore measurement, children use the data from *Math Masters,* page 196 to make up stories for others to solve. Encourage children to make up number stories that are 2-step or comparison problems. They record their work on *Math Masters,* page TA7.

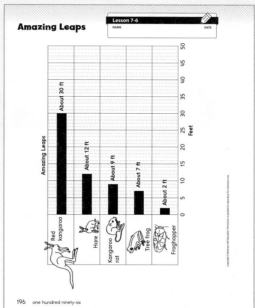

Extra Practice
5–15 min

WHOLE CLASS
SMALL GROUP
PARTNER
INDEPENDENT

Comparing Arm Span to Height

Math Masters, p. 197; per group: tape measure

For additional practice with measurement, children compare their arm span measurements to their height measurements. Say or display the following statement: *A person's height is about the same length as a person's arm span.* Ask: *Is this true?* Have children measure their height and arm span to the nearest inch to find out. They then explain why they think the statement is true or false. **GMP3.1** Encourage children to compare their findings with their classmates. **GMP3.2**

English Language Learners Support

Beginning ELL Model touching your arm as you say *arm*. Then ask children to touch their arms. With a broad gesture, show your arms outstretched and say: *This is my arm span.* Have a child stand in front of you with his or her arms outstretched. Touch one fingertip and then draw a line in the air to the other fingertip, saying: *This is your arm span.* Ask partners to demonstrate *arm span* with each other.

Go Online   ELL English Learners Support

Standards and Goals for
Mathematical Practice

SMP5 Use appropriate tools strategically.
> GMP5.2 Use tools effectively and make sense of your results.

SMP6 Attend to precision.
> GMP6.2 Use an appropriate level of precision for your problem.

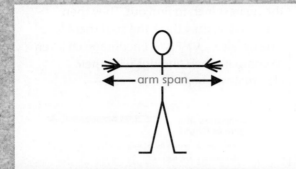

NOTE Remind children of the "2-inch, no zap" rule from Unit 4: you may "zap" the tape measure only when no more than 2 inches are showing. Following this rule will extend the life of the tape measures.

1 Warm Up 15–20 min [Go Online] ePresentations eToolkit

▶ Mental Math and Fluency

Dictate 2- and 3-digit numbers and have children mentally add or subtract 100, recording their answers on slates. *Leveled exercises:*

● ○ ○ Add 100 to 30 130, to 55 155, and to 88 188.
> Subtract 100 from 160 60, from 125 25, and from 176 76.

● ● ○ Add 100 to 500 600, to 640 740, and to 890 990.
> Subtract 100 from 500 400, from 220 120, and from 830 730.

● ● ● Add 100 to 97 197 to 193 293, and to 398 498.
> Subtract 100 from 801 701, from 696 596, and from 1,200 1,100.

▶ Daily Routines

Have children complete daily routines.

2 Focus 30–40 min [Go Online] ePresentations eToolkit

▶ Math Message

Math Masters, p. TA8

One friend measures your arm span in inches. Another friend measures your arm span in centimeters.

Who do you think will report the larger number? Why? Record your answer on My Exit Slip.

▶ Measuring Arm Spans

Math Masters, p. TA8

| WHOLE CLASS | SMALL GROUP | PARTNER | INDEPENDENT |

Math Message Follow-Up Ask children to share their answers. Expect most children to know that the measurement in centimeters will have a larger number because the centimeter is a smaller unit. Then have a volunteer and two helpers demonstrate for the class how to measure arm span.

The volunteer stands with his or her arms fully extended. **Arm span** is the distance from fingertip to fingertip across outstretched arms. Have the class make an estimate of the length of the volunteer's arm span in inches. Then one helper holds the end of the tape measure at the tip of the volunteer's right middle finger as the second helper pulls the tape tight across the child's chest. The second helper then holds the tape at the tip of the volunteer's left middle finger and reads the tape to the nearest inch. GMP5.2, GMP6.2 Turn the tape over, repeat the procedure, and have the second helper read the tape to the nearest centimeter. GMP5.2, GMP6.2

Ask: *Is the number larger when you measure in inches or in centimeters?* Centimeters *Why?* Centimeters are smaller units of length than inches. You can fit more centimeters than inches in the same arm span.

Repeat this routine several times with different volunteers and helpers. Check that the measurements they report are correct to the nearest inch and the nearest centimeter. **GMP5.2, GMP6.2**

Tell children that they will collect arm span and standing jump data to use in later lessons. Divide the class into groups of four. Children will remain in these groups for both data-collection activities. For each activity children take measurements in both centimeters and inches.

▶ Collecting and Recording Arm Span Data

Math Journal 2, p. 180

| WHOLE CLASS | **SMALL GROUP** | PARTNER | INDEPENDENT |

Children follow the procedure for collecting arm span data described in the Math Message Follow-Up. Circulate and make sure all children take a turn reading measurements. **GMP5.2** Remind them that all arm span measurements should be to the nearest inch or centimeter. When a measurement falls exactly on the half-inch or half-centimeter mark, they should report the next larger number. **GMP6.2** Children record their own arm span measurements in their journals. They will need to copy their own arm span measurement in inches onto their Home Link.

> **NOTE** The inch arm span data recorded on journal page 180 will also be used in Lesson 7-8. Problem 3 will be completed in Lesson 7-8.

Academic Language Development To teach the meaning of "measuring to the nearest unit," build on children's understanding of what it means to be near something or someone. Use a yardstick or a tape measure to find the length of something, such as a desk, to the nearest inch. Ask: *Which two numbers does the length of the desk fall between? Which number would you use in the measurement? Why?* Sample answer: It falls between the 35 and the 36. I would say the desk is 36 inches long because the edge is closer to 36. Stress that 36 is the nearest inch mark. Repeat with other objects, using both inches and centimeters. Help children generalize that "measuring to the nearest unit" means choosing the number nearest to the length of the object. Encourage them to use this expression when reporting out about measurements. Provide sentence frames such as the following: "This _____ measures _____ to the nearest _____."

Math Masters, p. TA8

TA8

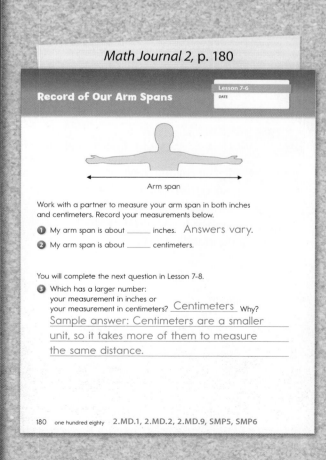

Math Journal 2, p. 180

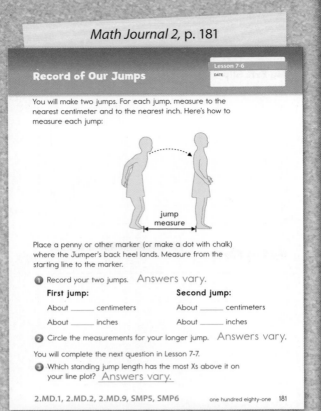

Math Journal 2, p. 181

▶ Collecting and Recording Standing Jump Data

Math Journal 2, p. 181

| WHOLE CLASS | **SMALL GROUP** | PARTNER | INDEPENDENT |

Children make two jumps and record the length of each one in both centimeters and inches on journal page 181. Follow these steps to explain how the data for the jumps will be gathered:

1. Assign a job to each group member.
   - The Jumper jumps.
   - The Line Judge makes sure the Jumper's toes don't cross the line.
   - The Marker marks where the Jumper lands.
   - The Measurer measures the length of the jump with the Jumper's help.
     Show children how to place a chalk dot, a penny, or other marker where the Jumper's back heel lands and how to measure from the starting line to the marker. After the Measurer measures the jump in one unit, he or she should turn the tape measure over to read the measurement in the other unit. **GMP5.2** Jumps are recorded to the nearest centimeter and inch. **GMP6.2**

2. Demonstrate a jump. The toes of both feet should be just touching the starting line. No running start is allowed. Neither is stepping back.

3. Let each child take several practice jumps before measuring a jump.

4. Each Jumper makes two jumps that are measured. They record the lengths of their own jumps in their journals. They also circle the measurement of the longer jump.

5. When the first Jumper has recorded two jumps, group members rotate jobs so that each child eventually performs all of the different jobs.

   NOTE The standing jump data measured in inches will also be used in the next lesson. Problem 3 will be completed in that lesson.

Common Misconception

Differentiate If children have difficulty with following the directions, make a display of the job flow diagram to help group members know how to switch jobs for the standing jumps.

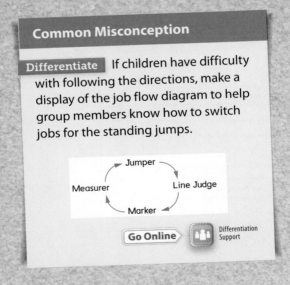

Go Online ▷   Differentiation Support

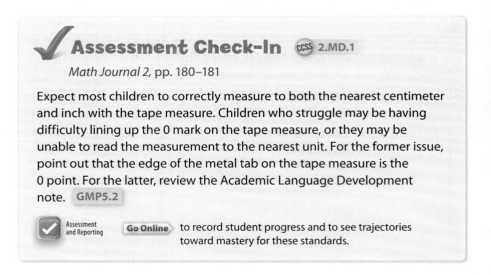

✔ Assessment Check-In   CCSS 2.MD.1

Math Journal 2, pp. 180–181

Expect most children to correctly measure to both the nearest centimeter and inch with the tape measure. Children who struggle may be having difficulty lining up the 0 mark on the tape measure, or they may be unable to read the measurement to the nearest unit. For the former issue, point out that the edge of the metal tab on the tape measure is the 0 point. For the latter, review the Academic Language Development note. **GMP5.2**

Assessment and Reporting   Go Online ▷ to record student progress and to see trajectories toward mastery for these standards.

Summarize Ask: *When measuring in inches, how did you determine the nearest inch?*

ePresentations   eToolkit   Home Connections

③ Practice   10–15 min   [Go Online]

▶ Comparing Measurements

Math Journal 2, p. 182

| WHOLE CLASS | SMALL GROUP | **PARTNER** | INDEPENDENT |

Partners choose measuring tools and measure their height, head size, and shoe length in centimeters. **GMP5.1** Then they find the differences between their measurements.

▶ Math Boxes 7-6: Preview for Unit 8

Math Journal 2, p. 183

| WHOLE CLASS | **SMALL GROUP** | PARTNER | **INDEPENDENT** |

Mixed Practice Math Boxes 7-6 are paired with Math Boxes 7-10. These problems focus on skills and understandings that are prerequisite for Unit 8. You may want to use information from these Math Boxes to plan instruction and grouping in Unit 8.

▶ Home Link 7-6

Math Masters, p. 198

Homework Remind children to copy the length of their arm span in inches from journal page 198 onto their Home Link. Children compare their arm spans to the arm span of someone else at home. They also find objects around their homes that are about the same length as their own arm spans.

Comparing Arm Spans

Home Link 7-6

NAME   DATE

Family Note

In today's lesson your child measured his or her standing jump and arm span in both centimeters and inches. Help your child compare his or her arm span to someone else's arm span at home. Also help your child find objects around the house that are about the same length as his or her arm span.

Please return this Home Link to school tomorrow.

My arm span is about _____ inches long.   Answers vary.

① Tell someone at home about how long your arm span is in inches.

② Compare your arm span to the arm span of someone at home. Can you find someone who has a longer arm span than you do? Is there someone at home who has a shorter arm span?

_____ has a longer arm span than I have.

_____ has a shorter arm span than I have.

③ List some objects that are about the same length as your arm span.

_____ _____ _____

④ Explain how you know the objects you listed in Problem 3 are about the same length as your arm span.

Answers vary.

Practice
Solve.   Unit

⑤ 57 + 3 = __60__   ⑥ 4 + 71 = __75__

⑦ __43__ = 34 + 9   ⑧ 48 + __8__ = 56

198 one hundred ninety-eight   2.NBT.5, 2.MD.1

Math Masters, p. 198

Math Journal 2, p. 182

Comparing Measurements

Lesson 7-6
DATE

Work with a partner. Measure your height, head size, and shoe length to the nearest centimeter. For each measurement, choose a tool to use. You may use a ruler, a meterstick, or a tape measure.

① Height Answers vary.

I am about _____ centimeters tall.

My partner is about _____ centimeters tall.

Who is taller? _____

How much taller? _____ centimeters

② Head size (the distance around your head) Answers vary.

My head size is about _____ centimeters.

My partner's head size is about _____ centimeters.

Who has the larger head size? _____

How much larger? _____ centimeters

③ Shoe length Answers vary.

My shoe is about _____ centimeters long.

My partner's shoe is about _____ centimeters long.

Who has the longer shoe length? _____

How much longer? _____ centimeters

182   one hundred eighty-two   2.NBT.4, 2.NBT.5, 2.MD.1, 2.MD.4, SMP5

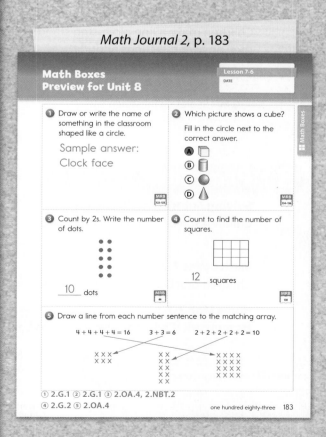

Math Journal 2, p. 183

Math Boxes Preview for Unit 8

Lesson 7-6
DATE

① Draw or write the name of something in the classroom shaped like a circle.

Sample answer: Clock face

MRB 122-125

② Which picture shows a cube?

Fill in the circle next to the correct answer.

Ⓐ
Ⓑ
Ⓒ
Ⓓ

MRB 134-136

③ Count by 2s. Write the number of dots.

__10__ dots

MRB 98

④ Count to find the number of squares.

__12__ squares

MRB 198

⑤ Draw a line from each number sentence to the matching array.

4 + 4 + 4 + 4 = 16   3 + 3 = 6   2 + 2 + 2 + 2 + 2 = 10

① 2.G.1 ② 2.G.1 ③ 2.OA.4, 2.NBT.2
④ 2.G.2 ⑤ 2.OA.4

one hundred eighty-three   183

Representing Data: Standing Jumps

Overview Children discuss the shortest and longest standing jumps and create a line plot for the data.

▶ **Before You Begin**
For the optional Enrichment activity, prepare a number line for children to use to organize data into a line plot. Include the label "Hand Span (cm)" below the line. For the optional Extra Practice activity, collect objects ranging in length from 1 to 12 inches. Some of the objects should be the same length.

▶ **Vocabulary**
line plot

Common Core State Standards

Focus Clusters
- Use place value understanding and properties of operations to add and subtract.
- Relate addition and subtraction to length.
- Represent and interpret data.

1 Warm Up 15–20 min

	Materials	
Mental Math and Fluency Children compare numbers.	slate	2.NBT.4
Daily Routines Children complete daily routines.	See pages 4–43.	See pages xiv–xvii.

2 Focus 20–30 min

Math Message Children record their standing-jump length in inches.	*Math Journal 2,* p. 181; stick-on notes	
Discussing the Data Children discuss the standing-jump data.	tape measure (for demonstration); stick-on notes from Math Message	2.NBT.5, 2.MD.6
✓ **Assessment Check-In** See page 657.		2.NBT.5
Making a Class Line Plot Children make a line plot using standing-jump data and then answer questions about it.	*Math Journal 2,* p. 181; *My Reference Book,* p. 114; stick-on notes from Math Message	2.MD.9 SMP4

2.MD.9 **Spiral Snapshot**

GMC Represent measurement data on a line plot.

7-7 Focus	7-8 Focus Practice	8-1 Practice	8-3 Practice	8-5 Practice	9-6 Practice

/// Spiral Tracker   **Go Online**   to see how mastery develops for all standards within the grade.

3 Practice 10–15 min

Solving Subtraction Problems Children solve subtraction problems.	*Math Journal 2,* p. 184; number grid or number line (optional); base-10 blocks (optional)	2.NBT.5, 2.NBT.7
Math Boxes 7-7 Children practice and maintain skills.	*Math Journal 2,* p. 185	See page 659.
Home Link 7-7 **Homework** Children solve problems based on data.	*Math Masters,* p. 201	2.NBT.5

connectED.mheducation.com

Plan your lessons online with these tools.

ePresentations   Student Learning Center   Facts Workshop Game   eToolkit   Professional Development   Home Connections   Spiral Tracker   Assessment and Reporting   English Learners Support   Differentiation Support

Differentiation Options

RtI

Readiness   5–15 min

WHOLE CLASS
SMALL GROUP
PARTNER
INDEPENDENT

Collecting and Organizing Data

Math Masters, p. 199; 6-sided die

For experience collecting and organizing data, children use a table to record the rolls of a die. Children roll a die and record the number rolled by drawing an X above the corresponding number.

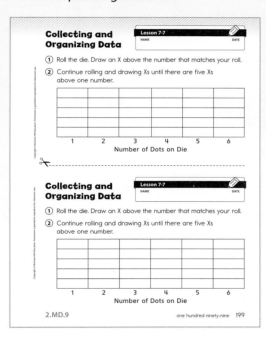

CCSS 2.MD.1, 2.MD.9, SMP4

Enrichment   5–15 min

WHOLE CLASS
SMALL GROUP
PARTNER
INDEPENDENT

Making a Line Plot

Activity Card 92, tape measure, stick-on notes, number line drawn by teacher

To apply their knowledge of collecting data and creating a display, children work in small groups to gather data to make a line plot. Classmates measure each other's hand spans and record the measurements on stick-on notes that they use to make a line plot. GMP4.1, GMP4.2

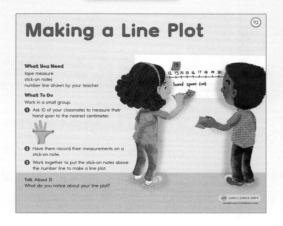

CCSS 2.MD.1, 2.MD.9, SMP4

Extra Practice   5–15 min

WHOLE CLASS
SMALL GROUP
PARTNER
INDEPENDENT

Measuring Objects for a Line Plot

Activity Card 93; *Math Masters,* p. 200; 12-inch ruler; objects from classroom varying in length from 1 to 12 inches

For additional experience gathering and organizing data, children measure objects and make a line plot. GMP4.1 They discuss what they notice about the data. GMP4.2

English Language Learners Support

Beginning ELL Use concrete objects and think-alouds to review how descriptive words in English take the comparative -*er* and superlative -*est* endings. *For example:* Hold up two pencils of different lengths and say: *This pencil is short. And this one is even shorter.* Repeat with different pairs of classroom objects: show an object, describe the attribute, and prompt children to use the -*er* ending with this sentence frame: "And that one is even _____-*er*." Show the pencils again and introduce -*est* by showing a third pencil. Say: *And this pencil is the shortest.* Use a variety of objects and show-me commands so that children can practice comparing first two and then three objects and using the -*er* and -*est* endings.

 Go Online   ELL   English Learners Support

Standards and Goals for
Mathematical Practice

SMP4 Model with mathematics.

  GMP4.1 Model real-world situations using graphs, drawings, tables, symbols, numbers, diagrams, and other representations.

  GMP4.2 Use mathematical models to solve problems and answer questions.

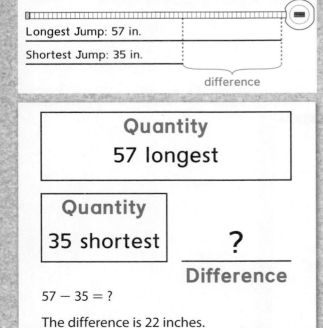

Longest Jump: 57 in.

Shortest Jump: 35 in.

difference

Quantity
57 longest

Quantity	
35 shortest	?

Difference

$57 - 35 = ?$

The difference is 22 inches.

① Warm Up 15–20 min 〔Go Online〕 ePresentations eToolkit

▶ Mental Math and Fluency

Dictate pairs of numbers for children to write on their slates and compare, recording the results with $>$, $<$, and $=$. Ask children to explain their answers in terms of place value. *Leveled exercises:*

● ○ ○ 123 and 123 $=$
   234 and 334 $<$
   456 and 457 $<$

● ● ○ 989 and 971 $>$
   445 and 454 $<$
   877 and 788 $>$

● ● ● 1,054 and 1,154 $<$
   1,243 and 1,233 $>$
   1,522 and 1,622 $<$

▶ Daily Routines

Have children complete daily routines.

② Focus 20–30 min 〔Go Online〕 ePresentations eToolkit

▶ Math Message

Math Journal 2, p. 181

Turn to journal page 181. Write your name and the length of your longer jump in inches on a stick-on note.

▶ Discussing the Data

WHOLE CLASS   SMALL GROUP   PARTNER   INDEPENDENT

Math Message Follow-Up As children read their jump lengths in inches, list the data in order from shortest to longest. Tape an actual tape measure to the board and mark the longest and shortest jump lengths. (*See margin.*)

  NOTE Explain that another name for the shortest jump is *minimum* and another name for the longest jump is *maximum.*

Partners work together to calculate the difference between the longest and shortest jumps and then share their solution strategies. Draw a comparison diagram on the board and fill in the known quantities—the lengths of the longest and shortest jumps. Because the difference is what you want to find, write a question mark for the difference. (*See margin.*)

Professional Development

A line plot is a quick and easy way to organize and display data. You can think of it as a rough sketch of a bar graph. Line plots are also called *pictographs* (when the marks used on the graph are pictures), *sketch graphs,* or *dot plots.* Line plots work best with numerical data. Check marks, Xs, stick-on notes, or other marks above a labeled line show the frequency of each value. In a line plot, three data values can be clearly identified: maximum, minimum, and range.

Go Online Professional Development

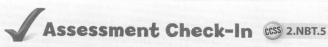

✓ Assessment Check-In CCSS 2.NBT.5

Observe children as they find the difference between the longest and shortest jumps. Expect that most will successfully find the difference without using a tool. If children struggle, suggest that they use a number grid or a number line for support.

 Assessment and Reporting   **Go Online** to record student progress and to see trajectories toward mastery for these standards.

▶ Making a Class Line Plot

Math Journal 2, p. 181; *My Reference Book,* p. 114

| WHOLE CLASS | SMALL GROUP | PARTNER | INDEPENDENT |

Display a long number line with tick marks at 6-inch intervals. Write the shortest standing-jump length below the leftmost tick mark and the longest length under the rightmost tick mark. Fill in all the intervening numbers. For example, if the shortest jump is 35 inches and the longest is 57 inches, write 35, 36, 37, . . . , 54, 55, 56, 57 beneath the tick marks. Write "Jump Length (inches)" below the line. Then follow the steps below to guide children to create a **line plot** of the data. GMP4.1

1. Children come to the display in small groups.

2. Children find the numbers on the number line that match their stick-on notes. They post their stick-on notes just above those tick marks. *For example:* Darius's jump is 38 inches, so he wrote 38 on his stick-on note. He posts his stick-on note just above the tick mark labeled 38.

3. If there is already a note above a tick mark, the new stick-on note goes directly above it. *For example:* Jun's jump is also 38 inches, so he posts his stick-on note directly above Darius's. Be sure that children leave enough room between the notes to be able to count the number of stick-on notes at each interval. GMP4.1

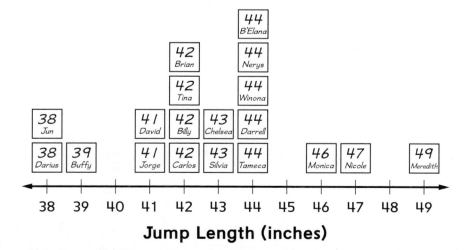

Jump Length (inches)

After all of the stick-on notes have been posted, remove them one by one and replace each note with an X. GMP4.1

Academic Language Development

Children may have heard the term *plot* used to mean "the events of a story," "a scheme to do something bad," or "to make a secret plan." Explain that in math, a *line plot* is a special kind of graph that uses a number line to show data or information, such as how often something occurs or how much of something there is.

NOTE It is important for you to replace the stick-on notes with Xs rather than asking children to do so because the Xs should be uniform in size and an equal distance apart to help children use the graph to answer questions.

Label the shortest and longest lengths on the line plot. Ask children questions about the set of data displayed in the line plot: **GMP4.2**

- *What does it mean when there are a lot of Xs above a number?* Many children have that standing-jump length.
- *How many children have a jump of 42 inches?* Answers vary.

Have children answer Problem 3 on journal page 181 and share their answers. Answers vary.

Summarize Read about line plots on *My Reference Book,* page 114 as a class. Be sure to discuss the Try It Together question at the bottom of the page.

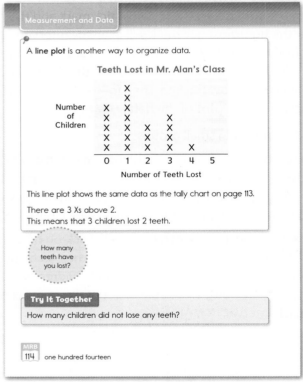

My Reference Book, p. 114

Math Journal 2, p. 184

Solving Subtraction Problems

Lesson 7-7

DATE

Make a ballpark estimate. Then solve.

① 42 − 19 = 23   Unit

② 64 − 39 = 25   Unit

Sample answer:
Ballpark estimate:
40 − 20 = 20

Sample answer:
Ballpark estimate:
60 − 40 = 20

Try This

③ 86 − 57 = 29   Unit

④ 103 − 58 = 45   Unit

Sample answer:
Ballpark estimate:
90 − 60 = 30

Sample answer:
Ballpark estimate:
100 − 60 = 40

184   one hundred eighty-four   2.NBT.5, 2.NBT.7

3 Practice   10–15 min   Go Online

ePresentations   eToolkit   Home Connections

▶ Solving Subtraction Problems

Math Journal 2, p. 184

WHOLE CLASS | SMALL GROUP | PARTNER | INDEPENDENT

Have children solve subtraction problems using whichever strategy they prefer. Encourage them to use number grids, number lines, or base-10 blocks.

▶ Math Boxes 7-7

Math Journal 2, p. 185

| WHOLE CLASS | SMALL GROUP | PARTNER | INDEPENDENT |

Mixed Practice Math Boxes 7-7 are paired with Math Boxes 7-9.

▶ Home Link 7-7

Math Masters, p. 201

Homework Children organize and answer questions about a set of data.

Interpreting Data

Home Link 7-7

NAME   DATE

Family Note

In this lesson your child examined classroom data on the length of classmates' standing jumps. The class found the shortest jump length and the longest jump length and calculated the difference between the lengths. They also made a line plot based on the data.

Please return this Home Link to school tomorrow.

The track team collected these standing-jump data:

Jumper	Standing-Jump Length
Fran	68 inches
Arturo	72 inches
Louise	57 inches
Kelsey	71 inches
Keisha	60 inches
Ray	64 inches
Maria	64 inches
Ben	62 inches

① List the inches for each jump in order from shortest to longest.

57   60   62   64   64   68   71   72

② What is the shortest jump length? __57__ inches

③ What is the longest jump length? __72__ inches

④ What is the difference between the longest jump length and the shortest jump length? __15__ inches

Practice

⑤ __98__ = 1 + 97   ⑥ 23 + 6 = __29__

2.NBT.5   two hundred one   201

Math Masters, p. 201

Math Journal 2, p. 185

Math Boxes

Lesson 7-7
DATE

① During the picnic I counted 14 grasshoppers, 16 flies, and 25 ants. How many insects did I count in all?

Answer: __55__ insects

② Write each number in expanded form.

591 __500 + 90 + 1__

311 __300 + 10 + 1__

Sample: 702 __700 + 0 + 2__

Sample: 920 __900 + 20__

③ The green snake is 10 cm long. The brown snake is 26 cm long. How much longer is the brown snake than the green snake? Draw a diagram to help.

Sample answer:
Number model:

10 + ? = 26; 26 − 10 = ?

Answer: __16__ cm longer.

④ Dan has 20 red blocks and 30 blue blocks. He gives 12 blocks to his sister. How many blocks does he have left?

Sample answer:
Write one or two number models:

__20 + 30 − 12 = ?__

Dan has __38__ blocks left.

⑤ **Writing/Reasoning** Look at Problem 3. If the green snake grows 10 cm, will it be longer than the brown snake? Explain.

No. Sample answer: If the green snake grows 10 cm, it will be 20 cm long. It will still be shorter than the brown snake.

① 2.OA.1, 2.NBT.5, 2.NBT.6
② 2.NBT.1, 2.NBT.3 ③ 2.OA.1, 2.NBT.5, 2.MD.5 ④ 2.OA.1, 2.NBT.5 ⑤ 2.OA.1, 2.NBT.5, 2.MD.5, SMP3   one hundred eighty-five 185

Representing Data: Arm Spans

Overview Children make a frequency table and a line plot for a set of data.

▶ **Before You Begin**
Place stick-on notes next to the Math Message. For the frequency table and line plot activities in Part 2, decide how you will display *Math Masters,* pages 204–205.

▶ **Vocabulary**
frequency table • line plot

Common Core State Standards

Focus Clusters
• Use place value understanding and properties of operations to add and subtract.
• Represent and interpret data.

1 Warm Up 15–20 min

	Materials	
Mental Math and Fluency Children mentally add and subtract 100.	slate	**2.NBT.8**
Daily Routines Children complete daily routines.	See pages 4–43.	See pages xiv–xvii.

2 Focus 30–40 min

Math Message Children explain how measurements relate to unit size.	*Math Journal 2,* p. 180; stick-on notes	**2.MD.2**
Comparing Arm Span Measures Children find the difference between the longest and shortest arm spans in the class.	*Math Journal 2,* p. 180; *My Reference Book,* p. 113; tape measure; stick-on notes from Math Message	**2.NBT.5, 2.MD.2, 2.MD.6**
Making a Frequency Table of Arm Span Data Children make a frequency table to show arm span data.	*Math Journal 2,* p. 186; *Math Masters,* p. 204; stick-on notes	**2.NBT.3** **SMP4**
Making a Line Plot of Arm Span Data Children make a line plot and then answer questions about it.	*Math Journal 2,* p. 187; *Math Masters,* p. 205	**2.MD.9** **SMP2, SMP4**
✓ **Assessment Check-In** See page 664.	*Math Journal 2,* p. 187	**2.MD.9**

CCSS 2.MD.9 **Spiral Snapshot**

GMC Represent measurement data on a line plot.

| 7-7
Focus | 7-8
Focus
Practice | 8-1
Practice | 8-3
Practice | 8-5
Practice | 9-6
Practice |

▦ Spiral Tracker   (Go Online)   to see how mastery develops for all standards within the grade.

3 Practice 10–15 min

Playing *Beat the Calculator* **Game** Children practice addition facts.	See Lesson 5-1.	**2.OA.2**
Math Boxes 7-8 Children practice and maintain skills.	*Math Journal 2,* p. 188; 10-centimeter ruler	See page 665.
Home Link 7-8 **Homework** Children answer questions based on a line plot.	*Math Masters,* p. 206	**2.NBT.5** **SMP4**

(connectED.mheducation.com >

Plan your lessons online with these tools.

 ePresentations
 Student Learning Center
 Facts Workshop Game
 eToolkit
 Professional Development
 Home Connections
 Spiral Tracker
 Assessment and Reporting
 English Learners Support
 Differentiation Support

Differentiation Options

RtI

Readiness
5–15 min

Recording Tally Marks

Math Masters, p. 155
(1 copy per partnership);
per partnership: 1 six-sided die

WHOLE CLASS
SMALL GROUP
PARTNER
INDEPENDENT

For experience recording tally marks,
partners take turns rolling a die and using
a tally mark to record the number they
rolled on *Math Masters*, page 155. See the
Readiness activity in Lesson 6-1 for details.

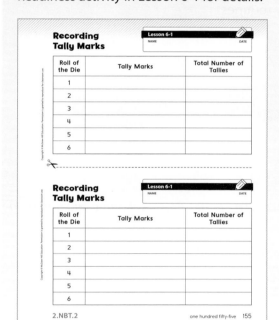

Enrichment
5–15 min

Questions for Arm Span Line Plot

Activity Card 94;
Math Journal 2, p. 187; paper

WHOLE CLASS
SMALL GROUP
PARTNER
INDEPENDENT

To extend their understanding of line
plots, children independently write five
questions based on the information on
journal page 187. Then partners trade
papers and answer each other's questions.
GMP4.2

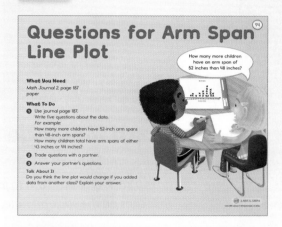

Extra Practice
5–15 min

Line Plot of Heights

Activity Card 95;
Math Masters, pp. 202–203

WHOLE CLASS
SMALL GROUP
PARTNER
INDEPENDENT

For additional experience with
measurement data, children use a
frequency table of children's heights on
Math Masters, page 202 to make a line plot
on *Math Masters*, page 203. GMP4.1

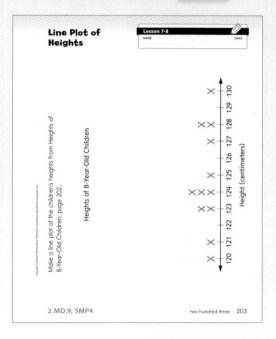

English Language Learners Support

Beginning ELL Use demonstrations to help children understand the terms *frequent*
and *frequency* in terms of counting and totaling. Display a two-column table with a list
of all the children's names in the left column; then have children indicate their arm span
measurements in the right column. Circle a measurement and ask: *How many children have
an arm span of _____? How frequent is the measurement _____? Let's count to see how many
times we find the measurement _____. Let's count to find the frequency of the measurement
_____.* After modeling a few such examples, ask questions like this one: *What is the
frequency of the measurement _____?*

Go Online   ELL English Learners Support

Standards and Goals for
Mathematical Practice

SMP2 **Reason abstractly and quantitatively.**
  GMP2.3 Make connections between representations.

SMP4 **Model with mathematics.**
  GMP4.1 Model real-world situations using graphs, drawings, tables, symbols, numbers, diagrams, and other representations.
  GMP4.2 Use mathematical models to solve problems and answer questions.

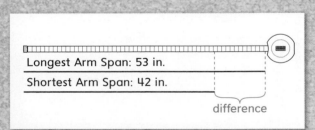

Longest Arm Span: 53 in.
Shortest Arm Span: 42 in.

difference

Quantity
53 longest

Quantity	
42 shortest	**?**
	Difference

$$53 - 42 = ?$$
$$53 - 42 = 11$$

1 Warm Up   15–20 min   Go Online
ePresentations   eToolkit

▶ Mental Math and Fluency

Dictate 2- and 3-digit numbers. Have children mentally add or subtract 100 and record their answers on slates. *Leveled exercises:*

●○○ Add 100 to 60 160, to 15 115, and to 93 193.
     Subtract 100 from 110 10, from 195 95, and from 134 34.

●●○ Add 100 to 300 400, to 820 920, and to 780 880.
     Subtract 100 from 100 0, from 630 530, and from 910 810.

●●● Add 100 to 91 191, to 188 288, and to 599 699.
     Subtract 100 from 999 899, from 798 698, and from 1,111 1,011.

▶ Daily Routines

Have children complete daily routines.

2 Focus   30–40 min   Go Online
ePresentations   eToolkit

▶ Math Message

Math Journal 2, p. 180

Take 1 stick-on note. Turn to journal page 180. Print your name and arm span in inches on the stick-on note.

Then answer Problem 3 on journal page 180.

▶ Comparing Arm Span Measures

Math Journal 2, p. 180; *My Reference Book*, p. 113

WHOLE CLASS	SMALL GROUP	PARTNER	INDEPENDENT

Math Message Follow-Up Circulate to be sure that children record their arm span data in inches. Discuss whether the measure—the number of units in the measurement—was smaller when they measured in centimeters or when they measured in inches. Inches Have children share their thinking as to why one measure was a smaller number than the other. Inches are a larger unit, so there are fewer of them.

Determine who has the shortest and longest arm spans and what their lengths are. Tape an actual tape measure to the board and mark the longest and shortest arm spans. (*See margin.*) Draw a comparison diagram on the board, filling in the known quantities and writing a question mark for the difference. (*See margin.*) With the class, find the difference between the two arm spans. Explain that children will use the class arm span data to make a **frequency table** and a line plot. You may want to read about tally charts on *My Reference Book*, page 113 with the class.

▶ Making a Frequency Table of Arm Span Data

Math Journal 2, p. 186; *Math Masters*, p. 204

WHOLE CLASS | **SMALL GROUP** | PARTNER | INDEPENDENT

Display a copy of the table on *Math Masters*, page 204 and work as a class to fill in the frequency table of arm spans. **GMP4.1** Record children's data on the display as children do so on journal page 186. Follow these steps:

1. Fill in the Arm Span column. In the first row, write the length of the shortest arm span in the class. Fill in the subsequent rows with all of the possible arm spans to the nearest inch, up to the length of the longest arm span in the class. For example, if the shortest span is 42 inches and the longest span is 53 inches, write 42, 43, 44, . . . , 51, 52, 53 in the arm span column. (*See margin.*)

2. Ask each child in turn to say his or her arm span. As children share their data, everyone (including the speaker) makes a tally mark next to the arm span length reported.

3. After all of the measurements have been tallied, write a number for each set of tallies.

4. To check that no measurements have been omitted, add the frequency numbers and compare the sum to the number of children in the class. Discuss the completed table. **GMP4.2** Ask: *How many children had an arm span of 50 inches? Of 45 inches? Which arm span lengths did nobody mention?* Sample answer: The ones with a 0 in the Number column

Academic Language Development The term *frequent* may not be familiar to children. Introduce the term using contextual information and restatements with more familiar everyday words to help children construct an understanding of the term. For example, say: *How many times have you gone to the dentist this year? Do you often go to the dentist? Do you make frequent visits to the dentist?* Point out to children that the words *frequent, frequently,* and *frequency* belong to the same word family.

Professional Development

Frequency means the number of times an event or a value occurs in a set of data. The arm span table shows frequency two ways: with tally marks and with numbers (counts). This filled-in table is called a *frequency table* because it shows the frequencies for all arm spans in your data set. Use the terms *frequency* and *frequency table*. Encourage, but do not expect, children to use these words at this time.

Go Online ▶ Professional Development

Our Arm Spans		
Arm Span (Inches)	Frequency	
	Tallies	Number
42	//	2
43	/	1
44		0
45	//	2
46	////	4
47	//	2
48	̶H̶H̶	5
49		0
50	/	1
51	/	1
52		0
53	/	1
	Total =	19

Math Journal 2, p. 186

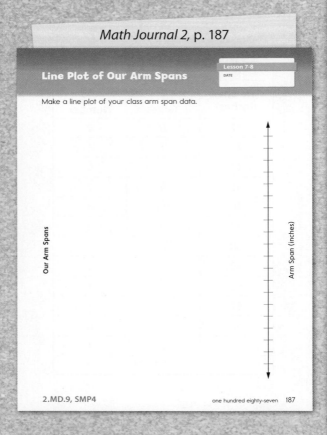

Math Journal 2, p. 187

► Making a Line Plot of Arm Span Data

Math Journal 2, p. 187; *Math Masters*, p. 205

WHOLE CLASS   **SMALL GROUP**   PARTNER   INDEPENDENT

Have children use the information in the frequency table to draw a line plot on journal page 187. **GMP4.1** A **line plot** is type of display that shows data organized above a labeled line. Display *Math Masters*, page 205 and show how to draw the scale of possible arm span lengths. Discuss how you know with which numbers to start and end. Children record the scale on their line plots. For each tally mark next to an arm span length in the frequency table, they draw an X above the tick mark for the corresponding length on the line plot. For example, if there are 2 tally marks next to 42 in the table, children draw two Xs above 42 on the line plot. You may wish to model a few examples for children. Remind them that every tally should be represented with an X on the line plot. To make sure that all of the data from the class are represented, prompt children to count the number of Xs and compare the total to the number of children in the class.

Discuss children's completed line plots. Ask: **GMP4.2**

- *What does it mean when there are a lot of Xs above a number?* Many children have arm spans that are that number of inches.
- *Which arm span is the most common?* Answers vary, but children should identify the number with the most Xs above it.
- *What do you know about the numbers that have no Xs above them?* No child has an arm span that is that long.
- *How many children have an arm span of 51 inches? Of 46 inches?* Answers vary.

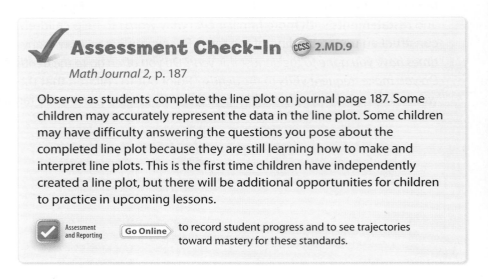

✓ Assessment Check-In   ccss 2.MD.9

Math Journal 2, p. 187

Observe as students complete the line plot on journal page 187. Some children may accurately represent the data in the line plot. Some children may have difficulty answering the questions you pose about the completed line plot because they are still learning how to make and interpret line plots. This is the first time children have independently created a line plot, but there will be additional opportunities for children to practice in upcoming lessons.

✓ Assessment and Reporting   〈Go Online〉 to record student progress and to see trajectories toward mastery for these standards.

Summarize Children compare the frequency table and the line plot on journal pages 186–187. Ask: *How are the frequency table and the line plot similar?* **GMP2.3** Sample answer: Both show the lengths of children's arm spans. *How are they different?* **GMP2.3** Sample answer: The frequency table uses tallies to show how many children have each length. The line plot uses Xs.

③ **Practice** 10–15 min

Go Online
ePresentations eToolkit Home Connections

▶ Playing *Beat the Calculator*

Assessment Handbook, pp. 98–99

| WHOLE CLASS | **SMALL GROUP** | PARTNER | INDEPENDENT |

Have small groups play the game as introduced in Lesson 5-1. As you circulate and observe, consider using *Assessment Handbook*, pages 98–99 to monitor children's progress with addition facts. By the end of Grade 2, children are expected to know from memory all sums of two 1-digit numbers.

Observe

- Which facts do children know from memory?
- Which children need additional support to play the game?

Discuss

- *What strategies did you use to solve the facts you did not know?*
- *Why is knowing addition facts helpful?*

▶ Math Boxes 7-8

Math Journal 2, pp. 188

| WHOLE CLASS | SMALL GROUP | **PARTNER** | **INDEPENDENT** |

Mixed Practice Math Boxes 7-8 are paired with Math Boxes 7-5.

▶ Home Link 7-8

Math Masters, p. 206

Homework Children answer questions based on a line plot showing basketball players' heights. **GMP4.2**

Math Journal 2, p. 188

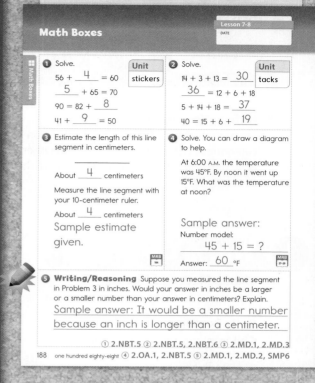

Math Masters, p. 206

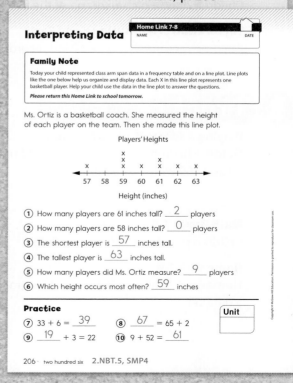

Exploring Shape Attributes, Graphs, and Measurements

Overview Children sort shapes, draw a picture graph, and measure body parts.

► **Before You Begin**
For both the Math Message and Exploration B, make a large tally chart on poster paper with the categories Apple, Banana, Grapes, and Other. Display it near the Math Message. Have paper clips and envelopes available for each child to store the Shape Cards they cut out from *Math Journal 2,* Activity Sheets 7–12.

Common Core State Standards

Focus Clusters
- Measure and estimate lengths in standard units.
- Represent and interpret data.
- Reason with shapes and their attributes.

---

1 Warm Up 15–20 min

	Materials	
Mental Math and Fluency Children add three or more numbers.	slate	2.NBT.5, 2.NBT.6
Daily Routines Children complete daily routines.	See pages 4–43.	See pages xiv–xvii.

2 Focus 30–40 min

Math Message Children draw tally marks next to their favorite fruit. They also discuss shapes with partners.	*Math Journal 2,* Activity Sheets 7–12	2.G.1 SMP4, SMP7
Discussing Shapes Children discuss what they notice about shapes.	*Math Journal 2,* Activity Sheets 7–12	2.G.1 SMP7
Exploration A: Sorting Shapes Children cut out shapes and sort them.	Activity Card 96; *Math Journal 2,* pp. 190–191 and Activity Sheets 7–12; scissors; envelope; paper clip	2.G.1 SMP7
Exploration B: Drawing a Picture Graph Children draw a picture graph of the favorite-fruit data.	Activity Card 97; *Math Journal 2,* p. 192; Our Favorite Fruits data	2.MD.10 SMP4
Exploration C: Measuring Body Parts Children measure body parts and record data.	Activity Card 98; *Math Journal 2,* p. 193; tape measure; yardstick; meterstick	2.MD.1, 2.MD.2, 2.MD.3, 2.MD.9 SMP5

3 Practice 10–15 min

Playing *Addition/Subtraction Spin* **Game** Children practice mentally adding and subtracting 10 and 100 with 3-digit numbers.	*My Reference Book,* pp. 138–139; *Math Masters,* p. G23 and p. G24 (optional); per partnership: paper clip, pencil, die (prepared in Lesson 5-6); calculator	2.NBT.5, 2.NBT.7, 2.NBT.8 SMP8
Math Boxes 7-9 Children practice and maintain skills.	*Math Journal 2,* p. 189	See page 671.
Home Link 7-9 **Homework** Children draw a picture graph to represent a data set.	*Math Masters,* p. 209	2.NBT.5, 2.MD.10

Differentiation Options

RtI

CCSS 2.MD.10

Readiness   5–15 min

Discussing Picture Graphs

My Reference Book, p. 115

| WHOLE CLASS |
| SMALL GROUP |
| PARTNER |
| INDEPENDENT |

For experience with picture graphs, children discuss *My Reference Book*, page 115. Ask: *What do you notice about the graph?* Sample answers: It has a title. It uses pictures of books to show the data. *What does each book symbol in the graph stand for?* Sample answer: Each book stands for one book a child read. *What does the graph show about the number of books read?* Sample answer: It shows how many books were read by each child. *How is a picture graph helpful?* Sample answers: It is easy to read. You can compare the number of pictures above or next to each label.

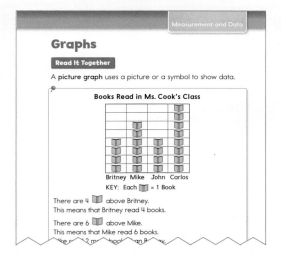

Measurement and Data

Graphs

Read It Together

A **picture graph** uses a picture or a symbol to show data.

Books Read in Ms. Cook's Class

Britney Mike John Carlos
KEY: Each 📖 = 1 Book

There are 4 📖 above Britney.
This means that Britney read 4 books.

There are 6 📖 above Mike.
This means that Mike read 6 books.

CCSS 2.MD.10, SMP4

Enrichment   5–15 min

Drawing a Bar Graph

Math Masters, p. 207

| WHOLE CLASS |
| SMALL GROUP |
| PARTNER |
| INDEPENDENT |

To apply their understanding of representing data, children survey classmates to gather data to display in their own bar graphs. Children choose a survey question that has up to four answers. For those who have difficulty choosing a question, suggest a topic, such as favorite seasons or sports. Children collect their data and then draw their bar graphs on *Math Masters*, page 207. **GMP4.1**

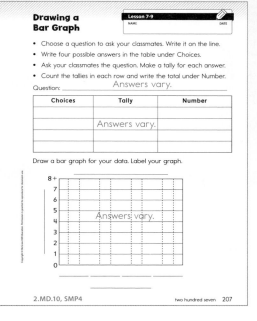

Drawing a Bar Graph
Lesson 7-9
NAME   DATE

- Choose a question to ask your classmates. Write it on the line.
- Write four possible answers in the table under Choices.
- Ask your classmates the question. Make a tally for each answer.
- Count the tallies in each row and write the total under Number.

Question: _____ Answers vary.

Choices	Tally	Number
	Answers vary.	

Draw a bar graph for your data. Label your graph.

Answers vary.

2.MD.10, SMP4   two hundred seven   207

CCSS 2.MD.10, SMP4

Extra Practice   5–15 min

Drawing a Favorite Fruits Bar Graph

Math Journal 2, p. 192;
Math Masters, p. 208

| WHOLE CLASS |
| SMALL GROUP |
| PARTNER |
| INDEPENDENT |

For additional practice representing data, children use the picture graph they made in Exploration B on journal page 192 to draw a bar graph on *Math Masters*, page 208. **GMP4.1**

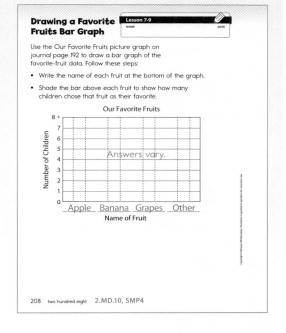

Drawing a Favorite Fruits Bar Graph
Lesson 7-9
NAME   DATE

Use the Our Favorite Fruits picture graph on journal page 192 to draw a bar graph of the favorite-fruit data. Follow these steps:

- Write the name of each fruit at the bottom of the graph.
- Shade the bar above each fruit to show how many children chose that fruit as their favorite.

Our Favorite Fruits

Number of Children

Answers vary.

Apple Banana Grapes Other
Name of Fruit

208   two hundred eight   2.MD.10, SMP4

English Language Learners Support

Beginning ELL Use the verbs *sort* and *group* interchangeably to build children's understanding that *sorting* results in groups of objects that share a common attribute. Use a think-aloud to sort objects into two different groups according to an attribute, such as color. For example, say: *I am sorting things by color. This is red. I will sort it into the red group.* Put one object in the wrong group and say: *This object doesn't belong in this group. Where does it belong? How should I group, or sort, it?* For practice with the terms *sort* and *group*, display sets of mixed objects and ask children to sort or group them by a given attribute.

CCSS Standards and Goals for
Mathematical Practice

SMP4 **Model with mathematics.**
GMP4.1 Model real-world situations using graphs, drawings, tables, symbols, numbers, diagrams, and other representations.

SMP5 **Use appropriate tools strategically.**
GMP5.2 Use tools effectively and make sense of your results.

SMP7 **Look for and make use of structure.**
GMP7.1 Look for mathematical structures such as categories, patterns, and properties.

1 Warm Up 15–20 min

Go Online

▶ Mental Math and Fluency

Display these problems. Ask children to write their answers on their slates and then share their strategies for solving. *Leveled exercises:*

● ○ ○ **? = 9 + 3 + 11** 23; Sample strategy: I took 1 away from 11 to add to 9 to make 10. Then I added another 10 and 3 more to get 23.
8 + 12 + 9 = ? 29; Sample strategy: I know that 8 and 2 are easy numbers to add to make 10. Then I added 10 more to get 20, and 20 + 9 = 29.

● ● ○ **? = 9 + 3 + 11 + 17** 40; Sample strategy: First I added the 1s: 3 + 7 = 10, and 9 + 1 = 10. Then I added the 10s: 10 + 10 + 10 + 10 = 40.
? = 12 + 13 + 17 + 8 50; Sample strategy: First I added the 1s: 2 + 8 = 10, and 3 + 7 = 10. Then I added the 10s: 10 + 10 + 10 + 10 + 10 = 50.

● ● ● **25 + 4 + 16 + 15 = ?** 60; Sample strategy: I added 4 + 16 = 20. Then I added 25 + 15 = 40. Finally I added 20 + 40 = 60.
33 + 25 + 25 + 17 = ? 100; Sample strategy: I know that 2 quarters equals 50¢, so 25 + 25 = 50. I added 33 + 17 = 50, and 50 + 50 = 100.

▶ Daily Routines

Have children complete daily routines.

Math Journal 2, Activity Sheet 7

Shape Cards 1

Activity Sheet 7

---

2 Focus 30–40 min

Go Online

▶ Math Message

Activity Sheets 7–12

Place a tally mark by your favorite fruit. **GMP4.1**

Then find Activity Sheets 7–12 in the back of your journal. Remove the pages from your journal and look at the shapes. Talk with your partner about things you notice about the shapes. **GMP7.1**

▶ Discussing Shapes

Math Journal 2, Activity Sheets 7–12

| WHOLE CLASS | SMALL GROUP | PARTNER | INDEPENDENT |

Math Message Follow-Up Have children share what they notice about the shapes. Sample answers: They have straight lines. Some shapes have the same number of sides. There are triangles, rectangles, a square, and some shapes that look like kites.

Have children find a shape with three sides. Tell them to look for other shapes with the same number of sides. GMP7.1 Ask: *Are the three-sided shapes the same or different? Explain your answer.* Sample answers: They are the same because they have three sides. They are different because some of their sides are different lengths.

Explain that in today's lesson children will be cutting out the Shape Cards and sorting them, making a picture graph, and measuring body parts. After explaining the Explorations activities, assign groups to each one. Plan to spend most of your time with children working on Exploration A.

▶ Exploration A: Sorting Shapes

Activity Card 96; *Math Journal 2*, pp. 190–191 and Activity Sheets 7–12

| WHOLE CLASS | **SMALL GROUP** | **PARTNER** | INDEPENDENT |

Children cut out the Shape Cards from Activity Sheets 7–12 and write their names or initials on the back of each card. After they have finished, have them think of a way to sort them into groups. GMP7.1 As children sort their shapes, ask: *How are you sorting your shapes?* Sample answer: I am putting all the shapes that have three sides together. *Which shapes are in this group?* Answers vary. *Which shapes are in that group?* Answers vary. *Why did you put them there?* Answers vary.

Academic Language Development Children may be more familiar with the term *sort* to mean *type* or *sort of*. As an example, say: *What sort of games do you like—board games or video games?* Point out that *sort* also describes the action of putting objects into like groups. Use the word *group* interchangeably with *sort* to reinforce children's understanding of this use of the term as an action. For example, say: *You can sort shapes by the number of sides. How else might you group the shapes?*

Children trace three shapes from two different sorts on journal pages 190–191 and label the sorts.

> **NOTE** Have children clip their Shape Cards together and store them in envelopes for use in Unit 8.

Activity Card 96

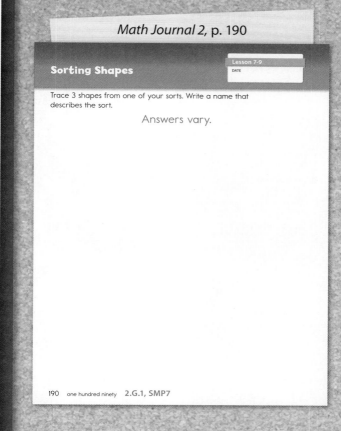

Math Journal 2, p. 190

Math Journal 2, p. 191

Lesson 7-9 **669**

Activity Card 97

Drawing a Picture Graph

Name of Fruit	
apple	ⅢⅠ
banana	ⅢⅠ-ⅢⅠ
grapes	ⅢⅠ
other	Ⅲ

What You Need
Math Journal 2, page 192
Our Favorite Fruits data

What To Do
❶ Count the tallies on the class chart.
❷ Use the tallies to complete the tally chart on journal page 192.
❸ Draw a picture graph.
• Write the name of each fruit on a line at the bottom of the graph.
• Draw smiley faces above each fruit to show the number of children who picked that fruit.

Talk About It
What does the picture graph show?
Do you think the graph would look different if another class was asked to pick their favorite fruits?

Drawing a Picture Graph

Lesson 7-9
DATE

Tally chart: Answers vary.

Name of Fruit	Number of Children
Apple	
Banana	
Grapes	
Other	

Picture graph: Graphs vary.

Our Favorite Fruits

 Apple Banana Grapes Other
Name of Fruit

KEY: 😊 = 1 child

192   one hundred ninety-two   **2.MD.10, SMP4**

▶ **Exploration B:** Drawing a Picture Graph

Activity Card 97; *Math Journal 2*, p. 192

| WHOLE CLASS | **SMALL GROUP** | **PARTNER** | INDEPENDENT |

Display the favorite-fruit data collected during the Math Message for all children to see while they complete Exploration B. Children complete a tally chart by counting the tallies from the Math Message and then use the data to draw a picture graph on journal page 192. GMP4.1

Note: The last category on the Favorite Fruits graph is "other," and it may include data from several different types of fruits. If the other category has many "smiley faces," it might be helpful to discuss the different fruits that comprise the category and why the category has so many smiley faces.

▶ **Exploration C:** Measuring Body Parts

Activity Card 98; *Math Journal 2*, p. 193

| WHOLE CLASS | **SMALL GROUP** | **PARTNER** | INDEPENDENT |

Children work as partners or in a small group to measure specified body parts. GMP5.2 They record their measurements on journal page 193.

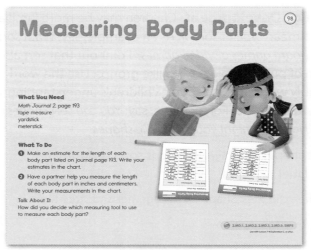

Activity Card 98

Summarize Have children share how they sorted their shapes in Exploration A.

3 Practice 10–15 min

Go Online · ePresentations · eToolkit · Home Connections

▶ Playing *Addition/Subtraction Spin*

My Reference Book, pp. 138–139; *Math Masters*, p. G23

| WHOLE CLASS | SMALL GROUP | **PARTNER** | INDEPENDENT |

Partners play several rounds of *Addition/Subtraction Spin*. See Lesson 5-6 for details.

Observe
- Which children are engaged in the game?
- Which children need additional support to play the game?

Discuss
- *What shortcut or rule did you use to help you add or subtract?*
- *How do you know your rule works?* **GMP8.1**

▶ Math Boxes 7-9

Math Journal 2, p. 189

| WHOLE CLASS | **SMALL GROUP** | **PARTNER** | **INDEPENDENT** |

Mixed Practice Math Boxes 7-9 are paired with Math Boxes 7-7.

▶ Home Link 7-9

Math Masters, p. 209

Homework Children draw a picture graph to represent a data set.

Vegetable Picture Graph

Home Link 7-9
NAME   DATE

Family Note
Today your child drew a picture graph, which uses pictures or symbols to show data. The key on a picture graph tells what each picture is for. Have your child use the data table to draw the graph.
Please return this Home Link to school tomorrow.

Favorite Vegetables

Name of Vegetable	Number of People
Carrots	4
Peas	5
Corn	3
Other	6

Favorite Vegetables Picture Graph

Carrots Peas Corn Other
Name of Vegetable
KEY: ☺ = 1 child

Practice
① __26__ = 21 + 5   ② 63 + 4 = __67__
③ __2__ + 88 = 90   ④ 7 + 35 = __42__

2.NBT.5, 2.MD.10   two hundred nine 209

Math Masters, p. 209

Math Journal 2, p. 193

Measuring Body Parts Lesson 7-9 DATE

Complete the chart. Answers vary.

Body Part	Centimeters	Inches
Wrist	Estimate: About ___ centimeters / Measurement: About ___ centimeters	Estimate: About ___ inches / Measurement: About ___ inches
Arm span	Estimate: About ___ centimeters / Measurement: About ___ centimeters	Estimate: About ___ inches / Measurement: About ___ inches
Around head	Estimate: About ___ centimeters / Measurement: About ___ centimeters	Estimate: About ___ inches / Measurement: About ___ inches
Waist	Estimate: About ___ centimeters / Measurement: About ___ centimeters	Estimate: About ___ inches / Measurement: About ___ inches
Height	Estimate: About ___ centimeters / Measurement: About ___ centimeters	Estimate: About ___ inches / Measurement: About ___ inches

2.MD.1, 2.MD.2, 2.MD.3, 2.MD.9, SMP5   one hundred ninety-three 193

Math Journal 2, p. 189

Math Boxes Lesson 7-9 DATE

① The first-grade class has 23 children. The second-grade class has 21 children. The third-grade class has 17 children. How many children are there in all?

Answer: __61__ children

② Write each number in expanded form.
463 __400 + 60 + 3__
Sample: 205 __200 + 5__
Sample: 640 __600 + 40 + 0__
888 __800 + 80 + 8__

③ The red rope is 45 feet long. The yellow rope is 30 feet long. How much longer is the red rope than the yellow rope? You may draw a diagram to help.
Number models: Sample: 45 − 30 = ?; 30 + ? = 45
Answer: __15__ feet

④ Emma had 30 stickers. Her teacher gave her 12 more. Then Emma gave 9 stickers to her friend. How many stickers does Emma have now?
Number model: Sample: 30 + 12 − 9 = ?
Emma has __33__ stickers.

⑤ **Writing/Reasoning** For Problem 2, Luka wrote 40 + 60 + 3 as the expanded form of 463. Do you agree with Luka? Explain.
Sample answer: No. The 4 is in the hundreds place and has a value of 400, not 40.

① 2.OA.1, 2.NBT.5, 6 ② 2.NBT.1, 3 ③ 2.OA.1, 2.NBT.5, 2.MD.5
④ 2.OA.1, 2.NBT.5 ⑤ 2.NBT.1, 2.NBT.3, SMP3   one hundred eighty-nine 189

Unit 7 Progress Check

Overview   **Day 1:** Administer the Unit Assessments.
Day 2: Administer the Open Response Assessment.

2-Day Lesson

 Student Learning Center
Students may take
assessments digitally.

 Assessment and Reporting
Record results and track
progress toward mastery.

Day 1: Unit Assessments

1 Warm Up   5–10 min

Materials

Self Assessment
Children complete the Self Assessment.

Assessment Handbook, p. 46

2a Assess   35–50 min

Unit 7 Assessment
These items reflect mastery expectations to this point.

Assessment Handbook, pp. 47–49

Unit 7 Challenge (Optional)
Children may demonstrate progress beyond expectations.

Assessment Handbook, p. 50

CCSS Common Core State Standards	**Goals for Mathematical Content (GMC)**	Lessons	Self Assessment	Unit 7 Assessment	Unit 7 Challenge
2.NBT.5	Add within 100 fluently.	7-1, 7-3, 7-8	1	1a–1d, 2	1a, 1b, 2
2.NBT.6	Add up to four 2-digit numbers.	7-2, 7-3	2	2	2
2.NBT.9	Explain why addition and subtraction strategies work.	7-1, 7-3	1		2
2.MD.1	Measure the length of an object.	7-4 to 7-6, 7-9	3	4a, 4b	
	Select appropriate tools to measure length.	7-4, 7-5, 7-9	6	3a–3d	
2.MD.2	Measure an object using 2 different units of length.	7-4, 7-6		4a, 4b	
	Describe how length measurements relate to the size of the unit.*	7-6, 7-8		4c	
2.MD.3	Estimate lengths.	7-4 to 7-6, 7-9	5	4a, 4b, 5a–5d	
2.MD.9	Represent measurement data on a line plot.	7-7, 7-8	4	6	
	Goals for Mathematical Practice (GMP)				
SMP5	Choose appropriate tools. GMP5.1	7-4, 7-5		3a–3d	
	Use tools effectively and make sense of your results. GMP5.2	7-4 to 7-6, 7-9	6		
SMP6	Explain your mathematical thinking clearly and precisely. GMP6.1				1c
SMP7	Use structures to solve problems and answer questions. GMP7.2	7-2, 7-3			2

*Instruction and most practice on this content is complete.

 Spiral Tracker   **Go Online** to see how mastery develops for all standards within the grade.

1 Warm Up 5–10 min

▶ Self Assessment

Assessment Handbook, p. 46

| WHOLE CLASS | SMALL GROUP | PARTNER | **INDEPENDENT** |

Children complete the Self Assessment to reflect on their progress in Unit 7.

NAME	DATE

Put a check in the box that tells how you do each skill.

Skills	I can do this. I can explain how to do this.	I can do this by myself.	I can do this with help.
① Play *Hit the Target.* MJ2 165			
② Add 3 or more numbers. MJ2 171			
③ Measure objects to the nearest inch and centimeter. MJ2 180–181			
④ Complete a line plot. MJ2 187			
⑤ Use personal references to help estimate length. MJ2 178			
⑥ Use measuring tools correctly. MJ2 178			

Unit 7 Self Assessment — Lesson 7-10

Copyright © McGraw-Hill Education. Permission is granted to reproduce for classroom use.

Assessment Handbook, p. 46

2a Assess 35–50 min

[Go Online] ✓ Assessment and Reporting   ▦▦ Differentiation Support

▶ Unit 7 Assessment

Assessment Handbook, pp. 47–49

| WHOLE CLASS | SMALL GROUP | PARTNER | **INDEPENDENT** |

Children complete the Unit 7 Assessment to demonstrate their progress on the Common Core State Standards covered in this unit.

[Go Online] for generic rubrics in the *Assessment Handbook* that can be used to evaluate children's progress on the Mathematical Practices.

Assessment Handbook, p. 47

NAME	DATE	Lesson 7-10 ✓

Unit 7 Assessment

① Fill in the missing numbers.

   a. $33 + \underline{7} = 40$ b. $90 = 88 + \underline{2}$

   c. $\underline{5} + 45 = 50$ d. $30 = \underline{6} + 24$

② Two teams are playing *Basketball Addition.* Here are their scores for the first half:

	Team A	Team B
	Player 1: 10	Player 1: 20
	Player 2: 12	Player 2: 5
	Player 3: 6	Player 3: 15
	Player 4: 8	Player 4: 3

   a. Find each team's score for the first half.

     Team A: $\underline{36}$ points   Team B: $\underline{43}$ points

   b. Which team is winning at halftime? $\underline{\text{Team B}}$

③ a. What measuring tool would you use to measure the length of your Math Journal? $\underline{\text{Answers vary.}}$

   b. Explain your choice. Answers vary but should include an explanation that is justifiable.

   c. What unit would you use when measuring the length of your Math Journal? $\underline{\text{Answers vary.}}$

   d. Explain your choice. Answers vary but should include an explanation that is justifiable.

Copyright © McGraw-Hill Education. Permission is granted to reproduce for classroom use.

Assessment Masters **47**

Assessment Handbook, p. 48

NAME	DATE	Lesson 7-10 ✓

Unit 7 Assessment (continued)

④ Estimate and then measure the length of each picture below in inches and centimeters.

   a.

   Estimates: Answers vary. Measures:

   about _____ inches   about $\underline{6}$ inches

   about _____ cm   about $\underline{15}$ cm

   b.

   Estimates: Answers vary. Measures:

   about _____ inches   about $\underline{4}$ inches

   about _____ cm   about $\underline{9}$ cm

   c. Did you get a smaller number when you measured in inches or centimeters? $\underline{\text{Inches}}$ Why?

   Sample answer: Inches are larger units than centimeters.

Copyright © McGraw-Hill Education. Permission is granted to reproduce for classroom use.

48 Assessment Handbook

Assessment Handbook, p. 49

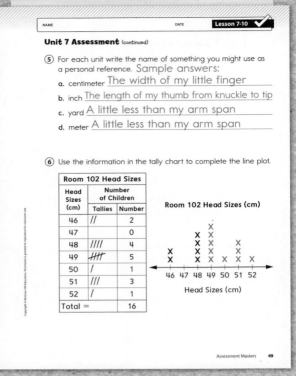

NAME   DATE   Lesson 7-10 ✓

Unit 7 Assessment (continued)

⑤ For each unit write the name of something you might use as a personal reference. Sample answers:

a. centimeter The width of my little finger

b. inch The length of my thumb from knuckle to tip

c. yard A little less than my arm span

d. meter A little less than my arm span

⑥ Use the information in the tally chart to complete the line plot.

Room 102 Head Sizes

Head Sizes (cm)	Number of Children	
	Tallies	Number
46	//	2
47		0
48	////	4
49	//////	5
50	/	1
51	///	3
52	/	1
Total =		16

Room 102 Head Sizes (cm)

Head Sizes (cm)

Assessment Masters   49

Assessment Handbook, p. 50

NAME   DATE   Lesson 7-10 ✓

Unit 7 Challenge

① You are playing *Hit the Target*. Your target number is 70 and your starting number is 36.

a. Show how you can hit the target number in two changes.

Target number: **70**   Sample answer:

Starting number	Change	Result	Change	Result
36	+ 4	40	+ 30	70

b. Show how you can hit the target number in one change.

Target number: **70**

Starting number	Change	Result
36	+ 34	70

c. How can you use Problem 1a to help you solve Problem 1b?
Sample answer: I added my two changes and made one change.

② Solve 6 + 19 + 11 + 4. **40** Explain which numbers you added first and why. Sample explanation:
First I added 6 + 4 = 10. Then I added 19 + 11 because 9 + 1 = 10 and that's 30 altogether. 10 + 30 = 40.

50   Assessment Handbook

Item(s)	Adjustments
1a–1d	To scaffold item 1, have children use a number grid to find the difference to the next multiple of 10.
2	To extend item 2, have children find the total halftime score for both teams and explain their strategy.
3	To scaffold item 3, provide visuals of a ruler, tape measure, yardstick, and meterstick.
4	To extend item 4, have children measure each item to the nearest $\frac{1}{2}$ inch.
5	To extend item 5, have children use their personal references to estimate the length of their desk.
6	To scaffold item 6, have children make a bar graph using the same data.

Advice for Differentiation

All instruction and most practice is complete for the content that is marked with an asterisk (*) on page 672.

Use the online assessment and reporting tools to track children's performance. Differentiation materials are available online to help you address children's needs.

> **NOTE** See the Unit Organizer on pages 608–609 or the online Spiral Tracker for details on Unit 7 focus topics and the spiral.

▶ Unit 7 Challenge (Optional)

Assessment Handbook, p. 50

WHOLE CLASS	SMALL GROUP	PARTNER	**INDEPENDENT**

Children can complete the Unit 7 Challenge after they complete the Unit 7 Assessment.

Day 2: Open Response Assessment

▶ **Before You Begin**
Make base-10 blocks available with at least 3 flats, 15 longs, and 20 cubes.

		Materials
2b Assess 50–55 min		
Solving the Open Response Problem Children decide if two different sets of base-10 blocks represent the same number and explain their thinking.		*Assessment Handbook,* p. 51; base-10 blocks
Discussing the Problem Children discuss the representations and their thinking.		*Assessment Handbook,* p. 51

CCSS Common Core State Standards	**Goals for Mathematical Content (GMC)**	**Lessons**
2.NBT.1	Understand 3-digit place value.	7-1
	Represent whole numbers as hundreds, tens, and ones.	7-1
2.NBT.1a	Understand exchanging tens and hundreds.	7-1
	Goal for Mathematical Practice (GMP)	
SMP2	Make sense of the representations you and others use. **GMP2.2**	7-1

Spiral Tracker **Go Online** ▷ to see how mastery develops for all standards within the grade.

▶ **Evaluating Children's Responses**
Evaluate children's abilities to apply place-value understanding to solve a problem. Use the rubric below to evaluate their work based on **GMP2.2**.

Goal for Mathematical Practice	**Not Meeting Expectations**	**Partially Meeting Expectations**	**Meeting Expectations**	**Exceeding Expectations**
GMP2.2 Make sense of the representations you and others use.	May state that one or both of the representations are correct, but provides no evidence of interpreting the base-10 blocks as ones, tens, and hundreds.	States that one or both of the representations are correct and provides evidence of interpreting base-10 blocks as ones, tens, and hundreds, but does not provide evidence of interpreting 10 ones as a ten and 10 tens as a hundred.	States that both of the representations are correct, and provides evidence of interpreting 10 ones as a ten and 10 tens as a hundred.	Meets expectations and provides evidence in at least two forms, each of which provides evidence on its own of interpreting 10 ones as a ten and 10 tens as a hundred. **OR** Meets expectations and refers to Maria's representation as most efficient (e.g., saying that it shows the number with the fewest blocks).

		Materials
3 Look Ahead 10–15 min		
Math Boxes 7-10 Children practice and maintain skills.		*Math Journal 2,* p. 194
Home Link 7-10 Children take home the Family Letter that introduces Unit 8.		*Math Masters,* pp. 210–213

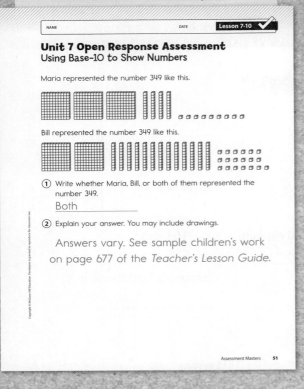

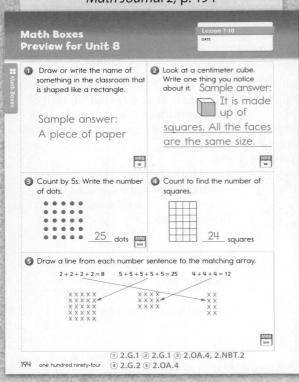

2b Assess   50–55 min   Go Online ✓
Assessment and Reporting

▶ Solving the Open Response Problem

Assessment Handbook, p. 51

| WHOLE CLASS | SMALL GROUP | **PARTNER** | **INDEPENDENT** |

This open response problem requires children to apply skills and concepts from Unit 7 to make sense of two base-10 block representations of the number 349. The focus of this task is on **GMP2.2:** Make sense of the representations you and others use.

Distribute *Assessment Handbook,* page 51. Make base-10 blocks available. Point out to children that their job in Problem 1 is to write who is correct: Maria, Bill, or both. Then they should explain their thinking in Problem 2. They should use words, drawings, or both to show what they understand about Maria's and Bill's representations. GMP2.2

Observe children as they work. You may wish to take notes. For example, note the different ways that children group 10 cubes into a long and 10 longs into a flat in Bill's representation. Encourage children to use base-10 blocks to decide who is correct and to make drawings of their trades.

> **Differentiate**   **Adjusting the Assessment**
>
> For children who have difficulty explaining their answer on paper, in words or drawings, ask them to explain orally and show you what they did using base-10 blocks. Then have them make a drawing. You may either write what they dictate or ask them to write down what they said to you.

▶ Discussing the Problem

Assessment Handbook, p. 51

| WHOLE CLASS | SMALL GROUP | PARTNER | INDEPENDENT |

After children have had a chance to complete their work, have volunteers demonstrate their strategies and explain their thinking. Include successful and unsuccessful strategies. In particular, have a volunteer show a strategy that includes exchanging blocks to make a hundred from 10 longs and a long from 10 cubes. Remind children that all correct representations show the same number, 349.

> **Differentiate**   **Common Misconception**
>
> Some children may state that Bill is incorrect because he did not exchange 10 cubes for a long and 10 longs for a flat. These children are showing important understanding of base-10 block exchanges and efficient ways of representing numbers, but they are confusing efficiency with being correct. Bill's representation is no less valid than Maria's. In fact, it is sometimes important to be able to think about and show numbers with more than the fewest number of blocks (such as when subtracting).

Evaluating Children's Responses 2.NBT.1

Collect children's work and review any of your notes about children's performance. For the content standard, expect most children to show evidence of interpreting the number 349 as 3 hundreds, 4 tens, and 9 ones. You can use the rubric on page 675 to evaluate children's work for **GMP2.2**.

See the sample in the margin. This work meets expectations for the content standard because the explanation indicates an understanding that the base-10 blocks represent hundreds, tens, and ones, and both representations show 349. The work meets expectations for the mathematical practice because the child states that both representations are correct, and the drawing shows that 10 ones can be exchanged for a ten, and 10 tens can be exchanged for a hundred. This work would exceed expectations if the child had explained trading 10 tens for a hundred to make 300 and 10 ones for a ten to make 40. **GMP2.2**

Go Online ✓ Assessment and Reporting

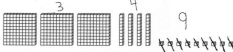

Maria represented the number 349 like this.

Bill represented the number 349 like this.

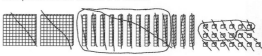

① Write whether Maria, Bill, or both of them represented the number 349. _____ **Both**

② Explain your answer. You may include drawings.

I Think Both Because Bill. has 349 and Maria has 349. Bill has 2 Hudrids and 13 tens and 19 ones that egles 349. I made a ten of ones and I made a Hudrid out of tens.

---

3 Look Ahead 10–15 min

Go Online 🏠 Home Connections

▶ Math Boxes 7-10: Preview for Unit 8

Math Journal 2, p. 194

| WHOLE CLASS | SMALL GROUP | PARTNER | INDEPENDENT |

Mixed Practice Math Boxes 7-10 are paired with Math Boxes 7-6. These problems focus on skills and understandings that are prerequisite for Unit 8. You may want to use information from these Math Boxes to plan instruction and grouping in Unit 8.

▶ Home Link 7-10: Unit 8 Family Letter

Math Masters, pp. 210–213

Home Connection The Unit 8 Family Letter provides information and activities related to Unit 8 content.

Home Link 7-10

Unit 8: Family Letter

NAME   DATE

Geometry and Arrays

In Unit 8 children explore 2-dimensional shapes, including triangles, quadrilaterals, pentagons, and hexagons. They describe and sort the shapes according to their attributes, such as number of sides, length of sides, number of angles, and whether they have right angles or parallel sides.

These shapes each have at least one right angle.   These shapes have no right angles.

Children also look for 2-dimensional shapes in 3-dimensional objects. For example, they look at a cube and notice that each face, or side, of the cube is a square.

After these shape activities, children also explore techniques for partitioning rectangles into rows and columns of same-size squares. These activities lay the foundation for area measurement in Grade 3.

This rectangle is partitioned into 2 rows and 6 columns of squares.

In the last part of the unit, children solve number stories involving equal groups of objects. In some cases equal groups are small clusters of objects, such as petals on flowers. In other cases the equal groups are the rows or columns of rectangular arrays.

Equal groups of petals: 3 flowers with 5 petals on each flower is 15 petals in all.   An array of chairs: 3 rows with 5 chairs in each row is 15 chairs in all.

Children build equal groups and arrays with counters and explore strategies for finding how many counters there are in all. These activities lay the foundation for work with multiplication in Grade 3.

Please keep this Family Letter for reference as your child works through Unit 8.

210   two hundred ten

Unit 8 Organizer
Geometry and Arrays

In this unit, children explore 2- and 3-dimensional shapes and their attributes. They partition rectangles into rows and columns of same-size squares. At the end of the unit, they explore strategies for determining the total number of objects in equal groups and rectangular arrays. Children's learning will focus on two clusters of the Common Core's content standards, as well as in-depth work on two of the Mathematical Practices.

 Standards for Mathematical Content

Domain	Cluster
Operations and Algebraic Thinking	Work with equal groups of objects to gain foundations for multiplication.
Geometry	Reason with shapes and their attributes.

Because the standards within each domain can be broad, *Everyday Mathematics* has unpacked each standard into Goals for Mathematical Content GMC . For a complete list of Standards and Goals, see page EM1.

For an overview of the CCSS domains, standards, and mastery expectations in this unit, see the **Spiral Trace** on pages 684–685. See the **Mathematical Background** (pages 686–688) for a discussion of the following key topics:

- 2- and 3-Dimensional Shapes
- Partitioning Rectangles
- Equal Groups and Arrays

 Standards for Mathematical Practice

SMP1 Make sense of problems and persevere in solving them.

SMP7 Look for and make use of structure.

For a discussion about how *Everyday Mathematics* develops these practices and a list of Goals for Mathematical Practice GMP , see page 689.

  **Virtual Learning Community** **Go Online** to **vlc.cemseprojects.org** to search for video clips on each practice.

©Lane Oatey/Blue Jean Images/Getty Images

Go Digital with these tools at **connectED.mheducation.com**

 ePresentations Student Learning Center Facts Workshop Game eToolkit Professional Development Home Connections Spiral Tracker Assessment and Reporting English Learners Support Differentiation Support

Contents

*The standards listed here are addressed in the **Focus** of each lesson. For all the standards in a lesson, see the Lesson Opener.

Unit 8 Materials

VLC Virtual Learning Community

See how *Everyday Mathematics* teachers organize materials. Search "Classroom Tours" at **vlc.cemseprojects.org**.

Lesson	Math Masters	Activity Cards	Manipulative Kit	Other Materials
8-1	p. 214	99–100	per group: 4 each of number cards 0–10; pattern blocks; geoboards; rubber bands; ruler	slate; Shape Cards (from Lesson 7-9); Two-Dimensional Shapes poster
8-2	pp. 215; TA8; TA31; G28	101	10-centimeter ruler	slate; Activity Sheet 13; Shape Cards; Attribute Cards; scissors; Fact Triangles
8-3	pp. 216; G19; G20	99; 102	straws; twist ties; per partnership: base-10 blocks (5 flats, 30 longs, and 30 cubes); 4 each of number cards 0–9	slate; straw-and-twist-tie triangles (from Math Message); small bag or box; Shape Cards; pretzel sticks or toothpicks
8-4	pp. 217–219; TA5		1 square pattern block	slate; chart paper; straightedge, colored pencils (optional); Two-Dimensional Shapes poster; Guidelines for Discussions Poster; children's work from Day 1; selected samples of children's work
8-5	pp. 128; 220–221; TA32	103	base-10 thousand cube; inch ruler	slate; stick-on note; Class Data Pad; rectangular prism, cylinder, cone, sphere, and pyramid (for demonstration); 3-dimensional shapes (brought in by children); stick-on notes from the Math Message; Pattern-Block Template; per group: 27 centimeter cubes, small stickers (optional)
8-6	pp. 222–228; TA3; G17		1-inch square pattern block; counters; cards from the Everything Math Deck; calculator; per partnership: 4 each of number cards 0–9	slate; 20 centimeter cubes; Pattern-Block Template; scissors
8-7	pp. 229–233		1-inch square pattern blocks; base-10 blocks (optional); centimeter ruler; cards from the Everything Math Deck	slate; number grid or number line (optional); Pattern-Block Template; index card; scissors
8-8	pp. 234; TA7; *Assessment Handbook*, pp. 98–99		per group: 4 each of number cards 0–9, calculator; 24 counters; inch ruler	slate; per group: large paper triangle (optional)
8-9	pp. 235–236; G26–G27	104	36 counters; per partnership: one 6-sided die; per group: one 20-sided polyhedral die or three 6-sided dice	slate; magazine pictures (optional)
8-10	pp. 237; TA25; G29–G31	105–106	per partnership: 12 counters; base-10 blocks (optional)	scissors; *Array Concentration* Array Cards and Number Cards; number grid or number line, slate (optional); per partnership: 50 centimeter cubes
8-11	pp. 238–240; TA9 (optional); TA10; G31	107–110	trapezoid pattern blocks; geoboards; rubber bands	slate; Shape Cards; several small opaque bags or boxes; Pattern-Block Template; *Array Concentration* Array Cards and Number Cards; large piece of paper; scissors; tape or glue; folder
8-12	pp. 241–244; *Assessment Handbook*, pp. 52–61			1 centimeter cube

Literature Link
8-1 *The Greedy Triangle* (optional) 8-3 *Shape Up* (optional)
8-8 *Each Orange Had 8 Slices: A Counting Book* (optional) 8-10 *One Hundred Hungry Ants* (optional)

Go Online for a complete literature list for Grade 2.

Problem Solving  Professional Development

Everyday Mathematics emphasizes equally all three of the Common Core's dimensions of **rigor:**

- conceptual understanding
- procedural skill and fluency
- applications

Math Messages, other daily work, Explorations, and Open Response tasks provide many opportunities for children to apply what they know to solve problems.

▶ Math Message

Math Messages require children to solve a problem they have not previously been shown how to solve. Math Messages provide almost daily opportunities for problem solving.

▶ Daily Work

Journal pages, Home Links, Writing/Reasoning prompts, and Differentiation Options often require children to solve problems in mathematical contexts and real-life situations. **Minute Math+** offers Number Stories for transition times and spare moments throughout the day. See Routine 6, pages 38–43.

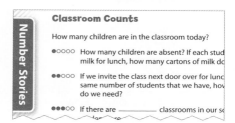

▶ Explorations

In Exploration A children identify hidden shapes based on their attributes. In Exploration B children build shapes with trapezoids based on given attributes. In Exploration C children partition shapes on a geoboard into equal parts.

▶ Open Response and Reengagement

In Lesson 8-4 children draw quadrilaterals with specified attributes. Children explain how they know their shapes are quadrilaterals and have the specified attributes. The reengagement discussion on Day 2 could focus on determining whether or not other children's shapes meet the requirements for the problem or helping each other write better explanations. Making mathematical conjectures and arguments is the focus practice for the lesson. Making conjectures, or educated guesses, can help children extend their mathematical thinking. Making arguments, or complete and convincing explanations, can help children organize their thoughts and communicate them to others. GMP3.1

VLC **Virtual Learning Community** Go Online to watch an Open Response and Reengagement lesson in action. Search "Open Response" at **vlc.cemseprojects.org**.

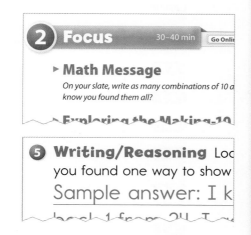

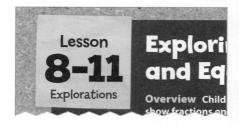

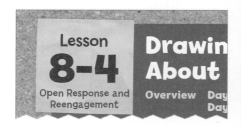

Assessment and Differentiation

Assessment and Reporting

See pages xxii–xxv to learn about a comprehensive online system for recording, monitoring, and reporting children's progress using core program assessments.

 **Virtual Learning Community** **Go Online** to **vlc.cemseprojects.org** for tools and ideas related to assessment and differentiation from *Everyday Mathematics* teachers.

✓ Ongoing Assessment

In addition to frequent informal opportunities for "kid watching," every lesson (except Explorations) offers an **Assessment Check-In** to gauge children's performance on one or more of the standards addressed in that lesson.

Lesson	Task Description	CCSS Common Core State Standards
8-1	Sort shapes according to whether or not they have parallel sides and whether or not they have right angles.	2.G.1, SMP7
8-2	Draw and name a triangle, a shape with a right angle, and a shape with at least 1 pair of parallel sides.	2.G.1
8-3	Build and draw 5- and 6-sided polygons.	2.G.1
8-4	Attempt to improve the drawing of Linda's garden and the explanation of why the drawing works.	2.G.1, SMP3
8-5	Describe a cube as having 6 equal-size square faces.	2.G.1, SMP6
8-6	Show evidence of a strategy to partition a rectangle, such as tracing the pattern block multiple times, drawing rows or columns of squares, or drawing squares along the edges of the rectangles.	2.G.2
8-7	Partition a square into two rows with two same-size squares in each row and count the total number of squares.	2.G.2
8-8	Solve number stories using drawings and write addition number models.	2.OA.1, 2.OA.4, SMP4
8-9	Use counters to create arrays, draw the arrays, and record addition number models.	2.OA.4, SMP2
8-10	Match number cards and array cards and write addition number models for the arrays.	2.OA.4

▶ Periodic Assessment

Unit 8 Progress Check This assessment focuses on the CCSS domains of *Operations and Algebraic Thinking* and *Geometry*. It also contains a Cumulative Assessment to help monitor children's learning and retention of content that was the focus of Units 1–7.

NOTE Odd-numbered units include an **Open Response Assessment.** Even-numbered units include a **Cumulative Assessment.**

▶ Unit 8 Differentiation Activities

 Differentiation Support   English Learners Support

Differentiation Options Every regular lesson provides **Readiness, Enrichment, Extra Practice,** and **English Language Learners Support** activities that address the Focus standards of that lesson.

Activity Card 104

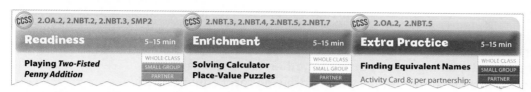

Activity Cards These activities, written to the children, enable you to differentiate Part 2 of the lesson through small-group work.

English Language Learners Activities and point-of-use support help children at different levels of English language proficiency succeed.

Differentiation Support Two online pages for most lessons provide suggestions for game modifications, ways to scaffold lessons for children who need additional support, and language development suggestions for Beginning, Intermediate, and Advanced English language learners.

Differentiation Support online pages

For **ongoing distributed practice,** see these activities:
- Mental Math and Fluency
- Differentiation Options: Extra Practice
- Part 3: Journal pages, Math Boxes, *Math Masters,* Home Links
- Print and online games

Ongoing Practice   Differentiation Support

▶ Games

Games in *Everyday Mathematics* are an essential tool for practicing skills and developing strategic thinking.

Lesson	Game	Skills and Concepts	CCSS Common Core State Standards
8-1	Subtraction Top-It	Practicing subtraction facts	2.OA.2
8-2	Shape Capture	Identifying shapes by their attributes	2.G.1, SMP7
8-3	Target	Using base-10 blocks to model addition and subtraction	2.NBT.1, 2.NBT.1a, 2.NBT.1b, 2.NBT.3, 2.NBT.7, SMP2
8-6	The Number-Grid Difference Game	Practicing subtraction with 2-digit numbers	2.NBT.5, 2.NBT.7, SMP2
8-8	Beat the Calculator	Practicing addition facts	2.OA.2
8-9	Basketball Addition	Adding three or more numbers	2.NBT.5, 2.NBT.6, 2.NBT.9, SMP1, SMP7
8-10 8-11	Array Concentration	Finding the total number of objects in arrays and writing matching number models	2.OA.4, 2.NBT.2, SMP2, SMP6
8-10	Array Bingo	Finding the total number of objects in arrays	2.OA.4, SMP2

VLC Virtual Learning Community **Go Online** to look for examples of *Everyday Mathematics* games at **vlc.cemseprojects.org**.

Ⓒ Spiral Trace: Skills, Concepts, and Applications

★ **Mastery Expectations** This Spiral Trace outlines instructional trajectories for key standards in Unit 8. For each standard, it highlights opportunities for Focus instruction, Warm Up and Practice activities, as well as formative and summative assessment. It describes the **degree of mastery**— as measured against the entire standard—expected at this point in the year.

Operations and Algebraic Thinking

2.OA.1 Use addition and subtraction within 100 to solve one- and two-step word problems involving situations of adding to, taking from, putting together, taking apart, and comparing, with unknowns in all positions, e.g., by using drawings and equations with a symbol for the unknown number to represent the problem.

| 5-12 Progress Check | 6-1 Warm Up Practice | 6-2 through 6-5 Warm Up Focus Practice | 6-7 Practice | 6-9 Focus | 7-2 Focus Practice | 8-8 Focus | 8-12 Progress Check | 9-9 Focus Practice | 9-10 Focus Practice | 9-12 Progress Check |

★ By the end of Unit 8, expect children to **use addition and subtraction within 100 to solve 1-step word problems involving situations of adding to, taking from, putting together, taking apart, and comparing, with unknowns in all positions, e.g., by using drawings and equations with a symbol for the unknown to represent the problem; use addition and subtraction within 100 to solve 2-step word problems involving situations of adding to, taking from, putting together, taking apart, and comparing, with unknowns in all positions, e.g., by using drawings to represent the problem.**

2.OA.4 Use addition to find the total number of objects arranged in rectangular arrays with up to 5 rows and up to 5 columns; write an equation to express the total as a sum of equal addends.

| 4-11 Focus | 5-5 Focus | 6-10 Focus Practice | 8-8 through 8-10 Focus Practice | 8-11 Practice | 8-12 Progress Check | 9-1 Practice | 9-5 through 9-8 Practice | 9-10 Warm Up Focus | 9-12 Progress Check |

★ By the end of Unit 8, expect children to **use counting strategies or addition to find the total number of objects arranged in rectangular arrays with up to 5 rows and up to 5 columns; write an equation to express the total as a sum of equal addends.**

Number and Operations in Base Ten

2.NBT.2 Count within 1000; skip-count by 5s, 10s, and 100s.

| 5-2 through 5-4 Warm Up Focus Practice | 5-6 Warm Up Focus | 5-10 Warm Up Focus | 6-1 Focus Practice | 8-8 through 8-10 Focus Practice | 8-11 Practice | 9-1 Practice | 9-8 Focus Practice | 9-11 Focus Practice | 9-12 Progress Check |

★ By the end of Unit 8, expect children to **count within 1,000; skip count by 5s, 10s, and 100s; and apply these skills to find the total number of objects in arrays.**

Go to **connectED.mheducation.com** for comprehensive trajectories that show how in-depth mastery develops across the grade.

Spiral Tracker

Geometry

2.G.1 Recognize and draw shapes having specified attributes, such as a given number of angles or a given number of equal faces. Identify triangles, quadrilaterals, pentagons, hexagons, and cubes.

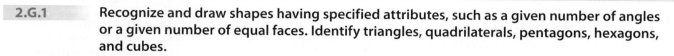

| 5-5 Focus | 6-10 Focus | 7-6 Practice | 7-9 Focus | 7-10 Practice | 8-1 through 8-5 Focus Practice | 8-11 Focus Practice | 8-12 Progress Check | 9-5 Practice |

By the end of Unit 8, expect children to **recognize, identify, and draw shapes having specified attributes, such as with a given number of angles or sides or a given number of pairs of parallel sides; identify triangles, quadrilaterals, hexagons, and cubes.**

2.G.2 Partition a rectangle into rows and columns of same-size squares and count to find the total number of them.

| 1-12 Focus | 3-11 Focus | 8-6 Focus Practice | 8-7 Focus Practice | 8-12 Progress Check | 9-1 Practice | 9-3 Practice | 9-6 Practice | 9-11 Practice |

By the end of Unit 8, expect children to **partition a rectangle into 2 or 4 rows and columns of same-size squares and count to find the total number of them.**

2.G.3 Partition circles and rectangles into two, three, or four equal shares, describe the shares using the words *halves, thirds, half of, a third of,* etc., and describe the whole as two halves, three thirds, four fourths. Recognize that equal shares of identical wholes need not have the same shape.

| 2-8 Focus | 8-11 Focus | 9-1 through 9-4 Focus Practice | 9-5 through 9-11 Practice | 9-12 Progress Check |

By the end of Unit 8, expect children to **partition simple shapes into two equal shares using geoboards and rubber bands.**

Key ✓ = Assessment Check-In    = Progress Check Lesson   = Current Unit   = Previous or Upcoming Lessons

Mathematical Background: Content

 This discussion highlights the major content areas and the Common Core State Standards addressed in Unit 8. See the online Spiral Tracker for complete information about the learning trajectories for all standards.

▶ 2- and 3-Dimensional Shapes

(Lessons 8-1 through 8-5)

The study of geometry has many connections to the real world. Children encounter and make sense of 2- and 3-dimensional shapes in their daily lives, so they have a wealth of informal knowledge about shapes when they come to school. Through their study of geometry, children will organize and formalize their knowledge of shapes and the relationships of these shapes to each other.

Just as numbers can be related to one another in various ways—for example, $5 > 2$ and $\frac{1}{2} = 0.5$—geometric objects can be related to one another. Children discuss two types of geometric relationships in this unit: parallel and congruent.

Two lines are *parallel* if they lie in the same plane and are always the same distance apart. Many objects in the real world resemble parallel lines, including railroad tracks, highway lane markings, and lines on notebook paper. Two sides of a 2-dimensional shape are parallel if they lie along parallel lines. That is, if the sides could be extended to form parallel lines, then the sides themselves are parallel.

Two geometric figures are *congruent* if they are exactly the same size and the same shape. Two line segments or sides are congruent if they are the same length. Two 2-dimensional shapes are congruent if one shape can be flipped, turned, or slid to line up exactly with the other.

Two 3-dimensional figures can also be congruent if they are exactly the same size and shape. Likewise, the various faces of a 3-dimensional shape can be congruent. For example, all the faces of a cube are congruent to one another because they are all squares and all the same size.

Second graders are not expected to use the word *congruent*. Lesson 8-5 discusses the idea of congruent faces but uses the more kid-friendly language of *same size and same shape*. Because congruence is the geometric equivalent of numerical equality, it is natural and appropriate for children to say that two congruent faces are "equal." However, it is mathematically imprecise to say that geometric figures are equal, so children should be guided toward describing faces as being the same size and the same shape.

In Lessons 8-1 through 8-5, children describe shapes according to their attributes. **2.G.1** An *attribute* is a feature of an object or a common feature of a set of objects. The attributes of a 2-dimensional shape include the number and lengths of the sides, the number and sizes of the angles, and the presence or absence of parallel sides. The attributes of a 3-dimensional shape include the number, shape, and size of the faces. Comparison and sorting activities are used to draw children's attention to particular attributes. For more information, see the discussion of Mathematical Practice 7 on page 689.

 Standards and Goals for Mathematical Content

Because the standards within each domain can be broad, *Everyday Mathematics* has unpacked each standard into Goals for Mathematical Content GMC. For a complete list of Standards and Goals, see page EM1.

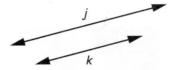

Line *j* is parallel to line *k*.

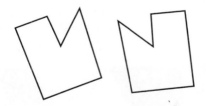

Congruent 2-dimensional figures

▶ Partitioning Rectangles (Lessons 8-6 and 8-7)

To *partition* an object or space is to divide it into parts. For example, temporary walls are often used to partition a workspace into cubicles, and food trays and to-go boxes are sometimes partitioned so that foods can be separated.

In the real world, partitions may not all be the same size. In mathematics, it is often useful to partition objects into same-size pieces. Fractions, for example, are based on the idea of partitioning a whole into equal parts. Similarly, a ruler represents the partition of a length into same-size pieces, with one piece being the length unit.

In Lessons 8-6 and 8-7, children partition rectangles into rows and columns of same-size squares. **2.G.2** These lessons build on earlier work of covering rectangles with same-size squares by having children visualize how this covering would look and divide the rectangle into squares to show where the pattern-block faces would be.

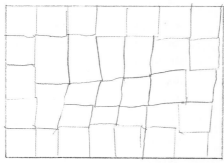

One child's attempt at partitioning a rectangle

Visualizing the row-and-column structure of a partitioned or covered rectangle is not trivial for children. Often children are able to draw squares around the edges of a rectangle but lose track of the structure in the middle, particularly for larger rectangles. Children are encouraged to monitor their own thinking and keep trying because spatial thinking develops through practice.

After children have successfully partitioned a rectangle, they count the resulting squares. **2.G.2** Accurately counting the squares is also not a trivial task. When counting squares one-by-one, some children may miss some squares or count some twice. The row-and-column structure can be helpful in this regard. If children know that there are 5 squares in each row, for example, they can count by 5s as they point to the rows to help them find the total. In this case, a row is being used as a *composite unit,* or a larger unit made up of smaller units. However, this strategy won't be apparent to all children, and Lessons 8-6 and 8-7 do not emphasize this idea. Children will gain more practice with "thinking in rows and columns" when they work with arrays in Lessons 8-8 through 8-10.

The partitioning work in Grade 2 lays the foundation for area measurement in Grade 3. The *area* of a figure is a count of how many identical square units are needed to cover the figure without gaps or overlaps. To understand the meaning of an area measurement in square units, children need to be able to visualize the row-and-column structure of square units covering a surface. In third grade, children will build on this work and discover, for example, that the length of a rectangle in inches corresponds to the number of inch squares that fit in one row, just as the width of a rectangle in inches corresponds to the number of inch squares that fit in one column. Making these connections is an important step toward understanding why multiplying the length by the width of a rectangle produces its area.

▶ Equal Groups and Arrays

(Lessons 8-8 through 8-10)

One of the clusters in the Grade 2 Common Core State Standards requires children to "Work with equal groups of objects to gain foundations for multiplication." Just as addition and subtraction can be interpreted in several ways (for example, adding to, taking from, putting together, taking apart, and comparing), so can multiplication. Two interpretations of multiplication are explored in Lessons 8-8 through 8-10: equal groups and rectangular arrays.

Four equal groups of eggs

Starting from a very young age, children encounter *equal-groups* situations. For example, crayons and markers come in packages with equal numbers in each package. Tables in a classroom often all have the same number of chairs. In Lesson 8-8 children solve number stories involving equal groups of objects by using their own invented strategies. Children will often solve these problems by modeling them with concrete objects or drawings, skip counting, or adding in two or more steps. **2.OA.1, 2.NBT.2, 2.NBT.7**

Many real-world situations involve rectangular *arrays,* or rectangular arrangements of objects in rows and columns. For example, the seats in a movie theater are often arranged in rectangular arrays, and a checkerboard is a rectangular array of squares. Because each row has the same number of objects, the rows in an array can be thought of as equal groups. Likewise, the columns of an array can also be thought of as equal groups. In Lesson 8-8 children solve number stories about finding the total number of objects in rectangular arrays, often by modeling, skip counting, or using repeated addition. **2.OA.4, 2.NBT.2, 2.NBT.7**

Array situations could be considered special cases of equal-groups situations. When equal groups are not arranged in an array, the equal-groups situation is not as structured. Array situations also have the added element of showing two sets of equal groups simultaneously (the rows and the columns). For these as well as other reasons, arrays are often more difficult for children to interpret compared with other equal-groups situations. Therefore, in this unit arrays are treated as related to but separate from other equal-groups situations. Children will encounter both types of situations and look for connections between them. For example, in Lesson 8-9 children use counters to show that 3 groups of 5 objects have the same total number of objects as an array with 3 rows of 5 objects in each row.

While solving number stories, as well as when working with more abstract groups of counters, children write number models to represent their equal groups and arrays. All children are expected to write addition models. For example, the number model $5 + 5 + 5 = 15$ represents 3 groups of 5, or an array with 3 rows of 5. Some children may be ready to write multiplication number models, such as $3 \times 5 = 15$, but this is not required in Grade 2. Regardless of whether children write addition or multiplication number models, they are encouraged to read them using equal-groups language, such as "3 groups of 5 is 15 altogether." Using this language, rather than "5 plus 5 plus 5 is 15" or "3 times 5 is equal to 15," helps children build a conceptual understanding of multiplication. It is particularly important for children to develop a solid grasp of the foundational concepts of multiplication in Grade 2. The Common Core State Standards expects mastery of all multiplication facts by the end of Grade 3.

Mathematical Background: Practices

 *In Everyday Mathematics, children learn the **content** of mathematics as they engage in the **practices** of mathematics. As such, the Standards for Mathematical Practice are embedded in children's everyday work, including hands-on activities, problem-solving tasks, discussions, and written work. Read here to see how Mathematical Practices 1 and 7 are emphasized in this unit.*

▶ Standard for Mathematical Practice 1

According to Mathematical Practice 1, mathematically proficient students "monitor and evaluate their progress [when solving a problem] and change course if necessary." In Lessons 8-6 and 8-7, children partition rectangles into rows and columns of same-size squares. This task is difficult for most children because it requires a type of spatial sense that takes practice to develop.

In Lesson 8-6 children share their first attempts at partitioning a rectangle. They discuss the difficulties they encountered and the mistakes they may have made. Then they use what they learned to try again, thereby learning to "keep trying when [a] problem is hard." GMP1.3 Before they try to partition a new rectangle, in Lesson 8-7 children discuss things that they should think about while they partition, such as making same-size squares and making rows that all have the same number of squares. This discussion prepares children to "reflect on [their] thinking as [they] solve [their] problem." GMP1.2

▶ Standard for Mathematical Practice 7

According to Mathematical Practice 7, mathematically proficient students "look closely to discern a pattern or structure. Young students, for example, . . . may sort a collection of shapes according to how many sides the shapes have." In Lessons 8-1 through 8-5, children describe shapes according to their properties. For example, children might describe a rectangle as having 4 sides, 4 angles, 2 pairs of equal-length sides, 2 pairs of parallel sides, and/or 4 right angles. However, children may not notice all of these attributes by looking at a single shape.

Comparison and sorting activities are used in these lessons to draw children's attention to specific attributes. For example, comparing a rectangle to a square can focus children's attention on the lengths of the sides. Comparing a cube to a pyramid can focus children's attention on the shapes of the faces. Sorting shapes according to the presence or the absence of a right angle focuses children's attention on right angles. In this way children "look for mathematical structures such as . . . properties" and "use structures to solve problems" in Unit 8. GMP7.1, GMP7.2 These activities help prepare children to recognize and draw shapes having specified attributes, or properties, as required by the Grade 2 content standards.

 Standards and Goals for Mathematical Practice

SMP1 Make sense of problems and persevere in solving them.

> **GMP1.1** Make sense of your problem.

> **GMP1.2** Reflect on your thinking as you solve your problem.

> **GMP1.3** Keep trying when your problem is hard.

> **GMP1.4** Check whether your answer makes sense.

> **GMP1.5** Solve problems in more than one way.

> **GMP1.6** Compare the strategies you and others use.

SMP7 Look for and make use of structure.

> **GMP7.1** Look for mathematical structures such as categories, patterns, and properties.

> **GMP7.2** Use structures to solve problems and answer questions.

Go Online to the *Implementation Guide* for more information about the Mathematical Practices.

For children's information on the Mathematical Practices, see *My Reference Book*, pages 1–22.

▶ **Before You Begin**
Be sure to have available the Shape Cards that children cut out and sorted in Lesson 7-9.
For the optional Extra Practice activity, obtain the book *The Greedy Triangle* by Marilyn Burns
(Scholastic Inc., 2008).

▶ **Vocabulary**
attribute • side • angle • vertex • parallel • right angle

 Common Core State Standards

Focus Cluster
Reason with shapes and their attributes.

---

1 Warm Up 15–20 min

	Materials	
Mental Math and Fluency Children solve basic facts.	slate	2.OA.2
Daily Routines Children complete daily routines.	See pages 4–43.	See pages xiv–xvii.

---

2 Focus 20–30 min

Math Message Children identify 3- and 4-sided shapes.	Shape Cards (from Lesson 7-9)	2.G.1
Describing Shapes Children describe 3-, 4-, 5-, and 6-sided shapes.	Shape Cards (for demonstration), Two-Dimensional Shapes poster, Shape Cards	2.G.1 SMP7
Discussing Attributes Children sort shapes based on parallel sides and right angles.	Shape Cards, ruler	2.G.1 SMP7
Assessment Check-In See page 694.		2.G.1, SMP7

CCSS 2.G.1 Spiral Snapshot

GMC Identify 2- and 3-dimensional shapes.

| 5-5
Focus | 6-10
Focus | 8-1
Focus
Practice | 8-2 through 8-5
Focus
Practice | 8-11
Focus
Practice | 9-5
Practice |

Spiral Tracker   **Go Online** to see how mastery develops for all standards within the grade.

---

3 Practice 10–15 min

Playing Subtraction Top-It **Game** Children practice subtraction facts while playing *Subtraction Top-It*.	*My Reference Book*, pp. 170–172; per group: 4 each of number cards 0–10	2.OA.2
Math Boxes 8-1 Children practice and maintain skills.	*Math Journal 2*, p. 195	See page 695.
Home Link 8-1 **Homework** Children find shapes in newspapers and magazines.	*Math Masters*, p. 214	2.NBT.5, 2.G.1

---

connectED.mheducation.com

Plan your lessons online with these tools.

 ePresentations Student Learning Center Facts Workshop Game eToolkit Professional Development Home Connections Spiral Tracker Assessment and Reporting English Learners Support Differentiation Support

Differentiation Options RtI

CCSS 2.G.1

Readiness   5–15 min

Sorting Pattern Blocks

WHOLE CLASS
SMALL GROUP
PARTNER
INDEPENDENT

pattern blocks

To explore 2-dimensional shapes, children sort pattern blocks. Partnerships need one of each pattern block. Partners decide together how to sort the blocks into two groups and then they carry out the sort. Ask children to describe how they sorted the blocks. Sample answers: Blocks with four sides and blocks that do not have four sides; shapes you can make with other shapes, and shapes you cannot make with other shapes

CCSS 2.G.1

Enrichment   5–15 min

Solving Shape Riddles

WHOLE CLASS
SMALL GROUP
PARTNER
INDEPENDENT

Activity Card 99, Shape Cards, paper

To apply their understanding of attributes of 2-dimensional shapes, children pose and solve shape riddles. Display a set of Shape Cards for children to see and handle. Each child chooses a shape and makes up a riddle based on its attributes. Partners then solve each other's riddles. Children may make up additional riddles as time allows. Consider making a class collection of shape riddles to post on a bulletin board or assemble in a class book.

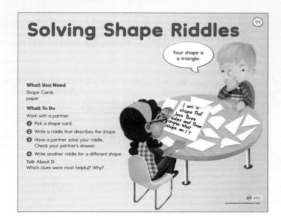

CCSS 2.G.1, SMP4

Extra Practice   15–30 min

Identifying Attributes of Shapes

WHOLE CLASS
SMALL GROUP
PARTNER
INDEPENDENT

Activity Card 100, *The Greedy Triangle*, geoboards, rubber bands

Literature Link For practice identifying attributes of shapes, children read ***The Greedy Triangle*** by Marilyn Burns (Scholastic Inc., 2008) in small groups, as partners, or on their own. This book focuses on the characteristics of different shapes. A triangle, dissatisfied with its life, asks for more angles and sides, but it eventually realizes that staying its own shape is best. Children each make a triangle on a geoboard with a rubber band. GMP4.1 As the story is read, children make changes to their shapes' angles and sides to fit the story. After the story children discuss the different attributes of the shapes in the story.

English Language Learners Support

Beginning ELL To prepare children to discuss the attributes of shapes, preteach the terms *side, straight, parallel, angle,* and *vertex* by using Total Physical Response directions and think-aloud statements. Construct shapes out of straws and twist ties, pointing out attributes such as *straight sides, parallel sides,* and *angles.* Invite children to make their own shapes and imitate your actions, repeating the names of the attributes. Assess children's understanding of the terms by displaying a variety of shapes and asking them to name the attributes orally or, depending on their oral proficiency, point them out.

Go Online   ELL English Learners Support

CCSS Standards and Goals for
Mathematical Practice

SMP7 **Look for and make use of structure.**
GMP7.1 Look for mathematical structures such as categories, patterns, and properties.
GMP7.2 Use structures to solve problems and answer questions.

1 Warm Up   15–20 min   Go Online   ePresentations   eToolkit

▶ Mental Math and Fluency

Pose facts one at a time. Children write the sum or difference on their slates.
Leveled exercises:

●○○ 3 + 4 7   ●●○ 3 + 8 11   ●●● 5 − 2 3
       5 + 6 11   7 + 9 16   6 − 4 2
       4 + 5 9   6 + 8 14   7 − 3 4

▶ Daily Routines

Have children complete daily routines.

2 Focus   20–30 min   Go Online   ePresentations   eToolkit

▶ Math Message

Look at your Shape Cards. Pick one shape with 3 sides and another shape with 4 sides. Be prepared to describe the shapes.

▶ Describing Shapes

| **WHOLE CLASS** | SMALL GROUP | PARTNER | INDEPENDENT |

Math Message Follow-Up First discuss the 3-sided shapes. Ask:

• *Which Shape Cards have 3 sides?* Shapes A, B, and C
• *Describe a 3-sided shape.* Sample answers: It has 3 sides. It has 3 angles. It has 3 vertices. **GMP7.1**

Display Shape A and point to the sides and the angles.

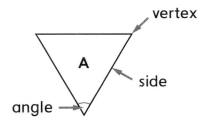

• *What do we call 3-sided shapes?* Triangles

Next discuss the 4-sided shapes. Ask:

• *Which Shape Cards have 4 sides?* Shapes D, G, H, I, J, K, and L
• *Describe a 4-sided shape.* Sample answers: It has 4 sides. It has 4 angles. It has 4 vertices. **GMP7.1**

Have children look at the Two-Dimensional Shapes poster and locate the 4-sided shapes. Explain that all 4-sided shapes belong to a family of shapes called *quadrilaterals*.

NOTE Although all of the 4-sided shapes can be classified as quadrilaterals, some quadrilaterals have special features and names. Shapes G, H, I, J, and K are trapezoids. Shapes I and J are rhombuses. Shape J is a square. Shapes J and K are rectangles. Shapes I, J, and L are kites.

It is important that children know all 4-sided polygons are quadrilaterals but do not expect them to know all the names of various 4-sided shapes.

Now have children find the Shape Cards that have 5 sides. Shapes E, F, and O Have children describe a 5-sided shape. Sample answers: It has 5 sides. It has 5 angles. It has 5 vertices. **GMP7.1** Have children look at the Two-Dimensional Shapes poster and locate the 5-sided shapes. Explain that all 5-sided shapes are called *pentagons*. Repeat with the 6-sided shapes. Shapes M, N, and P; 6 sides, 6 angles, and 6 vertices Have children look at the Two-Dimensional Shapes poster and locate the 6-sided shapes. Explain that all 6-sided shapes are called *hexagons*. Tell children they will explore other shape **attributes.**

▶ Discussing Attributes

WHOLE CLASS	SMALL GROUP	PARTNER	INDEPENDENT

Ask children to examine their Shape Cards to find things that are the same about all the shapes. You might take this opportunity to assess children's knowledge of shape attributes. Expect answers like the following:

- All of the shapes are made up of straight **sides** (line segments).
- Any two sides that meet form an **angle.** All the shapes have angles.
- The point at which two sides of a shape meet is called the **vertex.**

Explain that another attribute some shapes have is **parallel** sides. Have children place a ruler on blank paper and draw lines along its top edge and its bottom edge. Explain that the two line segments are parallel because they are the same distance apart.

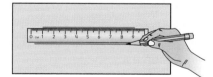

Using a ruler to draw parallel line segments

Tell children to place a pencil on the top line segment. Have them slowly slide it down toward the bottom line segment without turning or angling it at all. Have them repeat the action starting at the bottom segment and sliding the pencil to the top segment. Explain that because the pencil can be slid from one segment to the other without any turning or angling, they are parallel.

Display Shape G and point to the pair of parallel sides. Ask: *Do you think these two sides are parallel?* Yes. Have children slide a pencil from one side of the shape to the other to check if the sides are parallel. Remind them that they should not turn or angle their pencil. If children are still unsure, have them trace the opposite sides of the shape. Children should see that the sides are parallel.

Now ask whether the other pair of opposite sides is parallel. No. Have children slide a pencil from one of these sides to the other and observe that they cannot make the pencil line up with the other side without turning it. Therefore, these sides are not parallel.

Discuss real-world examples of parallel line segments, such as railroad tracks, shelves on bookcases, and so on. Ask children to identify examples of parallel lines or parallel line segments from the classroom or the hallway. Sample answers: Opposite edges of doors, opposite edges of books, opposite edges of whiteboards

Now display Shape K. Ask volunteers to point to the sides that are parallel. Both pairs of opposite sides are parallel. Have children work as partners to examine all of the 4-sided shapes and sort the shapes into two piles: shapes that have parallel sides and shapes that do not have parallel sides. Parallel sides: G, H, I, J, and K; no parallel sides: D and L GMP7.1, GMP7.2 Children share their results.

Display Shape K. Ask children to look at all of the angles and point out the angles that form a square corner. Explain that the name for this type of angle is a **right angle.**

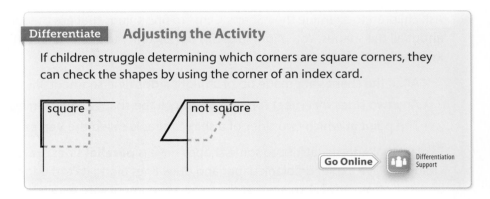

Differentiate   **Adjusting the Activity**

If children struggle determining which corners are square corners, they can check the shapes by using the corner of an index card.

Go Online   Differentiation Support

Working in partnerships, children find and then sort all of the 3- and 4-sided shapes into two piles: shapes with right angles and shapes that have no right angles. Right angles: B, J, and K; no right angles: A, C, D, G, H, I, and L GMP7.1, GMP7.2 Children share their results. Have children point to all of the right angles on each shape.

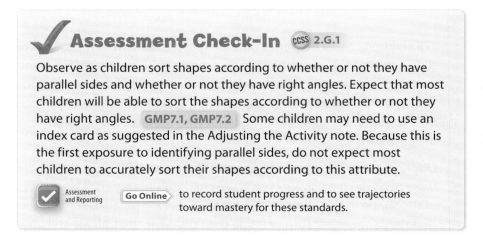

✓ **Assessment Check-In** CCSS 2.G.1

Observe as children sort shapes according to whether or not they have parallel sides and whether or not they have right angles. Expect that most children will be able to sort the shapes according to whether or not they have right angles. GMP7.1, GMP7.2 Some children may need to use an index card as suggested in the Adjusting the Activity note. Because this is the first exposure to identifying parallel sides, do not expect most children to accurately sort their shapes according to this attribute.

Assessment and Reporting   Go Online to record student progress and to see trajectories toward mastery for these standards.

My Reference Book, p. 170

Games

Top-It

Materials ☐ number cards 0–15 (2 of each)
Players 2 or more
Skill Comparing numbers
Object of the Game To collect more cards.

Directions

1 Shuffle the cards. Place the deck number-side down on the table.

2 Each player turns over 1 card and says the number on it.

3 The player with the larger number takes all the cards. If two cards show the same number, those players turn over another card. The player with the larger number then takes all the cards for that round.

4 The game is over when all of the cards have been turned over.

5 The player with the most cards wins.

MRB
170 one hundred seventy

Summarize Have children name some attributes of shapes. Sample answers: Number of sides, number of angles, parallel sides, right angles

③ Practice   10–15 min

Go Online

ePresentations   eToolkit   Home Connections

▶ Playing *Subtraction Top-It*

My Reference Book, pp. 170–172

| WHOLE CLASS | **SMALL GROUP** | **PARTNER** | INDEPENDENT |

Children play Subtraction *Top-It* to practice subtraction facts. See *My Reference Book,* pages 170–172 or Lesson 3-6 for additional instructions.

Differentiate **Adjusting the Activity**

Make counters or a number line available for children as needed.

Go Online   Differentiation Support

Observe

• Which children automatically know the differences?
• What strategies are the remaining children using to determine the differences?

Discuss

• *How did you figure out the differences?*
• *How did you know which comparison symbol to write on the record sheet?*

▶ Math Boxes 8-1

Math Journal 2, p. 195

| WHOLE CLASS | SMALL GROUP | PARTNER | **INDEPENDENT** |

Mixed Practice Math Boxes 8-1 are paired with Math Boxes 8-3.

▶ Home Link 8-1

Math Masters, p. 214

Homework Children find shapes in newspapers and magazines. They bring the pictures to school to use in a bulletin-board display or class shapes book. You may wish to extend this assignment for more than one day. If so, be sure to remind children to bring in other examples as they find them.

Math Journal 2, p. 195

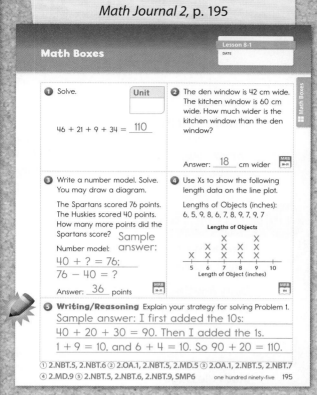

Math Boxes
Lesson 8-1
DATE

① Solve.   Unit

$46 + 21 + 9 + 34 =$ __110__

② The den window is 42 cm wide. The kitchen window is 60 cm wide. How much wider is the kitchen window than the den window?

Answer: __18__ cm wider

③ Write a number model. Solve. You may draw a diagram.

The Spartans scored 76 points. The Huskies scored 40 points. How many more points did the Spartans score?

Number model: Sample answer:
$40 + ? = 76;$
$76 - 40 = ?$

Answer: __36__ points

④ Use Xs to show the following length data on the line plot.

Lengths of Objects (inches): 6, 5, 9, 8, 6, 7, 8, 9, 7, 9, 7

⑤ **Writing/Reasoning** Explain your strategy for solving Problem 1. Sample answer: I first added the 10s. $40 + 20 + 30 = 90.$ Then I added the 1s. $1 + 9 = 10,$ and $6 + 4 = 10.$ So $90 + 20 = 110.$

① 2.NBT.5, 2.NBT.6 ② 2.OA.1, 2.NBT.5, 2.MD.5 ③ 2.OA.1, 2.NBT.5, 2.NBT.7 ④ 2.MD.9 ⑤ 2.NBT.5, 2.NBT.6, 2.NBT.9, SMP6   one hundred ninety-five 195

Math Masters, p. 214

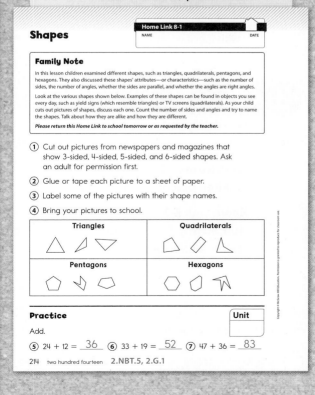

Shapes
Home Link 8-1
NAME   DATE

Family Note

In this lesson children examined different shapes, such as triangles, quadrilaterals, pentagons, and hexagons. They also discussed these shapes' attributes—or characteristics—such as the number of sides, the number of angles, whether the sides are parallel, and whether the angles are right angles.

Look at the various shapes shown below. Examples of these shapes can be found in objects you see every day, such as yield signs (which resemble triangles) or TV screens (quadrilaterals). As your child cuts out pictures of shapes, discuss each one. Count the number of sides and angles and try to name the shapes. Talk about how they are alike and how they are different.

Please return this Home Link to school tomorrow or as requested by the teacher.

① Cut out pictures from newspapers and magazines that show 3-sided, 4-sided, 5-sided, and 6-sided shapes. Ask an adult for permission first.

② Glue or tape each picture to a sheet of paper.

③ Label some of the pictures with their shape names.

④ Bring your pictures to school.

| Triangles | Quadrilaterals |
| Pentagons | Hexagons |

Practice   Unit

Add.

⑤ $24 + 12 =$ __36__ ⑥ $33 + 19 =$ __52__ ⑦ $47 + 36 =$ __83__

214 two hundred fourteen   2.NBT.5, 2.G.1

Lesson 8-2

Playing *Shape Capture*

Overview Children identify shapes that have certain attributes while playing the game *Shape Capture*.

▶ **Before You Begin**

Have available the Shape Cards that children cut out and sorted in Lesson 7-9. For the Math Message, display shapes A and K. In Part 2 each child will cut out the Attribute Cards from *Math Journal 2*, Activity Sheet 13, but each group or partnership will need only one set of cards to play *Shape Capture*.

 Common Core State Standards

Focus Cluster
Reason with shapes and their attributes.

1 Warm Up 15–20 min	**Materials**	
Mental Math and Fluency Children solve basic facts.	slate	2.OA.2
Daily Routines Children complete daily routines.	See pages 4–43.	See pages xiv–xvii.

2 Focus 20–30 min		
Math Message Children identify attributes of shapes.	Shape Cards A and K (for demonstration), slate	2.G.1 SMP7
Identifying Attributes Children describe attributes of shapes.	Shape Cards A and K (for demonstration), ruler (for demonstration)	2.G.1 SMP7
Demonstrating and Playing *Shape Capture* Game Children identify shapes by their attributes.	*Math Journal 2*, Activity Sheet 13 (see Before You Begin); *Math Masters*, p. G28; scissors; Shape Cards (for demonstration); per group: Shape Cards	2.G.1 SMP7
✔ **Assessment Check-In** See page 700.	*Math Masters*, p. TA8	2.G.1

CCSS 2.G.1 **Spiral Snapshot**

GMC Recognize and draw shapes with specified attributes.

◀ | 6-10 Focus | 7-9 Focus | 8-1 Focus Practice | 8-2 Focus Practice | 8-3 through 8-5 Focus Practice | 8-11 Focus Practice | 9-5 Practice |

▓ Spiral Tracker   (Go Online) to see how mastery develops for all standards within the grade.

3 Practice 10–15 min		
Practicing with Fact Triangles Children practice facts using fact triangles.	*Math Journal 2*, pp. 250–253; Fact Triangles	2.OA.2
Math Boxes 8-2 Children practice and maintain skills.	*Math Journal 2*, p. 196; 10-centimeter ruler	See page 701.
Home Link 8-2 **Homework** Children identify shapes that have certain attributes.	*Math Masters*, p. 215	2.G.1

connectED.mheducation.com ▶

Plan your lessons online with these tools.

 ePresentations   Student Learning Center   Facts Workshop Game   eToolkit   Professional Development   Home Connections   Spiral Tracker   Assessment and Reporting   English Learners Support   Differentiation Support

696 Unit 8 | Geometry and Arrays

Differentiation Options RtI

Readiness
5–15 min

Identifying Attributes

Shape Cards

| WHOLE CLASS |
| SMALL GROUP |
| **PARTNER** |
| INDEPENDENT |

To explore attributes of 2-dimensional shapes using a concrete model, children identify different attributes of their Shape Cards. Prompt them to choose a shape, such as a square (Shape J). Name a specific attribute, such as a right angle, and have children point to that attribute on their shape. **GMP7.1** Continue the activity with various shapes and attributes.

Enrichment
5–15 min

Comparing Shapes

Math Masters, p. TA31; Shape Cards

| WHOLE CLASS |
| SMALL GROUP |
| **PARTNER** |
| INDEPENDENT |

To extend their work with attributes of 2-dimensional shapes, children compare two shapes using a Venn diagram on *Math Masters*, page TA31. Explain to children that a Venn diagram is a tool for organizing information.

Children select two different shapes from their Shape Cards, such as a triangle and a square, and write the name of each shape above one of the circles in the diagram. In the center of the diagram, where the circles overlap, children list attributes shared by both shapes. **GMP7.1** Then children list attributes unique to each shape in the nonoverlapping parts of the circles. Encourage children to use the vocabulary from the *Shape Capture* Attribute Cards in their comparisons.

Extra Practice
5–15 min

Drawing Shapes

Activity Card 101, Attribute Cards, paper

| WHOLE CLASS |
| SMALL GROUP |
| **PARTNER** |
| INDEPENDENT |

For additional practice with attributes of 2-dimensional shapes, children use their Attribute Cards to draw a shape that matches the attribute(s) listed on the card. **GMP7.2** Children compare their drawings and discuss similarities and differences.

English Language Learners Support

Beginning ELL Use role play to introduce the word *capture*. Place several pattern blocks between yourself and a child. Take some of the child's blocks and say: *I have taken some of your blocks. I have captured some of your blocks.* Model several rounds of the *Shape Capture* game. As you capture each shape, say the following (filling in the shape name and the attribute, respectively): *I have captured the _____ shape because it has _____.* Then have children who are at the English production stage practice a round using similar sentence frames: "I have captured the _____ shape. The shape I captured has _____." For pre–English production children, use gestures and Total Physical Response prompts to direct them to capture specific blocks, certain numbers of blocks, or blocks with specific attributes.

 Go Online | ELL English Learners Support

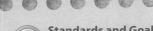

1 Warm Up 15–20 min | Go Online | ePresentations | eToolkit |

▶ Mental Math and Fluency

Pose facts one at a time. Children write the sum or difference on their slates.
Leveled exercises:

● ○ ○ 　7 + 1 　8 　　　● ● ○ 　2 + 4 　6 　　　● ● ● 　8 − 2 　6
　　　 6 + 1 　7 　　　　　　 3 + 5 　8 　　　　　　 7 − 2 　5
　　　 8 + 1 　9 　　　　　　 5 + 4 　9 　　　　　　 9 − 6 　3

▶ Daily Routines

Have children complete daily routines.

2 Focus 20–30 min | Go Online | ePresentations | eToolkit |

▶ Math Message

Display Shape Cards A and K.

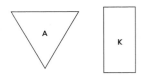

On your slate, describe the shapes using the following words: side, angle, vertex, parallel, *and* right angle. **GMP7.1**

▶ Identifying Attributes

| WHOLE CLASS | SMALL GROUP | PARTNER | INDEPENDENT |

Math Message Follow-Up Ask children to share their descriptions of Shape A, making sure they use the words *side, vertex, angle, parallel,* and *right angle* as suggested in the Math Message. **GMP7.1** Sample answers: It has 3 sides. It has 3 angles. it has 3 vertices. It has no parallel sides. It has no right angles.

If no children mention the lengths of Shape A's sides, ask: *What do you notice about the lengths of the sides?* All the sides are the same length. You or a volunteer can measure the lengths of the sides to show that they are the same length. Ask: *What is the name of the shape?* Triangle

Have children share their descriptions of Shape K. **GMP7.1** Sample answers: It has 4 sides. It has 4 vertices. It has 4 angles. It has 2 pairs of parallel sides. It has 4 right angles. It has 2 pairs of equal-length sides. Ask: *What is the name of the shape?* Rectangle, quadrilateral

Explain that today children will identify shapes that have specified attributes. **GMP7.1, GMP7.2**

Academic Language Development Provide sentence frames for children to use to describe shapes:

- This is a _____.
- It has _____ sides.
- It has _____ vertices.
- It has _____ angles.
- It has _____ pair(s) of parallel sides.
- It has _____ right angle(s).

Differentiate | **Adjusting the Activity**

To support children who struggle to identify shapes with specified attributes, provide cards that illustrate each attribute. For example, on one card show a right triangle with the right angle labeled. On another card, show a rectangle with the parallel sides labeled. **Go Online** | Differentiation Support

▶ **Demonstrating and Playing**
Shape Capture

Math Journal 2, Activity Sheet 13; *Math Masters,* p. G28

WHOLE CLASS | SMALL GROUP | PARTNER | INDEPENDENT

Have children carefully cut apart the Attribute Cards from Activity Sheet 13 in their journals. Children will identify attributes of shapes as they play *Shape Capture.* **GMP7.1, GMP7.2** The game is played with two players or two teams of two players each. Each partnership or group will use one set of Shape Cards and one set of Attribute Cards.

Play several rounds of *Shape Capture* with the class to help children learn the rules. Consider displaying a set of Shape Cards while the children arrange their shapes on their desks.

Math Journal 2, **Activity Sheet 13**

Shape Capture Attribute Cards

All of the angles are right angles.	There are no right angles.	There is only 1 right angle.
There are no parallel sides.	Only 1 pair of sides is parallel.	All opposite sides are parallel.
There are 3 sides, 3 vertices and 3 angles.	There are 4 sides, 4 vertices and 4 angles.	There are 5 sides, 5 vertices and 5 angles.
There are 6 sides, 6 vertices and 6 angles.	All sides are the same length.	Only some of the sides are the same length.

Activity Sheet 13

Math Masters, p. G28

**Shape Capture
Record Sheet**

NAME   DATE

Round	Attribute(s)	Name(s) of Shape(s)	Number of Shapes Captured
1			
2			
3			
4			
5			
			Total:

G28

Directions

Play with a partner or in two teams of two.

1. Spread out the Shape Cards on a flat surface. Shuffle the Attribute Cards and place the pile facedown.

2. Players take turns. When it is your turn, do the following:
   - Turn over the top card from the Attribute Card pile.
   - Take, or capture, all the shapes that have the attributes shown on the Attribute Card. Name each shape as you capture it.
   - If no shapes have the attribute named on the card, your turn is over.
   - At the end of your turn, if you have not captured a shape that you could have taken, the other player or team may name and capture it.

3. If you run out of Attribute Cards, reshuffle and continue play.

4. The game ends when there are no shapes left. The winner is the player or the team with more captured shapes.

Have children record their first five rounds of play on *Math Masters,* page G28. Encourage them to abbreviate attributes in a few words instead of copying all the words on the card.

Observe
- Which children can correctly find shapes with specified attributes?
- Which children are checking the other team or player's selections?

Discuss
- *How did you check to be sure the other team or player was capturing shapes that matched the Attribute Cards?*
- *Which shapes were easier to capture? Why? Which shapes were harder to capture? Why?*

Summarize Using their *Shape Capture* record sheet for reference, have children choose one Attribute Card and name the shape(s) they captured.

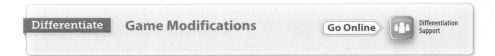

Differentiate **Game Modifications** **Go Online** Differentiation Support

✓ **Assessment Check-In** CCSS 2.G.1

Math Masters, p. TA8

Have children draw three shapes on an Exit Slip (*Math Masters,* page TA8)— one with 3 sides, 3 vertices, and 3 angles; one with at least 1 right angle; and one with at least 1 pair of parallel sides. Then have children name the shapes. Expect most children to successfully draw a triangle and a shape with a right angle. Some children may be able to draw a shape with parallel sides. If children struggle, have them use a Pattern-Block Template to select and draw shapes.

 Assessment and Reporting   **Go Online** to record student progress and to see trajectories toward mastery for these standards.

3 Practice 10–15 min

Go Online   ePresentations   eToolkit   Home Connections

▶ Practicing with Fact Triangles

Math Journal 2, pp. 250–253

| WHOLE CLASS | **SMALL GROUP** | **PARTNER** | INDEPENDENT |

As children practice with Fact Triangles, have them use the Addition Facts Inventory Record, Parts 1 and 2 (*Math Journal 2,* pages 250–253) to take inventory of the addition facts they know and the facts that need more practice. See Lesson 3-3 for additional details.

▶ Math Boxes 8-2

Math Journal 2, p. 196

| WHOLE CLASS | **SMALL GROUP** | **PARTNER** | **INDEPENDENT** |

Mixed Practice Math Boxes 8-2 are paired with Math Boxes 8-4.

▶ Home Link 8-2

Math Masters, p. 215

Homework Children identify shapes that have certain attributes.

Math Journal 2, p. 196

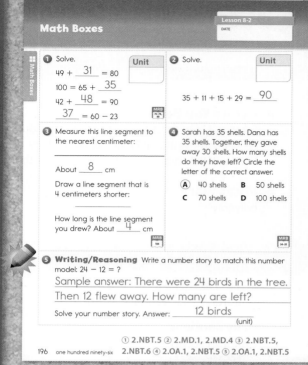

Math Masters, p. 215

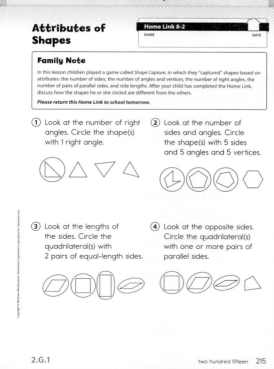

Comparing Triangles, Pentagons, and Hexagons

Overview Children build and compare various polygons.

▶ **Before You Begin**
For Part 2 have twist ties and straws of various lengths available. Each child will need at least 22 of each. For the optional Extra Practice activity, obtain the book **Shape Up** by David Adler (Holiday House, 2008) and have pretzel sticks or toothpicks available.

▶ **Vocabulary**
polygon

 **Common Core State Standards**

Focus Cluster
Reason with shapes and their attributes.

	Materials	
① Warm Up 15–20 min		
Mental Math and Fluency Children make ballpark estimates.	slate	**2.NBT.5, 2.NBT.7**
Daily Routines Children complete daily routines.	See pages 4–43.	See pages xiv–xvii.

② Focus 20–30 min		
Math Message Children use straws and twist ties to make 3-sided shapes.	straws and twist ties	**2.G.1**
Comparing Triangles Children compare triangles and discuss polygons.	straw-and-twist-tie triangles (from the Math Message)	**2.G.1** **SMP7**
Comparing Pentagons and Hexagons Children build and compare 5- and 6-sided shapes.	*Math Journal 2*, p. 197; straws and twist ties	**2.G.1** **SMP7**
✓ **Assessment Check-In** See page 706.	*Math Journal 2*, p. 197	**2.G.1**

CCSS 2.G.1   **Spiral Snapshot**

GMC Identify 2- and 3-dimensional shapes.

| 6-10 Focus | 8-1 Focus Practice | 8-2 Focus Practice | 8-3 Focus Practice | 8-4 Focus | 8-5 Focus Practice | 8-11 Focus Practice | 9-5 Practice |

 Spiral Tracker   (Go Online)   to see how mastery develops for all standards within the grade.

③ Practice 10–15 min		
Playing *Target* to 200 **Game** Children use base-10 blocks to model addition and subtraction.	per player: *Math Masters,* pp. G19–G20; per partnership: base-10 blocks (5 flats, 30 longs, and 30 cubes); 4 each of number cards 0–9	**2.NBT.1, 2.NBT.1a, 2.NBT.1b,** **2.NBT.3, 2.NBT.7** **SMP2**
Math Boxes 8-3 Children practice and maintain skills.	*Math Journal 2*, p. 198	See page 707.
Home Link 8-3 **Homework** Children draw and label polygons. They gather 3-dimensional shapes for the Shapes Museum.	*Math Masters*, p. 216	**2.G.1**

connectED.mheducation.com

Plan your lessons online with these tools.

 ePresentations   Student Learning Center   Facts Workshop Game   eToolkit   Professional Development   Home Connections   Spiral Tracker   Assessment and Reporting   English Learners Support   Differentiation Support

Differentiation Options

RtI

CCSS 2.G.1, SMP7

Readiness
5–15 min

Playing Touch and Match with Shapes

WHOLE CLASS
SMALL GROUP
PARTNER
INDEPENDENT

2 each of Shape Cards A–C, E–F, and M–P; small bag or box

For experience identifying similarities and differences among triangles, pentagons, and hexagons, children identify shapes by feel. Divide Shape Cards into two sets, each made up of 1 each of Shape Cards A–C, E–F, and M–P. Place one set of the specified Shape Cards in full view. Without children seeing, place Shape A from the other set in a bag or a box. Have a volunteer reach inside the container, feel the shape without looking at it, and find the matching shape from the displayed set. **GMP7.1**

Ask: *How did you find the match?* Sample answer: The shape I was holding had 3 sides, 3 vertices and 3 angles, and all the sides felt like the same length. Repeat with other shapes and other volunteers.

CCSS 2.G.1

Enrichment
5–15 min

Solving Shape Riddles

WHOLE CLASS
SMALL GROUP
PARTNER
INDEPENDENT

Activity Card 99, Shape Cards, paper

To apply their understanding of the attributes of 2-dimensional shapes, children continue to pose and solve shape riddles. See the Enrichment activity in Lesson 8-1 for directions.

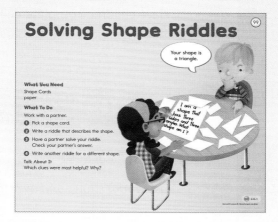

CCSS 2.G.1

Extra Practice
5–15 min

Making Shapes

WHOLE CLASS
SMALL GROUP
PARTNER
INDEPENDENT

Activity Card 102, *Shape Up*, pretzel sticks or toothpicks

Literature Link For practice with different attributes of shapes, children read ***Shape Up*** by David Adler (Holiday House, 2008) in small groups or as partners. This book introduces triangles and other polygons while showing children how to make shapes using different food items, such as cheese, slices of bread, and pretzels. After reading the book, children build shapes that have attributes such as parallel lines, right angles, and specified numbers of sides. Then they discuss and compare their shapes.

English Language Learners Support

Beginning ELL Contrast the term *in common* with the term *different*. Draw a T-chart on paper and label one column In Common and the other column Different. Place it on a flat surface. Place several Shape Cards next to the T-chart, some having a specific and obvious attribute (such as having 3 sides) and others that don't. Use think-alouds as you place the triangles in the In Common column and place the other polygons in the Different column, saying: *This polygon goes in the In Common column. This polygon goes in the Different column.* After placing a few polygons, point to each group and say the following: *These polygons have _____ in common. These polygons are different.* Next have children place other polygons in the column in which they think they belong. Affirm correct choices by using the target terms *in common* and *different* for children to hear and repeat.

 Go Online   ELL English Learners Support

Lesson 8-3   **703**

1 Warm Up   15–20 min   Go Online   ePresentations   eToolkit

▶ Mental Math and Fluency

Children make a ballpark estimate for each sum and record their estimates as number models on their slates. *Leveled exercises:*

- ●○○ 48 + 46 = ?   Sample answers: 50 + 50 = 100; 50 + 46 = 96
  13 + 59 = ?   Sample answers: 10 + 60 = 70; 13 + 60 = 73
- ●●○ 76 + 188 = ?   Sample answers: 80 + 190 = 270; 70 + 200 = 270
  85 + 165 = ?   Sample answers: 80 + 170 = 250; 90 + 160 = 250
- ●●● 183 + 211 = ?   Sample answers: 180 + 210 = 390; 200 + 200 = 400
  296 + 373 = ?   Sample answers: 300 + 370 = 670; 300 + 373 = 673

▶ Daily Routines

Have children complete daily routines.

2 Focus   20–30 min   Go Online   ePresentations   eToolkit

▶ Math Message

Using your straws and twist ties, make at least two different 3-sided shapes.

▶ Comparing Triangles

Math Message Follow-Up Display some examples of children's triangles. Ask: *What do these shapes have in common?* They have 3 sides, 3 vertices, and 3 angles. *What name describes all of these shapes?* Triangle

Compare some of the different triangles, focusing on the sides and the angles. Be sure to display one triangle with a right angle and one with all equal-length sides and discuss these attributes. **GMP7.1, GMP7.2** Have children describe each triangle's attributes and then discuss the differences between the triangles.

Explain to children that another attribute all the triangles share is that they are all **polygons.** As you discuss these traits of polygons, point them out on one or more straw triangles:

- Polygons are made up of all straight sides (line segments).
- The sides of a polygon do not cross.
- Polygons are "closed" figures: you can trace their sides and come back to where you started without retracing or crossing any part.

Ask children to trace their straw-and-twist-tie triangles and confirm that they meet all the criteria to be polygons. Display some examples of nonpolygons and discuss why these shapes are not polygons.

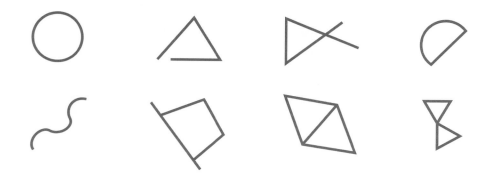

Children should be able to determine why these shapes are not polygons. Have a volunteer trace some of the shapes with a finger. For shapes made up of two polygons, ask: *Can you return to the starting point without retracing or crossing any part of the shape?* No. *Is this a polygon?* No.

Explain that children will compare various polygons by examining their attributes.

▶ Comparing Pentagons and Hexagons

Math Journal 2, p. 197

| WHOLE CLASS | SMALL GROUP | PARTNER | INDEPENDENT |

Children work with partners to use their straws and twist ties to build 5- and 6-sided polygons as directed on journal page 197. Then each child draws their polygons on the page.

When children are finished drawing the polygons, ask: *What is the name for any 5-sided polygon?* Pentagon *What is the name for any 6-sided polygon?* Hexagon *Which of the shapes you drew are polygons?* All of them *How do you know?* Sample answer: They are closed figures, they have straight sides, and their sides don't cross. Have children trace the sides of each shape on their journal page with their finger to be sure that all their polygons are closed, have straight sides, and have sides that do not cross.

Display two different straw-and-twist-tie pentagons made by children, making sure to choose examples that have different attributes. As a class, compare and contrast the shapes based on their attributes. Encourage children to examine the number of angles, the number of vertices, the number of sides, side lengths, the number of right angles, and the number of parallel sides. GMP7.1, GMP7.2

Guide children to notice the attributes by asking the following questions:

• *How are these two pentagons alike?* Answers vary.

• *How are they different?* Answers vary.

Repeat this comparison with two examples of children's straw-and-twist-tie hexagons.

Math Journal 2, p. 197

Comparing Pentagons and Hexagons

Lesson 8-3
DATE

Use straws and twist ties.

1. Build a 5-sided shape. Draw the shape below.

Sample answer:

Shape name: Pentagon

2. Build a different 5-sided shape. Draw the shape below.

Sample answer:

Shape name: Pentagon

3. How are the shapes you drew for Problems 1 and 2 alike? How are they different?
Answers vary.

4. Build a 6-sided shape. Draw the shape below.

Sample answer:

Shape name: Hexagon

5. Build a different 6-sided shape. Draw the shape below.

Sample answer:

Shape name: Hexagon

6. How are the shapes you drew for Problems 4 and 5 alike? How are they different?
Answers vary.

2.G.1, SMP7   one hundred ninety-seven 197

If children struggle comparing and contrasting polygons, provide them with sentence frames in which they fill in the attributes that the shapes do or do not share:

- These two shapes are alike because they both have _____.
- They are different because _____.

Allow children to practice in small groups before the large-group discussion.

 Differentiation Support

 Assessment Check-In CCSS 2.G.1

Math Journal 2, p. 197

Expect that children can build 5- and 6-sided polygons and draw them on journal page 197. Some children may be able to compare and contrast the polygons. If children struggle building or drawing pentagons and hexagons, refer them to *My Reference Book,* page 123 or the Two-Dimensional Shapes poster. You may also have children find the 5- and 6-sided Shape Cards and describe them.

Assessment and Reporting | Go Online | to record student progress and to see trajectories toward mastery for these standards.

Summarize Children describe the attributes of various polygons.

3 Practice 10–15 min

Go Online | ePresentations | eToolkit | Home Connections

▶ Playing *Target* to 200

Math Masters, pp. G19–G20

| WHOLE CLASS | **SMALL GROUP** | **PARTNER** | INDEPENDENT |

Have children play *Target* to 200 to apply their understanding of place value. The rules are the same as *Target* to 50 (see Lesson 4-7 for detailed directions), but the object of the game is to have two flats on the game mat. GMP2.1, GMP2.2

Observe

- Which children are correctly representing their numbers with base-10 blocks?
- Which children seem to have a strategy for deciding whether to make a 1- or 2-digit number? To add or subtract their numbers?

Discuss

- *How did you decide what number to make? Whether to add or subtract?*
- *How did you know when to make an exchange?*

▶ Math Boxes 8-3 ✏️

Math Journal 2, p. 198

| WHOLE CLASS | SMALL GROUP | PARTNER | INDEPENDENT |

Mixed Practice Math Boxes 8-3 are paired with Math Boxes 8-1.

▶ Home Link 8-3

Math Masters, p. 216

Homework Children draw and label 4-, 5-, and 6-sided shapes. They gather 3-dimensional shapes for the Shapes Museum.

Planning Ahead

Gather models of 3-dimensional shapes—such as a cylinder, a cone, a sphere, a cube, a pyramid, and a rectangular prism—for demonstration for Lesson 8-5.

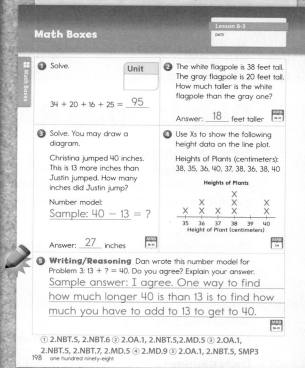

Math Journal 2, p. 198

Math Boxes   Lesson 8-3
DATE

① Solve.

Unit

$34 + 20 + 16 + 25 =$ __95__

② The white flagpole is 38 feet tall. The gray flagpole is 20 feet tall. How much taller is the white flagpole than the gray one?

Answer: __18__ feet taller

③ Solve. You may draw a diagram.

Christina jumped 40 inches. This is 13 more inches than Justin jumped. How many inches did Justin jump?

Number model:
Sample: $40 - 13 = ?$

Answer: __27__ inches

④ Use Xs to show the following height data on the line plot.

Heights of Plants (centimeters): 38, 35, 36, 40, 37, 38, 36, 38, 40

Heights of Plants

Height of Plant (centimeters)
35 36 37 38 39 40

⑤ **Writing/Reasoning** Dan wrote this number model for Problem 3: $13 + ? = 40$. Do you agree? Explain your answer.
Sample answer: I agree. One way to find how much longer 40 is than 13 is to find how much you have to add to 13 to get to 40.

① 2.NBT.5, 2.NBT.6 ② 2.OA.1, 2.NBT.5,2.MD.5 ③ 2.OA.1,
2.NBT.5, 2.NBT.7, 2.MD.5 ④ 2.MD.9 ⑤ 2.OA.1, 2.NBT.5, SMP3
198   one hundred ninety-eight

Math Masters, p. 216

Shapes Museum   Home Link 8-3
NAME   DATE

Family Note

In today's lesson children used straws and twist ties to build polygons and then drew the shapes. Children learned that polygons are closed figures made up of all straight sides that do no cross.

Polygons: ▢ ◇ △   NOT polygons: 〜 ○ △

Please return the top part of this Home Link to school tomorrow.

For Problems 1–3, draw the polygon and write its name on the line. Sample drawings:

① 6-sided polygon: ② 5-sided polygon: ③ 4-sided polygon:

⬡   ⬠   ▢

Sample answers:

__Hexagon__   __Pentagon__   __Quadrilateral__

④ Are these three shapes all polygons? Explain.
Yes. Sample answer: They are all closed shapes with straight sides that don't cross.

Shapes Museum

For the next few days our class will collect items to put into a Shapes Museum. Starting tomorrow, bring items such as boxes, soup cans, party hats, pyramids, and balls to school. Ask an adult for permission to bring in these items.

216   two hundred sixteen   2.G.1

Drawing and Reasoning About Quadrilaterals

2-Day Lesson

Overview **Day 1:** Children draw quadrilaterals with given attributes.
Day 2: The class discusses solutions, and children revise their work.

Day 1: Open Response

▶ **Before You Begin**
Prepare three copies of *Math Masters,* page 218 per child. You may wish to read with the class *My Reference Book,* pages 10–11 to preview the ideas of conjecture and argument. If possible, schedule time to review children's work and plan for Day 2 of this lesson with your grade-level team.

▶ **Vocabulary**
attribute · quadrilateral · side · angle · parallel sides · right angle

 **Common Core State Standards**

Focus Cluster
Reason with shapes and their attributes.

---

1 Warm Up   15–20 min

	Materials	
Mental Math and Fluency Children record a number model and make a ballpark estimate.	slate	2.NBT.5
Daily Routines Children complete daily routines.	See pages 4–43.	See pages xiv–xvii.

2a Focus   45–55 min

Math Message Children examine examples and non-examples of quadrilaterals and make conjectures about the attributes of a quadrilateral.	*Math Journal 2,* p. 199	2.G.1 SMP3, SMP7
What Is a Quadrilateral? Children discuss their conjectures about the attributes of a quadrilateral.	*Math Journal 2,* p. 199; per partnership: 1 square pattern block; chart paper; Two-Dimensional Shapes Poster	2.G.1 SMP3, SMP7
Solving the Open Response Problem Children draw quadrilaterals with given attributes to plan a garden and argue that their shapes have the attributes.	*Math Masters,* pp. 217–218; straightedge (optional); 1 square pattern block (optional)	2.G.1 SMP1, SMP3, SMP7

Getting Ready for Day 2 →

Review children's work and plan discussion for reengagement.   *Math Masters,* p. TA5; children's work from Day 1

CCSS 2.G.1 **Spiral Snapshot**

GMC Recognize and draw shapes with specified attributes.

6-10 Focus	7-9 Focus	8-1 through 8-3 Focus Practice	8-4 Focus Practice	8-5 Focus	8-11 Focus Practice	9-5 Practice

Spiral Tracker **Go Online** to see how mastery develops for all standards within the grade.

---

connectED.mheducation.com

Plan your lessons online with these tools.

 ePresentations   Student Learning Center   Facts Workshop Game   eToolkit   Professional Development   Home Connections   Spiral Tracker   Assessment and Reporting   English Learners Support   Differentiation Support

1 Warm Up 15–20 min

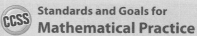

Go Online

ePresentations   eToolkit

▶ Mental Math and Fluency

Children write number models on their slates to show their estimates.

●○○ $72 + 21 = ?$ Sample answer: $70 + 20 = 90$
    $33 + 49 = ?$ Sample answers: $30 + 50 = 80$; $33 + 50 = 83$

●●○ $64 + 129 = ?$ Sample answer: $60 + 130 = 190$
    $55 + 125 = ?$ Sample answers: $60 + 130 = 190$; $50 + 120 = 170$

●●● $174 + 210 = ?$ Sample answers: $170 + 200 = 370$; $200 + 200 = 400$
    $396 + 473 = ?$ Sample answers: $400 + 473 = 873$; $400 + 500 = 900$

▶ Daily Routines

Have children complete daily routines.

2a Focus 45–55 min

Go Online

ePresentations   eToolkit

▶ Math Message

Math Journal 2, p. 199

Look at journal page 199. Discuss the questions with a partner.
GMP3.1, GMP7.1

Differentiate **Adjusting the Activity**

If children have difficulty identifying attributes, review examples of possible attributes of shapes. Ask: *What attributes of shapes have we talked about before?* Sample answer: number of sides, number of vertices (corners), number of angles, parallel sides, straight or curved sides Have children look for these attributes in the quadrilaterals.

Standards and Goals for Mathematical Practice

SMP1 Make sense of problems and persevere in solving them.
   GMP1.3 Keep trying when your problem is hard.

SMP3 Construct viable arguments and critique the reasoning of others.
   GMP3.1 Make mathematical conjectures and arguments.

SMP7 Look for and make use of structure.
   GMP7.1 Look for mathematical structures such as categories, patterns, and properties.

Professional Development

The focus of this lesson is **GMP3.1**. A conjecture is an educated guess. In math class, it is a good guess that uses mathematics. Children think about possible attributes of a quadrilateral as they examine examples and non-examples. The process of making a conjecture often includes testing that conjecture. As children have ideas about possible attributes, encourage them to check their ideas against each of the given shapes. In the open response problem, children shift from making conjectures to arguments. Children will use arguments that draw on evidence to explain why their quadrilateral drawings have the given attributes.

Go Online for information about **SMP3** in the *Implementation Guide.*

What Is a Quadrilateral?

Lesson 8-4

DATE

These shapes are quadrilaterals:

A B C

D E F

These shapes are not quadrilaterals:

G H I

J K L

Look at the shapes.

What do the quadrilaterals have in common?

What do you think are the attributes of a quadrilateral?

Tell your partner your ideas.

2.G.1, SMP3, SMP7   one hundred ninety-nine   199

▶ What Is a Quadrilateral?

Math Journal 2, p. 199

WHOLE CLASS | SMALL GROUP | PARTNER | INDEPENDENT

Math Message Follow-Up As partners discuss their ideas about the shapes, suggest they look for **attributes** that are common to all of the quadrilaterals. GMP7.1 Have children share their ideas about attributes of **quadrilaterals.** When a child suggests an attribute, ask other children to check whether each of the shapes in the first group has that attribute. GMP3.1 Patterns that children find in the first group of shapes should reveal a few attributes, such as four **sides,** four **angles,** and four corners. Counting the angles in concave shapes can be challenging, so do not focus on the number of angles in shape B. As children agree on attributes for quadrilaterals, record them on a class chart.

Examining the shapes in the second group may help identify more attributes of quadrilaterals. GMP3.1 These shapes reveal that quadrilaterals have four straight (and not just four) sides, and that the figure must be closed. Add these attributes to the chart.

If it does not come up in discussion, ask children if one or two pairs of **parallel sides** are attributes of quadrilaterals. If needed, review the definition and test for parallel sides. Have children look on the Two-Dimensional Shapes Poster and find shapes that have parallel sides. Because parallel sides only occur in some quadrilaterals, children should conclude that parallel sides are not attributes of all quadrilaterals.

When you are satisfied with your list of attributes, help children see that all the quadrilaterals have all (and not just some) of the attributes. Review each of the shapes in the second group and ask children to explain why each of these shapes is not a quadrilateral. GMP3.1, GMP7.1

Give each partnership a square pattern block. Ask: *What is the name of the angles in this shape?* Right angle Review how to use the square pattern block or the corner of a sheet of paper to decide whether an angle is a **right angle.** Give children a moment to test whether any of the shapes from the Math Message has one or more right angles. GMP7.1

Tell children that they are going to draw their own quadrilaterals and write about their attributes.

Academic Language Development This lesson relies on vocabulary that can be challenging for all children, including English language learners. Children will learn the words as they hear them and especially as they use them in various contexts. As you use the vocabulary, regularly use gestures and point to diagrams, objects, and written words that are connected with the vocabulary. If there are children in your class who speak Spanish, you may want to make a connection between the word *quadrilateral* and the Spanish words *cuatro* (four) and *lado* (side).

▶ Solving the Open Response Problem

Math Masters, pp. 217–218

| WHOLE CLASS | SMALL GROUP | **PARTNER** | **INDEPENDENT** |

Distribute *Math Masters,* page 217 and one copy of *Math Masters,* page 218 to all children. Read Problem 1 as a class. Tell children to make drawings for Juan's garden on the dot paper. Remind children to use their square pattern block or the corner of a piece of paper to check their drawings for the number of right angles. Tell children that when they decide a shape does not work for Juan's garden, they can cross it out, but they should not erase it.

When most children have created drawings for Juan's garden, hand out a second copy of *Math Masters,* page 218 to all children. Read Problem 2 as a class. Have children make drawings for Linda's garden on the second page of dot paper. Tell children they should expect to make several drawings before they find one that may work for Linda. **GMP1.3** Provide extra dot paper to children who need it. Read Problem 3 as a class. Tell children that they should move on to Problem 3 only after they have created one or more successful drawings for Linda. Remind children that they may use the list of quadrilateral attributes from the Math Message as they write their explanations for Problem 3. **GMP3.1**

> **NOTE** Creating a quadrilateral with exactly two right angles is challenging. The most important element of this activity is not that children create a successful drawing for Linda on Day 1, but that they get practice drawing figures, examining them, and deciding whether they are quadrilaterals with two right angles. **GMP7.1** In this way, unsuccessful drawings are as important as successful ones.

Observe children as they work. Encourage them to create multiple drawings that fit Juan's or Linda's plan. **GMP1.3** Expect on Day 1 that only some children will successfully create shapes that satisfy Problem 2. During the reengagement discussion on Day 2, children will discuss successful examples and have time to complete a drawing and write an argument for Problem 3.

Differentiate   **Adjusting the Activity**

For children who get frustrated with their efforts to draw a garden for Linda, ask them to draw one right angle and then a second that is connected to it. You will usually see them create a figure similar to the one shown below. When you see this, ask: *Can you draw the fourth side directly across this opening? Why or why not?* Sample answer: No, because that would be four right angles. Let children keep working to see if they can make progress without more help.

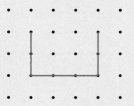

Math Masters, p. 217

Drawing and Reasoning About Quadrilaterals   Lesson 8-4

The members of Juan's family are making plans for a garden that will be in the shape of a quadrilateral.

① Juan wants a quadrilateral with four right angles. Try drawing shapes on a sheet of dot paper that will work for Juan's plan. Circle the one you think Juan should use for the garden. Answers vary. See sample work for Child A on page 713.

② Juan's sister, Linda, wants a quadrilateral with just two right angles. Try drawing shapes on a second sheet of dot paper that will work for Linda's plan. Circle the one you think Linda should use. Answers vary. See sample children's work on page 717.

③ Explain how you know your circled shape for Problem 2 works for Linda's plan.

Answers vary. See sample children's work on page 717 of the *Teacher's Lesson Guide.*

2.G.1, SMP1, SMP3, SMP7   two hundred seventeen 217

Math Masters, p. 218

Garden Plans   Lesson 8-4

Garden Plans for _____

218 two hundred eighteen

Partners can work together to examine each other's shapes and share ideas, but children should complete their own drawings and arguments.

Summarize Ask: *Why do you think it is important to keep trying and not give up when your problem is hard?* GMP1.3 Answers vary.

Collect children's work so that you can evaluate it and prepare for Day 2.

Getting Ready for Day 2

Math Masters, p. TA5

Planning a Follow-Up Discussion

Review children's work. Use the Reengagement Planning Form (*Math Masters*, page TA5) and the rubric on page 714 to plan ways to help children meet expectations for both the content and practice standards. Look for shapes that include the given attributes for the gardens as well as those that do not. Look for interesting ways children explained how they knew their shapes worked for Linda's plans.

Reengagement Planning Form

Common Core State Standard (CCSS):
2.G.1 Recognize and draw shapes having specified attributes, such as a given number of angles or a given number of equal faces. Identify triangles, quadrilaterals, pentagons, hexagons, and cubes.

Goal for Mathematical Practice (GMP): *GMP3.1 Make mathematical conjectures and arguments.*

Organize the discussion in one of the ways below or in another way you choose. If children's work is unclear or if you prefer to show work anonymously, rewrite the work for display.

Go Online for sample children's work that you can use in your discussion.

1. Review Problem 1 and display a child's drawings for Juan's plan. Select work that has shapes that do and do not work for Juan's plan. See sample work from Child A. Ask:

   • *What needs to be true about drawings that fit Juan's plan?* GMP7.1
   Sample answer: They should be closed and have four straight sides and four right angles.

   • *How can we check to see whether a shape has four right angles?* GMP3.1, GMP7.1 Sample answer: Put the corner of a square pattern block or piece of paper inside each angle and see if it matches.

   • *Which of these drawings do you think will work for Juan's plan? Explain why you think they will work.* GMP3.1 Answers vary.

   • *Which of these drawings do you think will not work for Juan's plan? Explain why you think they won't work.* GMP3.1 Answers vary.

2. Display examples of a child's drawings for Linda's plan. Include successful, unsuccessful, and incomplete drawings. See sample work from Child B. Ask:

   • *What needs to be true about drawings that fit Linda's plan?* GMP7.1 Sample answer: They should be closed and have four straight sides and four angles. Only two angles should be right angles.

   • *It looks like this child decided that several of these shapes don't work. Do you agree? Why or why not?* GMP3.1, GMP7.1 Answers vary.

   • *How could we change one of the drawings that doesn't work to make it work?* GMP1.3, GMP3.1, GMP7.1 Answers vary.

   • *This child didn't cross out two of the shapes. Do you think one or both of these drawings work for Linda's garden? Why or why not?* GMP3.1, GMP7.1 Sample answer: Both work because there are four straight sides and just two right angles.

3. Display a child's successful shape for Problem 2 and response to Problem 3. Consider using a response that is incomplete to generate more discussion. See sample work from Child C. Ask:

   • *What do you think about this child's explanation?* GMP3.1, GMP7.1 Sample answer: It is correct that the shape has four sides. It is incorrect to say the shape has two angles. It has four angles.

   • *What could this child do to improve the explanation?* GMP1.3, GMP3.1, GMP7.1 Sample answer: The child could add that the figure is closed, that there are two right angles, and show which angles are right angles and how he or she knows they are right angles.

Planning for Revisions

Have extra copes of Math Masters, page 218 available for additional drawings. You might want to ask children to use colored pencils so you can see what they revised.

Sample child's work, Child A

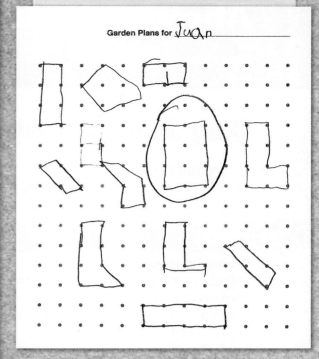

Garden Plans for Juan

Sample child's work, Child B

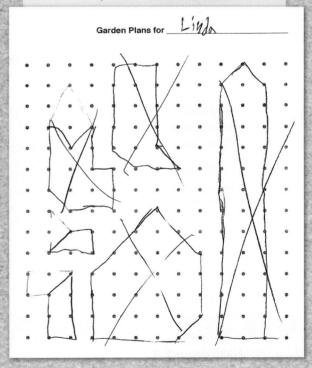

Garden Plans for Linda

Sample child's work, Child C

3 Explain how you know your circled shape for Problem 2 works for Linda's plan.

It has for corners and four sides and two angles

Drawing and Reasoning About Quadrilaterals

Day 2: Reengagement

▶ **Before You Begin**
Have extra copies available of *Math Masters,* page 218 for children to revise their work.

CCSS Common Core State Standards

Focus Cluster
Reason with shapes and their attributes.

2b Focus   50–55 min

Materials

Setting Expectations Children review the open response problem and discuss what would make a good garden plan and a complete explanation. They review how to discuss other children's work respectfully.	*My Reference Book,* p. 127 (optional); Guidelines for Discussions Poster	2.G.1 SMP3
Reengaging in the Problem Children discuss their drawings and evaluate their written arguments that their drawings have the given attributes.	selected samples of children's work	2.G.1 SMP3, SMP7
Revising Work Children revise drawings and arguments from Day 1.	*Math Masters,* p. 218 (optional); children's work from Day 1; colored pencils (optional)	2.G.1 SMP1, SMP3, SMP7
✓ **Assessment Check-In** See page 717 and the rubric below.		2.G.1 SMP3

Goal for Mathematical Practice **GMP3.1** Make mathematical conjectures and arguments.	Not Meeting Expectations	Partially Meeting Expectations	Meeting Expectations	Exceeding Expectations
	Does not provide an argument that a shape is appropriate for Linda's garden, or provides an argument that does not include any of the 3 elements required to meet expectations.	Provides a partial argument that a shape is appropriate for Linda's garden that includes one or two of the three elements required to meet expectations.	Provides an argument that a shape is appropriate for Linda's garden because • it has 2 right angles, • it is a closed figure, and • it has one or more of the following: 4 sides, 4 angles, or 4 corners.	Meets expectations and includes an explanation of how the child tested and confirmed that there were two right angles.

3 Practice   10–15 min

Math Boxes 8-4 Children practice and maintain skills.	*Math Journal 2,* p. 200	See page 716.
Home Link 8-4 **Homework** Children draw a triangle and quadrilateral and compare their attributes.	*Math Masters,* p. 219	2.G.1, 2.NBT.5

 Focus
50–55 min Go Online
ePresentations   eToolkit

NOTE These Day 2 activities will ideally take place within a few days of Day 1. Prior to beginning Day 2, see Planning a Follow-Up Discussion from Day 1.

▶ Setting Expectations

| **WHOLE CLASS** | SMALL GROUP | PARTNER | INDEPENDENT |

Review the open response problem from Day 1. Ask: *What do you think a complete answer to this problem needs to include?* GMP3.1 Sample answer: It needs drawings of gardens for Juan and Linda and an explanation for why the circled shape has the attributes for Linda's plan. You may wish to review attributes of quadrilaterals using *My Reference Book,* page 127.

Tell children that they are going to look at others' work, decide if the shapes work for Juan's and Linda's gardens, and think about the explanations. Point out that the shapes and explanations will be different. Some explanations will be correct and complete, and others will need more work. Remind children that they should help each other draw shapes that work and write complete explanations. Refer to your list of discussion guidelines and encourage children to use these sentence frames:

- That shape works because _____.
- I would add _____.

▶ Reengaging in the Problem

| **WHOLE CLASS** | SMALL GROUP | **PARTNER** | INDEPENDENT |

Children reengage in the problem by analyzing and critiquing other children's work in pairs and in a whole-group discussion. Have children discuss with partners before sharing with the whole group. Guide this discussion based on the decisions you made in Getting Ready for Day 2. GMP1.3, GMP3.1, GMP7.1

▶ Revising Work

WHOLE CLASS | SMALL GROUP | **PARTNER** | **INDEPENDENT**

Pass back children's work from Day 1. Before children revise anything, ask them to examine their drawings and arguments and decide how to improve them. Ask the following questions one at a time. Have partners discuss their responses and give a thumbs-up or thumbs-down based on their own work.

- *Did you try to draw several shapes that will work for Juan's plan?*
- *Did you circle one shape that worked for Juan's garden?*
- *Did you try to draw several shapes that will work for Linda's plan?* GMP1.3, GMP7.1
- *Did you circle one shape that worked for Linda's garden?*
- *Did you write how you know your circled shape for Problem 2 works for Linda's plan?*
- *Is your explanation clear enough that your partner can understand it?* GMP3.1

Tell children they now have a chance to revise their work. Tell children to add to their earlier work using colored pencils or make additional drawings on a new sheet of dot paper, instead of erasing their original work.

Differentiate | **Adjusting the Activity**

For children who did not complete a correct drawing for Problem 2, have them try new drawings based on the class discussion.

For children who completed correct drawings and strong explanations on Day 1, ask them to explore whether they can create a quadrilateral garden with exactly one right angle or exactly three right angles. Give them a new sheet of dot paper. They may create successful examples with one right angle, but it is impossible to create a quadrilateral with three right angles. GMP1.3, GMP7.1

Math Journal 2, p. 200

Math Boxes

Lesson 8-4
DATE

① Solve.   Unit

21 + __29__ = 50

100 = 52 + __48__

__51__ = 70 − 19

38 + __52__ = 90

② Solve.   Unit

Fill in the circle next to the correct answer.

23 + 18 + 12 + 19 = __72__

Ⓐ 72   Ⓑ 62

Ⓒ 73   Ⓓ 83

③ Measure this line segment in centimeters.

_____ About __2__ cm

Draw a line segment 4 centimeters longer.

_____

How long is the line segment you drew?

About __6__ cm

④ There are 15 second graders at the lunch table. Then 12 more sit down. After a while, 10 children leave. How many are at the lunch table now?

Number model(s): Sample answer:

15 + 12 = ?, 27 − 10 = ?

There are __17__ children.

⑤ Write a number story to match the number model: 35 + 15 = ?

Sample answer: Sarah did 35 jumping jacks. Amy did 15 jumping jacks. How many did they do all together?

Solve your number story.

Answer: 50 jumping jacks (Unit)

200   two hundred

① 2.NBT.5 ② 2.MD.1, 2.MD.4 ③ 2.NBT.5, 2.NBT.6
④ 2.OA.1, 2.NBT.5 ⑤ 2.OA.1, 2.NBT.5.

Assessment Check-In CCSS 2.G.1

Collect and review children's revised work. Expect children to improve their drawings and explanations based on the class discussion. For the content standard, expect most children to show in their drawings and in the reengagement discussion that they can recognize and draw quadrilaterals with specified attributes. You can use the rubric on page 714 to evaluate children's revised work for Problem 3 for **GMP3.1**.

✓ Assessment and Reporting Go Online▷ to record student progress and to see trajectories toward mastery for these standards.

Go Online▷ for optional generic rubrics in the *Assessment Handbook* that can be used to assess any additional GMPs addressed in the lesson.

Sample Children's Work—Evaluated

See the sample in the margin. This work meets expectations for the content standard because the child recognized and drew a quadrilateral with four right angles for Problem 1 (not shown) and two right angles for Problem 2 (shown), and in class discussion. This work meets expectations for the mathematical practice because the argument that the circled shape works for Linda's garden has all three of the required elements: "it is closed," "it has 4 corners or angles," and "it works for Linda because it has 2 right [angles]." **GMP3.1**

Go Online▷ for other samples of evaluated children's work.

Summarize Ask children to reflect on their work and revisions. Ask: *What did you do to improve your work?* Answers vary.

3 Practice 10–15 min Go Online▷ ePresentations eToolkit Home Connections

▶ Math Boxes 8-4

Math Journal 2, p. 200

| WHOLE CLASS | SMALL GROUP | PARTNER | INDEPENDENT |

Mixed Practice Math Boxes 8-4 are paired with Math Boxes 8-2.

▶ Home Link 8-4

Math Masters, p. 219

Homework Children draw a triangle and quadrilateral and compare their attributes.

Sample child's work, "Meeting Expectations"

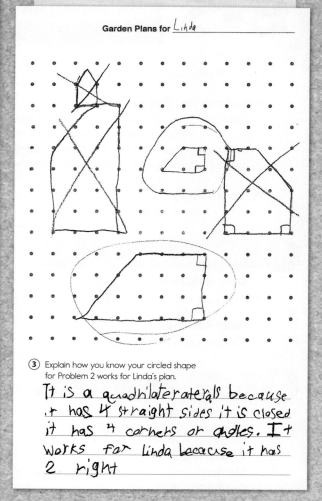

Garden Plans for Linda

③ Explain how you know your circled shape for Problem 2 works for Linda's plan.

It is a quadhilateraterals because it has 4 straight sides it is closed it has 4 corners or angles. It works for Linda because it has 2 right

---

Math Masters, p. 219

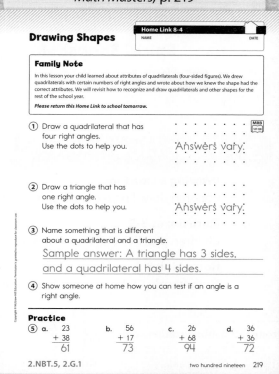

Drawing Shapes Home Link 8-4
NAME DATE

Family Note
In this lesson your child learned about attributes of quadrilaterals (four-sided figures). We drew quadrilaterals with certain numbers of right angles and wrote about how we knew the shape had the correct attributes. We will revisit how to recognize and draw quadrilaterals and other shapes for the rest of the school year.

Please return this Home Link to school tomorrow.

① Draw a quadrilateral that has four right angles. Use the dots to help you. Answers vary.

② Draw a triangle that has one right angle. Use the dots to help you. Answers vary.

③ Name something that is different about a quadrilateral and a triangle.
Sample answer: A triangle has 3 sides, and a quadrilateral has 4 sides.

④ Show someone at home how you can test if an angle is a right angle.

Practice
⑤ a. 23
+ 38
─────
61

b. 56
+ 17
─────
73

c. 26
+ 68
─────
94

d. 36
+ 36
─────
72

2.NBT.5, 2.G.1 two hundred nineteen 219

Lesson 8-5

Attributes of 3-Dimensional Shapes

Overview Children sort and compare 3-dimensional shapes according to their attributes.

▶ **Before You Begin**

Distribute stick-on notes to children and place a base-10 thousand cube near the Math Message. For Part 2 gather models of 3-dimensional shapes. Decide where you will display the shapes for the Shapes Museum. Write the terms *cube, rectangular prism, cylinder, cone, sphere, pyramid,* and *other* on index cards. For Part 3 designate a place where children can display their wrist measurements.

▶ **Vocabulary**

cube • face • apex

Common Core State Standards

Focus Cluster
Reason with shapes and their attributes.

---

1 Warm Up 15–20 min

	Materials	
Mental Math and Fluency Children solve basic facts.	slate	2.OA.2
Daily Routines Children complete daily routines.	See pages 4–43.	See pages xiv–xvii.

2 Focus 30–40 min

Math Message Children find examples of cubes.	*Math Journal 2,* p. 193; stick-on note; base-10 thousand cube	2.G.1
Describing Cubes Children discuss their descriptions of a cube.	Class Data Pad; per partnership: centimeter cube	2.G.1 SMP2, SMP6
Discussing Attributes of 3-Dimensional Shapes Children discuss attributes of 3-dimensional shapes.	base-10 thousand cube, rectangular prism, cylinder, cone, sphere, and pyramid (for demonstration)	2.G.1 SMP6
Comparing 3-Dimensional Shapes Children compare and discuss 3-dimensional shapes.	3-dimensional shapes (brought in by children)	2.G.1 SMP2, SMP6
✓ **Assessment Check-In** See page 722.		2.G.1, SMP6

CCSS 2.G.1 **Spiral Snapshot**

GMC Identify 2- and 3-dimensional shapes.

5-5 Focus	6-10 Focus	8-1 through 8-4 Focus Practice	8-5 Focus Practice	8-11 Focus Practice	9-5 Practice

/// **Spiral Tracker**   〈 **Go Online** 〉 to see how mastery develops for all standards within the grade.

3 Practice 10–15 min

Drawing a Line Plot Children draw line plots to represent class wrist-size data.	*Math Masters,* p. TA32; stick-on notes from the Math Message	2.MD.9
Math Boxes 8-5 Children practice and maintain skills.	*Math Journal 2,* p. 201; inch ruler	See page 723.
Home Link 8-5 **Homework** Children list examples of 3-dimensional shapes.	*Math Masters,* p. 221	2.NBT.5, 2.G.1

〉 **connectED.mheducation.com**

Plan your lessons online with these tools.

 ePresentations   Student Learning Center   Facts Workshop Game   eToolkit   Professional Development   Home Connections   Spiral Tracker   Assessment and Reporting   English Learners Support   Differentiation Support

Differentiation Options

 RtI

Readiness
5–15 min

Identifying Pattern-Block Template Shapes

Math Masters, p. 128; Pattern-Block Template

| WHOLE CLASS |
| SMALL GROUP |
| PARTNER |
| INDEPENDENT |

To explore 3-, 4-, and 5-sided polygons using a concrete model, have children revisit the Readiness activity in Lesson 5-5. They draw shapes with their Pattern-Block Templates and identify them on *Math Masters*, page 128. When children have completed the page, encourage them to describe similarities and differences among the shapes. **GMP7.1**

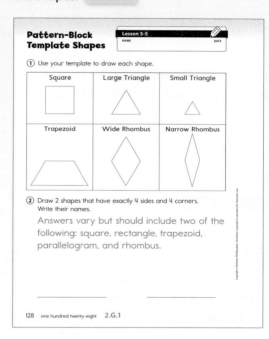

Pattern-Block Template Shapes — Lesson 5-5
NAME   DATE

① Use your template to draw each shape.

| Square | Large Triangle | Small Triangle |
| Trapezoid | Wide Rhombus | Narrow Rhombus |

② Draw 2 shapes that have exactly 4 sides and 4 corners. Write their names.

Answers vary but should include two of the following: square, rectangle, trapezoid, parallelogram, and rhombus.

128 one hundred twenty-eight 2.G.1

Enrichment
15–30 min

Describing Faces on a Cube

Math Masters, p. 220; per group: 27 centimeter cubes, small stickers (optional)

| WHOLE CLASS |
| SMALL GROUP |
| PARTNER |
| INDEPENDENT |

To apply their understanding of 3-dimensional figures, children complete *Math Masters*, page 220. Provide plenty of time for children to think about and try out their solution strategies. Children might build the larger cube with centimeter cubes and use small stickers to represent the painted sides, or they might draw each layer of the larger cube and figure out how many faces on each block will be painted.

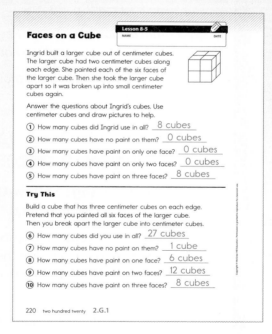

Faces on a Cube — Lesson 8-5
NAME   DATE

Ingrid built a larger cube out of centimeter cubes. The larger cube had two centimeter cubes along each edge. She painted each of the six faces of the larger cube. Then she took the larger cube apart so it was broken up into small centimeter cubes again.

Answer the questions about Ingrid's cubes. Use centimeter cubes and draw pictures to help.

① How many cubes did Ingrid use in all? __8 cubes__
② How many cubes have no paint on them? __0 cubes__
③ How many cubes have paint on only one face? __0 cubes__
④ How many cubes have paint on only two faces? __0 cubes__
⑤ How many cubes have paint on three faces? __8 cubes__

Try This

Build a cube that has three centimeter cubes on each edge. Pretend that you painted all six faces of the larger cube. Then you break apart the larger cube into centimeter cubes.

⑥ How many cubes did you use in all? __27 cubes__
⑦ How many cubes have no paint on them? __1 cube__
⑧ How many cubes have paint on one face? __6 cubes__
⑨ How many cubes have paint on two faces? __12 cubes__
⑩ How many cubes have paint on three faces? __8 cubes__

220 two hundred twenty 2.G.1

Extra Practice
5–15 min

Sorting Shapes

Activity Card 103, shapes from the Shapes Museum

| WHOLE CLASS |
| SMALL GROUP |
| PARTNER |
| INDEPENDENT |

For practice identifying attributes of shapes, children sort shapes from the Shapes Museum. Encourage them to sort by attributes, such as having all equal-size faces, having rectangular faces, having curved surfaces, having only flat surfaces, and so on.

Sorting Shapes (103)

I am going to find all the shapes with curved surfaces.

What You Need
shapes from the Shapes Museum

What To Do
Work with a partner or small group.
❶ Think about ways to sort the shapes from the Shapes Museum.
❷ Sort the shapes.
❸ Think of another way to sort the shapes.

Talk About It
How are the shapes alike? How are they different?

English Language Learners Support

Beginning ELL Children may be confused by use of the term *face* to describe an attribute of a geometric solid because they think of the term in reference to the front of their head. To help children learn the mathematical meaning of this term, introduce it by showing children a variety of solids with facial features drawn on their faces. Assess children's understanding of the term by showing them solids without the drawings and asking them to point to or count all the faces.

Go Online > **ELL** English Learners Support

SMP2 Reason abstractly and quantitatively.

GMP2.1 Create mathematical representations using numbers, words, pictures, symbols, gestures, tables, graphs, and concrete objects.

GMP2.2 Make sense of the representations you and others use.

SMP6 Attend to precision.

GMP6.3 Use clear labels, units, and mathematical language.

NOTE This lesson exposes children to many geometry terms that may be unfamiliar to them. At this time they are expected to recognize shapes having specified attributes, such as a given number of angles or a given number of "equal" or congruent faces. Children are not expected to know all the related vocabulary.

1 Warm Up 15–20 min   Go Online ePresentations eToolkit

▶ Mental Math and Fluency

Pose facts one at a time. Children write the sum or difference on their slates. *Leveled exercises:*

●○○		●●○		●●●	
5 + 2	7	9 + 2	11	9 − 5	4
7 + 2	9	6 + 3	9	7 − 4	3
6 + 2	8	3 + 8	11	8 − 5	3

▶ Daily Routines

Have children complete daily routines.

2 Focus 30–40 min   Go Online ePresentations eToolkit

▶ Math Message

Math Journal 2, p. 193

Turn to journal page 193. Record the measurement of your wrist size in centimeters on a stick-on note. Write large enough so others can see.

Then look at the base-10 thousand cube. Find examples of other cubes around the room.

▶ Describing Cubes

WHOLE CLASS SMALL GROUP PARTNER INDEPENDENT

Math Message Follow-Up On the Class Data Pad, list examples of cubes that children found. **GMP2.1**

Distribute a centimeter cube to each partnership. Have children share with their partners what they notice about the **cube.** After a few minutes, bring the class together to discuss children's observations. Expect them to note the following:

- There are six flat surfaces, or **faces.**
- All the faces are the same size.
- Each face is a square. **GMP6.3**

Ask children to point to each of the six faces on their cubes. Revisit the examples of cubes listed on the Class Data Pad, asking volunteers to confirm that each item fits the description of a cube. **GMP2.2** Adjust the list as needed.

▶ Discussing Attributes of 3-Dimensional Shapes

WHOLE CLASS | **SMALL GROUP** | PARTNER | INDEPENDENT

Use your models of 3-dimensional shapes to point out the following attributes:

- Cylinders, cones, and spheres all have curved surfaces.
- Rectangular prisms, cubes, pyramids, cylinders, and cones all have flat surfaces called *faces*.
- An edge of a cube, a prism, or a pyramid is a line segment where two faces meet.
- An edge of a cone or a cylinder is a curve where a flat face meets a curved surface.
- A vertex on a 3-dimensional shape such as a cube, a prism, or a pyramid is a point at which at least 3 edges meet. (The plural of *vertex* is *vertices*.)
- The **apex** of a cone is the point that is opposite the flat face.

Draw children's attention to the faces, edges, and vertices on the base-10 thousand cube. Have partners take turns running their fingers along the edges of their centimeter cubes and pointing out the faces and the vertices. GMP6.3

Academic Language Development Explain that *3-D* is an abbreviation for *3-dimensional,* which is a term used to describe solid shapes. Children may be familiar with the term *3-D* from watching movies or from pop-up books. Build on children's prior experiences to describe how in 3-D movies or pop-up pages, objects look like they pop out, as opposed to lying flat on the screen or the page.

▶ Comparing 3-Dimensional Shapes

WHOLE CLASS | **SMALL GROUP** | PARTNER | INDEPENDENT

Explain to children that they are going to make a Shapes Museum so they can examine different kinds of shapes. Help them set up the museum by placing the items they brought from home near the corresponding name cards. (*See Before You Begin.*) Shapes that do not fit into any of the six categories are placed near the "other" card. Add some of your own items to the museum.

Display models of pairs of shapes as specified below. GMP2.2 As you display each pair, ask: *How are these alike? How are they different?* GMP6.3 Sample observations that children might make include the following:

Cube and Rectangular Prism

- They have the same number of faces, vertices, and edges.
- Each face on both shapes has 4 sides and 4 angles.
- All of the faces of the cube are squares.
- The faces of the rectangular prism can be squares or rectangles.
- Rectangular prisms that have all square faces are called cubes. A cube is a special kind of rectangular prism.

Models of a cube and a rectangular prism

Models of a cube and a cylinder

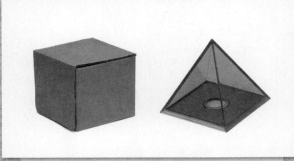

Models of a cube and a pyramid

Models of a cube and a cone

Cube and Cylinder

- The cylinder can roll when pushed. The cube can't.
- The cylinder has a curved surface. The cube doesn't.
- The cylinder has 2 flat faces. The cube has 6 flat faces.
- The cube's faces are squares. The flat faces on the cylinder are circles.

Cube and Pyramid

- Most of the faces of a pyramid are triangles. (Sometimes one of the faces is not a triangle.) All of the faces of a cube are square.
- A cube and a pyramid both have vertices where edges come together. A pyramid has a special vertex called an *apex* where the triangle faces come together.

Cube and Cone

- The cone can roll when pushed. The cube can't.
- The cone has a curved surface and 1 flat face in the shape of a circle. The cube has 6 flat faces that are all squares.
- The cone has a point, or apex, opposite the circular face. The cube has 8 vertices.

As time allows, compare and contrast other pairs of shapes.

> **Differentiate** **Adjusting the Activity**
>
> To help children describe the faces of 3-dimensional shapes, have them select a 3-dimensional shape and trace all of its flat faces on paper. They identify the shapes of the faces, record the names on the paper, and then use that information to describe the 3-dimensional shape. For example, a cube has 6 faces that are all squares. **Go Online** Differentiation Support

Over the next several days, allow small groups of children to visit the Shapes Museum. Have them examine the shapes and describe them in terms of their attributes. GMP6.3 For example, they may describe a cube by saying, "This shape has faces that are all equal. Each face is a square." They may describe a cylinder by saying, "This shape has a curved surface and two flat faces that are circles."

> ✔ **Assessment Check-In** CCSS 2.G.1
>
> Expect most children to be able to describe a cube as having 6 equal-size square faces. GMP6.3 If children struggle describing a cube by its attributes, have them trace the faces of a cube on paper as suggested in the Adjusting the Activity note. Some children may be able to describe other shapes by their attributes.
>
> ✔ Assessment and Reporting   **Go Online** to record student progress and to see trajectories toward mastery for these standards.

Summarize Have children share one or two things they learned about 3-dimensional shapes.

③ Practice 10–15 min

Go Online | ePresentations | eToolkit | Home Connections

▶ Drawing a Line Plot

Math Masters, p. TA32

| WHOLE CLASS | SMALL GROUP | PARTNER | INDEPENDENT |

With the class, organize and display the stick-on notes with wrist measurements (see the Math Message) in order from smallest to largest. Explain that children will draw line plots to show all the wrist measurements for the class.

Display *Math Masters*, page TA32 (or display a line for the line plot) and distribute copies to children. Discuss the horizontal scale. The wrist-size data are measurements to the nearest centimeter. The scale should begin with the smallest wrist size in the class, increase in 1-centimeter increments, and end with the measurement of the largest wrist size in the class. Explain that children do not need to use all the lines below the axis. Model writing the scale as children do so on page TA32.

Ask children to suggest a label for the horizontal axis and write it on the line. Sample answer: Wrist Size in Centimeters Then ask children to suggest a title for the line plot. Sample answer: Class Wrist Sizes

Have children draw Xs on their line plots to represent the class data. Remind them that each X represents one child. Model how to draw Xs one above the other for stick-on notes with the same measurements. Below is a sample line plot:

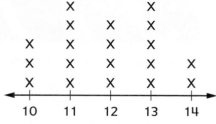

Class Wrist Sizes

```
        X           X
        X     X     X
  X     X     X     X
  X     X     X     X     X
  X     X     X     X     X
◄──┼─────┼─────┼─────┼─────┼──►
  10    11    12    13    14
      Wrist Size in Centimeters
```

▶ Math Boxes 8-5

Math Journal 2, p. 201

| WHOLE CLASS | SMALL GROUP | PARTNER | INDEPENDENT |

Mixed Practice Math Boxes 8-5 are paired with Math Boxes 8-7. For Problem 2, explain to children that the width of their journal is the distance across it from left to right, not top to bottom.

▶ Home Link 8-5

Math Masters, p. 221

Homework Children look for examples of 3-dimensional shapes at home.

Math Journal 2, p. 201

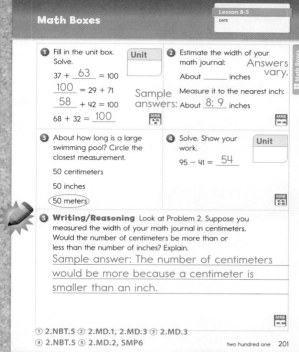

Math Boxes

Lesson 8-5
DATE

① Fill in the unit box. Solve.

Unit

$37 + \underline{63} = 100$
$\underline{100} = 29 + 71$
$\underline{58} + 42 = 100$
$68 + 32 = \underline{100}$

② Estimate the width of your math journal: About _____ inches

Answers vary.

Measure it to the nearest inch: About $\underline{8; 9}$ inches

Sample answers:

③ About how long is a large swimming pool? Circle the closest measurement.

50 centimeters
50 inches
(50 meters)

④ Solve. Show your work.

$95 - 41 = \underline{54}$

Unit

⑤ **Writing/Reasoning** Look at Problem 2. Suppose you measured the width of your math journal in centimeters. Would the number of centimeters be more than or less than the number of inches? Explain.

Sample answer: The number of centimeters would be more because a centimeter is smaller than an inch.

① 2.NBT.5 ② 2.MD.1, 2.MD.3 ③ 2.MD.3
④ 2.NBT.5 ⑤ 2.MD.2, SMP6

two hundred one 201

Math Masters, p. 221

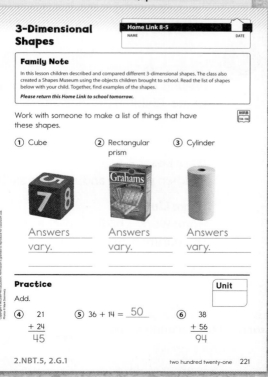

3-Dimensional Shapes

Home Link 8-5
NAME   DATE

Family Note

In this lesson children described and compared different 3-dimensional shapes. The class also created a Shapes Museum using the objects children brought to school. Read the list of shapes below with your child. Together, find examples of the shapes.

Please return this Home Link to school tomorrow.

Work with someone to make a list of things that have these shapes.

① Cube   ② Rectangular prism   ③ Cylinder

Answers vary.   Answers vary.   Answers vary.

Practice

Unit

Add.

④   21
   + 24
   ‾‾‾‾
    45

⑤ $36 + 14 = \underline{50}$

⑥   38
   + 56
   ‾‾‾‾
    94

2.NBT.5, 2.G.1

two hundred twenty-one 221

Partitioning Rectangles, Part 1

Overview Children use manipulatives to partition rectangles into same-size squares.

▶ **Before You Begin**
For Part 2 you will need a 1-inch square pattern block and 20 centimeter cubes per child.

▶ **Vocabulary**
row • column • partition

Common Core State Standards

Focus Cluster
Reason with shapes and their attributes.

---

1 Warm Up   15–20 min

	Materials	
Mental Math and Fluency Children practice place-value skills.	slate	2.NBT.1, 2.NBT.1a, 2.NBT.1b
Daily Routines Children complete daily routines.	See pages 4–43.	See pages xiv–xvii.

2 Focus   30–40 min

Math Message Children tile a rectangle.	*Math Journal 2*, p. 202; 20 centimeter cubes	2.G.2
Introducing Partitioning Children partition a rectangle into same-size squares.	*Math Journal 2*, p. 202; *Math Masters*, p. 225; 20 centimeter cubes	2.G.2 SMP1
Partitioning Rectangles Children use manipulatives to partition rectangles into same-size squares.	*Math Journal 2*, pp. 203–204; *Math Masters*, p. 226; 1-inch square pattern block	2.G.2 SMP3, SMP5
✓ **Assessment Check-In** See page 728.	*Math Journal 2*, p. 204	2.G.2

CCSS 2.G.2 Spiral Snapshot

GMC Partition a rectangle into rows and columns of same-size squares and count to find the total number of squares.

| 1-12
Focus | 3-11
Focus | 8-6
Focus
Practice | 8-7
Focus
Practice | 9-1
Practice | 9-3
Practice | 9-6
Practice | 9-11
Practice |

Spiral Tracker   **Go Online** to see how mastery develops for all standards within the grade.

3 Practice   10–15 min

Playing *The Number-Grid Difference Game* **Game** Children practice subtraction with 2-digit numbers.	*My Reference Book*, pp. 156–157; *Math Masters*, pp. TA3 and G17; counters; calculator; per partnership: 4 each of number cards 0–9	2.NBT.5, 2.NBT.7 SMP2
Math Boxes 8-6 Children practice and maintain skills.	*Math Journal 2*, p. 205	See page 729.
Home Link 8-6 **Homework** Children cover a rectangle with paper squares and partition an identical rectangle.	*Math Masters*, pp. 227–228	2.G.2

---

connectED.mheducation.com

Plan your lessons online with these tools.

 ePresentations   Student Learning Center   Facts Workshop Game   eToolkit   Professional Development   Home Connections   Spiral Tracker   Assessment and Reporting   English Learners Support   Differentiation Support

724   Unit 8 | Geometry and Arrays

Differentiation Options

RtI

Readiness
5–15 min

WHOLE CLASS
SMALL GROUP
PARTNER
INDEPENDENT

Covering Surfaces with Nonstandard Units

Math Masters, p. 222; 1-inch square pattern blocks; Pattern-Block Template; slate; paper; scissors; cards from the Everything Math Deck

To prepare for partitioning rectangles into same-size units, children use nonstandard units to cover various surfaces in the classroom. They record their work on *Math Masters*, page 222.

Covering Areas with Shapes

Lesson 8-6
NAME
DATE

Follow these steps to cover one Everything Math Deck card with square pattern blocks:

Answers vary.

- Lay the blocks flat on the card.
- Don't leave any spaces between the blocks.
- Keep the blocks inside the edges of the card. There may be open spaces along the edges.

① Count the blocks on the card. If a space could be covered by more than half of a block, count the space as one block. Do not count spaces that could be covered by less than half of a block.

I used _____ square pattern blocks to cover the card.

Use Everything Math Deck cards to cover your slate. Then use them to cover your Pattern-Block Template.

② Count the cards on the slate and the Pattern-Block Template.

I used _____ cards to cover my slate.

I used _____ cards to cover my Pattern-Block Template.

Fold a piece of paper into four equal pieces. Cut the pieces apart. Use the pieces to cover your desktop. Then use them to cover another large surface.

③ Count the pieces of paper on the desk and the other surface.

I used _____ pieces of paper to cover my desk.

I used _____ pieces of paper to cover _____.

222 two hundred twenty-two 2.G.2

Enrichment
5–15 min

WHOLE CLASS
SMALL GROUP
PARTNER
INDEPENDENT

Partitioning Rectangles without Tools

Math Masters, p. 223

To apply their understanding of partitioning rectangles, children partition rectangles into rows and columns of same-size squares on *Math Masters*, page 223. They determine the total number of squares it takes to cover each rectangle without using a tool.

Partition Rectangles without Tools

Lesson 8-6
NAME
DATE

① Look at the picture of the small square. How many small squares can cover the rectangle without gaps or overlaps? Partition the rectangle to show your answer.

How many rows of squares are there? 3
How many squares are in each row? 4
How many squares does it take to cover the rectangle? 12

② Partition this rectangle into 4 rows with 5 same-size squares in each row. There should be no gaps or overlaps.

How many squares does it take to completely cover the rectangle? 20

2.G.2
two hundred twenty-three 223

Extra Practice
5–15 min

WHOLE CLASS
SMALL GROUP
PARTNER
INDEPENDENT

Partitioning Rectangles into Squares

Math Masters, p. 224; 1-inch square pattern block; centimeter cube

For additional practice partitioning rectangles into same-size squares, children use a square pattern block and centimeter cube to help them partition rectangles on *Math Masters*, page 224.

Partition Rectangles into Squares

Lesson 8-6
NAME
DATE

① Partition this rectangle into squares that are the same size as a square pattern block. You may use a pattern block to help.

How many rows? 3 How many squares in each row? 6
How many squares in all? 18

② Partition this rectangle into squares that are the same size as one face of a centimeter cube. You may use a centimeter cube to help.

How many rows? 3
How many squares in each row? 8
How many squares in all? 24

224 two hundred twenty-four 2.G.2

English Language Learners Support

Beginning ELL Use think-alouds and folding activities to demonstrate the meaning of *partition* and connect it to the term *parts*. For example, show a sheet of paper and say: *I have a piece of paper. I want 4 smaller pieces. I will make 4 parts. I will partition the paper into 4 parts.* Model folding the paper into 4 equal parts and then have children do the same, counting the parts afterward. Mention that the term *partition* begins with *part-*. For additional practice, instruct children to partition paper into other numbers of parts, such as halves and eighths. Use directions such as the following: *Partition your paper into _____ parts.* Ask children to tell how many parts they have.

Go Online ELL English Learners Support

Professional Development

The centimeter cube can be used as a unit of *volume* (how much 3-dimensional space it takes up), a unit of *area* (how much 2-dimensional space one face takes up in covering a rectangle), and a unit of *length* (how many edges of cubes are needed to measure the length of a segment or path). It is not the object that is the unit. We use a feature of the object in our measurement work.

You can help children understand this by (1) being careful how you describe the attributes of the manipulatives and (2) listening for children who confuse one such unit for another.

Go Online Professional Development

1 Warm Up — 15–20 min — Go Online — ePresentations — eToolkit

▶ Mental Math and Fluency

Display the following place-value exercises. Have children write the numbers on their slates. *Leveled exercises:*

- ●○○ 4 hundreds, 10 tens, 2 ones 502
  - 8 hundreds, 10 tens 900
- ●●○ 10 tens, 7 hundreds, 9 ones 809
  - 7 ones, 6 hundreds, 10 tens 707
- ●●● 12 tens, 5 hundreds, 8 ones 628
  - 16 ones, 7 hundreds, 10 tens 816

▶ Daily Routines

Have children complete daily routines.

2 Focus — 30–40 min — Go Online — ePresentations — eToolkit

▶ Math Message

Math Journal 2, p. 202

Take 20 centimeter cubes. Complete Problem 1 on journal page 202.

▶ Introducing Partitioning

Math Journal 2, p. 202; *Math Masters*, p. 225

| WHOLE CLASS | SMALL GROUP | PARTNER | INDEPENDENT |

Math Message Follow-Up Display *Math Masters*, page 225 (which is identical to journal page 202) and have a volunteer cover Rectangle A with centimeter cubes. Ask children to share what they noticeut the cubes covering the rectangle.

They may observe that it took 15 cubes to completely cover the rectangle and that the cubes are arranged in **rows** and **columns.** Remind children that a rectangle is a 2-dimensional (flat) shape and a cube is a 3-dimensional shape. Ask: *What part of each cube actually covers the rectangle?* One of the faces *What shape is the face?* Square

Tell children to complete Problem 2 on journal page 202 by drawing squares on Rectangle B to show how they covered Rectangle A with centimeter cubes. Explain that when children draw same-size shapes to cover a shape, they are **partitioning,** or dividing, the shape into smaller shapes.

NOTE Some children may not be able to draw squares very well, resulting in different-size "squares" throughout the figure. As they practice more with partitioning, expect that their squares will become more uniform.

When most children are finished, have volunteers share their drawings. Identify a drawing that has 3 rows with 5 close-to-same-size squares in each row and ask children what they notice. Guide children to connect the equal rows of squares on Rectangle B to the equal rows of centimeter cubes that covered Rectangle A. Have children check that the squares they drew on Rectangle B match the arrangement of centimeter cubes that covered Rectangle A.

Discuss the challenges children faced in Problem 2. Ask: *How could you tell if you made a mistake?* GMP1.2 Sample answers: My squares were not in straight rows and columns. Some rows had more squares than others. *How did you fix your mistake?* GMP1.3 Sample answers: I erased my squares and redrew them. I checked that I had 5 squares in each row.

Academic Language Development Have children activate prior knowledge of the word *part* to help them understand the terms *partition* and *partitioning.* Point out that when they *partition* a figure, they are dividing it into equal-size *parts.* This is also called *partitioning.*

▶ Partitioning Rectangles

Math Journal 2, pp. 203–204; *Math Masters,* p. 226

| WHOLE CLASS | SMALL GROUP | PARTNER | INDEPENDENT |

Display a 1-inch square pattern block and *Math Masters,* page 226. Point to Problem 3. Ask children to think about how they might use a single square pattern block to find the total number of square pattern blocks needed to completely cover Rectangle C. Invite children to share their ideas with a partner and encourage them to make sense of their partner's ideas. GMP3.2 Bring the class together and explain that they will get a chance to try out their ideas.

Distribute a square pattern block to each child. Have them place the pattern block flat on the rectangle. Ask: *What part of the pattern block is actually on the rectangle?* A square face Tell children that to partition the rectangle, they need to draw squares to show where all of the square faces of the blocks would be if they covered the rectangle completely. Explain that their drawings should show where they put their square each time they moved it. Demonstrate drawing two or three squares on the display of *Math Masters,* page 226. Have partners use a single square pattern block to partition Rectangle C into same-size squares. GMP5.2

When they are finished, tell them to count the squares and answer the questions below Rectangle C. Bring the class together. Ask: *Into how many squares did you partition Rectangle C?* 35 *How did you use the square block to help partition the rectangle?* Expect strategies to include the following:

- I traced the pattern block multiple times to cover the rectangle.
- I traced a complete row or column of square pattern blocks and then filled in the other rows and columns.
- I traced the square pattern block to fill the space along all four edges of the rectangle and then filled in the middle.

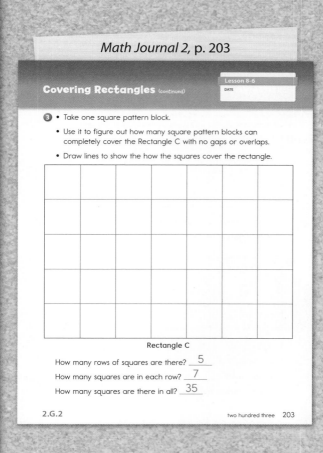

Math Journal 2, p. 202

Math Journal 2, p. 203

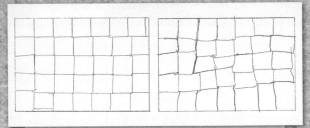

Examples of correct partitioning

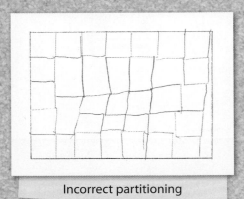

Incorrect partitioning

Math Journal 2, p. 204

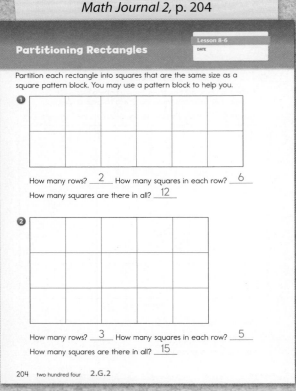

Partitioning Rectangles

Lesson 8-6

DATE

Partition each rectangle into squares that are the same size as a square pattern block. You may use a pattern block to help you.

❶

How many rows? __2__ How many squares in each row? __6__
How many squares are there in all? __12__

❷

How many rows? __3__ How many squares in each row? __5__
How many squares are there in all? __15__

One problem children may have executing these strategies is losing track of how many squares they should trace or draw in the middle of the rectangle.

Explain that one way to make partitioning easier is to first draw one row and one column of squares. Demonstrate by placing a square pattern block in the upper-left corner of Rectangle C and tracing a mark along its right edge. Then move the block to align its left edge with your mark. Continue making marks and moving the block to complete the row, pointing out that there are no gaps or overlaps between the squares. Ask: *How many squares will be in each row?* 7 Count the spaces to verify that 7 blocks will fit in a row.

Repeat the process for a column, starting in the upper-left corner and tracing marks along the bottom edge of the block. Ask: *How many rows will there be?* 5 Extend the lines for each row and column until the rectangle is completely partitioned into squares. Ask: *How many rows are there?* 5 *How many squares are in each row?* 7 Point out that the number of squares per row is the same as the number of columns. Ask: *How many same-size squares cover the rectangle?* 35

> **NOTE** Using a straightedge or a ruler to draw the partitioning lines may distract from the activity's focus, so it is preferred that neither you nor the children do so. Children's lines do not need to be completely straight as long as squares can be seen inside the rectangle. Have partners work together to use one square pattern block to help partition the rectangles on journal page 204.

Differentiate   **Adjusting the Activity**

For children who struggle with partitioning, provide enough square pattern blocks to completely cover the rectangle in Problem 1 on journal page 204. Children count and record the number of pattern blocks they used and then remove the blocks. They then use their recorded numbers as a guideline to partition the rectangle.   Go Online  Differentiation Support

✔ **Assessment Check-In** (CCSS) **2.G.2**

Math Journal 2, p. 204

Because this is their first exposure to partitioning, do not expect children to accurately partition the rectangles on journal page 204 into same-size squares. Expect their attempts to show evidence of a strategy, such as tracing the pattern block multiple times, drawing rows or columns of squares, or drawing squares along the edges of the rectangles. The "squares" each child draws may vary in size and shape. Some children will have a harder time drawing squares in the middle of the rectangle than on the edges. For those who struggle, use the suggestion in the Adjusting the Activity note above. Some children may be able to accurately partition the rectangle into same-size squares with or without the square pattern block.

 Assessment and Reporting   Go Online to record student progress and to see trajectories toward mastery for these standards.

Summarize Have children share their strategies for partitioning the rectangles in Problems 1 and 2 into same-size squares.

3 Practice 10–15 min

Go Online

ePresentations eToolkit Home Connections

▶ Playing *The Number-Grid Difference Game*

My Reference Book, pp. 156–157; *Math Masters*, pp. TA3 and G17

| WHOLE CLASS | SMALL GROUP | PARTNER | INDEPENDENT |

Partners play *The Number-Grid Difference Game* to practice subtracting with 2-digit numbers. For directions see *My Reference Book,* pages 156–157. Explain the directions and model a round or two.

Observe

• How are children using the number grid to calculate differences?

• Which children are using calculators to add their five scores?

Discuss

• *How did you decide on the order of the digits in your 2-digit numbers?*

• *What did you find easy about this game? Challenging?*

▶ Math Boxes 8-6

Math Journal 2, p. 205

| WHOLE CLASS | SMALL GROUP | PARTNER | INDEPENDENT |

Mixed Practice Math Boxes 8-6 are paired with Math Boxes 8-10.

▶ Home Link 8-6

Math Masters, pp. 227–228

Homework Children cover one rectangle with paper squares and partition an identical rectangle by drawing the same number of squares.

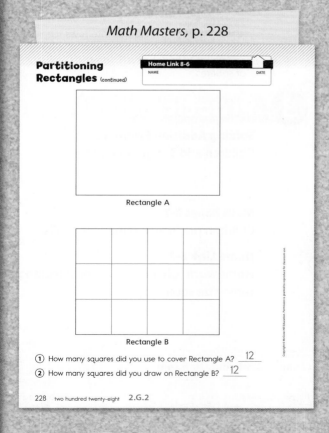

Math Journal 2, p. 205

Math Masters, p. 228

Partitioning Rectangles, Part 2

Overview Children partition rectangles into same-size squares.

▶ **Before You Begin**
For Part 2 place 1-inch square pattern blocks near the Math Message.

**Common Core
State Standards**

Focus Cluster
Reason with shapes and
their attributes.

1 **Warm Up** 15–20 min	**Materials**	
Mental Math and Fluency Children practice basic facts.	slate	**2.OA.2**
Daily Routines Children complete daily routines.	See pages 4–43.	See pages xiv–xvii.

2 **Focus** 30–40 min		
Math Message Children use a square pattern block to partition a rectangle into same-size squares.	*Math Journal 2*, p. 206; 1-inch square pattern block	**2.G.2**
Partitioning Strategies Children use a visual aid to partition rectangles into equal rows of same-size squares.	*Math Journal 2*, p. 206	**2.G.2** SMP2
Partitioning into Same-Size Squares Children partition rectangles into equal rows without manipulatives or visual aids.	*Math Journal 2*, p. 207	**2.G.2** SMP1
✓ **Assessment Check-In** See page 735.	*Math Journal 2*, p. 207	**2.G.2**

 CCSS 2.G.2 **Spiral Snapshot**

GMC Partition a rectangle into rows and columns of same-size squares and count to find the total number of squares.

| 1-12
Focus | 3-11
Focus | 8-6
Focus
Practice | 8-7
Focus
Practice | 9-1
Practice | 9-3
Practice | 9-6
Practice | 9-11
Practice |

Spiral Tracker   **Go Online**   to see how mastery develops for all standards within the grade.

3 **Practice** 15–20 min		
Solving Addition Problems Children add 3-digit numbers.	*Math Journal 2*, p. 208; number grid or number line (inside back cover of *Math Journal 2*, optional); base-10 blocks (optional)	**2.NBT.5, 2.NBT.7** SMP5
Math Boxes 8-7 Children practice and maintain skills.	*Math Journal 2*, p. 209; centimeter ruler	See page 735.
Home Link 8-7 **Homework** Children partition a rectangle into rows of same-size squares.	*Math Masters*, p. 233	**2.NBT.5, 2.G.2**

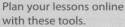 **connectED.mheducation.com**

Plan your lessons online
with these tools.

 ePresentations   Student Learning Center   Facts Workshop Game   eToolkit   Professional Development   Home Connections   Spiral Tracker   Assessment and Reporting   **ELL** English Learners Support   Differentiation Support

Differentiation Options

RtI

Covering More Surfaces with Nonstandard Units

WHOLE CLASS
SMALL GROUP
PARTNER
INDEPENDENT

Math Masters, p. 229; 1-inch square pattern blocks; Pattern-Block Template; index card; *Math Journal 2; My Reference Book;* paper; scissors; cards from the Everything Math Deck

To prepare for partitioning rectangles into same-size units, children use nonstandard units to cover various surfaces in the classroom. They record their work on *Math Masters,* page 229.

More Covering Areas with Shapes
Lesson 8-7
NAME   DATE

Cover one index card with square pattern blocks:

- Lay the blocks flat on the card. Answers vary.
- Don't leave any spaces between the blocks.
- Keep the blocks inside the edges of the card. There may be open spaces around the edges.
① Count the blocks on the card. If a space could be covered by more than half of a block, count the space as one block. Do not count spaces that could be covered by less than half of a block.

   I used _____ square pattern blocks to cover the card.

Trace the card onto a piece of paper. Use your Pattern-Block Template to show how you covered the card with blocks.
_____

Use Everything Math Deck cards to cover your journal and your *My Reference Book.*
② Count the cards on the journal and your *My Reference Book.*

   I used _____ cards to cover my journal.

   I used _____ cards to cover *My Reference Book.*

Fold a piece of paper into four equal pieces. Cut the pieces apart. Use the pieces to cover large surfaces other than your desktop.
③ I used _____ pieces of paper to cover _____.

   I used _____ pieces of paper to cover _____.

2.G.2   two hundred twenty-nine 229

Partitioning Polygons

WHOLE CLASS
SMALL GROUP
PARTNER
INDEPENDENT

Math Masters, p. 230

Children deepen their understanding of partitioning by partitioning rectilinear polygons into smaller rectangles. Then they partition each smaller rectangle into same-size squares. Children record their work on *Math Masters,* page 230.

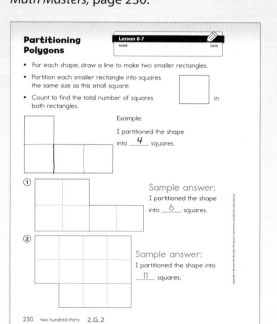

Partitioning Polygons
Lesson 8-7
NAME   DATE

- For each shape, draw a line to make two smaller rectangles.
- Partition each smaller rectangle into squares the same size as this small square:
- Count to find the total number of squares in both rectangles.

Example:

I partitioned the shape into __4__ squares.

① Sample answer:
I partitioned the shape into __6__ squares.

② Sample answer:
I partitioned the shape into __11__ squares.

230   two hundred thirty   2.G.2

Partitioning Rectangles

WHOLE CLASS
SMALL GROUP
PARTNER
INDEPENDENT

Math Masters, pp. 231–232

For additional practice partitioning rectangles into same-size squares, children complete *Math Masters,* pages 231–232. For Problem 3 children explain whether a given rectangle has been partitioned correctly. **GMP3.2**

Sample answer: I agree with Cathy. Jack's squares are not the same size. He has 7 squares in each row, but he has 5 squares in 6 columns and only 4 squares in 1 column.

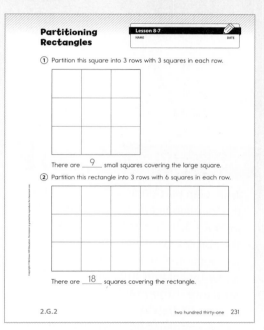

Partitioning Rectangles
Lesson 8-7
NAME   DATE

① Partition this square into 3 rows with 3 squares in each row.

There are __9__ small squares covering the large square.

② Partition this rectangle into 3 rows with 6 squares in each row.

There are __18__ squares covering the rectangle.

2.G.2   two hundred thirty-one 231

English Language Learners Support

Beginning ELL Introduce the adjective *same-size* using stick-on notes of different sizes and colors. Show how stick-on notes of the same size cover each other exactly, regardless of color. Ask *yes* or *no* questions to compare different stick-on notes. Ask: *Are these stick-on notes the same size? Are these same-size stick-on notes?* Provide practice hearing and responding to the term by directing children to organize the stick-on notes by putting specific same-size notes together.

Go Online   ELL **English Learners Support**

Standards and Goals for
Mathematical Practice

SMP1 **Make sense of problems and persevere in solving them.**
  GMP1.2 Reflect on your thinking as you solve your problem.

SMP2 **Reason abstractly and quantitatively.**
  GMP2.2 Make sense of the representations you and others use.

1 Warm Up   15–20 min   〔Go Online〕 ePresentations   eToolkit

▶ Mental Math and Fluency

Pose facts one at a time. Children write the sum or difference on their slates.
Leveled exercises:

●○○		●●○		●●●	
3 + 4	7	8 + 5	13	11 − 2	9
4 + 5	9	6 + 8	14	12 − 3	9
5 + 6	11	7 + 5	12	11 − 3	8

▶ Daily Routines

Have children complete daily routines.

2 Focus   30–40 min   〔Go Online〕 ePresentations   eToolkit

▶ Math Message

Math Journal 2, p. 206

Take one square pattern block. Complete Problem 1 on journal page 206. When you finish, return the square pattern block.

▶ Partitioning Strategies

Math Journal 2, p. 206

WHOLE CLASS	SMALL GROUP	PARTNER	INDEPENDENT

Math Message Follow-Up Have children share their strategies for partitioning the rectangle in Problem 1 on journal page 206 into same-size squares. Display a drawing that shows equal rows with equal numbers of close-to-same-size squares in each row.

Have children run a finger along each row on their rectangles. Ask:

- *How many rows does your drawing have?* 3
- *How many squares are in each row?* 2 Have children check that they have the same number of squares in each row.
- *Where are the columns? Point to them.* Children should indicate the columns on their rectangles.
- *How many columns are there?* 2
- *Why does this rectangle have 2 columns?* **GMP2.2** Because each row has 2 squares Count the squares in the first row aloud while pointing to each square: *1, 2.* Point out that each square in the first row is at the top of a new column. Count the columns aloud as you run your finger down the columns from top to bottom: *1, 2.*

- *How many squares are in each column?* 3
- *Why does this rectangle have 3 squares in each column?* GMP2.2

Because there are 3 rows of squares Count the squares in the first column. Point out that each square is at the beginning of a row.

Draw children's attention to the picture of the square to the right of the rectangle in Problem 2 on journal page 206. Explain that they will use the picture to help them figure out how many squares of that size are needed to cover the rectangle. Have children imagine that they are picking up the square and using it to partition the rectangle the same way they used the square block to partition the rectangle in Problem 1. As children work, check to make sure that they are drawing the same number of squares in each row and that the squares are about the same size.

Ask children to share their strategies for determining how many squares are needed to cover the rectangle. Some children may have visually estimated how many squares will fit in one row and one column, while others may have used their fingers or marks on paper to help them estimate. Ask: *How were you able to make sure that your squares were the same size?* Sample answers: I made a picture in my mind of the size of the square and then thought about how many squares it would take to cover the rectangle. I measured the length of the square in the picture with my finger and used my finger to mark the squares on the rectangle. I marked the edge of a paper with the length of the square and used the marks on the paper to help me know how big the squares needed to be.

Invite volunteers who drew equal rows of close-to-same-size squares to demonstrate how they drew their size squares.

Have children complete Problem 3. Bring the class together to share their strategies.

Differentiate **Adjusting the Activity**

If children struggle drawing the same number of squares in each row in Problem 3, suggest that they draw one row of squares at the top of the rectangle and then the first square on the left in each of the other rows. Then have them place their fingers on the first square in each row and run their fingers across the rectangle to help visualize each row. Ask: *How many rows are there?* 2 *How many squares should there be in each row?* 5

(Go Online) 📖 Differentiation Support

Math Journal 2, p. 206

Partitioning Strategies

Lesson 8-7
DATE

Math Message

❶ Use one square pattern block to partition the rectangle without any gaps or overlaps. Draw where you placed your square each time.

How many rows of squares? 3

How many squares in each row? 2

How many squares did you draw to cover the whole rectangle? 6

Partition each rectangle below into same-size squares. Use the small square to help you. Draw squares to show your partitioning.

❷

How many rows of squares are there? 2

How many squares are in each row? 4

How many squares did you draw to cover the rectangle? 8

❸ How many rows of squares are there? 2

How many squares are in each row? 5

How many squares did you draw to cover the rectangle? 10

206 two hundred six   2.G.2

Math Journal 2, p. 207

Partitioning into Same-Size Squares

Lesson 8-7
DATE

1. Partition this square into 2 rows with 2 same-size small squares in each row.

   How many small squares cover the large square? __4__

2. Partition this rectangle into 5 rows with 7 same-size squares in each row.

   How many squares cover the rectangle? __35__

2.G.2

Math Journal 2, p. 208

Solving Addition Problems

Lesson 8-7
DATE

Fill in the unit box. Then solve each problem. You can draw an open number line or use a number line, number grid, or base-10 blocks to help you. Show your work.

Unit

1. $\begin{array}{r} 38 \\ + 41 \\ \hline 79 \end{array}$

2. $\begin{array}{r} 26 \\ + 57 \\ \hline 83 \end{array}$

3. $\begin{array}{r} 454 \\ + 365 \\ \hline 819 \end{array}$

4. $\begin{array}{r} 258 \\ + 667 \\ \hline 925 \end{array}$

▶ Partitioning into Same-Size Squares

Math Journal 2, p. 207

| WHOLE CLASS | SMALL GROUP | PARTNER | INDEPENDENT |

Draw children's attention to journal page 207. Point out that there are no pictures to show the size of the squares that are supposed to cover each rectangle. Instead, children are given the number of rows and the number of squares in each row.

Display a rectangle and say: *I have to partition this rectangle into 2 rows with 3 same-size squares in each row. Suppose I make each row this tall.* (Make a mark too low.) *Will two rows fill up the rectangle?* No. *What about here?* (Make a mark too high.) No. *Where should the mark be?* About halfway between the top and bottom edges of the rectangle Make a mark halfway between the top and bottom edges of the rectangle and draw a line to partition it into 2 equal rows. Say: *Now I have to draw 3 squares in each row.* Invite a volunteer to make marks for the squares in the top row. Ask: *How can we check to make sure that these squares are the same size?* Sample answers: We can see that they are the same. We can use our thumb and forefinger to measure the length of one square and then measure the other squares to make sure they are the same length.

Before children begin work on journal page 207, ask them what they should think about as they partition the rectangles. **GMP1.2** Expect responses to include the following ideas:

- All the squares should be the same size.
- There should be the same number of squares in each row.
- There should be the same number of squares in each column.

Circulate as children complete the journal page and check that they are drawing the correct number of rows with the same number of squares in each row. Encourage them to help each other check whether their squares are the same size.

Differentiate **Common Misconception**

Watch for children who partition their rectangles into one too many rows or one too many columns. Suggest that they run their fingers along each row or column as they count. As they adjust their drawings, have them check that the squares are the same size. **Go Online** ▸ Differentiation Support

✓ Assessment Check-In ⒸⒸⓈⓈ 2.G.2

Math Journal 2, p. 207

Expect that most children will be able to partition the square in Problem 1 into two rows with two same-size squares in each row and count the total number of squares. If children struggle making same-size squares, suggest that they use a square pattern block as a reference. Some children may successfully partition the rectangle in Problem 2. Additional practice with partitioning will be provided in later lessons.

☑ Assessment and Reporting   ⟨Go Online⟩ to record student progress and to see trajectories toward mastery for these standards.

Summarize Have children discuss their strategies for partitioning the rectangles on journal page 207 into same-size squares.

③ Practice 15–20 min ⟨Go Online⟩ 🖥 🔧 🏠

ePresentations eToolkit Home Connections

▸ Solving Addition Problems

Math Journal 2, p. 208

| WHOLE CLASS | SMALL GROUP | **PARTNER** | **INDEPENDENT** |

Children add 2- and 3-digit numbers. As needed, encourage them to draw open number lines, use base-10 blocks, or use the number grids or number lines on the inside back covers of their journals. **GMP5.1, GMP5.2**

▸ Math Boxes 8-7 ✦

Math Journal 2, p. 209

| WHOLE CLASS | **SMALL GROUP** | **PARTNER** | **INDEPENDENT** |

Mixed Practice Math Boxes 8-7 are paired with Math Boxes 8-5.

▸ Home Link 8-7

Math Masters, p. 233

Homework Children partition a rectangle into rows of same-size squares.

Math Journal 2, p. 209

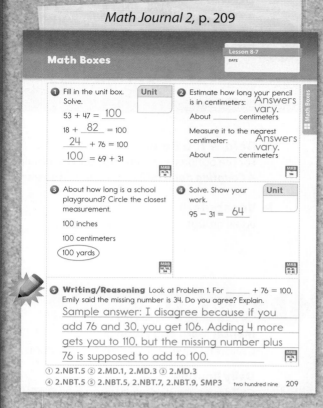

Math Boxes   Lesson 8-7
                                  DATE

1. Fill in the unit box. Solve.   Unit

   $53 + 47 = \underline{100}$

   $18 + \underline{82} = 100$

   $\underline{24} + 76 = 100$

   $\underline{100} = 69 + 31$

2. Estimate how long your pencil is in centimeters: Answers vary.

   About _____ centimeters

   Measure it to the nearest centimeter: Answers vary.

   About _____ centimeters

3. About how long is a school playground? Circle the closest measurement.

   100 inches

   100 centimeters

   ⟨100 yards⟩

4. Solve. Show your work.   Unit

   $95 - 31 = \underline{64}$

5. **Writing/Reasoning** Look at Problem 1. For _____ + 76 = 100, Emily said the missing number is 34. Do you agree? Explain.

   Sample answer: I disagree because if you add 76 and 30, you get 106. Adding 4 more gets you to 110, but the missing number plus 76 is supposed to add to 100.

① 2.NBT.5 ② 2.MD.1, 2.MD.3 ③ 2.MD.3
④ 2.NBT.5 ⑤ 2.NBT.5, 2.NBT.7, 2.NBT.9, SMP3   two hundred nine 209

Math Masters, p. 233

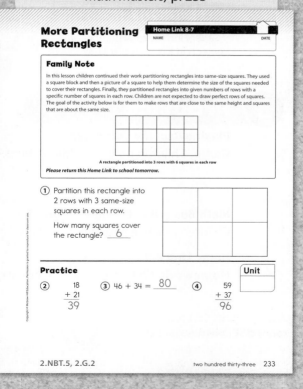

More Partitioning Rectangles   Home Link 8-7

                                    NAME   DATE

Family Note

In this lesson children continued their work partitioning rectangles into same-size squares. They used a square block and then a picture of a square to help them determine the size of the squares needed to cover their rectangles. Finally, they partitioned rectangles into given numbers of rows with a specific number of squares in each row. Children are not expected to draw perfect rows of squares. The goal of the activity below is for them to make rows that are close to the same height and squares that are about the same size.

A rectangle partitioned into 3 rows with 6 squares in each row

Please return this Home Link to school tomorrow.

① Partition this rectangle into 2 rows with 3 same-size squares in each row.

   How many squares cover the rectangle? __6__

Practice   Unit

② $\begin{array}{r} 18 \\ +\ 21 \\ \hline 39 \end{array}$ ③ $46 + 34 = \underline{80}$ ④ $\begin{array}{r} 59 \\ +\ 37 \\ \hline 96 \end{array}$

2.NBT.5, 2.G.2   two hundred thirty-three 233

Lesson 8-8
Equal-Groups and Array Number Stories

Overview Children solve number stories about equal groups and arrays.

▶ **Before You Begin**
For the optional Extra Practice activity, obtain one or more copies of *Each Orange Had 8 Slices: A Counting Book* by Paul Giganti Jr. (Greenwillow Books, 1999).

▶ **Vocabulary**
equal groups • array • row • column

 Common Core State Standards

Focus Clusters
• Represent and solve problems involving addition and subtraction.
• Work with equal groups of objects to gain foundations for multiplication.
• Understand place value.

1 Warm Up 15–20 min

	Materials	
Mental Math and Fluency Children solve addition and subtraction facts.	slate	2.OA.2
Daily Routines Children complete daily routines.	See pages 4–43.	See pages xiv–xvii.

2 Focus 20–30 min

Math Message Children draw pictures to help them solve a number story.		2.OA.1, 2.OA.4 SMP4
Discussing Equal Groups and Arrays Children discuss equal-groups and array number stories.		2.OA.1, 2.OA.4, 2.NBT.2 SMP2, SMP4
Solving Equal-Groups and Array Number Stories Children solve equal-groups and array number stories.	*Math Journal 2*, p. 210	2.OA.1, 2.OA.4, 2.NBT.2 SMP2, SMP4
✓ **Assessment Check-In** See page 741.	*Math Journal 2*, p. 210	2.OA.1, 2.OA.4, SMP4

CCSS 2.OA.4 Spiral Snapshot

GMC Find the total number of objects in a rectangular array.

| 5-5
Focus | 6-10
Focus
Practice | 8-8
Focus
Practice | 8-9
Focus
Practice | 8-10
Focus
Practice | 8-11
Practice | 9-1
Practice | 9-10
Warm Up
Focus |

Spiral Tracker **Go Online** to see how mastery develops for all standards within the grade.

3 Practice 10–15 min

Playing *Beat the Calculator* **Game** Children practice addition facts.	*Assessment Handbook*, pp. 98–99; per group: 4 each of number cards 0–9, calculator, large paper triangle (optional)	2.OA.2
Math Boxes 8-8: Preview for Unit 9 Children practice and maintain skills.	*Math Journal 2*, p. 211; inch ruler	See page 741.
Home Link 8-8 **Homework** Children work with equal groups and arrays.	*Math Masters*, p. 234	2.OA.4, 2.NBT.2

 connectED.mheducation.com ▶

Plan your lessons online with these tools.

 ePresentations   Student Learning Center   Facts Workshop Game   eToolkit   Professional Development   Home Connections   Spiral Tracker   Assessment and Reporting   English Learners Support   Differentiation Support

Differentiation Options

RtI

Readiness   10–15 min

Making Equal Rows

| WHOLE CLASS |
| SMALL GROUP |
| PARTNER |
| INDEPENDENT |

24 counters

For experience making equal rows using a concrete model, children play *Simon Says*. Distribute 24 counters to each child and have them make different arrangements of equal rows to represent the number 24. **GMP2.1** Give directions such as the following:

- *Simon says, "Put all of your counters in rows with 3 in each row." How many rows are there?* 8 rows
- *Simon says, "Put all of your counters in 4 equal rows." How many counters are in each row?* 6 counters

---

Enrichment   5–15 min

Writing Number Stories for Equal Groups and Arrays

| WHOLE CLASS |
| SMALL GROUP |
| PARTNER |
| INDEPENDENT |

Math Masters, p. TA7

Children apply their understanding of equal groups and arrays by writing and illustrating equal-groups or array number stories on *Math Masters,* page TA7. Children may use the number stories on journal page 210 as examples.

A Number Story

NAME   DATE

Unit

TA7

Extra Practice   5–15 min

Equal Groups in Literature

| WHOLE CLASS |
| SMALL GROUP |
| PARTNER |
| INDEPENDENT |

Each Orange Had 8 Slices: A Counting Book

Literature Link For practice with equal groups, children read ***Each Orange Had 8 Slices: A Counting Book*** by Paul Giganti Jr. (Greenwillow Books, 1999) in small groups, as partners, or on their own. Illustrations in the book show situations involving equal groups in the real world, and questions are posed about them. *For example:* If each orange has 8 slices and each slice has 2 seeds, how many seeds are there in all? 16 Have children answer the questions posed in the book.

English Language Learners Support

Beginning ELL Provide children with physical experiences to reinforce their understanding of the terms *column* and *row*. Using gestures and think-alouds, invite children to form a column by lining up one in front of each other. Next have children stand side by side to form a row. Post pictures of examples of children or objects arranged in a column and in a row and label them *column* and *row*. Use Total Physical Response directions to have children practice arranging themselves in a column and then in a row. Use show-me commands to prompt children to point to examples of columns and rows around the classroom or in pictures, such as an auditorium seating plan.

Go Online   ELL   English Learners Support

SMP2 **Reason abstractly and quantitatively.**
   GMP2.2 Make sense of the representations you and others use.
   GMP2.3 Make connections between representations.

SMP4 **Model with mathematics.**
   GMP4.1 Model real-world situations using graphs, drawings, tables, symbols, numbers, diagrams, and other representations.

Professional Development

Working with equal groups and arrays can help children build foundations for multiplication. If children leave second grade with a solid conceptual understanding of how, for example, finding the total number of objects in 3 equal groups of 5 relates to finding the product 3×5, they are more likely to successfully learn multiplication facts by the end of third grade.

Go Online   Professional Development

Sample pictures for the
Math Message problem

---

1 Warm Up   15–20 min   Go Online

ePresentations   eToolkit

▶ Mental Math and Fluency

Pose one fact at a time. Children write the sums or differences on their slates. *Leveled exercises:*

●○○	4 + 3 7	●●○	4 + 7 11	●●●	12 − 4 8
	6 + 5 11		8 + 4 12		13 − 8 5
	6 + 7 13		4 + 9 13		12 − 7 5

▶ Daily Routines

Have children complete daily routines.

---

2 Focus   20–30 min   Go Online

ePresentations   eToolkit

▶ Math Message

Jane bought 3 packs of gum. There are 5 sticks of gum in each pack. How many sticks of gum did she buy? Draw pictures to help find the answer.
GMP4.1

▶ Discussing Equal Groups and Arrays

WHOLE CLASS	SMALL GROUP	PARTNER	INDEPENDENT

Math Message Follow-Up Ask children to share their drawings and solution strategies. Expect a variety of representations, including drawings of groups, arrays, or tallies. **GMP4.1** (*See margin.*) Strategies may include counting the objects in the picture by 1s, counting by 5s, adding 5s, or doubling 5 and then adding 5 more.

> **NOTE** Whenever you ask children to represent a number story with a drawing, make a quick sketch to show them that the drawings need not be elaborate.

Ask: *What do all these drawings have in common?* **GMP2.3** Sample answer: They all show groups of 5. Tell children that groups with the same number of objects in them are called **equal groups.** Stories that involve finding the total number of objects in sets of equal groups are called *equal-groups number stories.* Ask volunteers to explain how their drawings show the equal groups from the story. Sample answers: I drew a square around each group of 5 sticks of gum. My tallies are in groups of 5. The rows in my array are groups of 5.

Ask children to suggest number models for the Math Message problem. **GMP4.1** Some children may suggest $5 + 5 + 5 = 15$. Ask: *How does this number model show what is happening in our drawings?* Sample answer: We are adding three 5s because we have three equal groups of 5. **GMP2.3**

Some children may suggest the number model 3 × 5 = 15 to represent the story. If so, explain that this is a *multiplication* number model and that multiplication as an operation involves finding the number of objects in equal groups or rows. Explain that when children solve equal-groups number stories, they are doing multiplication.

Write 5 + 5 + 5 = 15 and, if someone suggested it, 3 × 5 = 15. Have children practice reading the number models as "3 groups of 5 each is 15 in all." GMP2.2

> **NOTE** Some children may want to write multiplication number models for the problems in this lesson and in Lessons 8-9 and 8-10. Others may be more comfortable with addition number models. It is fine and appropriate for children to use whichever model they are comfortable with at this time. Multiplication models may come up in discussion, but they are not required at Grade 2. The activities in these lessons are intended to lay the foundation for multiplication. Children will do much more work with multiplication in Grade 3.

Look for children who drew arrays to represent the Math Message problem. Ask them to share their drawings, or, if no one drew an array, sketch one yourself. (*See margin.*) Remind the class that a rectangular **array** is an arrangement of objects or symbols in **rows** and **columns.** Point out that an array is one way to represent equal groups because all of the rows have the same number of objects and all of the columns have the same number of objects. Ask: *How are the equal groups from the gum problem represented in this array?* Sample answer: The rows are like the packs of gum because they each have 5. The equal groups in this problem could be represented by either the rows or the columns in an array, depending on whether children drew 3 rows of 5 or 3 columns of 5. But children should recognize that the number story calls for 3 groups of 5 each, *not* 5 groups of 3 each. The number model 5 + 5 + 5 = 15 is more appropriate for this problem than 3 + 3 + 3 + 3 + 3 = 15.

Explain that many real-life objects are arranged in arrays. Pose the following number story: *There are 2 rows of eggs in a carton. There are 6 eggs in each row. How many eggs are there in all?* Ask children to draw a picture and solve. GMP4.1 (*See margin.*)

Ask volunteers to share their drawings and answers. Expect most children to draw an array like the one shown in the margin. Ask: *What number model could we write for this story and drawing?* Sample answers: 6 + 6 = 12; 2 × 6 = 12 *How could we read this number model?* Sample answer: 2 rows of 6 each is 12 all together. GMP2.2

Differentiate **Adjusting the Activity**

Have children sketch the array, circle each row, and write 6 at the end of each row. This may help children see how 6 + 6 = 12 represents the array.

(Go Online) Differentiation Support

Tell children that the egg problem is an example of an array number story, which is one kind of equal-groups number story. In an array story the equal groups can be either the rows or the columns.

NOTE Although 3 × 5 is often read as "3 times 5," the word *times* does not communicate much about the meaning of multiplication. Rather than saying *times*, try referring to the idea of groups or rows—for example, "3 × 5 means 3 groups of 5 each." This will help children relate the numerical expression to the meaning of multiplication, which is important as they begin to invent ways to solve multiplication problems.

Academic Language Development

To help children remember how to distinguish *row* from *column,* have them write the words *row* horizontally and *column* vertically. Crossword puzzle grids, bingo game cards, and pictures of objects that can be identified by row and column (such as auditorium seats) may also be useful references for children as they practice using these terms.

Possible arrays for the Math Message problem

Sample array for the egg problem

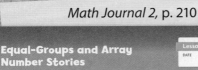

Equal-Groups and Array Number Stories

Lesson 8-8

DATE

For each each number story, draw a picture. Then answer the question and write a number model.

1. There are 3 rows of cans on the shelf with 4 cans in each row. How many cans are there in all?

Sample drawing:

××××
××××
××××

There are __12__ cans in all.

Number model: Samples:
4 + 4 + 4 = 12;
3 × 4 = 12

2. Mr. Yung has 4 boxes of markers. There are 6 markers in each box. How many markers does he have in all?

Sample drawing:

Mr. Yung has __24__ markers.

Number model: Samples:
6 + 6 + 6 + 6 = 24;
4 × 6 = 24

3. Sandy has 4 bags of marbles. Each bag has 4 marbles. How many marbles does she have in all? Sample drawing:

Sandy has __16__ marbles.

Number model: Samples:
4 + 4 + 4 + 4 = 16;
4 × 4 = 16

Try This

4. Mei folded his paper into 2 columns of 4 boxes each. How many boxes did he make?

Sample drawing:

Mei made __8__ boxes.

Number model: Samples:
4 + 4 = 8; 2 × 4 = 8

210   two hundred ten   2.OA.1, 2.OA.4, 2.NBT.2, SMP4

Adjusting the Activity

Differentiate Have children model the problems with counters before drawing pictures.

Go Online   Differentiation Support

Tell children that they will solve and write number models for more equal-groups and array number stories. Although it is not important for children to be able to distinguish between equal-groups and array number stories, it is important that they have experience with both.

▶ Solving Equal-Groups and Array Number Stories

Math Journal 2, p. 210

| WHOLE CLASS | SMALL GROUP | PARTNER | INDEPENDENT |

Pose number stories involving equal groups or arrays of objects. Tell children to work with their partners and use drawings to model and solve each problem. GMP4.1   After each number story, have volunteers share their strategies. Then work as a class to write an addition (and, if appropriate, a multiplication) number model to represent the number story. As children share number models, guide them to practice reading the number models aloud. GMP2.2   They should use language such as the following:

- 3 equal groups of 2 is 6.
- 2 columns of 4 each is 8 all together.
- 3 rows of 7 each makes 21 in all.

Suggested number stories:

- Your family has 3 bicycles. Each bicycle has 2 wheels. How many wheels are there in all? 6 wheels; Sample number models: $2 + 2 + 2 = 6$; $3 × 2 = 6$
  Sample strategies:
  - Make or draw 3 groups of 2 and count the objects by 1s.
  - Skip count by 2s, moving from group to group: 2, 4, 6.
- You see a pattern of floor tiles with 4 columns of tiles and 4 tiles in each column. How many tiles are there in all? 16 tiles; Sample number models: $4 + 4 + 4 + 4 = 16$; $4 × 4 = 16$
  Sample strategies:
  - Think about doubles facts: $4 + 4 = 8$ for the first two columns and $4 + 4 = 8$ for the last two columns. So the total is $8 + 8 = 16$.
  - Use an open number line to think about adding 4s. (*See margin.*)
- Part of a calendar shows 3 weeks with 7 days in each week. How many days are there in all? 21 days; Sample number models: $7 + 7 + 7 = 21$; $3 × 7 = 21$
  Sample strategies:
  - Make an array with 3 rows of 7 and then skip count by 3s while moving from column to column: 3, 6, 9, 12, 15, 18, 21.
  - First think $7 + 7 = 14$. Then add 7 more: $14 + 7 = 21$.

After the class has solved several problems, have children work in partnerships or small groups to complete journal page 210. Children should draw a picture or an array to model each number story. Encourage them to make quick, simple sketches using dots or Xs. Then they find the total number of objects and write a number model. GMP4.1

Assessment Check-In (CCSS) 2.OA.1, 2.OA.4

Math Journal 2, p. 210

Expect that most children will be able to correctly solve the number stories on journal page 210 using drawings and be able to write addition number models. **GMP4.1** If children struggle finding the totals, suggest that they use counters to model the number stories before drawing their pictures as suggested in the Adjusting the Activity note.

☑ Assessment and Reporting **Go Online** to record student progress and to see trajectories toward mastery for these standards.

Summarize Have children share with a partner one strategy they can use to find the total number of objects in equal groups or arrays. *Sample answer: I can find how many are in each row and then add up all the rows.*

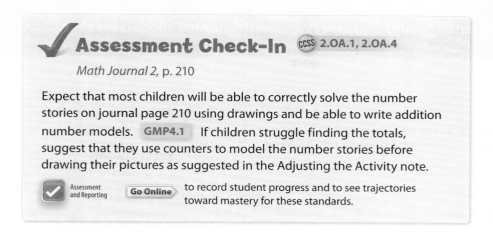

③ Practice 10–15 min

Go Online ePresentations eToolkit Home Connections

▶ Playing *Beat the Calculator*

Assessment Handbook, pp. 98–99

| WHOLE CLASS | **SMALL GROUP** | PARTNER | INDEPENDENT |

Have small groups play the game as introduced in Lesson 5-1. Use *Assessment Handbook,* pages 98–99 to monitor children's progress with addition facts. By the end of Grade 2, children are expected to know from memory all sums of two 1-digit numbers.

Observe

- Which facts do children know from memory?
- Which children need additional support to play the game?

Discuss

- *What strategies did you use to solve the facts you did not know?*
- *Why is it helpful to know addition facts?*

▶ Math Boxes 8-8: Preview for Unit 9

Math Journal 2, p. 211

| WHOLE CLASS | SMALL GROUP | **PARTNER** | **INDEPENDENT** |

Mixed Practice Math Boxes 8-8 are paired with Math Boxes 8-12. These problems focus on skills and understandings that are prerequisite for Unit 9. You may want to use information from these Math Boxes to plan instruction and grouping in Unit 9.

▶ Home Link 8-8

Math Masters, p. 234

Homework Children work with equal groups and arrays.

Math Journal 2, p. 211

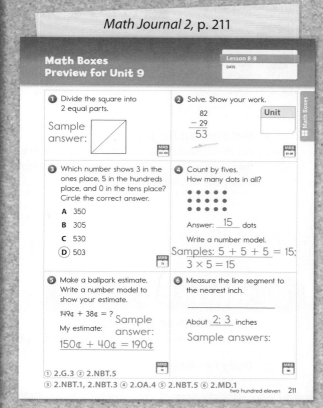

Math Masters, p. 234

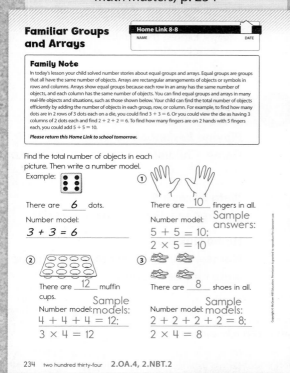

More Equal Groups and Arrays

Overview Children build equal groups and arrays and write number models for them.

▶ **Before You Begin**

For the optional Extra Practice activity, you may want to have several pictures from magazines that show equal groups or arrays (for example, packages of markers, window panes on a house, or a six-pack of bottled water).

Common Core State Standards

Focus Clusters
- Represent and solve problems involving addition and subtraction.
- Work with equal groups of objects to gain foundations for multiplication.
- Understand place value.

1 Warm Up 15–20 min

	Materials	
Mental Math and Fluency Children make ballpark estimates for subtraction problems.	slate	2.NBT.5
Daily Routines Children complete daily routines.	See pages 4–43.	See pages xiv–xvii.

2 Focus 20–30 min

Math Message Children write a number story to match a number model.	Math Journal 2, p. 212	2.OA.1, 2.OA.4 SMP2
Sharing Number Stories Children share and discuss their number stories.	Math Journal 2, p. 212	2.OA.1, 2.OA.4 SMP2
Building Equal Groups and Arrays Children build equal groups and arrays and then write and interpret number models.	Math Journal 2, p. 212; per child: 36 counters; per partnership: one 6-sided die, slate	2.OA.4, 2.NBT.2 SMP2
✓ **Assessment Check-In** See page 746.	Math Journal 2, p. 212	2.OA.4, SMP2

 2.OA.4 **Spiral Snapshot**

GMC Express the number of objects in an array as a sum of equal addends.

5-5 Focus	6-10 Focus Practice	8-8 Focus Practice	8-9 Focus Practice	8-10 Focus Practice	8-11 Practice	9-1 Practice	9-10 Warm Up Focus

/// **Spiral Tracker**   **Go Online** ▶ to see how mastery develops for all standards within the grade.

3 Practice 10–15 min

Playing Basketball Addition **Game** Children practice adding three or more numbers.	Math Masters, pp. G26–G27; per group: one 20-sided polyhedral die or three 6-sided dice	2.NBT.5, 2.NBT.6, 2.NBT.9 SMP1, SMP7
Math Boxes 8-9 Children practice and maintain skills.	Math Journal 2, p. 213	See page 747.
Home Link 8-9 **Homework** Children draw equal groups and arrays and write number models.	Math Masters, p. 236	2.OA.4, 2.NBT.2, 2.NBT.5 SMP2

connectED.mheducation.com ▶

Plan your lessons online with these tools.

 ePresentations   Student Learning Center   Facts Workshop Game   eToolkit   Professional Development   Home Connections   Spiral Tracker   Assessment and Reporting   English Learners Support Differentiation Support

Differentiation Options RtI

 CCSS 2.OA.4, 2.NBT.2, SMP2

Readiness   5–15 min

| WHOLE CLASS |
| SMALL GROUP |
| PARTNER |
| INDEPENDENT |

Connecting Arrays and Equal Groups

25 counters (for demonstration); per child or partnership: 25 counters

For experience connecting equal groups to arrays, children use counters to show equal groups and then arrange their groups into an array. **GMP2.3**

Instruct children to group their counters into 3 groups of 2 counters each as you do the same. Ask: *How many counters do you have in all?* 6 Point to each group as you skip count together: 2, 4, 6. Next model moving each group of 2 to form a row in an array, so that you end up with an array with 3 rows of 2 counters each. Have children do the same with their own counters. Ask: *How many counters do you have in all?* 6 Point to each row as you skip count together: 2, 4, 6.

Repeat the activity with other numbers as needed. *Suggestions:*

- 3 groups of 3 counters each
- 4 groups of 5 counters each
- 2 groups of 4 counters each

CCSS 2.OA.4, SMP1

Enrichment   5–15 min

| WHOLE CLASS |
| SMALL GROUP |
| PARTNER |
| INDEPENDENT |

Solving Equal-Groups and Array Riddles

Math Masters, p. 235

To further explore equal groups and arrays, children make sense of and solve challenging riddles on *Math Masters,* page 235. **GMP1.1** Children also write their own riddles.

> **Equal-Groups and Array Riddles**   Lesson 8-9
> NAME   DATE
>
> Solve the riddles in Problems 1–2.
>
> ① I have some counters. If I put them into 3 rows with 7 in each row, there are 5 counters left over. Draw a picture of my counters:
>
> Sample answer:
>
> How many counters do I have in all? __26__ counters
>
> ② I have between 20 and 30 counters. When I put them into 6 equal groups, there is an even number in each group and 1 is left over. Draw a picture of my counters:
>
> Sample answer:
>
> How many counters do I have in all? __25__ counters
> How many counters are in each group? __4__ counters
>
> ③ Write your own riddle about equal groups or arrays. Solve your riddle. Draw a picture to show your counters.
>
> Answers vary.
>
> 2.OA.4, SMP1   two hundred thirty-five   235

CCSS 2.OA.4, SMP2, SMP4

Extra Practice   5–15 min

| WHOLE CLASS |
| SMALL GROUP |
| PARTNER |
| INDEPENDENT |

Finding Equal Groups and Arrays

Activity Card 104, paper, magazine pictures (optional)

For additional practice writing number models for equal groups and arrays, children find equal groups and arrays that are in the classroom or in magazine pictures of real-life objects. They draw pictures and write number models to match the arrays or equal groups. **GMP4.1** They explain how their number models match their arrays or equal groups. **GMP2.3**

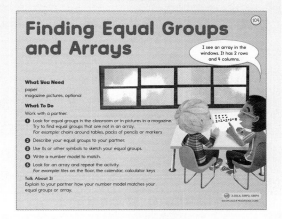

English Language Learners Support

Beginning ELL Build on English language learners' prior experiences with array arrangements by using the term *array* to describe items they may have seen in everyday life. Display everyday items such as an egg carton, a muffin pan, an ice-cube tray, or a checkerboard. Use think-alouds to name the objects and describe them using the terms *array, row,* and *column.* For example, describe the muffin pan this way: *This muffin pan shows an array. It has 3 rows, and it has 4 columns.* Point to each row and then each column and ask: *How many are in each row? How many are in each column? How many muffins in all would I get from this array?* Accept nonverbal or one-word responses by allowing children to either use number cards or tell you the number.

Go Online > **ELL** English Learners Support

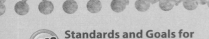

CCSS Standards and Goals for Mathematical Practice

SMP2 Reason abstractly and quantitatively.

GMP2.1 Create mathematical representations using numbers, words, pictures, symbols, gestures, tables, graphs, and concrete objects.

GMP2.2 Make sense of the representations you and others use.

GMP2.3 Make connections between representations.

Math Journal 2, p. 212

Equal Groups and Arrays

Lesson 8-9
DATE

Math Message

1 Write a number story to match the number model
4 + 4 + 4 = 12. Draw a picture of your story.

_____

Answers vary.

2 Draw the last set of equal groups you or your partner made from counters.

3 Draw the last array you or your partner made from counters.

Answers vary.   Answers vary.

4 Write a number model for the equal groups and the array you drew.

Answers vary.

212 two hundred twelve   2.OA.1, 2.OA.4, 2.NBT.2, SMP2

---

1 Warm Up   15–20 min

Go Online   ePresentations   eToolkit

▶ Mental Math and Fluency

Pose subtraction problems. Children make ballpark estimates and record number models for their estimates on their slates. *Leveled exercises:*

- ●○○ 60 − 18   Sample answer: 60 − 20 = 40
  70 − 39   Sample answer: 70 − 40 = 30
- ●●○ 98 − 42   Sample answer: 100 − 40 = 60
  87 − 31   Sample answer: 90 − 30 = 60
- ●●● 45 − 22   Sample answer: 45 − 20 = 25
  56 − 44   Sample answer: 55 − 45 = 10

▶ Daily Routines

Have children complete daily routines.

---

2 Focus   20–30 min

Go Online   ePresentations   eToolkit

▶ Math Message

Math Journal 2, p. 212

Complete Problem 1 on journal page 212. Share your number story with a partner. GMP2.2

▶ Sharing Number Stories

Math Journal 2, p. 212

| WHOLE CLASS | SMALL GROUP | PARTNER | INDEPENDENT |

Math Message Follow-Up Invite volunteers to share their number stories and drawings. Some children may have written equal-groups stories and others may have written array stories. *For example:*

- Amy has 3 bags of apples. There are 4 apples in each bag. How many apples does Amy have in all? She has 12 apples.

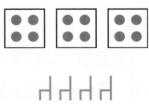

- Shawn set up 3 rows of chairs. There are 4 chairs in each row. How many chairs did Shawn set up all together? He set up 12 chairs all together.

- Omar brought 4 books to school. The next day he brought 4 more books. The day after that, he brought 4 more books. How many books did he bring all together? He brought 12 books all together.

Make sure a variety of stories and drawings are shared. If no one drew an array to represent their story, choose one story and ask children how they might represent it with an array.

Ask children to compare a drawing showing equal groups that are not represented in an array (such as the drawing for the apple story) with a drawing of an array (such as the drawing for the chair story). Ask: *How are these drawings similar?* **GMP2.3** They both show 3 groups of 4 each. *How are they different?* **GMP2.3** Sample answer: In one picture the groups are in different boxes, and in the other picture the groups are the rows. *Do both pictures match the number model? How can you tell?* **GMP2.3** Sample answer: Yes. They both show 3 different groups of 4, with 12 objects in all.

If your class used multiplication number models in Lesson 8-8, ask children to suggest a multiplication number model that matches the drawings. $3 \times 4 = 12$ Ask: *How can we read the number model in words?* **GMP2.2** Sample answers: 3 groups with 4 in each group is 12 in all. 3 rows of 4 chairs each make 12 all together.

Tell children that solving number stories like these depends on being able to think about equal groups. For more practice with this, they will use counters to build equal groups and arrays and then write number models to represent them.

▶ Building Equal Groups and Arrays

Math Journal 2, p. 212

| WHOLE CLASS | SMALL GROUP | PARTNER | INDEPENDENT |

Distribute 36 counters to each child and one die and one slate to each partnership. Explain the following directions:

1. Partner A rolls the die. This is the number of groups, rows, or columns.

2. Partner B rolls the die. This is the number in each group, row, or column.

3. Partner A uses counters to make equal groups (not arranged in an array) to match the numbers. Partner B uses counters to make an array to match the numbers. **GMP2.1**

4. Partner A finds the total number of counters in the equal groups, and Partner B finds the total number of counters in the array. Partners compare their totals to make sure they are the same.

5. Partner A writes a number model on the slate to match the counters. **GMP2.1** Partner B reads the number model in words. **GMP2.2**

6. Partners switch roles and repeat the activity.

Academic Language Development

Have children complete 4-Square Graphic Organizers (*Math Masters,* page TA42) for the terms *equal groups* and *array* to increase their understanding of the terms.

Adjusting the Activity

Differentiate Some children may benefit from completing the Building Equal Groups and Arrays activity in two teams of two children so that they can work together to build the representations.

Go Online Differentiation Support

Model a sample round for the class.

Sample Round

- Partner A rolls a 2. Partner B rolls a 3.
- Partner A makes 2 groups of 3 counters each. Partner B makes an array with 2 rows of 3 counters each.

Partner A writes $3 + 3 = 6$ or $2 \times 3 = 6$ on the slate. Partner B reads the number model aloud as "2 groups of 3 is 6 all together." Circulate and observe as children build equal groups and arrays and write and read number models. As appropriate, guide them to skip count or add to find the total number of counters rather than counting by 1s. Encourage children to read the number models using language about equal groups (or rows or columns). They should say "2 groups of 3 is 6 all together" rather than "3 plus 3 is 6" or "2 times 3 is 6." Using equal-groups language helps children build a conceptual foundation for multiplication.

When children have had several chances to practice both roles, tell them to each record their final set of equal groups, their final array, and the matching number model on the bottom of journal page 212. If children write multiplication number models, ask them to also write addition number models and discuss the connections between the two number models with their partners. GMP2.3

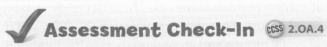

✔ Assessment Check-In CCSS 2.OA.4

Math Journal 2, p. 212

Expect that most children will be able use counters to create arrays, draw them on journal page 212, and record addition number models. GMP2.1 If children struggle to write addition number models, encourage them to circle each row or each column in their arrays to highlight the idea of equal groups. Then help them connect the groups to the equal addends in their number models. Some children may be able write multiplication number models for the arrays, though this is not expected at this time.

 Assessment and Reporting Go Online to record student progress and to see trajectories toward mastery for these standards.

Summarize Have children use counters to solve the following problem and share their answers. Ask: *Which will have more counters—an array with 3 rows and 5 in each row or an array with 5 rows and 3 in each row?* Sample answers: The arrays will have the same number of counters. The arrays are the same except that one is turned.

3 Practice 10–15 min

Go Online

ePresentations eToolkit Home Connections

▶ Playing *Basketball Addition*

Math Masters, pp. G26–G27

| WHOLE CLASS | **SMALL GROUP** | PARTNER | INDEPENDENT |

Divide children into teams of 3 to 5 to play *Basketball Addition* as introduced in Lesson 7-3.

Observe

- Which children can add the numbers to find the total score?
- Which children need additional support to play the game?

Discuss

- *Which numbers did you choose to add first? Why?* GMP7.2
- *Can you use another strategy to add the numbers?* GMP1.5

▶ Math Boxes 8-9

Math Journal 2, p. 213

| WHOLE CLASS | SMALL GROUP | PARTNER | INDEPENDENT |

Mixed Practice Math Boxes 8-9 are paired with Math Boxes 8-11.

▶ Home Link 8-9

Math Masters, p. 236

Homework Children draw equal groups and arrays and write number models to represent them. GMP2.1

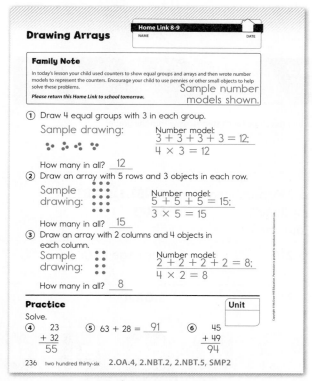

Math Masters, p. 236

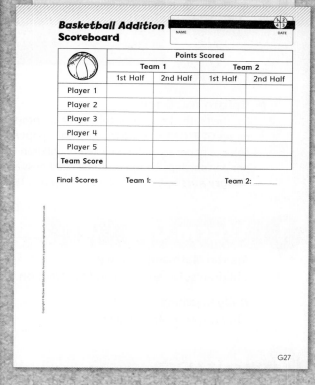

Math Journal 2, p. 213

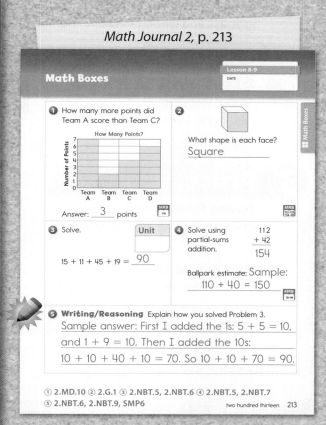

Playing *Array Concentration*

Overview Children play *Array Concentration* to practice finding the total number of objects in arrays and writing corresponding number models.

▶ **Before You Begin**

For the Math Message, distribute 12 counters per partnership. For Part 2 make copies of *Math Masters*, pages G29–G30 on cardstock or heavy paper. Consider copying one of the pages onto a different color of paper so that it is easier for children to keep the decks separate. You'll need one copy of each per partnership. For the optional Readiness activity, obtain one or more copies of **One Hundred Hungry Ants** by Elinor Pinczes (Sandpiper, 1999).

 Common Core State Standards

Focus Clusters
- Work with equal groups of objects to gain foundations for multiplication.
- Understand place value.

1 Warm Up   15–20 min

	Materials	
Mental Math and Fluency Children solve addition and subtraction facts.	slate (optional)	2.OA.2
Daily Routines Children complete daily routines.	See pages 4–43.	See pages xiv–xvii.

2 Focus   30–40 min

Math Message Children find different ways to arrange 12 desks in equal rows.	*Math Journal 2,* p. 214; per partnership: 12 counters	2.OA.4 SMP4
Arranging Desks Children share their arrays and find all possible arrays for 12.	*Math Journal 2,* p. 214	2.OA.4 SMP2, SMP4
Discussing the Array Cards Children cut out *Array Concentration* cards and discuss strategies for finding the total number of dots in arrays.	*Math Masters,* pp. G29–G30; scissors	2.OA.4, 2.NBT.2 SMP6
Introducing and Playing *Array Concentration* **Game** Children practice finding the total number of objects in arrays and writing matching number models.	*Math Masters,* p. G31; *Array Concentration* Array Cards and Number Cards	2.OA.4, 2.NBT.2 SMP2, SMP6
✓ **Assessment Check-In** See page 753.	*Math Masters,* p. G31	2.OA.4

CCSS 2.OA.4   **Spiral Snapshot**

GMC Express the number of objects in an array as a sum of equal addends.

5-5 Focus	6-10 Focus Practice	8-8 Focus Practice	8-9 Focus Practice	8-10 Focus Practice	6-11 Practice	9-1 Practice	9-10 Warm Up Focus

Spiral Tracker   **Go Online** to see how mastery develops for all standards within the grade.

3 Practice   10–15 min

Solving Subtraction Problems Children use strategies to subtract.	*Math Journal 2,* p. 215; number grid or number line (optional); base-10 blocks (optional)	2.NBT.5, 2.NBT.7 SMP5
Math Boxes 8-10 Children practice and maintain skills.	*Math Journal 2,* p. 216	See page 753.
Home Link 8-10 **Homework** Children complete *Array Concentration* rounds.	*Math Masters,* p. 237	2.OA.4

connectED.mheducation.com

Plan your lessons online with these tools.

ePresentations   Student Learning Center   Facts Workshop Game   eToolkit   Professional Development   Home Connections   Spiral Tracker   Assessment and Reporting   English Learners Support   Differentiation Support

Differentiation Options

RtI

Readiness
5–15 min

WHOLE CLASS
SMALL GROUP
PARTNER
INDEPENDENT

Arrays in Literature
One Hundred Hungry Ants

Literature Link For practice representing the same number with more than one array, children read **One Hundred Hungry Ants** by Elinor Pinczes (Sandpiper, 1999) in small groups, as partners, or on their own. In this book 100 ants are on their way to a picnic. To get there as fast as possible, they keep rearranging themselves into different arrays, such as 1 row of 100, 2 rows of 50, and 4 rows of 25. After children read the book, review the different arrays into which the ants arranged themselves.

Enrichment
10–15 min

WHOLE CLASS
SMALL GROUP
PARTNER
INDEPENDENT

Exploring Square Numbers

Activity Card 105;
Math Masters, p. TA25;
per partnership: 50 centimeter cubes

To further explore arrays, children build arrays with centimeter cubes. They build all the possible arrays for the numbers 1, 4, 9, and 16 to figure out why these numbers are called *square numbers*. Children are encouraged to find the next two or three square numbers (25, 36, and 49). They record their arrays on centimeter grid paper (*Math Masters*, page TA25) and write a number model below each array. **GMP2.1** Children discuss the patterns they notice in the arrays. **GMP7.2**

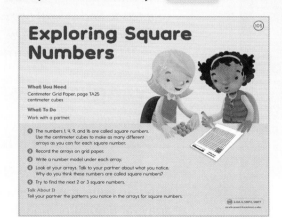

Extra Practice
5–15 min

WHOLE CLASS
SMALL GROUP
PARTNER
INDEPENDENT

Array Bingo

Activity Card 106,
Array Concentration Array Cards and Number Cards

For additional practice finding the total number of objects in arrays, children use their *Array Concentration* cards to play *Array Bingo*. They arrange the array cards to form a 3-by-3 bingo card and flip over the cards when the matching number is called. **GMP2.2** The first player to flip over three cards in a row, column, or diagonal wins.

English Language Learners Support

Beginning ELL Use concrete examples of equal groups of objects to illustrate the term *equal addends*. Start with a set of objects (that can be arranged in equal groups) and arrange the objects in equal groups as you make think-aloud statements like the following: *This group of _____ has the same number as this group. They both have _____ objects. They are equal groups.* Continue to use think-alouds as you write corresponding number models. Say: *Now I will write a number model.* Point to the addends, saying: *These are the same numbers. These are equal addends. Here is the answer. Here is the sum.* Assess children's understanding of the term *equal addends* by presenting different number models, such as $4 + 3 + 2 = 9$ and $3 + 3 + 3 = 9$, and asking children to point to the number model with equal addends.

Go Online | ELL English Learners Support

Standards and Goals for
Mathematical Practice

SMP2 **Reason abstractly and quantitatively.**
    GMP2.3 Make connections between representations.
SMP4 **Model with mathematics.**
    GMP4.2 Use mathematical models to solve problems and answer questions.
SMP6 **Attend to precision.**
    GMP6.4 Think about accuracy and efficiency when you count, measure, and calculate.

1 Warm Up   15–20 min   Go Online   ePresentations   eToolkit

▶ Mental Math and Fluency

Pose addition and subtraction facts. Children answer orally or on their slates. By the end of Grade 2, children should know from memory the facts in Level 1 and Level 2. *Leveled exercises:*

●○○ 8 + 7 15   ●●○ 5 + 9 14   ●●● 13 − 7 6
      9 + 8 17   5 + 8 13   14 − 9 5
      9 + 7 16   7 + 8 15   15 − 7 8

▶ Daily Routines

Have children complete daily routines.

2 Focus   30–40 min   Go Online   ePresentations   eToolkit

▶ Math Message

Math Journal 2, p. 214

Work with a partner. Suppose there are 12 desks in a classroom. Use your counters to find at least two ways to put the desks in rows with the same number of desks in each row. GMP4.2 *Draw your arrays on journal page 214. Write an addition number model for each array.*

▶ Arranging Desks

Math Journal 2, p. 214

| WHOLE CLASS | SMALL GROUP | PARTNER | INDEPENDENT |

Math Message Follow-Up Remind children that when they arrange things in equal rows, they are making arrays. Ask volunteers to share their arrays and number models. GMP4.2 If children wrote multiplication number models, ask them to suggest addition models as well.

Record children's arrays and number models. Ask children to continue sharing answers until no one has a different answer to share. Then have children look at all the number models. Ask: *How are these number models alike?* GMP2.3 Sample answers: They all have 12 as a sum. In each number model, the addends are all equal. Focus the discussion on the idea that there are several different ways to arrange the 12 desks in equal rows.

Have children use their counters to arrange the desks in equal rows of 5. Ask: *Can we make equal rows of 5? Why or why not?* GMP4.2 No. We can make only 2 rows of 5, and we still have 2 left over. *What would a number model for this arrangement look like?* $5 + 5 + 2 = 12$ *Does this number model have addends that are all equal?* No. *Is this arrangement an array?* No.

Math Journal 2, p. 214

Arranging Desks   Lesson 8-10
                                       DATE

Math Message

Draw at least two arrays to show how you can put 12 desks in rows with the same number of desks in each row. Then write an addition number model to match each array.   Sample answers:

$4 + 4 + 4 = 12$ ▪▪▪▪ $3 + 3 + 3 + 3 = 12$
              ▪▪▪▪
              ▪▪▪▪

$6 + 6 = 12$ ▪▪▪▪▪▪ $2 + 2 + 2 +$
            ▪▪▪▪▪▪ $2 + 2 + 2 = 12$

▪▪▪▪▪▪▪▪▪▪▪▪ $1 + 1 + 1 + 1 +$
$1 + 1 + 1 + 1 +$ $1 + 1 + 1 + 1 +$
$1 + 1 + 1 + 1 +$ $1 + 1 + 1 = 12$
$1 + 1 + 1 = 12$

Ask: *Did we find all the different ways to arrange 12 desks in equal rows?* *Answers vary.* *How could we check?* If no one mentions it, suggest the following strategy: check whether we can make rows of 1, then rows of 2, then rows of 3, and so on, until we have all possible arrays for 12.

Work together as a class to find any missing arrays and add them to the class list. Children can record additional arrays on journal page 214. Remind them to check that in each array, all the rows have the same number of desks. They should also check that in each number model, all the addends are equal.

Tell children that they will play a game to practice finding how many objects are in arrays and writing corresponding number models.

▶ Discussing the Array Cards

Math Masters, pp. G29–G30

| WHOLE CLASS | SMALL GROUP | PARTNER | INDEPENDENT |

Have each partnership cut out one set of *Array Concentration* Number Cards (*Math Masters,* page G29) and one set of *Array Concentration* Array Cards (*Math Masters,* page G30). If the cards are not copied onto different colors of paper (*see Before You Begin*), tell children to write an N on the back of each number card and an A on the back of each array card to help them keep the two decks separate.

Have children find the array card that says "2 by 3" at the bottom. Ask them what they think "2 by 3" might mean. If no one suggests it, explain that this is a short way to describe an array that has 2 rows and 3 columns.

Academic Language Development To reinforce children's understanding and use of the phrase _____ *by* _____ (for example, 3 by 2) to describe the rows and columns in an array, have them work in pairs to label everyday objects laid out in arrays. For example, give children different-size muffin pans that they might label 3 by 2, 3 by 4, or 4 by 6 (depending on size). Provide sentence frames that children can use to describe their arrays: "My _____ has an array of _____ by _____. My _____ has _____ rows and _____ columns."

Next have children find the 4-by-5 array card. Ask children to share strategies for finding the total number of dots in the array. 20 *Sample strategies:*

- Count all of the dots by 1s to get 20.
- Skip count by 5s as you point to each row: 5, 10, 15, 20. Or add 5s as you point to each row: $5 + 5 + 5 + 5 = 20$.
- Add 4s as you point to each column: $4 + 4 + 4 + 4 + 4 = 20$.

Ask: *Which strategies might help you find the total faster? Why?* **GMP6.4**
Sample answers: It's faster to skip count than to count by 1s because I don't have to say as many numbers. I can add fast because I know my addition facts.

Math Masters, p. G29

Array Concentration Number Cards

4	6	8	10
9	12	15	16
20	25	2	5

G29

Math Masters, p. G30

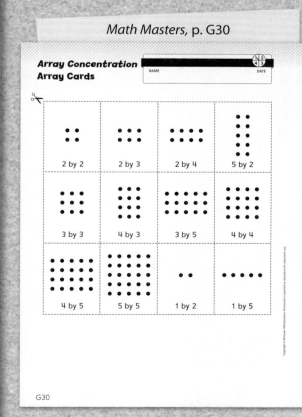

Array Concentration Array Cards

G30

Math Journal 2, p. 215

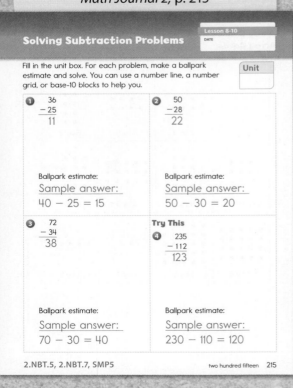

► Introducing and Playing *Array Concentration*

Math Masters, p. G31

WHOLE CLASS · SMALL GROUP · **PARTNER** · INDEPENDENT

Playing *Array Concentration* provides practice finding the total number of objects in arrays and writing corresponding addition number models. Distribute an *Array Concentration* Record Sheet (*Math Masters,* page G31) to each child. Play a few sample rounds to introduce the game.

Directions

1. Shuffle the deck of number cards. Place them facedown in a 4-by-3 array as shown below.

2. Shuffle the deck of array cards and place them facedown in another 4-by-3 array as shown below.

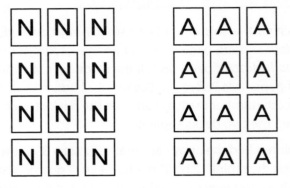

3. Player A flips over one number card and one array card and then finds the total number of dots in the array on the array card.
   - If the total number of dots matches the number on the number card, Player A collects the pair of cards and takes another turn.
   - If the total number of dots does not match the number on the number card, Player A turns the cards back over in the same positions, and play passes to Player 2.

4. When a player makes a match, that player records the number and the array on the record sheet and writes a matching addition number model. Some children may record a multiplication number model, too.

5. Play continues until all of the cards have been matched. The player with the most matches wins.

When children seem ready, have partnerships play the game.

> **NOTE** Most arrays have two possible addition number models. A 2-by-3 array, for example, can be represented with $3 + 3 = 6$ or $2 + 2 + 2 = 6$. But if an array has only one row or only one column, or if it has the same number of rows as columns, there is only one addition number model for it. For example, the only addition model for a 1-by-5 or 5-by-1 array is $1 + 1 + 1 + 1 + 1 = 5$. The only addition model for a 2-by-2 array is $2 + 2 = 4$.

Observe

- What strategies are children using to find the total number of dots in each array? Which children have efficient strategies?
- Which children need support to understand and play the game?

Discuss

- *How did you find the number of dots? Is there a faster way?* **GMP6.4**
- *How do you know your number model matches the array?* **GMP2.3**

Differentiate **Game Modifications** **Go Online** 👥 Differentiation Support

✓ Assessment Check-In **CCSS 2.OA.4**

Math Masters, p. G31

Expect most children to correctly match number cards and array cards and write correct addition number models for the arrays on *Math Masters,* page G31. If children struggle matching the arrays to the number cards or writing number models, have them copy the array onto a sheet of paper and mark rows, columns, or individual dots as they count to help them keep track. Some children may be able to write two addition number models for some of the arrays or write multiplication number models.

✓ Assessment and Reporting **Go Online** to record student progress and to see trajectories toward mastery for these standards.

Summarize Have children share the arrays for which they easily found the total numbers of dots and the arrays for which they had to use strategies to find the totals.

---

③ Practice 10–15 min

Go Online ePresentations eToolkit Home Connections

▶ Solving Subtraction Problems

Math Journal 2, p. 215

| WHOLE CLASS | SMALL GROUP | **PARTNER** | **INDEPENDENT** |

Children use strategies to subtract. As needed, encourage them to choose and use tools such as base-10 blocks or the number grids or number lines on the inside back covers of their journals. **GMP5.1, GMP5.2**

▶ Math Boxes 8-10

Math Journal 2, p. 216

| WHOLE CLASS | **SMALL GROUP** | **PARTNER** | **INDEPENDENT** |

Mixed Practice Math Boxes 8-10 are paired with Math Boxes 8-6.

▶ Home Link 8-10

Math Masters, p. 237

Homework Children complete sample rounds of a game of *Array Concentration.*

Math Journal 2, p. 216

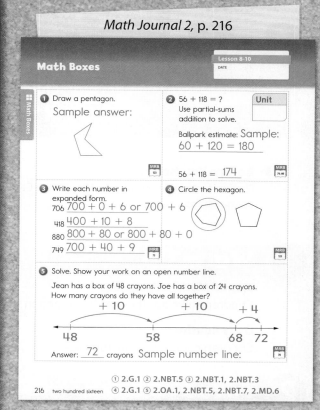

Math Boxes Lesson 8-10
DATE

① Draw a pentagon.
Sample answer:

② 56 + 118 = ?
Use partial-sums addition to solve.
Unit

Ballpark estimate: Sample:
60 + 120 = 180

56 + 118 = __174__

③ Write each number in expanded form.
706 __700 + 0 + 6 or 700 + 6__
418 __400 + 10 + 8__
880 __800 + 80 or 800 + 80 + 0__
749 __700 + 40 + 9__

④ Circle the hexagon.

⑤ Solve. Show your work on an open number line.
Jean has a box of 48 crayons. Joe has a box of 24 crayons. How many crayons do they have all together?

+ 10   + 10   + 4
48   58   68   72

Answer: __72__ crayons   Sample number line:

216 two hundred sixteen

① 2.G.1 ② 2.NBT.5 ③ 2.NBT.1, 2.NBT.3
④ 2.G.1 ⑤ 2.OA.1, 2.NBT.5, 2.NBT.7, 2.MD.6

Math Masters, p. 237

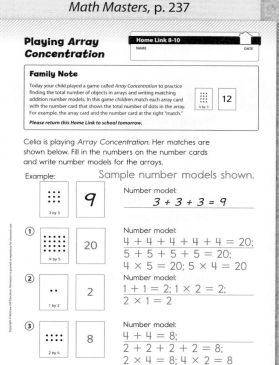

Playing *Array Concentration* Home Link 8-10
NAME   DATE

Family Note
Today your child played a game called *Array Concentration* to practice finding the total number of objects in arrays and writing matching addition number models. In this game children match each array card with the number card that shows the total number of dots in the array. For example, the array card and the number card at the right "match."
Please return this Home Link to school tomorrow.

12
4 by 3

Celia is playing *Array Concentration.* Her matches are shown below. Fill in the numbers on the number cards and write number models for the arrays.

Example: Sample number models shown.

9
3 by 3

Number model:
3 + 3 + 3 = 9

① 20
4 by 5

Number model:
4 + 4 + 4 + 4 + 4 = 20;
5 + 5 + 5 + 5 = 20;
4 × 5 = 20; 5 × 4 = 20

② 2
1 by 2

Number model:
1 + 1 = 2; 1 × 2 = 2;
2 × 1 = 2

③ 8
2 by 4

Number model:
4 + 4 = 8;
2 + 2 + 2 + 2 = 8;
2 × 4 = 8; 4 × 2 = 8

2.OA.4   two hundred thirty-seven 237

Exploring Mystery Shapes, Polygons, and Equal Parts

Overview Children describe attributes of shapes, build polygons with trapezoids, and show fractions on a geoboard.

▶ **Before You Begin**
For Exploration A you will need the Shape Cards cut out during Lesson 7-9. You will also need several opaque bags or boxes.

Common Core State Standards

Focus Cluster
Reason with shapes and their attributes.

	Materials	
1 Warm Up 15–20 min		
Mental Math and Fluency Children solve addition and subtraction facts.	slate	2.OA.2
Daily Routines Children complete daily routines.	See pages 4–43.	See pages xiv–xvii.
2 Focus 30–40 min		
Math Message Children draw shapes with different attributes.	slate	2.G.1
Comparing Shapes Children compare shapes.	slate	2.G.1
Exploration A: Identifying Mystery Shapes Children identify hidden shapes based on their attributes.	Activity Card 107; *Math Journal 2*, p. 217; Shape Cards; several small opaque bags or boxes	2.G.1 SMP2
Exploration B: Making Pattern-Block Worktables Children build shapes with trapezoids and use a Pattern-Block Template to record their work.	Activity Card 108; *Math Masters*, p. 239; trapezoid pattern blocks; Pattern-Block Template	2.G.1 SMP7
Exploration C: Partitioning Shapes into Equal Parts Children divide shapes on a geoboard into fractional parts.	Activity Card 109; *Math Masters*, p. TA10; *Math Masters*, p. TA9 (optional); geoboards; rubber bands	2.G.3 SMP2
3 Practice 10–15 min		
Playing *Array Concentration* **Game** Children play a game to practice finding the total number of objects in arrays and writing corresponding number models.	*Math Masters*, p. G31; *Array Concentration* Number Cards and Array Cards	2.OA.4, 2.NBT.2 SMP2, SMP6
Math Boxes 8-11 Children practice and maintain skills.	*Math Journal 2*, p. 218	See page 759.
Home Link 8-11 **Homework** Children write a shape riddle based on the attributes of a shape.	*Math Masters*, p. 240	2.G.1

Differentiation Options

RtI

CCSS 2.G.3

Readiness — 5–15 min

| WHOLE CLASS |
| SMALL GROUP |
| PARTNER |
| INDEPENDENT |

Showing Equal Parts

slate

For experience partitioning a shape into equal shares, children show equal parts on their slates. First instruct children to divide their slates into 2 equal parts. Then discuss the different ways children divided their slates (for example, by drawing a line vertically, horizontally, or diagonally). Discuss how they know the 2 parts are equal. Sample answer: Each part covers the same amount of space. Repeat the activity, prompting children to divide their slates into 3 and 4 equal parts.

CCSS 2.G.1

Enrichment — 15–30 min

| WHOLE CLASS |
| SMALL GROUP |
| PARTNER |
| INDEPENDENT |

Sorting Shape Words

Math Masters, p. 238; large piece of paper; scissors; tape or glue

To further explore shape attributes, partners cut out the shape-word cards on *Math Masters*, page 238, draw examples of the shapes or attributes named on the cards, and discuss ways to sort them. Encourage children to talk about which cards show shape names and which ones show shape attributes. After sorting the cards into groups, children tape or glue each group of cards onto paper and write a descriptive heading for each group. For example, a group including the trapezoid, square, and rectangle cards might have the heading Four Sides. Expect a wide variety of groupings. Children don't have to use all of the cards. Invite children to share and explain their groupings with the class.

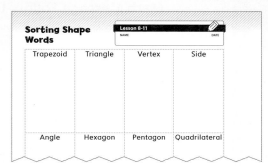

Sorting Shape Words — Lesson 8-11

Trapezoid	Triangle	Vertex	Side
Angle	Hexagon	Pentagon	Quadrilateral

CCSS 2.G.1, SMP2

Extra Practice — 5–15 min

| WHOLE CLASS |
| SMALL GROUP |
| PARTNER |
| INDEPENDENT |

Making My Shape

Activity Card 110, geoboard, rubber bands, folder

For additional practice recognizing attributes of shapes, children make shapes with specified attributes on a geoboard. One partner makes a shape on a hidden geoboard and orally describes it to the other partner, who tries to replicate the shape on another geoboard. GMP2.1 Partners compare the two geoboard shapes and discuss the attributes of each. They trade roles and repeat.

English Language Learners Support

Beginning ELL Scaffold the vocabulary needed to participate in Exploration A by having children draw pictures on the backs of cards that name shapes or shape attributes. You might use the Side, Angle, Right Angle, and Parallel Lines cards from *Math Masters,* page 238 to illustrate those terms. Allow children to use the cards as references when trying to create the drawings based on their partners' descriptions. When they must describe the mystery shapes to partners, allow them to show the word cards that describe the shape's attributes.

Go Online **ELL** English Learners Support

1 Warm Up 15–20 min

Go Online
ePresentations   eToolkit

▶ Mental Math and Fluency

Pose facts one at a time. Children write the sum or difference on their slates. *Leveled exercises:*

●○○	7 + 5	12		●●○	5 + 9	14		●●●	17 − 8	9
	6 + 8	14			7 + 9	16			16 − 7	9
	8 + 7	15			8 + 5	13			17 − 9	8

▶ Daily Routines

Have children complete daily routines.

2 Focus 30–40 min

Go Online
ePresentations   eToolkit

▶ Math Message

Draw these shapes on your slate:

- *Draw a shape with 3 sides and 3 angles.*
- *Draw a shape with 1 or 2 pairs of parallel sides.*
- *Draw a shape with 1 right angle.*

▶ Comparing Shapes

WHOLE CLASS	SMALL GROUP	PARTNER	INDEPENDENT

Math Message Follow-Up Have children share what they notice about the shapes they drew. Sample answers: Shapes with 3 sides are called triangles. Squares and rectangles have parallel sides. Squares, rectangles, and some triangles have right angles. **Ask:** *How are some of the shapes alike?* Sample answer: Some of the shapes have the same number of sides and the same number of angles. *How are some of the shapes different?* Sample answer: Some of the shapes have right angles, and some do not. They don't all have the same number of sides. Remind children that the number of sides, the number of angles, the number of vertices, the lengths of sides, the number of pairs of parallel sides, and the number of right angles are all different attributes of shapes.

Explain to children that in today's lesson they will describe shapes in terms of their attributes, build polygons out of trapezoids, and form equal shares on a geoboard.

After explaining the Explorations activities, assign groups to each one. Plan to spend most of your time with children working on Exploration A.

Activity Card 107

Mystery Shapes

107

I feel four sides, two of them are longer than the other two. The shape has right angles.

What You Need
Math Journal 2, page 217
Shape Cards
small bag or box
paper

What To Do
Work with a partner.

1. Put the shapes in a small bag or box.
2. Without looking, pick a shape.
3. Describe its attributes without saying its name.
4. Your partner draws the shape you describe on journal page 217 and names the shape.
5. Compare the shape to your partner's drawing.
6. Trade roles and repeat Steps 2–5.

Talk About It
Which clues were most helpful? Why?

▶ Exploration A: Identifying Mystery Shapes

Activity Card 107; *Math Journal 2,* p. 217

| WHOLE CLASS | **SMALL GROUP** | **PARTNER** | INDEPENDENT |

To explore attributes of shapes, partners take turns figuring out mystery shapes based on their attributes. Without looking, one partner reaches inside a bag or a box containing a Shape Card, feels the shape, and describes the shape's attributes without saying its name. Based on this description the other partner draws the shape on journal page 217.
GMP2.1 Partners compare the drawing to the shape. As children are engaged in the activity, ask questions such as the following: *What geometry words are you using to describe the shapes?* Sample answer: I'm telling the number of sides, angles, and vertices and whether they have parallel sides or right angles.

▶ Exploration B: Making Pattern-Block Worktables

Activity Card 108; *Math Masters,* p. 239

| WHOLE CLASS | **SMALL GROUP** | **PARTNER** | INDEPENDENT |

To explore building different polygon shapes from trapezoids, children pretend that trapezoid pattern blocks are small tables and that you, their teacher, want to make larger worktables by fitting the small trapezoid tables together. Children follow the directions on Activity Card 108 to make tables of varying sizes and shapes.

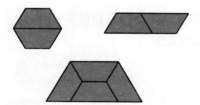

They use a Pattern-Block Template to record their shapes on *Math Masters,* page 239. Group members compare their results to find as many different worktable sizes and shapes as possible. **GMP7.1**

Math Journal 2, p. 217

Mystery Shapes

Lesson 8-11

DATE

Draw the mystery shapes in the space below. Write the name of each shape under your drawing.

2.G.1, SMP2   two hundred seventeen   217

Activity Card 108

Making Pattern-Block Worktables (108)

What You Need
Making Pattern-Block Worktables, page 239
Pattern-Block Template
trapezoid pattern blocks

What To Do
Work in a small group or with a partner.
1. Pretend the red trapezoid pattern block is a small table. Make a larger table by fitting these blocks together.
2. Record your work on page 239. Write the names for the shapes in Steps 3–5.
3. Make a worktable with three sides.
4. Make a worktable with six sides.
5. Make a worktable with at least one pair of parallel sides.
6. Make another worktable with at least one pair of parallel sides that is bigger than the one you just made.
7. Use more than one block to make a worktable shaped like a trapezoid.
8. Make other worktable shapes.

Talk About It
Find all the different-size and different-shape worktables that your group made.

2.G.1, SMP7

Making Pattern-Block Worktables

Lesson 8-11

NAME   DATE

Pretend that each red trapezoid pattern block is a small table. Your teacher wants to make larger worktables by fitting these small tables together. The directions for making the larger tables are on Activity Card 108.

① Name the shape of a worktable with three sides.
  Triangle

② Name the shape of a worktable with six sides.
  Hexagon

③ Name the shape of a worktable with at least one pair of parallel sides. Sample answer: Quadrilateral

④ Use a Pattern-Block Template to draw the worktables you make below. You can also draw worktables on the back of the paper.

2.G.1   two hundred thirty-nine 239

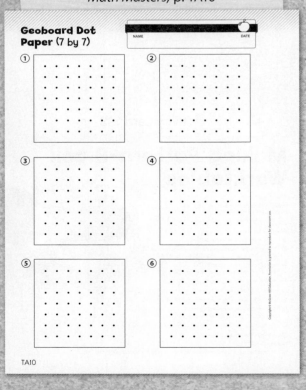

Geoboard Dot Paper (7 by 7)

NAME   DATE

① ② ③ ④ ⑤ ⑥

TA10

▶ # Exploration C: Partitioning Shapes into Equal Parts

Activity Card 109; *Math Masters*, p. TA10

| WHOLE CLASS | SMALL GROUP | PARTNER | INDEPENDENT |

To explore equal parts, partners partition shapes on geoboards into equal parts. One partner forms a shape on a geoboard with one rubber band. The other partner tries to divide the shape into 2 (or 3 or more) equal parts using additional rubber bands. Children record their results on the 7×7 geoboard dot paper on *Math Masters*, page TA10. GMP2.1 If your children are working with 5×5 geoboards, have them outline a 5×5 dot array or use *Math Masters*, page TA9.

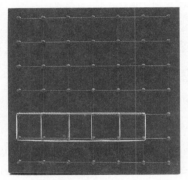

Children partition shapes into equal parts on geoboards.

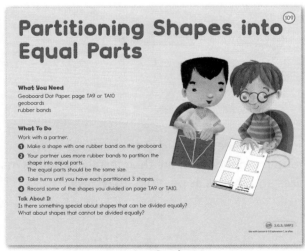

Partitioning Shapes into Equal Parts ⁽¹⁰⁹⁾

What You Need
Geoboard Dot Paper, page TA9 or TA10
geoboards
rubber bands

What To Do
Work with a partner.
❶ Make a shape with one rubber band on the geoboard.
❷ Your partner uses more rubber bands to partition the shape into equal parts.
  The equal parts should be the same size.
❸ Take turns until you have each partitioned 3 shapes.
❹ Record some of the shapes you divided on page TA9 or TA10.

Talk About It
Is there something special about shapes that can be divided equally?
What about shapes that cannot be divided equally?

Activity Card 109

Summarize Have children share two attributes they used in describing the shapes in Exploration A.

③ Practice 10–15 min | Go Online › | ePresentations | eToolkit |

▶ Playing *Array Concentration*

Math Masters, p. G31

| WHOLE CLASS | SMALL GROUP | **PARTNER** | INDEPENDENT |

Have partners play *Array Concentration* to practice finding the total number of objects in arrays and writing corresponding addition number models. See Lesson 8-10 for detailed directions.

Observe

- Do children have efficient strategies for finding the total number of dots in an array?
- Which children need additional support to understand and play the game?

Discuss

- *How did you find the total number of dots? Is there a faster way?* GMP6.4
- *How do you know your number model matches the array?* GMP2.3

▶ Math Boxes 8-11 ✸

Math Journal 2, p. 218

| WHOLE CLASS | **SMALL GROUP** | PARTNER | INDEPENDENT |

Mixed Practice Math Boxes 8-11 are paired with Math Boxes 8-9.

▶ Home Link 8-11

Math Masters, p. 240

Homework Children write a shape riddle based on a shape's attributes.

Math Journal 2, p. 218

Math Boxes | Lesson 8-11 | DATE |

① How many more pine trees are there than maple trees?

Kinds of Trees

Answer: __6__ trees

② What shape is each face? __Triangle__

③ Solve. | Unit |

$9 + 41 + 20 + 18 =$ __88__

④ Solve using partial-sums addition.

$$131 + 57 = 188$$

Ballpark estimate: Sample:
$130 + 60 + = 190$

⑤ **Writing/Reasoning** In Problem 1, Natalie says there are 17 trees in all, Allison says there are 27 trees in all. Who is correct? Explain. Sample answer: Natalie is correct.
$4 + 1 + 5 + 7 = 17$, not 27.

218 two hundred eighteen

① 2.MD.10 ② 2.G.1 ③ 2.NBT.5, 2.NBT.6
④ 2.NBT.5, 2.NBT.7 ⑤ 2.MD.10, SMP3

Math Masters, p. 240

Writing a Shape Riddle | Home Link 8-11 | NAME | DATE |

Family Note

In this lesson your child learned to recognize a 2-dimensional shape based on specific attributes, such as the following:

- number of angles
- number of sides
- number of pairs of parallel sides
- number of right angles

Using these attributes of 2-dimensional shapes, ask your child to write a shape riddle. *For example:* I am a shape that has 3 sides and 3 angles. I have no parallel sides. What shape am I? (The answer is "a triangle.") Your child can share the riddle with a family member or a friend.

Please return this Home Link to school tomorrow.

① Make up your own shape riddle. Give it to someone to solve.

Answers vary.

240 two hundred forty 2.G.1, SMP7

Unit 8 Progress Check

Overview   Day 1: Administer the Unit Assessments.
Day 2: Administer the Cumulative Assessment.

Day 1: Unit Assessments

1 Warm Up   5–10 min

Self Assessment
Children complete the Self Assessment.

Materials

Assessment Handbook, p. 52

2a Assess   35–45 min

Unit 8 Assessment
These items reflect mastery expectations to this point.

Unit 8 Challenge (Optional)
Children may demonstrate progress beyond expectations.

Assessment Handbook, pp. 53–55;
1 centimeter cube per child

Assessment Handbook, p. 56

CCSS Common Core State Standards	Goals for Mathematical Content (GMC)	Lessons	Self Assessment	Unit 8 Assessment	Unit 8 Challenge
2.OA.1	Use addition and subtraction to solve 1-step number stories.*	8-8		7b	
	Model 1-step problems involving addition and subtraction.*	8-8, 8-9		7c	
2.OA.4	Find the total number of objects in a rectangular array.	8-8 to 8-10	4, 6	7b	
	Express the number of objects in an array as a sum of equal addends.	8-8 to 8-10	5, 6	7c	2
2.G.1	Recognize and draw shapes with specified attributes.	8-1 to 8-5, 8-11	1	1–5	
	Identify 2- and 3-dimensional shapes.	8-1 to 8-5, 8-11	2	1, 2, 4, 5	
2.G.2	Partition a rectangle into rows and columns of same-size squares and count to find the total number of squares.	8-6, 8-7	3	6	1

	Goals for Mathematical Practice (GMP)				
SMP1	Make sense of your problem. GMP1.1				1
SMP4	Model real-world situations using graphs, drawings, tables, symbols, numbers, diagrams, and other representations. GMP4.1	8-8		7a, 7c	2
	Use mathematical models to solve problems and answer questions. GMP4.2	8-10		7b	2
SMP6	Use clear labels, units, and mathematical language. GMP6.3	8-5		4, 5	

*Instruction and most practice on this content is complete.

 Spiral Tracker   ⟨ **Go Online** ⟩   to see how mastery develops for all standards within the grade.

1 Warm Up 5–10 min

▶ Self Assessment

Assessment Handbook, p. 52

| WHOLE CLASS | SMALL GROUP | PARTNER | **INDEPENDENT** |

Children complete the Self Assessment to reflect on their progress in Unit 8.

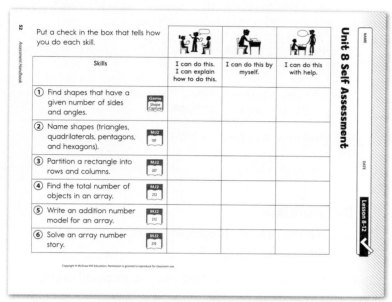

Assessment Handbook, p. 52

2a Assess 35–45 min

Go Online ✓ Assessment and Reporting   Differentiation Support

▶ Unit 8 Assessment

Assessment Handbook, pp. 53–55

| WHOLE CLASS | SMALL GROUP | PARTNER | **INDEPENDENT** |

Children complete the Unit 8 Assessment to demonstrate their progress on the Common Core State Standards covered in this unit. For Problem 5, provide each child with 1 centimeter cube.

Go Online for generic rubrics in the *Assessment Handbook* that can be used to evaluate children's progress on the Mathematical Practices.

Assessment Handbook, p. 53

NAME _____ **DATE** _____ **Lesson 8-12** ✓

Unit 8 Assessment

① Draw a 3-sided shape.

Answers vary but should be a triangle.

What is the name of your shape?
_____triangle_____

② Draw a 4-sided shape with 4 right angles.

Answers vary but should be a square or rectangle.

What is the name of your shape? Sample answers: quadrilateral, square, rectangle

③ Circle the shapes that have parallel sides.

④ Describe the shape below. Use the words *sides* and *angles*. Then write the name of the shape.

Sample answer: It has 6 sides and 6 angles.

Shape name: _____hexagon_____

Assessment Masters   **53**

Assessment Handbook, p. 54

NAME _____ **DATE** _____ **Lesson 8-12** ✓

Unit 8 Assessment (continued)

⑤ Your teacher will give you a shape. Describe the shape and write its name. Sample answer:
This is a cube. It has 6 faces that are the same size and the same shape.

⑥ Partition the rectangle into 2 rows with 2 same-size squares in each row.

How many squares did you draw? ____4____

54   Assessment Handbook

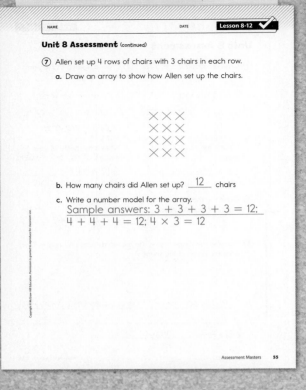

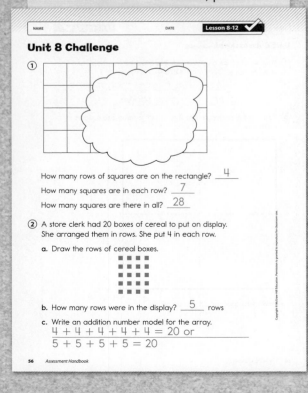

Differentiate — Adjusting the Assessment

Item(s)	Adjustments
1, 2	To extend items 1 and 2, have children draw 3- and 4-sided shapes with additional attributes specified, such as a number of right angles or pairs of parallel lines.
3	To scaffold item 3, provide children with a straightedge that they can move from one side to another to check whether the sides are parallel.
4	To scaffold item 4, provide a shapes poster with labels identifying the angles and sides on each of the shapes.
5	To scaffold item 5, provide children with a list of words they can use to describe the shape.
6	To extend item 6, have children partition the rectangle into 5 rows with 4 same-size squares in each row.
7	To scaffold items 6 and 7, provide children with 20 square pattern blocks.

Advice for Differentiation

All instruction and most practice is complete for the content that is marked with an asterisk (*) on page 760.

Use the online assessment and reporting tools to track children's performance. Differentiation materials are available online to help you address children's needs.

> **NOTE** See the Unit Organizer on pages 684–685 or online Spiral Tracker for details on Unit 8 focus topics and the spiral.

▶ Unit 8 Challenge (Optional)

Assessment Handbook, p. 56

| WHOLE CLASS | SMALL GROUP | PARTNER | **INDEPENDENT** |

Children can complete the Unit 8 Challenge after they complete the Unit 8 Assessment.

Unit 8 Progress Check ✔️

Overview   Day 2: Administer the Cumulative Assessment.

Day 2: Cumulative Assessment

2b Assess   35–45 min

Materials

⭐ **Cumulative Assessment**
These items reflect mastery expectations to this point.

Assessment Handbook,
pp. 57–61

Common Core State Standards	Goals for Mathematical Content (GMC)	Cumulative Assessment
2.OA.1	Use addition and subtraction to solve 1-step number stories.*	4a, 4b
	Model 1-step problems involving addition and subtraction.*	4a, 4b
2.NBT.5	Add within 100 fluently.	4a
	Subtract within 100 fluently.	4a, 4b
2.NBT.6	Add up to four 2-digit numbers.*	3
2.NBT.9	Explain why addition and subtraction strategies work.	3
2.MD.3	Estimate lengths.*	2
2.MD.5	Solve number stories involving length by adding or subtracting.*	4a, 4b
	Model number stories involving length.*	4a, 4b
2.MD.7	Tell and write time using analog and digital clocks.*	1
2.MD.10	Organize and represent data on bar and picture graphs.*	5
	Answer questions using information in graphs.*	6

	Goals for Mathematical Practice (GMP)	
SMP4	Use mathematical models to solve problems and answer questions. **GMP4.2**	6
SMP6	Explain your mathematical thinking clearly and precisely. **GMP6.1**	2a, 2b, 3
	Use an appropriate level of precision for your problem. **GMP6.2**	2a, 2b

*Instruction and most practice on this content is complete.

▨ **Spiral Tracker**   ⟩ **Go Online**   to see how mastery develops for all standards within the grade.

3 Look Ahead   10–15 min

Materials

Math Boxes 8-12
Children preview skills and concepts for Unit 9.

Math Journal 1, p. 219

Home Link 8-12
Children take home the Family Letter that introduces Unit 9.

Math Masters, pp. 241–244

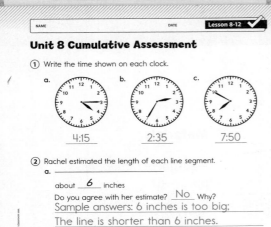

Unit 8 Cumulative Assessment

① Write the time shown on each clock.

a.   b.   c.

4:15   2:35   7:50

② Rachel estimated the length of each line segment.

a. _____

about __6__ inches

Do you agree with her estimate? __No__ Why?
Sample answers: 6 inches is too big;
The line is shorter than 6 inches.

How long do you think the line segment is?

About __3__ inches

b. _____

about __4 or 5__ centimeters   Sample
answers:
Do you agree with her estimate? __Yes__ Why?
The line looks like it is about 4 or 5 cm
long; I used my little finger to estimate.

How long do you think the line segment is?

About __4 or 5__ centimeters

Assessment Masters   57

NAME _____ DATE _____ Lesson 8-12 ✓

Unit 8 Cumulative Assessment (continued)

③ Solve. Try to make friendly numbers.

23 + 17 + 10 + 12 = __62__

16 + 31 + 14 + 19 = __80__

Pick one of the problems above. Explain how you added the numbers.

Sample answer: 16 + 31 + 14 + 19 = 80.
First I added 16 + 14 to get a friendly
number of 30. Then I added 31 + 19 and
got 50. I added 30 + 50 and got my final
answer of 80.

58   Assessment Handbook

---

② **Assess**   35–45 min   Go Online ▶

Assessment
and Reporting

Differentiation
Support

▶ Cumulative Assessment

Assessment Handbook, p. 57–61

| WHOLE CLASS | SMALL GROUP | PARTNER | **INDEPENDENT** |

Children complete the Cumulative Assessment. The problems in the Cumulative Assessment address content from Units 1–7.

Monitor children's progress on the Common Core State Standards using the online assessment and recording tools.

Go Online for generic rubrics in the *Assessment Handbook* that can be used to evaluate children's progress on the Mathematical Practices.

| **Differentiate** | **Adjusting the Assessment** |
Item(s)	Adjustments
1	To extend item 1, have children write the time 3 hours later than the time shown on each clock.
2	To scaffold Item 2, remind children of their personal measurement references for 1 inch and 1 centimeter.
3	To scaffold item 3, suggest that children look for combinations of 10.
4	To scaffold item 4, encourage children to use an open number line.
5, 6	To scaffold items 5 and 6, direct children to the data pages in *My Reference Book*.

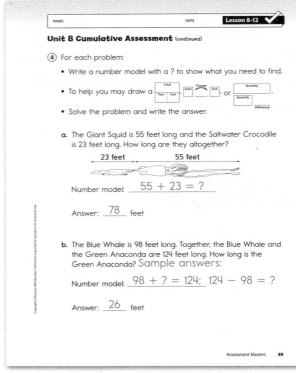

NAME _____ DATE _____ Lesson 8-12 ✓

Unit 8 Cumulative Assessment (continued)

④ For each problem:

• Write a number model with a ? to show what you need to find.

• To help you may draw a [Total / Part Part], [Start Change End], or [Quantity / Quantity Difference].

• Solve the problem and write the answer.

a. The Giant Squid is 55 feet long and the Saltwater Crocodile is 23 feet long. How long are they altogether?

23 feet   55 feet

Number model: __55 + 23 = ?__

Answer: __78__ feet

b. The Blue Whale is 98 feet long. Together, the Blue Whale and the Green Anaconda are 124 feet long. How long is the Green Anaconda? Sample answers:

Number model: __98 + ? = 124; 124 − 98 = ?__

Answer: __26__ feet

Assessment Masters   59

Assessment Handbook, p. 59

Advice for Differentiation

All instruction and most practice is complete for the content that is marked with an asterisk (*) on page 763.

Use the online assessment and reporting tools to track children's performance. Differentiation materials are available online to help you address children's needs.

3 Look Ahead 10–15 min   Go Online   🏠 Home Connections

▶ Math Boxes 8–12: Preview for Unit 9

Math Journal 2, p. 219

| WHOLE CLASS | SMALL GROUP | PARTNER | INDEPENDENT |

Mixed Practice Math Boxes 8-12 are paired with Math Boxes 8-8. These problems focus on skills and understandings that are prerequisite for Unit 9. You may want to use information from these Math Boxes to plan instruction and grouping in Unit 9.

Math Boxes
Preview for Unit 9   Lesson 8-12   DATE

① Divide the rectangle into 4 equal parts.

Sample answer: [grid]

② Solve. Show your work.   Unit

$$\begin{array}{r} 98 \\ -\ 75 \\ \hline 23 \end{array}$$

③ Which number shows 9 in the tens place, 8 in the hundreds place, and 1 in the ones place?

Fill in the circle next to the correct answer.

Ⓐ 891   Ⓑ 198
Ⓒ 819   Ⓓ 918

④ • • • • • • • • • •
• • • • • • • • • •

Count by 2s. How many dots in all?

Answer: __12__ dots

Write a number model.
$2 + 2 + 2 + 2 +$
$2 + 2 = 12$

⑤ Make a ballpark estimate. Write a number model to show your estimate.

$49¢ + 129¢ = ?$ Sample answer:

My estimate:
$50 + 130 = 180$ ¢

⑥ Measure the line segment to the nearest inch.

_____

Fill in the circle next to the correct answer.

Ⓐ about 2 inches
Ⓑ about 1 inch
Ⓒ about 4 inches

① 2.G.3 ② 2.NBT.5 ③ 2.NBT.1, 2.NBT.3
④ 2.OA.4 ⑤ 2.NBT.5 ⑥ 2.MD.1   two hundred nineteen 219

Math Journal 2, p. 219

▶ Home Link 8-12: Unit 9 Family Letter

Math Masters, pp. 241–244

Home Connection The Unit 9 Family Letter provides information and activities related to Unit 9 content.

Assessment Handbook, p. 60

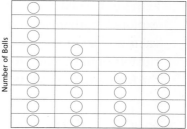

NAME   DATE   Lesson 8-12 ✓

Unit 8 Cumulative Assessment (continued)

⑤ The gym at McKenzie School has a basket of sports balls. Curtis sorted the balls. He made the tally chart below.

Type of Ball	Tallies	Number
Basketball	⊬⊬ ////	9
Soccer Ball	⊬⊬ /	6
Football	////	4
Softball	⊬⊬	5

Draw a picture graph to show his data.

Sports Balls at McKenzie School

[picture graph with circles]

Number of Balls

Basketball Soccer Ball Football Softball
Type of Balls

KEY: each ◯ = 1 ball

60 *Assessment Handbook*

Assessment Handbook, p. 61

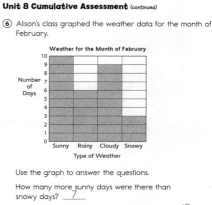

NAME   DATE   Lesson #-# ✓

Unit 8 Cumulative Assessment (continued)

⑥ Alison's class graphed the weather data for the month of February.

Weather for the Month of February

[bar graph: Number of Days vs Type of Weather: Sunny, Rainy, Cloudy, Snowy]

Use the graph to answer the questions.

How many more sunny days were there than snowy days? __7__

How many days were either cloudy or rainy? __15__

How many days in February were recorded on this graph? __28__

Write a question that can be answered with this graph.
Answers vary.

Write the answer to your question.
Answers vary.

Assessment Masters 61

Unit 9 Organizer
Equal Shares and Whole Number Operations

In this unit, children partition shapes into equal shares and apply these ideas to further explore length measurement. They also learn a new subtraction strategy based on place value and continue working with equal groups. Children's learning will focus on four clusters of the Common Core's content standards, as well as in-depth work on two of the Mathematical Practices.

CCSS Standards for Mathematical Content

Domain	Cluster
Operations and Algebraic Thinking	Work with equal groups of objects to gain foundations for multiplication.
Number and Operations in Base Ten	Understand place value.
	Use place value understanding and properties of operations to add and subtract.
Geometry	Reason with shapes and their attributes.

Because the standards within each domain can be broad, *Everyday Mathematics* has unpacked each standard into Goals for Mathematical Content **GMC** . For a complete list of Standards and Goals, see page EM1.

For an overview of the CCSS domains, standards, and mastery expectations in this unit, see the **Spiral Trace** on pages 772–773. See the **Mathematical Background** (pages 774–776) for a discussion of the following key topics:

- Equal Shares
- Place Value and Subtraction
- Money
- Multiples of 2, 5, and 10

CCSS Standards for Mathematical Practice

SMP1 Make sense of problems and persevere in solving them.

SMP2 Reason abstractly and quantitatively.

For a discussion about how *Everyday Mathematics* develops these practices and a list of Goals for Mathematical Practice **GMP** , see page 777.

 Virtual Learning Community  to **vlc.cemseprojects.org** to search for video clips on each practice.

Go Digital with these tools at **connectED.mheducation.com**

ePresentations Student Learning Center Facts Workshop Game eToolkit Professional Development Home Connections Spiral Tracker Assessment and Reporting English Learners Support Differentiation Support

Contents

*The standards listed here are addressed in the **Focus** of each lesson. For all the standards in a lesson, see the Lesson Opener.

Unit 9 Materials

 VLC Virtual Learning Community

See how *Everyday Mathematics* teachers organize materials. Search "Classroom Tours" at **vlc.cemseprojects.org**.

Lesson	Math Masters	Activity Cards	Manipulative Kit	Other Materials
9-1	pp. 245; G31	111	geoboard; rubber bands	paper squares; Class Data Pad or chart paper; *Array Concentration* Array Cards and Number Cards; circle templates; colored construction paper; scissors; slate
9-2	pp. 246–253	112–114	pattern blocks	Pattern-Block Template; Class Equal Shares posters; 2–3 paper strips; Fact Triangles; paper shapes; slate
9-3	pp. 254–256; TA5			Class Equal Shares posters; Standards for Mathematical Practice Poster; 15 cut out circles; scissors; glue; children's work from Day 1; selected samples of children's work; colored pencils (optional); slate; Guidelines for Discussions Poster
9-4	pp. 257–260; TA33	115		12-inch ruler; scissors; 3-inch paper strip; slate
9-5	pp. 261–263; G7–G8; G28	17	base-10 blocks; 4 each of number cards 0–9; calculator	per group: Attribute Cards, Shape Cards; quarter-sheets of paper; glue or tape; slate
9-6	pp. 264–265; TA22; TA32	116–117	base-10 blocks; 4 each of number cards 0–9	Class Data Pad; stick-on notes; slate
9-7	pp. 266; TA3; TA22; *Assessment Handbook,* pp. 98–99 (optional)	118–119	base-10 blocks; 4 each of number cards 0–9; per group: calculator	per group: large paper triangle; number grid; large poster paper; slate
9-8	pp. 267–269; 270 (optional); G25; 271		toolkit coins; toolkit bills; per partnership: 1 calculator	slate
9-9	pp. 272–275; TA5			Standards for Mathematical Practice Poster; scissors; glue; children's work from Day 1; selected samples of children's work; colored pencils (optional); Guidelines for Discussions Poster
9-10	pp. 276–277; TA6; TA7; TA25	120	10 counters	per partnership: 20 centimeter cubes; poster paper (or other large paper); slate
9-11	pp. 278–281; TA3 (optional)		Class Number Line; base-10 blocks; calculator	number line; number grid; 1 dime and 1 nickel (for demonstration); red, blue, and green colored pencils; slate
9-12	pp. 282–286; *Assessment Handbook,* pp. 62–69		base-10 blocks (optional); toolkit bills (optional)	scissors; number grid (optional)

Literature Link **9-1** *Ed Emberley's Picture Pie: A Cut and Paste Drawing Book* (optional)

Go Online for a complete literature list for Grade 2.

Problem Solving  Professional Development

Everyday Mathematics emphasizes equally all three of the Common Core's dimensions of **rigor:** conceptual understanding, procedural skill and fluency, and applications. Math Messages, other daily work, Explorations, and Open Response tasks provide many opportunities for children to apply what they know to solve problems.

▶ Math Message

Math Messages require children to solve a problem they have not been shown how to solve. Math Messages provide almost daily opportunities for problem solving.

▶ Daily Work

Journal pages, Home Links, Writing/Reasoning prompts, and Differentiation Options often require children to solve problems in mathematical contexts and real-life situations. **Minute Math+** offers Number Stories for transition times and for spare moments throughout the day. See Routine 6, pages 38–43.

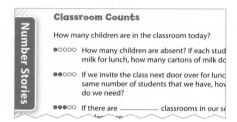

▶ Explorations

In Exploration A children divide shapes into equal parts and explain how they know the parts are equal.

▶ Open Response and Reengagement

In Lesson 9-3 children equally share 2 muffins between 2 people and 5 muffins among 4 people. They partition circles to show their shares and explain how they know their shares are equal. The reengagement discussion on Day 2 could focus on different ways to share a collection or how different partitions suggest different names for the shares. Creating mathematical models, such as drawings, can help children visualize the situation and solve a problem. GMP4.2

In Lesson 9-9 children select items to buy so that they spend as close to $100 as they can, without going over $100, and explain their strategies. The reengagement discussion on Day 2 could focus on different strategies for solving the problem or whether two different solutions can be correct. Writing clear and precise explanations can help children organize their thoughts, allowing them to become better problem solvers. GMP6.1

 **Virtual Learning Community** [Go Online] to watch an Open Response and Reengagement lesson in action. Search "Open Response" at **vlc.cemseprojects.org.**

▶ Open Response Assessment

In Progress Check Lesson 9-12, children determine the combination of 2-digit numbers that will produce the largest sum. GMP3.1

Look for GMP1.1–1.6 markers, which indicate opportunities for children to engage in SMP1: "Make sense of problems and persevere in solving them." Children also become better problem solvers as they engage in all of the CCSS Mathematical Practices. The yellow GMP markers throughout the lessons indicate places where you can emphasize the Mathematical Practices and develop children's problem-solving skills.

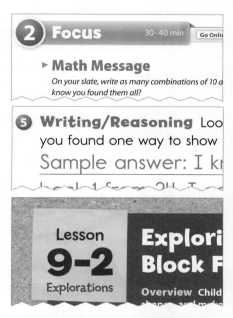

2 **Focus** 30–40 min [Go Onli]

▶ **Math Message**
On your slate, write as many combinations of 10 a know you found them all?

5 **Writing/Reasoning** Loo you found one way to show

Sample answer: I kn

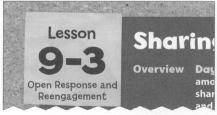

Lesson **9-2** Explorations — Explori Block F Overview Child

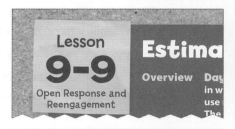

Lesson **9-3** Open Response and Reengagement — Sharin Overview Day amo shar and

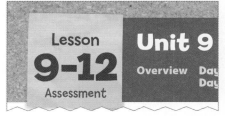

Lesson **9-9** Open Response and Reengagement — Estima Overview Day in w use The

Lesson **9-12** Assessment — Unit 9 Overview Day Day

Assessment and Differentiation

 Assessment and Reporting

See pages xxii–xxv to learn about a comprehensive online system for recording, monitoring, and reporting children's progress using core program assessments.

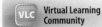 Virtual Learning Community Go Online to **vlc.cemseprojects.org** for tools and ideas related to assessment and differentiation from *Everyday Mathematics* teachers.

✔ Ongoing Assessment

In addition to frequent informal opportunities for "kid watching," every lesson (except Explorations) offers an **Assessment Check-In** to gauge children's performance on one or more of the standards addressed in that lesson.

Lesson	Task Description	CCSS Common Core State Standards
9-1	Show one way to partition rectangles into halves, fourths, and thirds; write one name for a part and one name for all of the parts together.	2.G.3, SMP2, SMP6
9-3	Partition one or all muffins into two and four equal shares; use appropriate language (halves, fourths, quarters) to talk about the shares. Attempt to complete or improve explanations of strategies based on the class discussion.	2.G.3, SMP4
9-4	Measure objects to the nearest inch.	2.MD.1, SMP5, SMP6
9-5	Write 3-digit numbers in expanded form and compare them.	2.NBT.3, 2.NBT.4, SMP2
9-6	Determine whether a trade is needed to solve a subtraction problem.	2.NBT.1a, 2.NBT.5, 2.NBT.7
9-7	Make reasonable estimates and write numbers in expanded form. Fluently subtract within 100 using strategies based on place value, properties of operations, and/or the relationship between addition and subtraction.	2.NBT.1a, 2.NBT.3, 2.NBT.5
9-8	Show at least one way to pay for food items.	2.MD.8, SMP1, SMP4
9-9	Choose at least three items, estimate or use mental math to find the total, and compare the total to $100. Improve work based on the class discussion.	2.OA.1, SMP6
9-10	Solve number stories and write an addition number model for each problem.	2.OA.1, 2.OA.3, 2.OA.4
9-11	Solve problems by skip counting or doubling.	2.NBT.2, 2.NBT.5

▶ Periodic Assessment

Unit 9 Progress Check This assessment focuses on the CCSS domains of *Operations and Algebraic Thinking, Number and Operations in Base Ten, Measurement and Data,* and *Geometry.* It also contains an Open Response Assessment to test children's ability to organize a set of digits to form the largest sum. GMP3.1

End-of-Year Assessment This benchmark test checks children's mastery of some of the important concepts and skills presented in *Second Grade Everyday Mathematics.* See the *Assessment Handbook.*

> **NOTE** Odd-numbered units include an **Open Response Assessment.** Even-numbered units include a **Cumulative Assessment.**

▶ Unit 9 Differentiation Activities

 Differentiation Support   English Learners Support

Differentiation Options Every regular lesson provides Readiness, Enrichment, **Extra Practice,** and **English Language Learners Support** activities that address the Focus standards of that lesson.

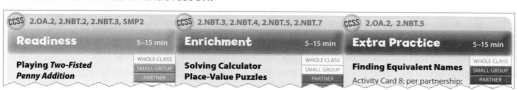

CCSS 2.OA.2, 2.NBT.2, 2.NBT.3, SMP2	CCSS 2.NBT.3, 2.NBT.4, 2.NBT.5, 2.NBT.7	CCSS 2.OA.2, 2.NBT.5
Readiness 5–15 min	**Enrichment** 5–15 min	**Extra Practice** 5–15 min
Playing *Two-Fisted Penny Addition* WHOLE CLASS SMALL GROUP PARTNER	Solving Calculator Place-Value Puzzles WHOLE CLASS SMALL GROUP PARTNER	Finding Equivalent Names WHOLE CLASS SMALL GROUP PARTNER Activity Card 8; per partnership;

Activity Cards These activities, written to the children, enable you to differentiate Part 2 of the lesson through small-group work.

English Language Learners Activities and point-of-use support help children at different levels of English language proficiency succeed.

Differentiation Support online pages

Differentiation Support Two online pages for most lessons provide suggestions for game modifications, ways to scaffold lessons for children who need additional support, and language development suggestions for Beginning, Intermediate, and Advanced English language learners.

Ongoing Practice Differentiation Support

For **ongoing distributed practice,** see these activities:
- Mental Math and Fluency
- Differentiation Options: Extra Practice
- Part 3: Journal pages, Math Boxes, *Math Masters,* Home Links
- Print and online games

▶ Games

Games in *Everyday Mathematics* are an essential tool for practicing skills and developing strategic thinking.

Lesson	Game	Skills and Concepts	CCSS Common Core State Standards
9-1	*Array Concentration*	Finding the total number of objects in an array and writing matching number models	2.OA.4, 2.NBT.2, SMP2, SMP6
9-5	*Shape Capture*	Identifying shapes by their attributes	2.G.1, SMP7
9-5	*Number Top-It*	Comparing multidigit numbers	2.NBT.1, 2.NBT.3, 2.NBT.4, SMP7
9-7	*Beat the Calculator*	Practicing addition facts	2.OA.2
9-8	*Hit the Target*	Finding differences between multiples of 10 and smaller 2-digit numbers	2.NBT.5, SMP1, SMP3

VLC Virtual Learning Community **Go Online** to look for examples of *Everyday Mathematics* games at **vlc.cemseprojects.org.**

⭐ **Mastery Expectations** This Spiral Trace outlines instructional trajectories for key standards in Unit 9. For each standard, it highlights opportunities for Focus instruction, Warm Up and Practice activities, as well as formative and summative assessment. It describes the **degree of mastery**—as measured against the entire standard—expected at this point in the year.

Operations and Algebraic Thinking

2.OA.4 Use addition to find the total number of objects arranged in rectangular arrays with up to 5 rows and up to 5 columns; write an equation to express the total as a sum of equal addends.

| 5-5 Focus | 6-10 Focus Practice | 8-8 through 8-10 Focus Practice | 8-11 Practice | 8-12 Progress Check | 9-1 Practice | 9-5 through 9-8 Practice | 9-10 Warm Up Focus | 9-11 Focus | 9-12 Progress Check |

⭐ By the end of Unit 9, expect children to **use addition to find the total number of objects arranged in rectangular arrays with up to 5 rows and up to 5 columns; write an equation to express the total as a sum of equal addends.**

Number and Operations in Base Ten

2.NBT.1 Understand that the three digits of a three-digit number represent amounts of hundreds, tens, and ones; e.g., 706 equals 7 hundreds, 0 tens, and 6 ones.

| 5-4 Practice | 5-5 Warm Up Practice | 5-10 Practice | 6-4 Warm Up Practice | 6-6 Practice | 6-11 Progress Check | 8-3 Practice | 8-6 Warm Up Practice | 9-1 Warm Up Practice | 9-5 through 9-7 Warm Up Focus Practice | 9-12 Progress Check |

⭐ By the end of Unit 9, expect children to **understand that the three digits of a 3-digit number represent amounts of hundreds, tens, and ones.**

2.NBT.3 Read and write numbers to 1000 using base-ten numerals, number names, and expanded form.

| 4-8 Practice | 4-12 Progress Check | 5-4 Practice | 5-10 Practice | 6-4 Focus Practice | 6-8 Focus Practice | 6-11 Progress Check | 7-8 Focus | 9-1 Warm Up Practice | 9-5 through 9-7 Warm Up Focus Practice | 9-12 Progress Check |

⭐ By the end of Unit 9, expect children to **read and write numbers to 1,000 using base-10 numerals, number names, and expanded form.**

2.NBT.5 Fluently add and subtract within 100 using strategies based on place value, properties of operations, and/or the relationship between addition and subtraction.

| 7-7 Focus Practice | 7-8 Focus Practice | 7-9 Warm Up Practice | 7-10 Progress Check | 8-1 Warm Up Practice | 8-3 Warm Up Practice | 9-6 Focus Practice | 9-7 Warm Up Focus Practice | 9-8 Practice | 9-9 Focus Practice | 9-11 Warm Up Focus Practice | 9-12 Progress Check |

⭐ By the end of Unit 9, expect children to **fluently add and subtract within 100 using strategies based on place value, properties of operations, and/or the relationship between addition and subtraction.**

2.NBT.9 Explain why addition and subtraction strategies work, using place value and the properties of operations.

| 5-11 Focus | 6-6 through 6-8 Focus Practice | 6-11 Progress Check | 7-1 Focus | 7-3 Focus | 7-4 Practice | 7-10 Progress Check | 8-9 Practice | 8-12 Progress Check | 9-6 Focus Practice | 9-7 Focus |

⭐ By the end of Unit 9, expect children to **explain why addition and subtraction strategies work, using place value and properties of operations.**

Spiral Tracker

Go to **connectED.mheducation.com** for comprehensive trajectories that show how in-depth mastery develops across the grade.

2.NBT.7 Add and subtract within 1000, using concrete models or drawings and strategies based on place value, properties of operations, and/or the relationship between addition and subtraction; relate the strategy to a written method. Understand that in adding or subtracting three-digit numbers, one adds or subtracts hundreds and hundreds, tens and tens, ones and ones; and sometimes it is necessary to compose or decompose tens or hundreds.

⭐ By the end of Unit 9, expect children to **add and subtract within 1,000 using concrete models or drawings and strategies based on place value, properties of operations, and/or the relationship between addition and subtraction; relate the strategy to a written method.** Expect children to understand that in adding or subtracting 3-digit numbers, one adds or subtracts hundreds and hundreds, tens and tens, ones and ones; and sometimes it is necessary to compose or decompose tens or hundreds.

Measurement and Data

2.MD.8 Solve word problems involving dollar bills, quarters, dimes, nickels, and pennies, using $ and ¢ symbols appropriately.

⭐ By the end of Unit 9, expect children to **solve word problems involving dollar bills, quarters, dimes, nickels, and pennies, using $ and ¢ symbols appropriately.**

Geometry

2.G.3 Partition circles and rectangles into two, three, or four equal shares, describe the shares using the words *halves, thirds, half of, a third of,* etc., and describe the whole as two halves, three thirds, four fourths. Recognize that equal shares of identical wholes need not have the same shape.

⭐ By the end of Unit 9, expect children to **partition circles and rectangles into two, three, or four equal shares using the words** *halves, thirds, half of, a third of,* **and so on, and describe the whole as** *two-halves, three-thirds, four-fourths.* **Expect children to recognize that equal shares of identical wholes need not have the same shape.**

Key ✓ = Assessment Check-In   ✓ = Progress Check Lesson   ▱ = Current Unit   ▰ = Previous Lessons

Mathematical Background: Content

 This discussion highlights the major content areas and the Common Core State Standards addressed in Unit 9. See the online Spiral Tracker for complete information about the learning trajectories for all standards.

▶ Equal Shares (Lessons 9-1 through 9-4)

In Lessons 9-1 and 9-2, children partition shapes into 2, 3, or 4 same-size parts. **2.G.3** Each part of a partitioned figure is called a *share*. When the shares are all the same size, they are called *equal shares*.

In Lesson 9-1 children fold paper squares to create equal shares. They use the folded squares to "prove" that their shares are equal by showing that the parts fit exactly on top of each other. They also explore the idea that there is more than one way to divide a particular shape into equal shares. For example, the figures below show three different ways that a square can be divided into 4 equal shares.

In Lesson 9-2 children extend their understanding of equal shares by exploring the idea that equal shares must be the same size but not necessarily the same shape. **2.G.3** (*See margin.*) Equal shares that are not the same shape do not line up exactly when they are placed on top of each other. Children "prove" that these shares are equal by picturing the smaller squares within each share. They count to make sure that each share is made up of the same number of small squares. Later in the unit, they also partition irregular shapes into equal shares and use the same square-counting strategy to show that the shares are equal. (*See margin.*)

This work with equal shares helps children build a conceptual understanding of fractions as numbers that represent a part of a whole. Children name the equal shares of the shapes using fraction words, such as *one-half, one-third of the shape*, and *1 out of 4 parts*. They name the whole using words such as *two-halves, three-thirds*, and *four-fourths*. **2.G.3**

Children are not introduced to the standard notation for fractions (e.g., $\frac{1}{2}$, $\frac{1}{3}$, and $\frac{4}{4}$) in Grade 2. Children often think of the standard notation as referring to two whole numbers and misapply whole-number reasoning to fractions. They may believe, for example, that $\frac{1}{4}$ is greater than $\frac{1}{2}$ because 4 is greater than 2. These misconceptions might be avoided if children begin by developing an understanding of fractions as *single numbers* rather than as two whole numbers. The fraction vocabulary used in this unit is intended to promote this kind of understanding.

In Lesson 9-4 children partition an inch into equal shares to explore measuring to the nearest half-inch. **2.MD.1**

 Standards and Goals for Mathematical Content

Because the standards within each domain can be broad, *Everyday Mathematics* has unpacked each standard into Goals for Mathematical Content GMC. For a complete list of Standards and Goals, see page EM1.

Each share is made up of 3 smaller squares.

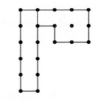

Each share is made up of 5 smaller squares.

╭────────────────────────────────────╮

Unit 9 Vocabulary

ballpark estimate	half-inch	precise
close-but-easier numbers	multiple	quarter-inch
equal share	one-fourth (1-fourth)	reasonable
expand-and-trade subtraction	one-half (1-half)	thousand cube
four-fourths (4-fourths)	one-quarter (1-quarter)	three-thirds (3-thirds)
fourth-inch	one-third (1-third)	two-halves (2-halves)

╰────────────────────────────────────╯

▶ Place Value and Subtraction

(Lessons 9-5 through 9-7)

In Lesson 9-5 children review place value for 3-digit numbers. They represent 3-digit numbers using base-10 blocks, place-value mats, and expanded form. **2.NBT.3** These representations reinforce the idea that the digits of a 3-digit number represent amounts of hundreds, tens, and ones. **2.NBT.1** Children extend their work with place value to numbers in the thousands and practice using expanded form to help them compare multidigit numbers.

In Lessons 9-6 and 9-7, children apply their understanding of place value to learn a new subtraction strategy called *expand-and-trade subtraction.* **2.NBT.5** In expand-and-trade subtraction, the numbers are first expanded to show the value of each digit. Then any necessary trades are made. Lastly, the subtraction is carried out. Children begin by modeling the numbers and trades with base-10 blocks in Lesson 9-6. **2.NBT.7** In Lesson 9-7 they transition to using expanded form to represent the numbers and write number sentences to show the trades. **2.NBT.3** See Lessons 9-6 and 9-7 for examples of how children use these representations.

Everyday Mathematics introduces expand-and-trade subtraction before the traditional subtraction algorithm for several reasons. By using representations that emphasize place value, children are able to think about the trades they are making, without losing track of what they are doing and which step they are on. For example, they can trade 1 ten for 10 ones and then verify that they are still working with the same number. Doing all the trading first also allows children to focus on one thing at a time, instead of switching back and forth between the trading and the subtraction. The diagrams in the margin show how a child might think about the trading involved in solving 51 − 24.

Expand-and-trade subtraction also allows children to start with any place-value position when they subtract. As in the example above, after all the necessary trades have been made, children can subtract the tens first or the ones first and get the same result. Children tend to operate from left to right, dealing with the largest place-value positions first. Starting on the right, as required by the traditional algorithm, can be counterintuitive to children and lead to computation errors. This is another reason that *Everyday Mathematics* provides experience with expand-and-trade subtraction before introducing the traditional algorithm in Grade 4.

A child's thinking about solving 51 − 24

> These blocks show 51. I need to take away 2 tens and 4 ones. Do I have enough ones to take away 4 ones? No, so I need to make a trade first.

> Now I have 4 tens and 11 ones. That's still 51, but now I have enough to take away 2 tens and 4 ones.

▶ Money (Lessons 9-8 and 9-9)

Money is a versatile topic that is closely connected to children's everyday lives. It provides a context for children to continue doing mathematics with their families during the break between grades. For this reason, work with money is included throughout *Everyday Mathematics* and reviewed during this last unit of Grade 2.

In Lesson 9-8 children review the values of pennies, nickels, dimes, and quarters. They make exchanges among coins and find different combinations of coins that can be used to pay for a particular item. **2.MD.8** This lesson provides contextualized practice with skip counting and concrete models for addition. **2.NBT.2, 2.NBT.7** The exploration with dimes and pennies also prepares children for an exploration of multiples of 10 and 5 later in the unit. (See the next section for details.)

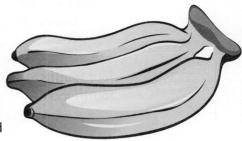

Bananas 59¢ lb

In Lesson 9-9 children use money amounts given in whole numbers of dollars to practice mental computation and estimation strategies and add three or more numbers. **2.NBT.5, 2.NBT.6, 2.NBT.8**

▶ Multiples of 2, 5, and 10 (Lessons 9-10 and 9-11)

In Lesson 9-10 children make connections among several different concepts they have studied during the year. They begin by solving number stories about the total number of objects in two equal groups or rows and writing addition number models to represent the stories. **2.OA.1, 2.OA.4** As they examine the number models, such as $2 + 2 = 4$, $6 + 6 = 12$, and $9 + 9 = 18$, they discover that all of them are doubles facts. **2.OA.2** They also notice that the total number of objects in two equal groups is always even and learn to use equal-groups number stories as a context for practice with writing even numbers as the sum of equal addends. **2.OA.3**

Two equal groups of 6;
$6 + 6 = 12$.

Lesson 9-10 is designed to help children build understanding of how doubles addition facts and even numbers relate to finding multiples of 2. A *multiple of 2* is the product of 2 and a counting number. For example, 6 is a multiple of 2 because $2 \times 3 = 6$. Lesson 9-11 uses money as a context for finding multiples of 5 and 10. Children skip count by 5s and 10s by counting collections of nickels and dimes. **2.NBT.2**

The 2s, 5s, and 10s multiplication facts are key groups of facts that will be used as helper facts to assist children with learning the rest of the multiplication facts by the end of Grade 3. The work with multiples and exposure to multiplication number models in Lessons 9-10 and 9-11 will build readiness for children to learn the key groups of facts quickly and move on to other, more challenging groups of facts.

Mathematical Background: Practices

 In Everyday Mathematics, *children learn the **content** of mathematics as they engage in the **practices** of mathematics. As such, the Standards for Mathematical Practice are embedded in children's everyday work, including hands-on activities, problem-solving tasks, discussions, and written work. Read here to see how Mathematical Practices 1 and 2 are emphasized in this unit.*

▶ Standard for Mathematical Practice 1

According to Mathematical Practice 1, mathematically proficient students "check their answers to problems using a different method" and "identify correspondences between different approaches." In Lesson 9-7 children solve subtraction problems using expand-and-trade subtraction and then solve at least one of the problems again using another method. They discuss how expand-and-trade subtraction is similar to and different from the other method they used, and they think about why it might be important to be able to "solve problems in more than one way." GMP1.5

Reasons to learn and practice a variety of strategies might include the ability to check one's own work and the idea that no single strategy is likely to be most efficient or useful for every problem. In Lesson 9-8 children continue to practice solving problems in more than one way by finding at least two combinations of coins that can be used to pay for a particular item. This activity situates solving problems in more than one way in a real-world context. If children know only one way to use coins to pay for an item but don't have the right coins in their pockets, they may think they cannot purchase the item(s).

▶ Standard for Mathematical Practice 2

According to Mathematical Practice 2, quantitative reasoning entails "knowing and flexibly using different properties of operations and objects." Children flexibly apply properties of equal shares and multidigit numbers as they "make sense of the representations [they] and others use" during Unit 9. GMP2.2

In Lessons 9-1 and 9-2, children partition shapes into equal shares. They reason that the shares must be equal because they have at least one property in common: they are the same size and shape, or they are composed of the same number of smaller squares.

In Lessons 9-5 through 9-7, children represent numbers using base-10 blocks and expanded form. Working with these representations, they focus on the place value of the digits in the numbers and use place-value properties to help them compare the numbers and make valid trades as they subtract.

 Standards and Goals for Mathematical Practice

SMP1 Make sense of problems and persevere in solving them.

GMP1.1 Make sense of your problem.

GMP1.2 Reflect on your thinking as you solve your problem.

GMP1.3 Keep trying when your problem is hard.

GMP1.4 Check whether your answer makes sense.

GMP1.5 Solve problems in more than one way.

GMP1.6 Compare the strategies you and others use.

SMP2 Reason abstractly and quantitatively.

GMP2.1 Create mathematical representations using numbers, words, pictures, symbols, gestures, tables, graphs, and concrete objects.

GMP2.2 Make sense of the representations you and others use.

GMP2.3 Make connections between representations.

Go Online to the *Implementation Guide* for more information about the Mathematical Practices.

For children's information on the Mathematical Practices, see *My Reference Book,* pages 1–22.

Creating and Naming Equal Parts

Overview Children divide shapes and use fraction vocabulary to name the shares.

▶ **Before You Begin**

For Part 2 each child will need eight 5-inch paper squares. You will need supplies to create three classroom posters. For the optional Enrichment activity, acquire one or more copies of *Ed Emberley's Picture Pie: A Cut and Paste Drawing Book* by Ed Emberley (Little, Brown and Company, 2006). You will also need to make circle templates by cutting different-size circles from tagboard.

▶ **Vocabulary**

equal share • one-half (1-half) • two-halves (2-halves) • one-fourth (1-fourth) • one-quarter (1-quarter) • four-fourths (4-fourths) • one-third (1-third) • three-thirds (3-thirds)

Common Core State Standards

Focus Cluster
Reason with shapes and their attributes.

---

1 Warm Up 15–20 min

	Materials	
Mental Math and Fluency Children write numbers in expanded form.	slate	2.NBT.1, 2.NBT.3
Daily Routines Children complete daily routines.	See pages 4–43.	See pages xiv–xvii.

---

2 Focus 30–40 min

Math Message Children show how to divide a sandwich into 2 equal shares.	paper squares (see Before You Begin)	2.G.3
Folding Squares into Equal Shares Children share ways to divide a square into 2 equal shares.	paper squares	2.G.3 SMP2
Naming 2, 4, and 3 Equal Shares Children use fraction words to name 2, 4, and 3 equal shares.	Class Data Pad or chart paper, paper squares	2.G.3 SMP2, SMP6
Partitioning Shapes Children partition shapes and name the parts.	*Math Journal 2,* pp. 220–221	2.G.3 SMP2, SMP6
✓ **Assessment Check-In** See page 783.	*Math Journal 2,* pp. 220–221	2.G.3, SMP2, SMP6

CCSS 2.G.3 **Spiral Snapshot**

GMC Partition shapes into equal shares.

| 2-8 Focus | 8-11 Focus | 9-1 Focus Practice | 9-2 Focus Practice | 9-3 Focus Practice | 9-4 through 9-11 Practice |

Spiral Tracker   **Go Online**   to see how mastery develops for all standards within the grade.

---

3 Practice 10–15 min

Playing *Array Concentration* **Game** Children find how many objects are in arrays.	*Math Masters,* p. G31; *Array Concentration* Array Cards and Number Cards	2.OA.4, 2.NBT.2 SMP2, SMP6
Math Boxes 9-1 Children practice and maintain skills.	*Math Journal 2,* p. 222	See page 783.
Home Link 9-1 **Homework** Children partition shapes and name the parts.	*Math Masters,* p. 245	2.G.3

connectED.mheducation.com

Plan your lessons online with these tools.

ePresentations | Student Learning Center | Facts Workshop Game | eToolkit | Professional Development | Home Connections | Spiral Tracker | Assessment and Reporting | English Learners Support | Differentiation Support

Differentiation Options

RtI

CCSS 2.G.3, SMP1

Readiness   5–15 min

WHOLE CLASS
SMALL GROUP
PARTNER
INDEPENDENT

Partitioning Slates into Equal Parts

slate

For experience partitioning shapes using a visual model, children divide their slates into equal parts. They start by dividing their slates into 2 equal parts and comparing their different strategies. GMP1.6

Emphasize that the 2 parts are equal and together make a whole. Have children look around the room for objects that are divided into 2 equal parts, such as bookcases or windows. Then have children divide their slates into 4 equal parts, compare their strategies, and look for classroom objects that are divided equally into 4 parts.

CCSS 2.G.3, SMP2

Enrichment   5–15 min

WHOLE CLASS
SMALL GROUP
PARTNER
INDEPENDENT

Naming Equal Parts Found in Literature

Activity Card 111,
Ed Emberley's Picture Pie: A Cut and Paste Drawing Book, colored construction paper, scissors, circle templates (see Before You Begin)

Literature Link To apply their understanding of names for equal parts, children read ***Ed Emberley's Picture Pie: A Cut and Paste Drawing Book*** by Ed Emberley (Little, Brown and Company, 2006) in small groups, as partners, or on their own. This book shows how a circle divided into equal parts can be used to make pictures of all kinds.

Children trace circles of different sizes and fold them to divide them into equal parts. Then they create their own circle designs and discuss the fractional circle parts they used. GMP2.3 Have children refer to the Equal Shares posters for help using different names of fractional parts.

Naming Equal Parts Found in Literature

What You Need
Ed Emberley's Picture Pie:
A Cut and Paste Drawing Book
construction paper of various colors
scissors

I will use fourths of the circle to make this part of the design.

CCSS 2.G.3, SMP1

Extra Practice   5–15 min

WHOLE CLASS
SMALL GROUP
PARTNER
INDEPENDENT

Dividing Shapes into Equal Parts

geoboard, rubber bands

To further explore equal shares, children divide a square on a geoboard into equal parts and name the parts. GMP1.5 Have children divide a square into 2 equal parts. Ask: *Is there more than one way to divide the square?* Sample answer: Yes. It can be divided from side to side or diagonally. *How can you name 1 part?* Sample answers: half; one-half; 1-half; 1 out of 2 equal parts *What is the name for both parts together?* Sample answers: two-halves; 2-halves; 2 out of 2; whole Repeat the activity with thirds and fourths.

English Language Learners Support

Beginning ELL Teach children some everyday meanings of the term *share* using teacher modeling, think-alouds, and commonly used classroom directions. For example, model sharing—in the sense of using something with someone else—by sharing a book with a child as you think aloud, saying: *I will share my book with* _____. Model what it means to *share*—in the sense of distributing—as you distribute pencils, saying: *Let's share the pencils so that everyone gets one.* Build on the latter meaning to demonstrate the mathematical meaning related to *equal shares.*

Go Online ELL English Learners Support

Standards and Goals for
Mathematical Practice

SMP2 **Reason abstractly and quantitatively.**
  GMP2.1 Create mathematical representations using numbers, words, pictures, symbols, gestures, tables, graphs, and concrete objects.
  GMP2.2 Make sense of the representations you and others use.

SMP6 **Attend to precision.**
  GMP6.3 Use clear labels, units, and mathematical language.

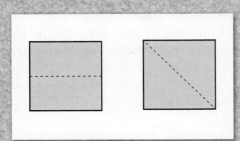

Sample "2 Equal Shares Poster"

Two Equal Shares

Name One Share

half
one-half
1-half
1 out of 2 parts

Name All Shares

two-halves
2-halves
2 out of 2 parts
whole

---

1 Warm Up   15–20 min   Go Online
ePresentations   eToolkit

▶ Mental Math and Fluency

Children write numbers in expanded form. *Leveled exercises*:

- ●○○ **124** 100 + 20 + 4; **283** 200 + 80 + 3
- ●●○ **399** 300 + 90 + 9; **207** Possible answers: 200 + 7 or 200 + 0 + 7
- ●●● **500** Possible answers: 500 or 500 + 0 + 0; **1,042** Possible answers: 1,000 + 40 + 2 or 1,000 + 0 + 40 + 2

▶ Daily Routines

Have children complete daily routines.

---

2 Focus   30–40 min   Go Online
ePresentations   eToolkit

▶ Math Message

Take 8 paper squares.

Two children want to share a sandwich equally. Fold a paper square to show how to divide the sandwich into 2 equal shares. Draw a line on the fold. Talk with a partner. Did you both fold the square the same way?

▶ Folding Squares into Equal Shares

| WHOLE CLASS | SMALL GROUP | PARTNER | INDEPENDENT |

Math Message Follow-Up Invite children to share the different ways they folded the squares. **GMP2.1** Expect children to fold a square into 2 **equal shares** in two ways: from side to side and diagonally. (*See margin.*)

Ask: *How can you show your partner that you have equal shares?* **GMP2.2** Sample answers: When folded, the 2 parts lie exactly on top of each other. We could cut along the fold, compare the 2 parts, and then see that they are the same size and shape.

Explain that children will divide squares into different numbers of equal shares and name the shares.

> **Professional Development** Children often apply whole-number reasoning to fractions. For example, they may say that $\frac{1}{2}$ is less than $\frac{1}{4}$ because 2 is less than 4. Using standard notation for fractions before development of conceptual understanding promotes such misconceptions. Children should begin by describing partitions of shapes in their own words while you model fraction vocabulary, such as *half, 1 out of 2 shares,* and *2 equal parts*. If children refer to standard notation (e.g., $\frac{1}{2}, \frac{1}{4}$), point out that they will see this notation in everyday life—but avoid using it in class.

| Differentiate | Go Online | for information about a common misconception.

▶ Naming 2, 4, and 3 Equal Shares

| WHOLE CLASS | SMALL GROUP | PARTNER | INDEPENDENT |

Begin a 2 Equal Shares poster on the Class Data Pad or chart paper. Have children refer to their paper squares from the previous activity. Ask the following questions. Record children's answers on the poster. (*See margin.*)

- *How can you name one child's share?* `GMP6.3` Sample answers: half; one-half; 1-half; 1 out of 2 equal parts

- *How can you name both shares together?* `GMP6.3` Sample answers: two-halves; 2-halves; 2 out of 2 equal parts; whole

If children do not mention all of the names shown on the sample poster, write one of the missing names and ask children why it should be on the poster. `GMP2.2` Repeat with any other missing names. As children share the names **one-half** and **two-halves,** point out that these can be written as number-and-word combinations: 1-half and 2-halves.

To complete the poster, attach examples of children's squares showing equal shares two different ways. As the unit progresses, add examples of other shapes partitioned into halves.

Draw a square divided into 2 unequal parts. Tell children it is a piece of toast. Ask: *Did I divide the toast into halves?* `GMP2.2` No. *Why or why not?* Sample answers: The 2 parts are not the same size. They are not equal shares. One piece is bigger than the other.

Next have children take another square. Ask: *How can 4 children share a sandwich equally?* Have partners fold several squares of paper to show as many different ways as possible to divide a sandwich into 4 equal parts. Tell them to draw lines along the folds to show the equal shares. `GMP2.1`

Observe partners as they work. If they do not suggest all of the solutions shown below, prompt them to look for other solutions.

Ask children how they can show that the shares of each paper square are equal. `GMP2.2` Sample answers: When folded, the parts will lie exactly on top of each other. We could cut along the folds, compare the shares, and see that they are the same size and shape.

Begin a 4 Equal Shares poster on the Class Data Pad. Ask the following questions. Record children's answers on the poster. (*See margin.*)

- *How can you name one child's share?* `GMP6.3` Sample answers: **one-fourth; 1-fourth; one-quarter; 1-quarter;** 1 out of 4 parts

- *How can you name all of the shares together?* `GMP6.3` Sample answers: **four-fourths; 4-fourths;** four-quarters; 4-quarters; 4 out of 4 parts; whole

---

Sample "4 Equal Shares Poster"

Four Equal Shares

Name One Share

one-fourth
1-fourth
one-quarter
1-quarter
1 out of 4 parts

Name All Shares

four-fourths
4-fourths
four-quarters
4-quarters
4 out of 4 parts
whole

NOTE If children use geometric terms such as *rectangle* or *triangle* to name the parts, make sure to acknowledge if they have correctly named the shape. Point out, however, that today the focus is on naming shares rather than naming shapes.

Sample "3 Equal Shares Poster"

Three Equal Shares

Name One Share

one-third
1-third
1 out of 3 parts

Name All Shares

three-thirds
3-thirds
3 out of 3 parts
whole

Showing Equal Parts

Lesson 9-1
DATE

① Divide each rectangle into 2 equal parts. Show two different ways.

Sample answers:

Write a name for 1 part.
one-half; 1-half; 1 out of 2 equal parts

Write a name for all of the parts together.
two-halves; 2-halves; 2 out of 2 equal parts; whole

② Divide each rectangle into 3 equal parts. Show two different ways.

Sample answers:

Write a name for 1 part.
one-third; 1-third; 1 out of 3 equal parts

Write a name for all of the parts together.
three-thirds; 3-thirds; 3 out of 3 equal parts; whole

③ Pick one rectangle from Problem 2. How could you show that the parts are equal? Sample answer: I can cut the rectangle into 3 parts and put them in a pile, and they will be the same size and shape.

220   two hundred twenty   2.G.3, SMP2, SMP6

Academic Language Development

To further children's understanding of the term *equal shares*, have partners complete 4-Square Graphic Organizers (*Math Masters*, page TA42) by including pictures, examples, nonexamples, and their own definitions of the term.

Showing Equal Parts (continued)

Lesson 9-1
DATE

④ Divide this rectangle into 4 equal parts.

Sample answers:

Write a name for 1 part.   one-fourth; 1-fourth;
1 out of 4 equal parts; one-quarter; 1-quarter

Write a name for all of the parts together.
four-fourths; 4-fourths; 4 out of 4 equal parts;
four-quarters; 4-quarters; whole

⑤ Ruthie divided this rectangle into 3 parts. She named one of the parts "one-third."

Do you agree with Ruthie? __No__
Explain your answer.
Sample answers: Two parts are the same size, but the third part is bigger. The parts are not equal.

Try This

⑥ Divide this circle into 4 equal parts.

2.G.3, SMP2, SMP6   two hundred twenty-one   221

If children do not mention all of the names shown on the sample poster, write one of the missing names and ask children why this name should be included on the poster. GMP2.2 Repeat with any other missing names.

To complete the poster, attach examples of children's squares showing 4 equal shares three different ways. As the unit progresses, add examples of other shapes partitioned into fourths.

Then discuss the idea of 3 equal shares. Ask: *How can 3 children share a sandwich equally?* Have partners use paper squares to show as many different ways as possible to divide a sandwich into 3 equal shares. GMP2.1 Before they begin, suggest that they think about what the 3 equal parts should look like.

Begin a 3 Equal Shares poster. Ask the following questions and record children's answers on the poster. (*See margin on previous page.*)

- *How can you name one child's share?* GMP6.3 Sample answers: **one- third; 1-third;** 1 out of 3 equal parts

- *How can you name all of the shares together?* GMP6.3 Sample answers: **three-thirds; 3-thirds;** 3 out of 3 equal parts; whole

If children do not mention all of the names shown on the sample poster, write one of the missing names and ask children why it should be included on the poster. GMP2.2 Repeat with any other missing names.

To complete the poster, trace a square and ask a volunteer to draw on it to show 3 equal shares. As the unit progresses, add examples of other shapes partitioned into thirds.

Next have children sort their squares into piles according to whether they show 2, 3, or 4 equal parts. Ask:

- *Is the whole the same for all of the squares you folded?* Yes. *How do you know?* GMP2.2 The squares are all the same size.

- *Are the shares the same for all of the squares you folded?* No. *How do you know?* GMP2.2 Sample answers: The shares are different shapes. The shares are different sizes when you share with more children. The more equal shares in the whole, the smaller the shares.

► ## Partitioning Shapes

Math Journal 2, pp. 220–221

WHOLE CLASS   SMALL GROUP   PARTNER   INDEPENDENT

Children partition shapes and, for each shape, name a single part and the whole. GMP2.1, GMP2.2, GMP6.3 Some children may benefit from folding $8\frac{1}{2}$"-by-11" sheets of paper to help them partition the shapes.

NOTE For Problem 4 some children may divide the rectangle into fourths using two diagonal lines. Others may say that the resulting parts are not equal. If this happens, tell children that even though they are different shapes, each part represents the same amount. Second-grade children do not yet have the tools to prove this formally.

✔ Assessment Check-In <small>CCSS</small> 2.G.3

Math Journal 2, pp. 220–221

Expect most children to be able to show one way to partition the rectangles on journal pages 220–221 into halves, fourths, and thirds and write one name for a part and one name for all of the parts together. **GMP2.1, GMP2.2, GMP6.3** Some children may be able to partition the rectangles in multiple ways and write more than one name for a part or for all of the parts together.

☑ Assessment and Reporting 〔Go Online〕 to record student progress and to see trajectories toward mastery for these standards.

Summarize Refer to the 2 Equal Shares, 4 Equal Shares, and 3 Equal Shares posters. Review the names on each poster for 1 share and for all shares together. Emphasize that these names can be used only for figures divided into parts that are equal. Have volunteers add rectangle drawings with appropriate partitions to each poster.

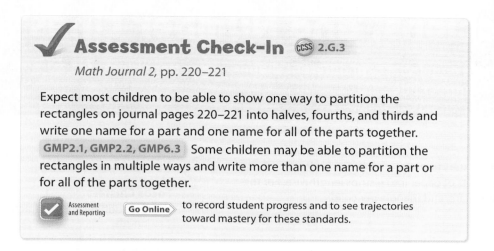

Math Journal 2, p. 222

③ Practice 10–15 min 〔Go Online〕

ePresentations   eToolkit   Home Connections

▶ Playing *Array Concentration*

Math Masters, p. G31

| WHOLE CLASS | SMALL GROUP | **PARTNER** | INDEPENDENT |

Children play *Array Concentration* to practice finding how many objects are in arrays and writing number models. See Lesson 8-10.

Observe

- Do children have efficient strategies for finding the total number of dots in an array? Which children need additional support?

Discuss

- *How did you find the total? Is there a faster way?* **GMP6.4**
- *How do you know that your number model matches the array?* **GMP2.3**

▶ Math Boxes 9-1

Math Journal 2, p. 222

| WHOLE CLASS | **SMALL GROUP** | **PARTNER** | **INDEPENDENT** |

Mixed Practice Math Boxes 9-1 are paired with Math Boxes 9-3.

▶ Home Link 9-1

Math Masters, p. 245

Homework Children partition shapes and name the equal parts.

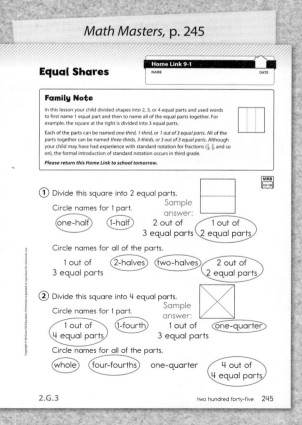

Math Masters, p. 245

Exploring Equal Shares, Pattern-Block Fractions, and Number Lines

Overview Children explore equal shares of different shapes, use pattern blocks to divide shapes, and make a number line.

▶ **Before You Begin**
For Explorations A and B, display the three Equal Shares posters from Lesson 9-1. For Exploration C, prepare 2 or 3 strips of paper for each child by cutting sheets of paper into 4 strips lengthwise. For the optional Readiness activity, cut out 1 large paper circle, square, and triangle per child.

Common Core State Standards

Focus Clusters
- Relate addition and subtraction to length.
- Reason with shapes and their attributes.

1 Warm Up 15–20 min

	Materials	
Mental Math and Fluency Children solve addition problems with three or four addends.	slate	**2.NBT.6**
Daily Routines Children complete daily routines.	See pages 4–43.	See pages xiv–xvii.

2 Focus 30–40 min

Math Message Children think about whether equal shares of a cracker can be different shapes.	*Math Journal 2*, p. 223	**2.G.3**
Explaining Equal Shares Children share their thinking about the Math Message.	*Math Journal 2*, p. 223; *Math Masters*, p. 248	**2.G.3**
Exploration A: Sharing Crackers Children divide crackers into equal parts and explain how they know the parts are equal.	Activity Card 112; *Math Masters*, pp. 249–250	**2.G.3** **SMP2**
Exploration B: Making Equal Parts Children use pattern blocks to divide shapes into equal parts.	Activity Card 113; *Math Masters*, pp. 251–252; Pattern-Block Template; pattern blocks; Equal Shares posters	**2.G.3** **SMP2**
Exploration C: Making a Number Line Children make number lines and label their halfway marks.	Activity Card 114; 2–3 paper strips (see Before You Begin)	**2.MD.6, 2.G.3** **SMP2**

3 Practice 10–15 min

Practicing with Fact Triangles Children practice facts using Fact Triangles.	*Math Journal 2*, pp. 250–253; Fact Triangles	**2.OA.2**
Math Boxes 9-2 Children practice and maintain skills.	*Math Journal 2*, p. 224	See page 789.
Home Link 9-2 **Homework** Children write names for parts of shapes.	*Math Masters*, p. 253	**2.G.3**

Differentiation Options

RtI

CCSS 2.G.3

Readiness
5–15 min

Folding Paper Pizzas

WHOLE CLASS
SMALL GROUP
PARTNER
INDEPENDENT

paper shapes
(see Before You Begin)

To explore dividing shapes into halves using a concrete model, children fold "pizza" circles, squares, and triangles into 2 equal parts each. Emphasize that for each pizza to be divided into halves, the 2 parts must be equal. Children decorate their pizzas with different ingredients on each half, for example, one-half sausage and one-half mushroom. Have children describe their pizzas and share their strategies for dividing shapes into 2 equal parts. Sample strategies: I folded the paper until the edges met all the way around. I looked at it and guessed and then folded to check my answer.

CCSS 2.G.3, SMP2, SMP7

Enrichment
5–15 min

Showing Fractions for One-Half

WHOLE CLASS
SMALL GROUP
PARTNER
INDEPENDENT

Math Masters, p. 246

To apply their understanding of equal shares that show one-half, children identify and create representations for one-half on *Math Masters,* page 246. **GMP2.1, GMP2.2** After children complete the page, have them discuss patterns they see in the figures. **GMP7.1**

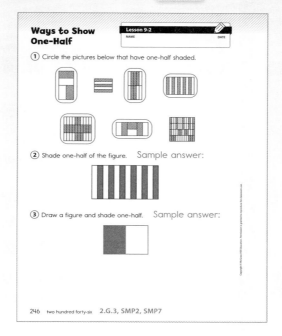

CCSS 2.G.3

Extra Practice
5–15 min

Finding Equal Parts of Shapes

WHOLE CLASS
SMALL GROUP
PARTNER
INDEPENDENT

Math Masters, p. 247

For practice finding equal parts of shapes, children partition irregular shapes on *Math Masters,* page 247. They explain how they know the parts are equal.

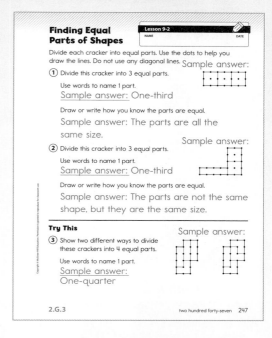

English Language Learners Support

Beginning ELL Provide additional experience with these terms from Lesson 9-1: *one-half, two-halves, one-third, three-thirds, one-fourth,* and *four-fourths.* Gather pictures of familiar items that are examples of each term and caption them accordingly (for example, one-half of an apple, two-halves of an orange, or four-fourths of a window pane). Distribute the pictures and use Total Physical Response prompts along with questions that do not require extended answers. Ask: *Who has two-halves of _____? Show me the card with one-fourth of _____. Show me a card with three-thirds of _____. What do you have three-thirds of?* Encourage children to repeat using short sentences, such as the following: *I have two-halves of an orange.*

> Go Online

> ELL English Learners Support

Standards and Goals for
Mathematical Practice

SMP2 Reason abstractly and quantitatively.

GMP2.1 Create mathematical representations using numbers, words, pictures, symbols, gestures, tables, graphs, and concrete objects.

GMP2.2 Make sense of the representations you and others use.

---

1 Warm Up · 15–20 min

Go Online — ePresentations · eToolkit

▶ Mental Math and Fluency

Display addition problems with three or four addends. Encourage children to look for combinations to make adding easier. *Leveled exercises:*

●○○ $15 + 15 + 10 = ?$ 40
$19 + 11 + 20 = ?$ 50
$10 + 17 + 23 = ?$ 50

●●○ $14 + 16 + 11 + 19 = ?$ 60
$16 + 24 + 22 + 28 = ?$ 90
$12 + 18 + 23 + 17 = ?$ 70

●●● $11 + 12 + 23 + 14 = ?$ 60
$21 + 11 + 23 + 24 = ?$ 79
$12 + 15 + 18 + 21 = ?$ 66

▶ Daily Routines

Have children complete daily routines.

---

2 Focus · 30–40 min

Go Online — ePresentations · eToolkit

▶ Math Message

Math Journal 2, p. 223

Solve the problem on journal page 223. Explain your thinking to a partner.

▶ Explaining Equal Shares

Math Journal 2, p. 223; *Math Masters,* p. 248

| WHOLE CLASS | SMALL GROUP | PARTNER | INDEPENDENT |

Math Message Follow-Up Invite children to share their thinking about whether Juan shared the cracker equally. Sample answer: Yes. Each person gets 3 cracker squares.

Some children may think that Juan didn't share the cracker equally because the pieces are different shapes. Guide the discussion so that children see that the 3 shares are equal in size. If no one mentions that each piece is composed of 3 smaller squares, display *Math Masters,* page 248 and use shading to show children the 3 small squares that make up each share.

After explaining the Explorations activities, assign groups to each one. Plan to spend most of your time with children working on Exploration A.

Math Journal 2, p. 223

Sharing a Cracker

Lesson 9-2
DATE

Math Message

Juan has a cracker he wants to share equally with 2 friends.

He divided the cracker in an unusual way:

Do you think Juan divided the cracker into 3 equal parts? Yes.

Explain your answer.
Sample answer: Each piece has 3 small squares of the same size, so everyone gets the same amount.

2.G.3

two hundred twenty-three 223

▶ # Exploration A: Sharing Crackers

Activity Card 112; *Math Masters,* pp. 249–250

| WHOLE CLASS | **SMALL GROUP** | **PARTNER** | INDEPENDENT |

Children divide shapes into equal parts. **GMP2.1** Explain that they can draw horizontal or vertical (but not diagonal) line segments to connect the dots on the crackers. After they divide the crackers, have children write fraction words to name parts of each cracker.

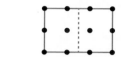

This line segment does not connect the dots.

This line segment is a diagonal line segment.

These line segments connect the dots properly.

Activity Card 112

Differentiate **Adjusting the Activity**

Some children may have difficulty partitioning the cracker in Problem 2 on *Math Masters,* page 250. Have them explain how they know the 2 halves of cracker in Problem 1 are equal. Children should see that each half of Cracker A is composed of 3 smaller squares.

Then have children compare the size of cracker in Problem 1 to the cracker in Problem 2. Ask: *Which cracker is larger?* Sample answers: They are the same size but different shapes. They are each made up of 6 smaller squares. Guide them to see that—just as in Cracker A—there are 3 small squares in each half of Cracker B.

Go Online ▶ Differentiation Support

Math Masters, p. 249

Exploration A: Sharing Crackers Lesson 9-2

NAME   DATE

① Divide the cracker below into 2 equal parts.

Sample answer:

Use words to name 1 part.
Sample answers: One-half; 1-half; 1 out of 2 equal parts
Draw or write how you know the parts are equal.
Sample answer: I can see that the top half is the same size as the bottom half.

② Divide the cracker below into 2 equal parts.

Sample answer:

Use words to name 1 part.
Sample answers: One-half; 1-half; 1 out of 2 equal parts
Draw or write how you know the parts are equal.
Sample answer: I see that there are 6 small squares inside the cracker. Each person would get 3 of the small squares.

2.G.3, SMP2   two hundred forty-nine   249

Math Masters, p. 250

Exploration A: Sharing Crackers (continued) Lesson 9-2

NAME   DATE

③ Divide the cracker below into 3 equal parts. Do it differently than Juan did in the Math Message.

Sample answer:

Use words to name ALL the parts.
Sample answers: Three-thirds; 3-thirds; 3 out of 3 equal parts; whole
Draw or write how you know the parts are equal.
Sample answer: I know that each part has to have 3 small squares, so I divided the shape into 3 parts with 3 squares in each part.

Try This

④ Divide the crackers below into 3 equal parts in two different ways.

Sample answers:

Use words to name 1 part.
Sample answers: One-third; 1-third; 1 out of 3 equal parts

250 two hundred fifty   2.G.3, SMP2

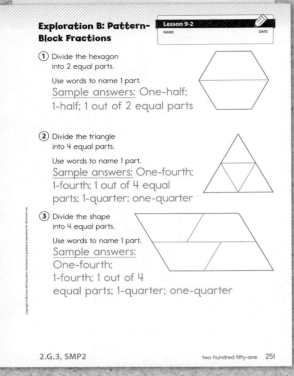

Exploration B: Pattern-Block Fractions

Lesson 9-2

NAME   DATE

① Divide the hexagon into 2 equal parts.

Use words to name 1 part.
<u>Sample answers:</u> One-half; 1-half; 1 out of 2 equal parts

② Divide the triangle into 4 equal parts.

Use words to name 1 part.
<u>Sample answers:</u> One-fourth; 1-fourth; 1 out of 4 equal parts; 1-quarter; one-quarter

③ Divide the shape into 4 equal parts.

Use words to name 1 part.
<u>Sample answers:</u> One-fourth; 1-fourth; 1 out of 4 equal parts; 1-quarter; one-quarter

2.G.3, SMP2   two hundred fifty-one   251

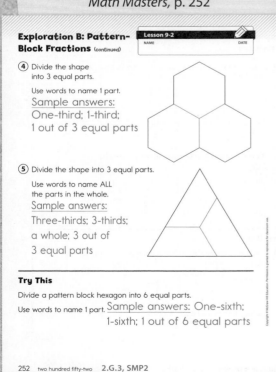

Exploration B: Pattern-Block Fractions (continued)

Lesson 9-2

NAME   DATE

④ Divide the shape into 3 equal parts.

Use words to name 1 part.
<u>Sample answers:</u> One-third; 1-third; 1 out of 3 equal parts

⑤ Divide the shape into 3 equal parts.

Use words to name ALL the parts in the whole.
<u>Sample answers:</u> Three-thirds; 3-thirds; a whole; 3 out of 3 equal parts

Try This

Divide a pattern block hexagon into 6 equal parts.
Use words to name 1 part. <u>Sample answers:</u> One-sixth; 1-sixth; 1 out of 6 equal parts

252   two hundred fifty-two   2.G.3, SMP2

► # Exploration B: Making Equal Parts

Activity Card 113; *Math Masters,* pp. 251–252

| WHOLE CLASS | **SMALL GROUP** | **PARTNER** | INDEPENDENT |

Children cover shapes with pattern blocks on *Math Masters,* pages 251–252 and use their Pattern-Block Template to record their work. GMP2.1

Remind children to use the Equal Shares posters from Lesson 9-1 to find words to name 1 part of each divided shape.

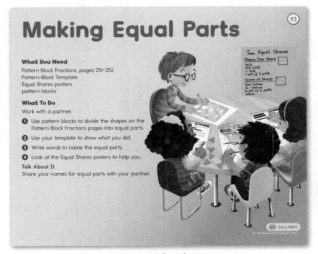

Activity Card 113

► # Exploration C: Making a Number Line

Activity Card 114

| WHOLE CLASS | **SMALL GROUP** | **PARTNER** | INDEPENDENT |

Children use paper strips to make number lines and label them with whole numbers. (*See Before You Begin.*) They discuss names for the numbers represented by tick marks between the whole numbers. GMP2.1, GMP2.2

Summarize Invite volunteers to discuss how they know that the crackers from Exploration A were divided into equal shares.

788   Unit 9 | Equal Shares and Whole Number Operations

③ Practice   10–15 min   〈Go Online〉   ePresentations   eToolkit   Home Connections

▶ Practicing with Fact Triangles

Math Journal 2, pp. 250–253

| WHOLE CLASS | SMALL GROUP | PARTNER | INDEPENDENT |

As children practice with Fact Triangles, have them use the Addition Facts Inventory Record, Parts 1 and 2 (journal pages 250–253) to take inventory of the addition facts they know and the facts that need more practice. See Lesson 3-3 for additional details.

▶ Math Boxes 9-2

Math Journal 2, p. 224

| WHOLE CLASS | SMALL GROUP | PARTNER | INDEPENDENT |

Mixed Practice Math Boxes 9-2 are paired with Math Boxes 9-4.

▶ Home Link 9-2

Math Masters, p. 253

Homework Children use fraction words to describe fractional parts of shapes.

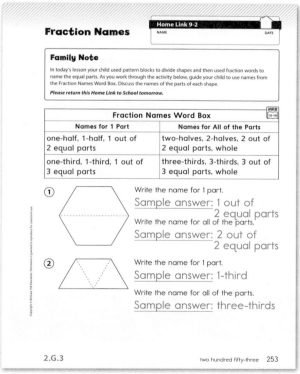

Math Masters, p. 253

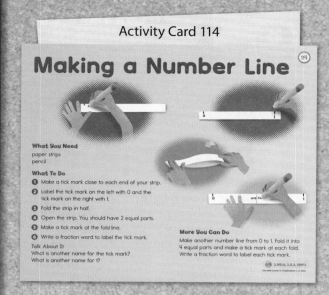

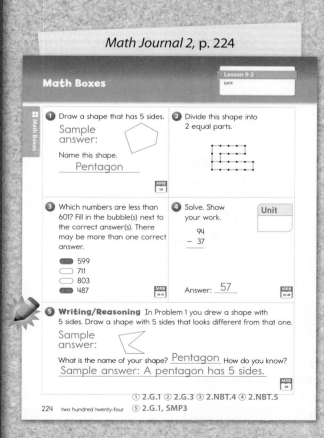

Math Journal 2, p. 224

Sharing Muffins

2-Day Lesson

Overview **Day 1:** Children decide how to share muffins equally and use words to name the shares. **Day 2:** The class discusses selected drawings and names, and children revise their work.

Day 1: Open Response

▶ **Before You Begin**
Cut out about 15 small circles and have them available for use during the open response problem. Display the Equal Shares posters from Lessons 9-1 and 9-2. If possible, schedule time to review children's work and plan for Day 2 of this lesson with your grade-level team.

▶ **Vocabulary**
equal shares • one-half • two-halves • one-fourth • four-fourths • one-quarter

CCSS Common Core State Standards

Focus Cluster
Reason with shapes and their attributes.

1 Warm Up 15–20 min	**Materials**	
Mental Math and Fluency Children solve addition problems with 3 or more addends.	slate	2.NBT.6
Daily Routines Children complete daily routines.	See pages 4–43.	See pages xiv–xvii.

2a Focus 45–55 min		
Math Message Children talk with a partner about ways of naming all the shares in a shape partitioned into thirds.	*Math Journal 2*, p. 225; Class Equal Shares posters	2.G.3 SMP1, SMP4
An Art Project Children discuss their ideas about ways of naming all the shares of a shape partitioned into thirds.	*Math Journal 2*, p. 225; Class Equal Shares posters	2.G.3 SMP1, SMP4
Solving the Open Response Problem Children show how to divide muffins equally among two and four children, describing each child's share.	*Math Masters*, pp. 254–255; Class Equal Shares posters; Standards for Mathematical Practice Poster; 15 cut out circles; scissors; glue	2.G.3 SMP4

Getting Ready for Day 2 →

Review children's work and plan discussion for reengagement. *Math Masters*, p. TA5; children's work from Day 1

CCSS 2.G.3 **Spiral Snapshot**

GMC Describe equal shares using fraction words.

| 9-1 Focus Practice | 9-2 Focus Practice | 9-3 Focus Practice | 9-4 Focus Practice | 9-5 through 9-7 Practice | 9-9 through 9-11 Practice |

 **Spiral Tracker** **Go Online** to see how mastery develops for all standards within the grade.

connectED.mheducation.com
Plan your lessons online with these tools.

 ePresentations Student Learning Center Facts Workshop Game eToolkit Professional Development Home Connections Spiral Tracker Assessment and Reporting ELL English Learners Support Differentiation Support

790 Unit 9 | Equal Shares and Whole Number Operations

1 Warm Up 15–20 min

Go Online

ePresentations   eToolkit

▶ Mental Math and Fluency

Display addition problems with three or four addends. Encourage children to look for combinations that make the addition easier.

● ○ ○ $10 + 12 + 10 = ?$ 32
$12 + 18 + 10 = ?$ 40
$13 + 17 + 12 = ?$ 42

● ● ○ $13 + 17 + 12 + 20 = ?$ 62
$14 + 23 + 21 + 27 = ?$ 85
$13 + 17 + 22 + 18 = ?$ 70

● ● ● $13 + 14 + 25 + 16 = ?$ 68
$24 + 14 + 26 + 27 = ?$ 91
$15 + 18 + 21 + 31 = ?$ 85

▶ Daily Routines

Have children complete daily routines.

2a Focus 45–55 min

Go Online

ePresentations   eToolkit

▶ Math Message

Math Journal 2, p. 225

Look at journal page 225. Talk with your partner about the problem.
GMP1.6, GMP4.2

▶ An Art Project

Math Journal 2, p. 225

| WHOLE CLASS | SMALL GROUP | PARTNER | INDEPENDENT |

Math Message Follow-Up Ask children to share their ideas about the Math Message problem. Discuss how both Jaylan and Leila had correct ways of naming the partitioned paper. **GMP1.6, GMP4.2** Use this discussion to review the representations and vocabulary on the Equal Shares posters.

Tell children that they are going to think more about how to make and name **equal shares.**

CCSS Standards and Goals for
Mathematical Practice

SMP1 Make sense of problems and persevere in solving them.
  GMP1.6 Compare the strategies you and others use.

SMP4 Model with mathematics.
  GMP4.1 Model real-world situations using graphs, drawings, tables, symbols, numbers, diagrams, and other representations.
  GMP4.2 Use mathematical models to solve problems and answer questions.

Professional Development

The focus of this lesson is **GMP4.2**. In the two previous lessons, children learned to partition a whole and name the equal shares. In this lesson, children make drawings to show equal shares of multiple wholes. Once they have made these visual models, they name the shares. Children should continue to use a variety of appropriate names for equal shares.

Go Online for information about **SMP4** in the *Implementation Guide.*

Math Journal 2, p. 225

An Art Project Lesson 9-3 DATE

A teacher is preparing to cut a piece of construction paper into three equal shares to use in an art project. The teacher drew dotted lines to guide the cuts.

Two children are talking about the paper. Jaylan says, "The paper is three-thirds." Leila says, "No, it's not. It's one whole."

Who is correct? Talk with a partner about your ideas.

2.G.3, SMP1, SMP4   two hundred twenty-five 225

▶ Solving the Open Response Problem

Math Masters, pp. 254–255

| WHOLE CLASS | SMALL GROUP | **PARTNER** | **INDEPENDENT** |

Distribute *Math Masters*, pages 254–255 to all children. Read the problem as a class and ask partners to discuss what the problem asks them to do. Encourage children to refer to the Equal Shares posters and use fraction vocabulary like that on the poster as they talk about and write responses to the problem. Review the terms **one-half, two-halves, one-fourth, four-fourths.** Tell children that an important part of the task is to write how much muffin is in one child's share.

NOTE Some children may describe a child's share in Problem 1 as "one-half" for at least two reasons. Children may incorrectly focus too narrowly on the amount that Anna or Sammy gets from one muffin that has been divided. On the other hand, children could describe a share as "one-half" because Anna and Sammy each get one-half of all of the muffins. This is correct if the children can express their thinking clearly. However, what we also want to know is, "How much muffin is one-half of all the muffins?" Appropriate responses are any equivalent way of saying one and one-half muffins (for example, three-halves or one whole and one-half). The same issue may arise in Problem 2 with children who describe a child's share as **"one-quarter"** or "one-fourth."

Math Masters, p. 255

Math Masters, p. 254

Circulate as children work. Ask children to explain their drawings and descriptions of one child's share, and encourage them to add details to clarify their responses. **GMP4.1, GMP4.2** You may also want to make notes about children's strategies.

Differentiate **Adjusting the Activity**

If children have trouble getting started, provide paper circles and scissors so they can act out the problem. Once they have acted it out, have them represent their actions with a drawing. If they still have difficulties, have them glue their circles and circle pieces to the paper. For children who struggle to use words to describe the shares, use the following sentence frame: "Each child gets _____." **GMP4.1**

Summarize *Ask: How did you show your work and thinking for this problem? Did you use words, symbols, or anything else?* **GMP4.1, GMP4.2** Answers vary. Refer children to the Standards for Mathematical Practice Poster for **GMP4.1** and **GMP4.2**.

Collect children's work so you can evaluate it and prepare for Day 2.

Getting Ready for Day 2

Math Masters, p. TA5

Planning a Follow-Up Discussion

Review children's work and any notes about their thinking. Use the Reengagement Planning Form (*Math Masters,* page TA5) and the rubric on page 795 to plan ways to help children meet expectations on both the content and practice standards. Look for misconceptions in children's descriptions of the equal shares as well as different correct ways children choose to share the muffins and name the shares.

Organize the discussion in one of the ways below or in another way you choose. If children's work is unclear or if you prefer to show work anonymously, rewrite the work for display.

Go Online for sample children's work that you can use in your discussion.

1. Display responses to Problem 1 that show different strategies for sharing the muffins and describing one child's share. See the sample work for Child A and Child B. Have children compare the strategies. Ask:

   - *How do you think Child A shared the muffins?* **GMP4.1** Sample answer: Child A gave Anna and Sammy each a whole muffin and split the other muffin into halves.

Academic Language Development

Use of the term "equal" within the context of equal shares may confuse students who often have used "equals" within the context of a number story, such as 3 plus 2 equals 5. It may be helpful to point out the varied uses of the term. In this lesson "equal shares" mean that the shares are the same size.

Sample child's work, Child A

Each peron gets one muffin and a half of a muffin.

Sample child's work, Child B

one half
one half

Sample child's work, Child C

one forth
one forth

- *Does Child B have the same strategy for sharing the muffins?* **GMP1.6, GMP4.1** Sample answer: No, Child B split each muffin into halves and gave a half of each muffin to Anna and the other 3 halves to Sammy.
- *How does the drawing for Child A show how much muffin Anna or Sammy gets?* **GMP4.1, GMP4.2** Sample answer: The arrows show how much muffin goes to each stick person.
- *How does the drawing for Child B show how much muffin Anna or Sammy gets?* **GMP4.1, GMP4.2** Sample answer: The letters show which pieces go to Anna and Sammy.
- *Do Child A and Child B agree or disagree about how much muffin goes to Anna or Sammy?* **GMP1.6, GMP4.2** Sample answer: They agree. One whole muffin a nd one-half muffin is the same amount as three muffin halves.
- *For Child A, do the words about a child's share match the drawing? How?* **GMP4.2** Sample answer: Yes, because the drawing and words show one muffin and a half of a muffin going to each child.
- *For Child B, do the words about a child's share match the drawing? How?* **GMP4.2** Sample answer: Not exactly. The words say "one-half," but the drawing shows three muffin halves go to Anna and three go to Sammy.

When discussing work like Child B's, talk about one child's share not only as one and one-half muffins, but also as "three-halves." In Problem 2, where a child may divide each of the five muffins into four pieces, talk about one child's share as "five-quarters" or "five-fourths."

2. Display responses to Problem 2. Discuss the strategies children use to share the muffins and how they describe one child's share. See the sample work for Child C. Have children interpret and compare the strategies. Ask:

- *What do you think this child is trying to show with this drawing?* **GMP4.2** Sample answer: The child puts the share for one child in a box. The circle in each box is a whole muffin, and the dark piece is a quarter from the split muffin in the middle.
- *Do you have any suggestions for how the drawing can be improved?* **GMP4.1** Sample answer: The muffin that is split is not split into equal pieces. The drawing should show equal quarters.
- *What do you think this child is trying to say with the words?* Sample answer: One-fourth is the amount of each of the pieces from the split muffin, but it isn't clear what the total share is for each child.
- *Do you have any suggestions for how the words can be improved?* **GMP4.2** Sample answer: The child needs to explain how much is in one of the boxes. It's one whole muffin and a quarter muffin.

Planning for Revisions

Have copies of *Math Masters*, pages 254–255 or extra paper available for children to use in revisions. You might want to ask children to use colored pencils so you can see what they revised.

Sharing Muffins

Overview   Day 2: The class discusses selected drawings and names, and children revise their work.

Day 2: Reengagement

▶ **Before You Begin**
Have extra copies available of *Math Masters*, pages 254–255 for children to revise their work.

Focus Cluster
Reason with shapes and their attributes.

2b Focus   50–55 min

Materials

Setting Expectations
Children review the open response problem and discuss what the drawing and descriptions should include. They review how to respectfully discuss others' work.

Guidelines for Discussions Poster

Reengaging in the Problem
Children compare strategies for partitioning the muffins and showing and describing each child's share.

selected samples of children's work

2.G.3
SMP1, SMP4

Revising Work
Children improve the clarity and completeness of their drawings and descriptions of each child's share.

Math Masters, pp. 254–255 (optional); children's work from Day 1; colored pencils (optional)

2.G.3
SMP4

✓ **Assessment Check-In**   See page 797 and the rubric below.

2.G.3
SMP4

Goal for Mathematical Practice	Not Meeting Expectations	Partially Meeting Expectations	Meeting Expectations	Exceeding Expectations
GMP4.2 Use mathematical models to solve problems and answer questions.	Does not provide a drawing that shows each child's share and does not name one share.	For both problems, provides drawings that show or attempt to show equal shares and uses words that name or attempt to name one share, but one of the following deficiencies may be present: • One or both drawings may not show shares of equal size, or are incomplete. • Appropriate names for one share are not given for one or both problems.	For both problems, provides drawings that show equal shares and words that name one share using appropriate, specific language.	Meets expectations and provides more than one appropriate way of naming one share.

3 Practice   10–15 min

Math Boxes 9-3
Children practice and maintain skills.

Math Journal 2, p. 226

See page 796.

Home Link 9-3
Homework Children partition a rectangle and give appropriate fraction names for one part and for the whole.

Math Masters, p. 256

2.NBT.5, 2.G.3

NOTE These Day 2 activities will ideally take place within a few days of Day 1. Prior to beginning Day 2, see Planning a Follow-Up Discussion from Day 1.

2a Focus

50–55 min   Go Online

ePresentations   eToolkit

▶ Setting Expectations

| WHOLE CLASS | SMALL GROUP | PARTNER | INDEPENDENT |

Briefly review the open response problem from Day 1. Ask: *What do you think a complete answer to this problem needs to include?* Sample answer: Problems 1 and 2 should each have a drawing showing how to share the muffins, and there should be words saying how much one child's share is.

Tell children that they are going to look at other children's work and think about different strategies for sharing the muffins and describing the shares. Point out that children can use different drawings and words for the shares and still be correct. Refer to your list of discussion guidelines and encourage children to use these sentence frames:

- Could you explain _____?
- I like how you showed _____.

▶ Reengaging in the Problem

| WHOLE CLASS | SMALL GROUP | PARTNER | INDEPENDENT |

Children reengage in the problem by analyzing and critiquing other children's work in pairs and in a whole-group discussion. Have children discuss with partners before sharing with the whole group. Guide this discussion based on the decisions you made in Getting Ready for Day 2. **GMP1.6, GMP4.1, GMP4.2**

▶ Revising Work

| WHOLE CLASS | SMALL GROUP | PARTNER | INDEPENDENT |

Pass back children's work from Day 1. Before children revise anything, ask them to look carefully at their responses and decide how to improve them. Ask the questions below one at a time. Have partners discuss their responses and give a thumbs-up or thumbs-down based on their own work.

- *Did you show how to share three muffins between Sammy and Anna?*
- *Did you explain how much muffin is in one child's share? Did you use words from the Equal Shares posters?*
- *Are your drawing and explanation clear enough that someone else could understand them?*
- *Did you show how to share five muffins among four children?*
- *Did you explain how much muffin is in one child's share?*
- *Are your drawing and explanation clear enough that someone else could understand them?* **GMP4.1, GMP4.2**

Tell children they now have a chance to revise their work. Tell them to add to their earlier work using colored pencils or to use another sheet of paper, instead of erasing their original work.

Math Journal 2, p. 226

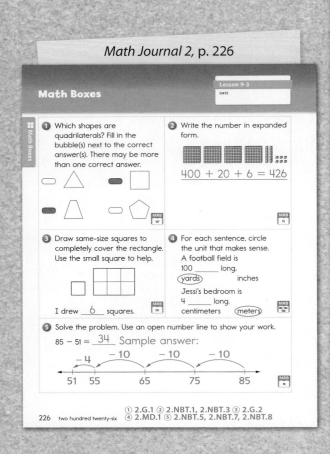

✓ Assessment Check-In CCSS 2.G.3

Collect and review children's revised work. Expect children to complete or improve their explanations based on the class discussion. For the content standard, expect most children to partition one or all muffins into two equal shares in Problem 1, and into four equal shares in Problem 2. Also expect children to use appropriate language (halves, fourths, quarters) as they talk about the shares in the discussion. You can use the rubric on page 795 to evaluate children's revised work for **GMP4.2.**

 Assessment and Reporting (Go Online) to record student progress and to see trajectories toward mastery for these standards.

(Go Online) for optional generic rubrics in the *Assessment Handbook* that can be used to assess any additional GMPs addressed in the lesson.

Sample Children's Work—Evaluated

See the sample in the margin. This work meets expectations for the content standard by showing one muffin appropriately partitioned into two parts in Problem 1 and four parts in Problem 2 (the latter is shown at the bottom right). The work meets expectations for the mathematical practice by including drawings for Problems 1 and 2 showing one child's share. In each case, the child clearly showed one muffin and the appropriate fraction of a muffin going to one person. The child used appropriate language to name a share as "one and a half" in Problem 1 and "one and a fourth" in Problem 2. GMP4.2

(Go Online) for other samples of evaluated children's work.

Summarize Ask children to reflect on their work and revisions. Ask: *What did you do to improve your drawing or names for equal shares?* Answers vary.

③ Practice 10–15 min

(Go Online) ePresentations eToolkit Home Connections

▶ Math Boxes 9-3

Math Journal 1, p. 226

| WHOLE CLASS | SMALL GROUP | PARTNER | INDEPENDENT |

Mixed Practice Math Boxes 9-3 are paired with Math Boxes 9-1.

▶ Home Link 9-3

Math Masters, p. 256

Homework Children partition a rectangle and give appropriate fraction names for one part and for the whole.

Sample child's work, "Meeting Expectations"

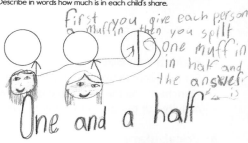

① Anna and Sammy want to share 3 muffins. Use the circles below to show how they can share the muffins fairly. Describe in words how much is in each child's share.

② Four children want to share 5 muffins. Show how they can share the muffins fairly. Describe in words how much is in each child's share.

Math Masters, p. 256

Naming Equal Shares Home Link 9-3

Family Note

In this lesson we continued making and naming equal shares of rectangles and circles. Your child showed and described how to share 3 muffins equally between 2 children, and 5 muffins equally among 4 children. By solving and discussing problems like these, your child will learn appropriate fraction vocabulary, such as 1 out of 2 equal shares, one-half, 1-third, one-quarter, 1-fourth, and one out of four equal shares. Practice making and naming fractional amounts will continue to the end of the year and will lead to a great deal of work with fractions in *Third Grade Everyday Mathematics.*

Please return this Home Link to school tomorrow.

① Divide the rectangle into 4 equal parts. Sample answer:

② How could you test that the parts are equal? Sample answer: Cut the rectangle out and fold it along the lines to see if the parts are the same size.

③ Use words to name one of the parts in at least two ways. Sample answer: 1-fourth, one-quarter

④ Use words to name all of the parts together. Sample answers: four out of four equal shares, 4-fourths

Practice Unit

⑤ 73
   + 34
   ‾‾‾
   107

⑥ 90
   − 43
   ‾‾‾
   47

⑦ 46
   + 36
   ‾‾‾
   82

256 two hundred fifty-six 2.NBT.5, 2.G.3

Fractional Units of Length

Overview Children measure lengths to the nearest half-inch.

▶ **Before You Begin**

For the Focus activities, label two different areas of the room with signs reading "1 inch" and "2 inches." You may want to copy *Math Masters*, page 259 onto cardstock. For the optional Readiness activity, cut out one 3-inch paper strip for each child.

▶ **Vocabulary**

half-inch • fourth-inch • precise • quarter-inch

CCSS Common Core State Standards

Focus Clusters
- Measure and estimate lengths in standard units.
- Relate addition and subtraction to length.

1 Warm Up   15–20 min

	Materials	
Mental Math and Fluency Children solve addition problems with three or four addends.	slate	2.NBT.6
Daily Routines Children complete daily routines.	See pages 4–43.	See pages xiv–xvii.

2 Focus   30–40 min

Math Message Children measure the length of their pinky finger.	12-inch ruler	2.MD.1 SMP5
Discussing the Need for Precise Measurements Children discuss how to make more-precise measurements.	12-inch ruler; *Math Journal 2*	2.MD.1 SMP5, SMP6
Introducing Half-Inches Children identify half- and quarter-inches on a 12-inch ruler.	*Math Masters*, p. TA33; 12-inch ruler	2.MD.6
Measuring to the Nearest Half-Inch Children measure objects to the nearest half-inch.	*Math Journal 2*, p. 227; *Math Masters*, p. 259; 12-inch ruler; scissors	2.MD.1, 2.MD.4 SMP5, SMP6
✓ **Assessment Check-In** See page 803.	*Math Journal 2*, p. 227	2.MD.1, SMP5, SMP6

CCSS 2.MD.1   **Spiral Snapshot**

GMC Measure the length of an object.

4-8 through 4-11 Focus Practice | 5-8 Practice | 6-10 Focus | 7-4 through 7-6 Focus Practice | 7-9 Focus | 9-4 Focus Practice

 Spiral Tracker   **Go Online** ▷ to see how mastery develops for all standards within the grade.

3 Practice   10–15 min

Partitioning Shapes into Equal Shares Children partition circles and describe the equal shares.	*Math Journal 2*, p. 228	2.G.3
Math Boxes 9-4 Children practice and maintain skills.	*Math Journal 2*, p. 229	See page 803.
Home Link 9-4 **Homework** Children measure line segments and objects to the nearest inch and half-inch.	*Math Masters*, p. 260	2.MD.1 SMP5, SMP6

 **connectED.mheducation.com** ▷

Plan your lessons online with these tools.

 ePresentations   Student Learning Center   Facts Workshop Game   eToolkit   Professional Development   Home Connections   Spiral Tracker   Assessment and Reporting   **ELL** English Learners Support   Differentiation Support

Differentiation Options

RtI

Readiness   5–15 min

WHOLE CLASS
SMALL GROUP
PARTNER
INDEPENDENT

Comparing Lengths of Objects

Activity Card 115;
Math Masters, p. 257;
3-inch paper strip; 12-inch ruler

For practice estimating and comparing lengths, children assemble a small group of objects and develop a strategy for comparing their lengths. After completing *Math Masters*, page 257, children share their comparisons. **GMP6.1**

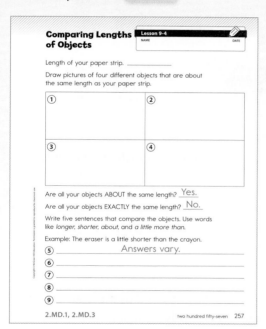

Comparing Lengths of Objects   Lesson 9-4
NAME   DATE

Length of your paper strip. _____

Draw pictures of four different objects that are about the same length as your paper strip.

| ① | ② |
| ③ | ④ |

Are all your objects ABOUT the same length? Yes.
Are all your objects EXACTLY the same length? No.

Write five sentences that compare the objects. Use words like *longer*, *shorter*, *about*, and *a little more than*.

Example: The eraser is a little shorter than the crayon.

⑤ _____ Answers vary. _____
⑥ _____
⑦ _____
⑧ _____
⑨ _____

2.MD.1, 2.MD.3   two hundred fifty-seven 257

Enrichment   5–15 min

WHOLE CLASS
SMALL GROUP
PARTNER
INDEPENDENT

Measuring a Crooked Path

Math Masters, p. 258;
12-inch ruler

To apply knowledge of linear measurement, children measure a crooked path to the nearest half-inch on *Math Masters*, page 258. Explain how they can put together pairs of half-inches to make a whole inch when calculating the total length of the path. After measuring the sections and calculating the total length, children create their own crooked paths for a partner to measure to the nearest half-inch.

Measuring a Crooked Path   Lesson 9-4
NAME   DATE

① A flea hopped along this path from one spot to another. Measure each part of the path to the nearest half-inch.

About 4 and 1-half inches
About 3 inches
About 1 inches
About 2 inches
About 4 and 1-half inches
About 3 and 1-half inches

② How far did the flea hop all together? Sample answer:
About 18 and 1-half inches

③ Draw your own crooked path. Have a partner measure the length of your crooked path to the nearest half-inch.
Answers vary.

The total length of my crooked path is about
_____ Answers vary. _____ inches.

258 two hundred fifty-eight 2.MD.1

Extra Practice   5–15 min

WHOLE CLASS
SMALL GROUP
PARTNER
INDEPENDENT

Drawing and Measuring to the Nearest Half-Inch

12-inch ruler, paper

For additional practice measuring length, children draw and measure a four-sided shape. Have each child draw a four-sided shape on paper and then trade drawings with a partner. The partner measures and records the length of each side of the shape to the nearest half-inch. Have children name the four-sided shapes.
Sample answers: Quadrilateral; square; rectangle; rhombus; trapezoid; kite

English Language Learners Support

Beginning ELL Scaffold for children to understand the meaning of *nearest* by beginning with a contrast of *near* and *far*. Identify children who are seated far from a point in the classroom, such as the door, and make think-aloud statements such as this: _____ *is seated far from the door.* Contrast with another child seated about half the distance from the door, saying: _____ *is nearer to the door.* Then point to another child seated very close to the door, saying: *But* _____ *is nearest to the door.* Ask questions that require children to make a choice, such as the following: *Who is nearest to the* _____: _____, _____, *or* _____?

Go Online   ELL English Learners Support

Standards and Goals for
Mathematical Practice

SMP5 Use appropriate tools strategically.
  GMP5.2 Use tools effectively and make sense of your results.
SMP6 Attend to precision.
  GMP6.2 Use an appropriate level of precision for your problem.
  GMP6.3 Use clear labels, units, and mathematical language.

1 Warm Up   15–20 min

Go Online   ePresentations   eToolkit

▶ Mental Math and Fluency

Display addition problems with three or four addends. Encourage children to look for combinations that will make the addition easier. *Leveled exercises:*

●○○ $12 + 18 + 20 = ?$ 50   ●●○ $13 + 27 + 21 + 19 = ?$ 80
    $17 + 13 + 10 = ?$ 40 $12 + 18 + 23 + 17 = ?$ 70
    $16 + 24 + 20 = ?$ 60 $26 + 24 + 32 + 18 = ?$ 100

●●● $11 + 22 + 23 + 33 = ?$ 89
    $17 + 21 + 28 + 11 = ?$ 77
    $19 + 16 + 32 + 27 = ?$ 94

▶ Daily Routines

Have children complete daily routines.

2 Focus   30–40 min

Go Online   ePresentations   eToolkit

▶ Math Message

Measure the length of your pinky finger. Is it about 1 inch or 2 inches long?
GMP5.2

▶ Discussing the Need for Precise Measurements

| WHOLE CLASS | SMALL GROUP | PARTNER | INDEPENDENT |

Math Message Follow-Up Have children stand in the area of the classroom with the label that corresponds to their pinky-finger measurement (*see Before You Begin*). GMP5.2 Ask: *Can we tell from the measurements which child has the longest pinky finger?* No. *The shortest?* No. *Why?* GMP5.2 Sample answer: Because more than one child has a pinky finger that measures about 1 inch, and more than one child has a pinky finger that measures about 2 inches. Some children might suggest determining who has the longest and the shortest pinky fingers by comparing them. Explain that this could work for the children in the class, but it wouldn't be practical if we were trying to determine the shortest and longest pinky fingers for children in a different school.

Explain that **precise** measurements can help determine which child has the shortest and which child has the longest pinky finger, regardless of where the children are located. Tell children that a measurement is more *precise* when it is made using a smaller unit. For example, a measurement to the nearest inch is more precise than a measurement to the nearest foot.

Ask children to measure the length of their journals (from top to bottom, not side to side) to the nearest foot and to the nearest inch. About 1 foot; about 11 inches Say: *Imagine that a child from another school tells you that their math books are about 1 foot tall, and another child from the same school tells you that the same math books are about 10 inches tall.* Ask: *Can you tell from the measurements to the nearest foot whether our math journals are taller than their math books?* GMP6.2 No. Both our journals and their books are about 1 foot tall. *Can you tell from the measurements to the nearest inch whether our math journals are taller than their math books?* GMP6.2 Yes. We know their books are shorter because they are only about 10 inches tall, but our journals are about 11 inches tall.

Point out that measurement tools are made by people. There is a limit to what people can observe and what tools can do, so all measurements are approximate—close to the actual measurement but not exact. However, some tools are better than others for making more-precise measurements because they are marked with smaller units, and measurements that are more precise tell us more about the "exact" measurement of an object than less-precise measurements do.

Tell children that they will explore a measurement unit that will allow them to make more-precise measurements.

▶ Introducing Half-Inches

Math Masters, p. TA33

| WHOLE CLASS | SMALL GROUP | PARTNER | INDEPENDENT |

Explain that measuring in half-inches, rather than inches or feet, produces more-precise measurements.

Display *Math Masters,* page TA33 and have children examine the inch ruler shown on it. Ask a volunteer to point to the mark that divides an inch on the ruler into 2 equal parts. Point out that this mark is called the "**half-inch** mark." Ask: *How long is the part between the 0 mark and the half-inch mark?* 1-half-inch *The part between the half-inch mark and the 1-inch mark?* 1-half-inch Use your fingers to trace the spaces between the 0 and half-inch marks and between the half-inch and 1-inch marks as children count the divisions chorally: 1-half, 2-halves. Ask: *How many half-inches make 1 whole inch?* 2-half-inches

Next ask: *How many spaces are marked between the 0 and 1-inch marks?* 4 *Are these spaces equal in length?* Yes. Have children look for the marks that divide the inch into 4 equal parts. Ask: *How long is the space between two such marks?* 1-**fourth-inch** or 1-**quarter-inch** Use your fingers to trace the quarter-inch spaces between the 0 and 1-inch marks and count the divisions of the inch in unison: *1-fourth, 2-fourths, 3-fourths, 4-fourths.* Ask: *How many fourth-inches make 1 whole inch?* 4-fourth-inches

You may want to add a ruler divided into quarter-inches to the 4 Equal Shares poster from Lesson 9-1.

Math Masters, p. TA33

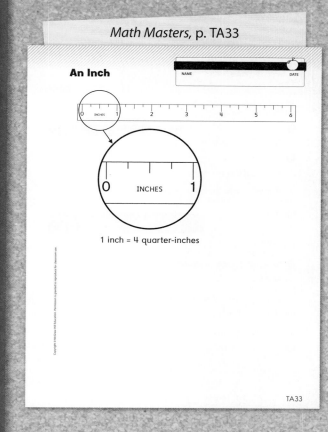

Math Masters, p. 259

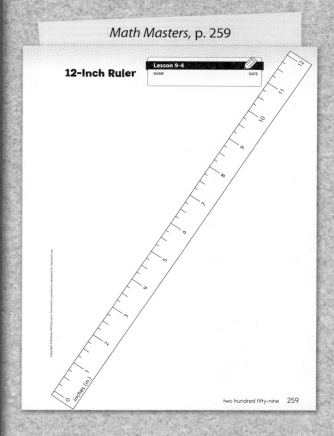

Math Journal 2, p. 227

Measuring Lengths
Lesson 9-4
DATE

Use your 12-inch ruler to measure the lengths of these objects to the nearest inch.

1. Crayon

   CRAYON

   About __4__ inches long

2. Eraser

   ERASER

   About __2__ inches long

Use your 12-inch ruler to measure the lengths of these objects to the nearest half-inch.

3. Marker

   About _____ inches long Possible
   answers: 4 and one-half; 4 and 1-half

4. Paintbrush

   About _____ inches long Possible
   answers: 5 and one-half; 5 and 1-half

Measure the length of your desk to the nearest half-inch.

5. My desk is about __Answers vary.__ inches long.

Use the measurements above to complete Problem 6.

6. The crayon is about __2__ inches longer than the eraser.

2.MD.1, 2.MD.4, SMP5, SMP6   two hundred twenty-seven 227

Math Journal 2, p. 228

Partitioning Circles into Halves, Thirds, and Fourths
Lesson 9-4
DATE

1. Divide this circle into 2 equal parts. Sample answers:

   Write a name for 1 part.
   1-half; 1 out of 2 equal parts

   Write a name for all of the parts together.
   two-halves; whole

2. Divide the circle into 4 equal parts. Sample answers:

   Write a name for 1 part.
   one-fourth; one-quarter

   Write a name for all of the parts together.
   4 out of 4 equal parts; 4 quarters

3. Which circle is divided into thirds (or 3 equal parts)? Circle A

   Sample
   answers:

   Circle A   Circle B

   How do you know? The 3 parts in Circle A are all the
   same size. Circle B is divided into 2 small parts
   and 1 bigger part.

   For the circle divided into thirds, write a name for all of the parts.
   three-thirds; 3 out of 3 equal parts

228 two hundred twenty-eight   2.G.3

▶ # Measuring to the Nearest Half-Inch

Math Journal 2, p. 227; Math Masters, p. 259

WHOLE CLASS | **SMALL GROUP** | **PARTNER** | INDEPENDENT

Have children cut out the 12-inch ruler on *Math Masters,* page 259.

Tell children that they can use the half-inch marks on the 12-inch ruler to help them measure to the nearest inch. Remind them to first line up one end of the object with the 0 mark. If the other end of the object goes past (or to the right of) the half-inch mark, the measurement is the next bigger number. If the end of the object stops before (or to the left) of the half-inch mark, the measurement is the next smaller number. Then explain that when measuring to the nearest half-inch, if the end of the object goes past the fourth-inch mark, the measurement is the next bigger half-inch. If the end of the object stops before the fourth-inch mark, the measurement is to the next smaller half-inch. Emphasize that nearest half-inch measurements may be whole numbers.

Both of the following are measurements to the nearest half-inch (the rulers are not actual size):

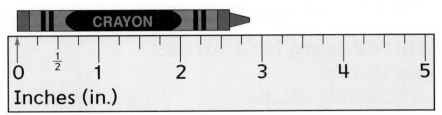

This crayon is about 3 inches long.

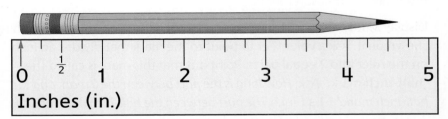

This pencil is about 4 and 1-half-inches long.

Have children use their 12-inch paper ruler to measure the lengths of objects on journal page 227 to the nearest inch and half-inch.
GMP5.2, GMP6.2, GMP6.3

Differentiate **Common Misconception**

Watch for children who mistakenly line up one end of the object with the end of the ruler instead of with the 0 mark. Consider highlighting the 0 mark with a colored marker to draw children's attention to it. Then remind children to carefully line up the 0 mark with one end of the object they are measuring.

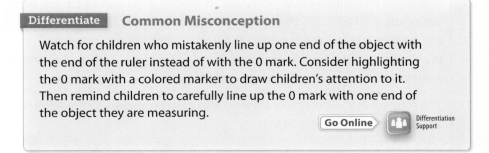

Go Online ▸ Differentiation Support

Have children measure their pinky fingers again, this time to the nearest half-inch. Ask them to raise their hands if their pinky is about 1-half inch, about 1 inch, about 1 and 1-half inches, or about 2 inches long. Ask: *Which measurement is more precise—to the nearest inch or to the nearest half-inch?* Nearest half-inch *Do the half-inch measurements tell who has the longest pinky? Explain.* Sample answers: No, two measurements to the nearest half-inch could be the same, with one pinky longer than the other.

✓ Assessment Check-In   CCSS 1.OA.6, 1.NBT.2, 1.OA.3

Math Journal 2, p. 227

Observe children as they measure the objects on journal page 227. Expect most children to correctly measure the objects in Problems 1–2 to the nearest inch. GMP5.2, GMP6.2, GMP6.3 If they struggle measuring to the nearest inch, consider using the suggestion in the Common Misconception note. Some children may be able to measure the objects in Problems 3–5 to the nearest half-inch. Others may connect the idea of a half-inch to the fraction work they did in previous lessons.

Assessment and Reporting   Go Online to record student progress and to see trajectories toward mastery for these standards.

Summarize Have volunteers share their measurements from journal page 227.

3 Practice   10–15 min   Go Online   ePresentations   eToolkit   Home Connections

▶ Partitioning Shapes into Equal Shares

Math Journal 2, p. 228

WHOLE CLASS | SMALL GROUP | PARTNER | INDEPENDENT

Refer children to the Equal Shares posters created in Lesson 9-1 and review the different ways of naming fractional parts. Then have them complete journal page 228.

▶ Math Boxes 9-4

Math Journal 2, p. 229

WHOLE CLASS | SMALL GROUP | PARTNER | INDEPENDENT

Mixed Practice Math Boxes 9-4 are paired with Math Boxes 9-2.

▶ Home Link 9-4

Math Masters, p. 260

Homework Children measure line segments and objects to the nearest inch and half-inch. GMP5.2, GMP6.2, GMP6.3

Math Journal 2, p. 229
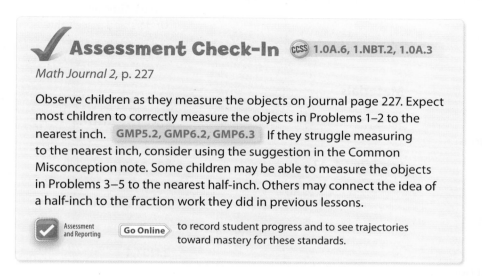

Math Masters, p. 260
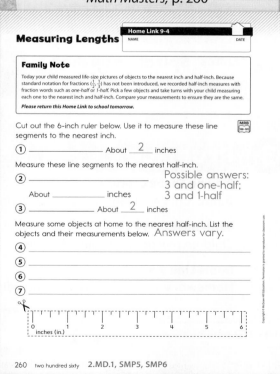

Reviewing Place Value

Overview Children write multidigit numbers in expanded form and compare them.

▶ **Before You Begin**
For the Focus activities, gather a set of base-10 blocks (1 thousand cube, 9 flats, 9 longs, and 9 small cubes) to use for demonstration purposes. Consider borrowing additional thousand cubes from other classrooms.

▶ **Vocabulary**
thousand cube

Common Core State Standards

Focus Cluster
Understand place value.

1 Warm Up   15–20 min

	Materials	
Mental Math and Fluency Children write numbers from place-value descriptions.	slate	2.NBT.1, 2.NBT.3
Daily Routines Children complete daily routines.	See pages 4–43.	See pages xiv–xvii.

2 Focus   30–40 min

Math Message Children identify the value of each digit in a 3-digit number.		2.NBT.1
Reviewing Place Value and Expanded Form Children use digit values to write a number in expanded form.		2.NBT.1, 2.NBT.3 SMP2
Representing Multidigit Numbers Children represent multidigit numbers with base-10 blocks and in expanded form.	*Math Masters*, p. 262; base-10 blocks for demonstration (see Before You Begin); 1 each of number cards 0–9; slate	2.NBT.1, 2.NBT.1a, 2.NBT.3 SMP2
Comparing Multidigit Numbers Children use expanded form to compare multidigit numbers.	*Math Journal 2*, p. 230; slate	2.NBT.1, 2.NBT.3, 2.NBT.4 SMP2, SMP6
✓ **Assessment Check-In** See page 809.	*Math Journal 2*, p. 230; base-10 blocks (optional)	2.NBT.3, 2.NBT.4, SMP2

CCSS 2.NBT.3 **Spiral Snapshot**

GMC Read and write numbers in expanded form.

4-4 through 4-6 Focus Practice | 4-8 Practice | 6-8 Focus | 9-1 Warm Up Practice | 9-5 Focus | 9-6 Warm Up Focus | 9-7 Focus Practice

⫶⫶⫶ Spiral Tracker   Go Online   to see how mastery develops for all standards within the grade.

3 Practice   10–15 min

Playing *Shape Capture* **Game** Children identify shapes by their attributes.	*Math Masters*, p. G28; per group: Attribute Cards, Shape Cards	2.G.1 SMP7
Math Boxes 9-5 Children practice and maintain skills.	*Math Journal 2*, p. 231	See page 809.
Home Link 9-5 **Homework** Children work with multidigit numbers.	*Math Masters*, p. 263	2.NBT.1, 2.NBT.4, 2.NBT.5

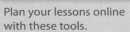
 ePresentations   Student Learning Center   Facts Workshop Game   eToolkit   Professional Development   Home Connections   Spiral Tracker   Assessment and Reporting   English Learners Support   Differentiation Support

Differentiation Options

RtI

CCSS 2.NBT.3, 2.NBT.7, 2.NBT.8, 2.NBT.9, SMP6   **CCSS** 2.NBT.1, 2.NBT.3, 2.NBT.7, SMP7   **CCSS** 2.NBT.1, 2.NBT.3, 2.NBT.4, SMP7

Readiness   5–15 min

WHOLE CLASS
SMALL GROUP
PARTNER
INDEPENDENT

Base-10 "Buildings"

Activity Card 17;
base-10 cubes, longs, and flats;
quarter-sheets of paper

To explore place value using a concrete model, children build base-10 "buildings" with flats, longs, and cubes. Partners explain to each other how they know they counted the blocks accurately. **GMP6.4** See Lesson 1-12 Exploration A for details.

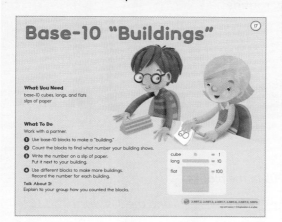

Enrichment   5–15 min

WHOLE CLASS
SMALL GROUP
PARTNER
INDEPENDENT

Exploring Place Value

Math Masters, p. 261; calculator

To extend their work with place value, children complete *Math Masters,* page 261. They add and subtract 10, 100, and 1,000 to and from numbers. **GMP7.1**

Exploring Place Value   Lesson 9-5

① Record the numbers that are 10 more, 100 more, and 1,000 more. You may use a calculator to help. Circle the digits that changed.

Number	10 More	100 More	1,000 More
135	1④5	②85	①135
307	3①7	④07	①307
2,369	2,3⑦9	2,④69	③,369
7,897	7,9⓪7	7,⑨97	⑧,897

② Record the numbers that are 10 less, 100 less, and 1,000 less. You may use a calculator to help. Circle the digits that changed.

Number	10 Less	100 Less	1,000 Less
4,572	4,5⑥2	4,④72	③,572
2,643	2,⑥3	2,⑤43	①,643
5,949	5,9③9	5,⑧49	④,949
Try This			
3,004	2,⑨94	2,⑨04	②,004

③ Write about some patterns that you notice. Sample answer: When I add or subtract 1,000, the thousands digit changes. When I add or subtract 100, the hundreds digit changes. When I add or subtract 10, the tens digit changes. Sometimes other digits change, too.

2.NBT.1, 2.NBT.3, 2.NBT.7, SMP7   two hundred sixty-one   261

Extra Practice   5–15 min

WHOLE CLASS
SMALL GROUP
PARTNER
INDEPENDENT

Playing *Number Top-It*

My Reference Book, pp. 170–172; *Math Masters,* pp. G7–G8; 4 each of number cards 0–9; glue or tape

For additional practice comparing multidigit numbers, children play *Number Top-It*. **GMP7.2** See Lesson 4-5 for details. As an extension, children can play the game by using 4 cards to make 4-digit numbers.

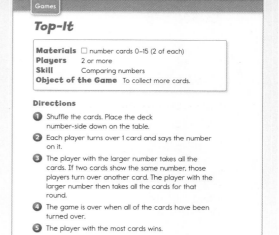

Games

Top-It

Materials	☐ number cards 0–15 (2 of each)
Players	2 or more
Skill	Comparing numbers
Object of the Game	To collect more cards.

Directions

① Shuffle the cards. Place the deck number-side down on the table.

② Each player turns over 1 card and says the number on it.

③ The player with the larger number takes all the cards. If two cards show the same number, those players turn over another card. The player with the larger number then takes all the cards for that round.

④ The game is over when all of the cards have been turned over.

⑤ The player with the most cards wins.

MRB 170   one hundred seventy

English Language Learners Support

Beginning ELL Review the terms *flat, long,* and *cube* as the names of base-10 blocks before introducing the term *thousand cube.* Use teacher modeling to name the blocks; then have children repeat the names. Use statements and questions such as these: *This is a flat. What is this?* Accept one-word responses and follow with teacher restatements using complete sentences for children to repeat. Assess children's knowledge of the block names by showing different blocks one at a time and asking questions requiring one-word or short-response answers.

 Go Online   ELL   English Learners Support

SMP2 Reason abstractly and quantitatively.

GMP2.1 Create mathematical representations using numbers, words, pictures, symbols, gestures, tables, graphs, and concrete objects.

GMP2.2 Make sense of the representations you and others use.

SMP6 Attend to precision.

GMP6.4 Think about accuracy and efficiency when you count, measure, and calculate.

1 Warm Up   15–20 min   Go Online   ePresentations   eToolkit

▶ Mental Math and Fluency

Read aloud the following descriptions of numbers. Have children write the numbers on their slates. *Leveled exercises:*

● ○ ○ Write a number with 5 in the hundreds place, 3 in the tens place, and 1 in the ones place.   531

● ● ○ Write a number with 7 in the hundreds place, 0 in the tens place, and 4 in the ones place.   704

● ● ● Write a number with 4 in the tens place, 9 in the hundreds place, and 5 in the ones place.   945

▶ Daily Routines

Have children complete daily routines.

2 Focus   30–40 min   Go Online   ePresentations   eToolkit

▶ Math Message

Tell a partner how you would fill in the blanks.

In 573, the 5 is worth _____, the 7 is worth _____, and the 3 is worth _____.

▶ Reviewing Place Value and Expanded Form

| WHOLE CLASS | SMALL GROUP | PARTNER | INDEPENDENT |

Math Message Follow-Up Have children share their answers. Ask:

- *What is the value of the 5?* 500 *How do you know?* Sample answer: The 5 is in the hundreds place.

- *What is the value of the 7?* 70 *How do you know?* Sample answer: The 7 is in the tens place.

- *What is the value of the 3?* 3 *How do you know?* Sample answer: The 3 is in the ones place.

Remind children that one representation of a number that shows the value of each digit is called *expanded form*. Ask a volunteer to write the expanded form for 573.   GMP2.1   500 + 70 + 3

Tell children that they will identify the values of digits and represent multidigit numbers with base-10 blocks and expanded form.

▶ Representing Multidigit Numbers

Math Masters, p. 262

WHOLE CLASS | **SMALL GROUP** | PARTNER | INDEPENDENT

Distribute a Place-Value Mat (*Math Masters,* page 262) to each child. Hold up a base-10 cube. Ask: *What is this base-10 block called?* A cube *What is its value?* 1 Repeat with a long and a flat. A long is worth 10. A flat is worth 100.

> **Differentiate** **Adjusting the Activity**
>
> For children who struggle to associate base-10 blocks with their names and values, make a classroom poster similar to the table on *My Reference Book,* page 71. Include a thousand cube if you are introducing it as well.
>
> **Go Online** ▸ [icon] Differentiation Support

Ask three volunteers to come to the front of the room. The first child, on the left as viewed by the class, holds 3 flats for all to see. The child in the middle holds up 5 longs, and the child on the right holds up 2 cubes. Ask: *What number do these blocks represent?* 352 Have children say the number aloud in unison and show it with number cards on their Place-Value Mats. Next ask: *How can these base-10 blocks help us write the expanded form for this number?* The 3 flats show 3 hundreds, so we write 300. The 5 longs show 5 tens, so we add 50. The 2 cubes show 2 ones, so we add 2. Have children write the expanded form on their slates. 300 + 50 + 2 Repeat this activity with several 3-digit numbers.

Reverse the procedure by displaying a 3-digit number and asking three volunteers to come to the front of the room and show the number with base-10 blocks. Then have all children show the number with number cards on their Place-Value Mats and write the number in expanded form on their slates. Include examples with 0 as a digit. *Suggestions:*

- 7 flats, 1 long, 6 cubes 716; 700 + 10 + 6
- 4 flats, 0 longs, 9 cubes 409; 400 + 0 + 9 or 400 + 9
- 150 1 flat, 5 longs, 0 cubes; 100 + 50 + 0 or 100 + 50
- 903 9 flats, 0 longs, 3 cubes; 900 + 0 + 3 or 900 + 3

If most children seem comfortable representing 3-digit numbers with base-10 blocks, have three volunteers at the front of the room represent 264 with base-10 blocks while the rest of the class shows this number on their Place-Value Mats. Hold up the thousand cube. Ask: *What do you think this block is worth?* 1,000 *How do you know?* Sample answers: It is made up of 1,000 cubes. It is 10 flats put together or 100 longs put together, and that's 1,000 in all. Tell children that this block is called a **thousand cube;** it can be used to help represent larger numbers.

Math Masters, p. 262

Place-Value Mat Lesson 9-5 NAME DATE

1s

10s

100s

1,000s

262 two hundred sixty-two

Academic Language Development

To deepen children's understanding of the term *expanded form,* have them complete a 4-Square Graphic Organizer (*Math Masters,* page TA42) for a given 4-digit number. Have children use base-10 shorthand to fill in the picture quadrant. For the definition quadrant, suggest this sentence starter: "When we expand a number, we _____."

Common Misconception

Differentiate Some children may think that the thousand cube is worth only 600 because they see only the six faces of the cube. Have children count by 100s as you stack 10 flats together to show them that the thousand cube consists of 10 flats, which is worth 1,000.

Go Online ▸ Differentiation Support

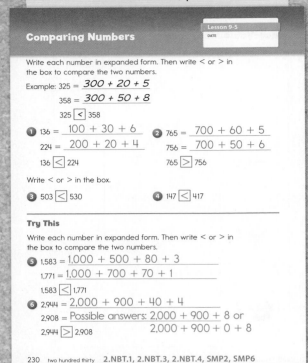

Math Journal 2, p. 230

Choose a fourth volunteer to come to the front of the room. Position this child at the left end of the line, holding up the thousand cube. Ask: *How many thousands did we add to our number?* 1 thousand Have children place a 1 in the thousands column of their Place-Value Mats. **GMP2.1** Ask: *What is the value of the 1?* **GMP2.2** 1,000 Prompt children to read the number aloud and write it in expanded form on their slates. **GMP2.1** One thousand two hundred sixty-four; $1,000 + 200 + 60 + 4$

To extend the activity for children who are ready, repeat it with other 4-digit numbers. Also reverse the procedure as you did for 3-digit numbers. The size of the numbers you can use will depend on the number of thousand cubes you have. If you have only one thousand cube, use numbers less than 2,000. If you have more thousand cubes, you can pose larger numbers. (*See Before You Begin.*) Suggestions:

- 1 thousand cube, 6 flats, 2 longs, 7 cubes 1,627; $1,000 + 600 + 20 + 7$
- **1,408** 1 thousand cube, 4 flats, 0 longs, 8 cubes; $1,000 + 400 + 8$ or $1,000 + 400 + 0 + 8$

▶ Comparing Multidigit Numbers

Math Journal 2, p. 230

| WHOLE CLASS | SMALL GROUP | PARTNER | INDEPENDENT |

Display the numbers 292 and 289. Ask children to write the expanded form for each number on their slates, with the hundreds, tens, and ones for each number aligned vertically. **GMP2.1** (*See margin.*) Ask: *How can we use the expanded form of each number to help us compare them?* **GMP2.2** Sample answer: First we can look at the hundreds. Both numbers have 2 hundreds, or 200, so next we look at the tens. We see that 292 has 9 tens, or 90, but 289 has 8 tens, or 80. So we know 292 is larger. Ask: *Do we need to look at the ones?* **GMP6.4** No. The tens told us that 292 is larger. Have children write a number sentence using < or > to compare the two numbers. $292 > 289$ Repeat this activity with other pairs of 3-digit numbers as needed.

If children seem comfortable comparing 3-digit numbers, display the numbers 3,948 and 3,953 and have children write the expanded form of each on their slates with the place-value parts aligned. **GMP2.1** (*See margin.*) Ask: *Where should we start if we want to compare these numbers?* With the thousands *How can we use the expanded form of each number to see which is larger?* **GMP2.2** Sample answer: Both numbers have 3 thousands and 9 hundreds. But 3,948 has 4 tens while 3,953 has 5 tens, so that means 3,953 is larger. Have children write a number sentence to record the comparison. $3,948 < 3,953$

To challenge children who are ready, pose additional problems involving comparisons of two 4-digit numbers. Then have all children complete journal page 230.

$292 = 200 + 90 + 2$

$289 = 200 + 80 + 9$

$3,948 = 3,000 + 900 + 40 + 8$

$3,953 = 3,000 + 900 + 50 + 3$

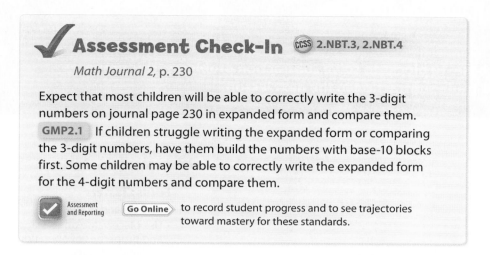

Math Journal 2, p. 231

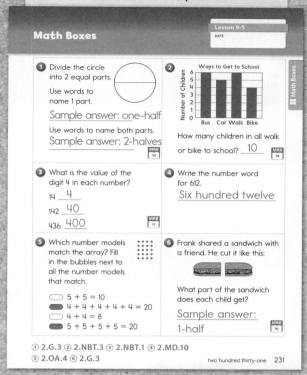

✔ Assessment Check-In CCSS 2.NBT.3, 2.NBT.4

Math Journal 2, p. 230

Expect that most children will be able to correctly write the 3-digit numbers on journal page 230 in expanded form and compare them.

GMP2.1 If children struggle writing the expanded form or comparing the 3-digit numbers, have them build the numbers with base-10 blocks first. Some children may be able to correctly write the expanded form for the 4-digit numbers and compare them.

✔ Assessment and Reporting   Go Online ▸ to record student progress and to see trajectories toward mastery for these standards.

Summarize Have children share their thoughts about which digit they should start with when they compare numbers. Sample answer: I should start with the digits in the largest place. If that digit is bigger in one number, then I know that number is the larger one right away.

③ Practice 10–15 min   Go Online ▸ ePresentations eToolkit Home Connections

▸ Playing *Shape Capture*

Math Masters, p. G28

Children identify attributes in shapes by playing *Shape Capture*.
GMP7.1, GMP7.2 The game is played with two players or two teams of two players each. See Lesson 8-2 for additional details.

Observe

- Which children can correctly find shapes with specified attributes?
- Which children are checking other team's or player's selections?

Discuss

- *How did you make sure the other team or player was capturing shapes that matched the Attribute Cards?*
- *Which shapes were easy to capture? Which were harder to capture? Why?*

▸ Math Boxes 9-5

Math Journal 2, p. 231

WHOLE CLASS | SMALL GROUP | PARTNER | INDEPENDENT

Mixed Practice Math Boxes 9-5 are paired with Math Boxes 9-7.

▸ Home Link 9-5

Math Masters, p. 263

Homework Children represent and compare multidigit numbers.

Math Masters, p. 263

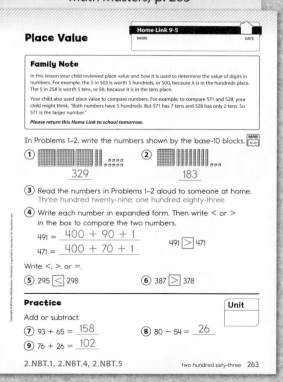

Lesson 9-6

Expand-and-Trade Subtraction, Part 1

Overview Children use base-10 blocks to solve subtraction problems. This prepares them to learn expand-and-trade subtraction in the next lesson.

Common Core State Standards

Focus Clusters
- Understand place value.
- Use place value understanding and properties of operations to add and subtract.

1 Warm Up 15–20 min

	Materials	
Mental Math and Fluency Children write numbers in expanded form.	slate	2.NBT.1, 2.NBT.3
Daily Routines Children complete daily routines.	See pages 4–43.	See pages xiv–xvii.

2 Focus 35–40 min

Math Message Children represent numbers using base-10 blocks.	base-10 blocks	2.NBT.1, 2.NBT.1a, 2.NBT.1b, SMP1
Representing Trades with Base-10 Blocks Children use base-10 blocks to represent place-value trades.	base-10 blocks	2.NBT.1, 2.NBT.1a, 2.NBT.3, 2.NBT.7, SMP2
Representing Subtraction without Trades Children use base-10 blocks to subtract without trades.	Class Data Pad; base-10 blocks	2.NBT.1, 2.NBT.1b, 2.NBT.5, 2.NBT.7, 2.NBT.9, SMP2
Representing Subtraction with Trades Children use base-10 blocks to subtract with trades.	Class Data Pad; base-10 blocks	2.NBT.1, 2.NBT.1a, 2.NBT.5, 2.NBT.7, 2.NBT.9, SMP2
Subtracting with Base-10 Blocks Children use base-10 blocks to subtract and use ballpark estimates to check the reasonableness of their answers.	*Math Journal 2*, p. 232; base-10 blocks	2.NBT.1, 2.NBT.1a, 2.NBT.1b, 2.NBT.5, 2.NBT.7, 2.NBT.9 SMP1
✓ **Assessment Check-In** See page 816.	*Math Journal 2*, p. 232	2.NBT.1a, 2.NBT.5, 2.NBT.7

CCSS 2.NBT.7 **Spiral Snapshot**

GMC Subtract multidigit numbers using models or strategies.

| 7-3
Warm Up | 7-7
Practice | 7-9
Practice | 8-3
Practice | 8-6
Practice | 8-10
Practice | 9-6
Focus
Practice | 9-7
Focus
Practice |

/// Spiral Tracker   Go Online to see how mastery develops for all standards within the grade.

3 Practice 10–15 min

Drawing a Line Plot Children represent class head-size measures on a line plot.	*Math Journal 2*, p. 193; *Math Masters*, p. TA32; stick-on notes	2.MD.9
Math Boxes 9-6 Children practice and maintain skills.	*Math Journal 2*, p. 233	See page 817.
Home Link 9-6 **Homework** Children practice subtraction with trades.	*Math Masters*, p. 265	2.NBT.1, 2.NBT.1a, 2.NBT.5, 2.NBT.7, 2.NBT.9

connectED.mheducation.com

Plan your lessons online with these tools.

 ePresentations   Student Learning Center   Facts Workshop Game   eToolkit   Professional Development   Home Connections   Spiral Tracker   Assessment and Reporting   English Learners Support   Differentiation Support

Differentiation Options

RtI

CCSS 2.NBT.1, 2.NBT.1a, SMP2

CCSS 2.NBT.1, 2.NBT.1a, 2.NBT.7, 2.NBT.9

CCSS 2.NBT.1, 2.NBT.5, 2.NBT.7, 2.NBT.9

Readiness · 10–15 min

| WHOLE CLASS |
| SMALL GROUP |
| **PARTNER** |
| INDEPENDENT |

Trading with Base-10 Blocks

Math Masters, p. TA22; base-10 blocks

For experience with trading using a concrete model, children use base-10 blocks to represent the same number in different ways. Have children display 3 flats, 2 longs, and 4 cubes on the Place-Value Mat on *Math Masters,* page TA22. Ask: *What number do these blocks represent?* 324 Then have them trade one of the longs for 10 cubes. Ask: *Do these blocks still represent the same number?* Sample answer: Yes, because 10 cubes have the same value as 1 long. Have children count all of the blocks to verify that they still represent 324. Ask: *Are there any other trades we could make and still show the same number?* Sample answers: Trade another long for 10 cubes; trade a flat for 10 longs.

Have children suggest other exchanges, make the trades, and count the blocks together to show that they represent the same number. Have them display multiple representations of base-10 blocks for other numbers, such as 365 and 289. **GMP2.2**

Enrichment · 10–20 min

| WHOLE CLASS |
| SMALL GROUP |
| **PARTNER** |
| INDEPENDENT |

Subtracting Multidigit Numbers

Activity Card 116;
Math Masters, p. 264;
base-10 blocks;
4 each of number cards 0–9

To further explore subtraction with multidigit numbers, children subtract 3-digit numbers from 4-digit numbers using base-10 blocks. They record their solutions to the problems on *Math Masters,* page 264 and explain how they used the base-10 blocks to solve the problems.

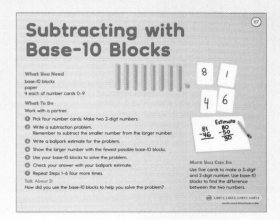

Extra Practice · 5–15 min

| WHOLE CLASS |
| SMALL GROUP |
| **PARTNER** |
| INDEPENDENT |

Subtracting with Base-10 Blocks

Activity Card 117,
base-10 blocks, paper,
4 each of number cards 0–9

For practice subtracting 2-digit numbers, partners use base-10 blocks to solve subtraction problems. Children make ballpark estimates to check the reasonableness of their answers and explain what they did.

English Language Learners Support

Beginning ELL To review the meaning of the term *trade,* present examples of trades using coins and bills. For example, trade 2 dimes and a nickel for 1 quarter or 10 one-dollar bills for 1 ten-dollar bill. Model first with real-world situations and think-aloud statements such as the following: *I need some quarters for the parking meter. I have only dimes and nickels. I need to make a trade.* Follow modeling with a request such as this: *Who can make a trade with me for a dime?* Provide sentence frames and encourage children to use them, such as "I can trade you _____ for _____."

  English Learners Support

Standards and Goals for
Mathematical Practice

SMP1 **Make sense of problems and persevere in solving them.**

GMP1.4 Check whether your answer makes sense.

GMP1.5 Solve problems in more than one way.

SMP2 **Reason abstractly and quantitatively.**

GMP2.1 Create mathematical representations using numbers, words, pictures, symbols, gestures, tables, graphs, and concrete objects.

GMP2.2 Make sense of the representations you and others use.

GMP2.3 Make connections between representations.

---

① Warm Up 15–20 min [Go Online] ePresentations eToolkit

▶ Mental Math and Fluency

Children write numbers in expanded form. *Leveled exercises:*

● ○ ○ **257** 200 + 50 + 7; **358** 300 + 50 + 8

● ● ○ **508** Possible answers: 500 + 8 or 500 + 0 + 8; **876** 800 + 70 + 6

● ● ● **1,090** Possible answers: 1,000 + 90 or 1,000 + 0 + 90 + 0; **2,007** Possible answers: 2,000 + 7 or 2,000 + 0 + 0 + 7

▶ Daily Routines

Have children complete daily routines.

---

② Focus 35–40 min [Go Online] ePresentations eToolkit

▶ Math Message

Use base-10 blocks to show 221 at least three different ways. Use base-10 shorthand to record your work. **GMP1.5** Sample answers:

☐ ☐ | | .

☐ | | | | | | | | | | | .

☐ ☐ | ....

▶ Representing Trades with Base-10 Blocks

| WHOLE CLASS | SMALL GROUP | PARTNER | INDEPENDENT |

Math Message Follow-Up Have children share their different representations. **GMP2.1** Record several examples, making sure to include the following two representations:

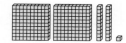

Guide a discussion about how children might translate from one representation to the other. **GMP2.3** For example, if they started with 2 flats, 2 longs, and 1 cube, they could trade 1 flat for 10 longs. They could also start with 1 flat, 12 longs, and 1 cube and trade in 10 longs for 1 flat. Have children model the trades with their base-10 blocks. Ask: *Which uses the fewest blocks possible? How do you know?* The example with 2 flats, 2 longs, and 1 cube uses the smallest number of blocks. In the other representations, we had to trade in 1 long for 10 cubes or 1 flat for 10 longs, so those used more blocks. *How could we use expanded form to show the representation with the smallest number of blocks?* **GMP2.3** 200 + 20 + 1

Display 200 + 20 + 1 and point out that each number in the expanded form shows the value of one type of block. For example, 200 represents the value of the 2 flats. Ask: *What number model could we write for the other representation?* **GMP2.3** 100 + 120 + 1 Display 100 + 120 + 1 and make the connection between the addends and each type of block.

▶ Representing Subtraction without Trades

| WHOLE CLASS | SMALL GROUP | PARTNER | INDEPENDENT |

Remind children that although there are many different ways to represent a number using base-10 blocks, they can use the fewest possible blocks by matching the number of each type of block to the digits in the number. Have children use the fewest possible base-10 blocks to represent the number 45. **GMP2.1** 4 longs and 5 cubes Record 4 longs and 5 cubes in base-10 shorthand for the class.

Ask: *Are there enough longs and cubes for me to remove 2 longs and 2 cubes?* Yes. *How do you know?* There are 4 longs, so there are enough to take away 2 longs. There are 5 cubes, so there are enough to take away 2 cubes. Then ask children how they would use their blocks to show 45 − 22. As they respond, record these steps on the Class Data Pad:

Show: 45 − 22		
Child says:	**Teacher records:**	**Teacher explains:**
First I show 45 with longs and cubes.	‖‖‖ ▪▪▪▪▪	
I take away 2 longs because I have to subtract 20.	Cross out 2 of the longs on your drawing. ‖‖⧅ ▪▪▪▪▪	We had 4 longs to begin with, so we have enough to remove 2 of the longs.
I also take away 2 cubes because I have to subtract 2.	Cross out 2 of the cubes. ‖‖⧅ ▪▪▪⊠⊠	We had 5 cubes to begin with, so we have enough to remove 2 of the cubes.
I have 2 longs and 3 cubes left, so the answer is 23.	Count the rest of the blocks to find the answer. Display this number model: 45 − 22 = 23	

Repeat this process with other subtraction problems that do not require a trade, such as 65 − 31 and 138 − 17. Discuss children's representations as a class as you record the steps on the Class Data Pad.

▶ Representing Subtraction with Trades

WHOLE CLASS   SMALL GROUP   PARTNER   INDEPENDENT

Tell children they will now use their base-10 blocks to solve 53 − 37. Ask children to represent 53 with base-10 blocks. When they have finished, record a sketch of 5 longs and 3 cubes.

‖‖‖‖‖ ▪▪▪

Ask: *Are there enough longs and cubes for me to remove 3 longs and 7 cubes?* No. There are only 3 cubes, so I can't remove 7 cubes. *How can I get more cubes so I can remove 7 cubes?* GMP2.3 Trade 1 long for 10 cubes Have children make the trade with their base-10 blocks. Represent this trade on your sketch by crossing out 1 long and adding 10 cubes.

Ask: *Do our blocks still show the number 53?* Yes. We made a trade, but the blocks still show the same number. *Do we have enough blocks so that we can remove 3 longs and 7 cubes (37) now?* GMP2.2 Yes. We now have enough cubes to take away 7 cubes. Complete the subtraction of 37 by removing 3 longs and 7 cubes.

Count the remaining blocks with children. Record the number sentence 53 − 37 = 16.

Repeat this process with other subtraction problems that require trades, such as 72 − 38 and 114 − 86. With the latter problem, make sure children understand that they will need to make two trades: 1 long for 10 cubes and 1 flat for 10 longs. Discuss children's representations as a class while recording the steps on the Class Data Pad.

▶ Subtracting with Base-10 Blocks

Math Journal 2, p. 232

| WHOLE CLASS | SMALL GROUP | PARTNER | INDEPENDENT |

Have children recall how they can check their answers for reasonableness. GMP1.4 By making ballpark estimates Remind them that making ballpark estimates can be helpful when they use any addition or subtraction method. If their answers are not close to their ballpark estimates, then children know they need to look back at their work and fix something.

Display the problem 143 − 65. As a class, make a ballpark estimate (either 140 − 60 = 80 or 140 − 70 = 70 because 65 is between 60 and 70). Then have children model and solve the problem with base-10 blocks. 78 Compare their answers to the estimate. Ask: *Is 80 (or 70) close to 78? Does the answer seem reasonable?* GMP1.4 Yes. The answer is close to the estimate, so it seems reasonable. I know the answer has to be much less than 143 because 65 is subtracted from it.

Have children complete the problems on journal page 232.

Math Journal 2, p. 232

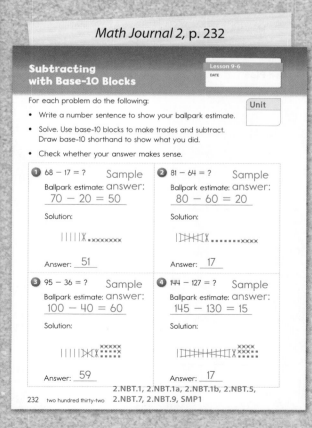

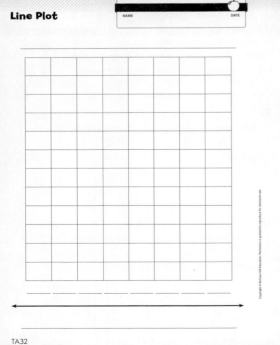

Math Journal 2, p. 193

Measuring Body Parts

Lesson 7-9

DATE

Complete the chart. Answers vary.

Body Part	Centimeters	Inches
Wrist	Estimate: About .... centimeters Measurement: About .... centimeters	Estimate: About .... inches Measurement: About .... inches
Arm span	Estimate: About .... centimeters Measurement: About .... centimeters	Estimate: About .... inches Measurement: About .... inches
Around head	Estimate: About .... centimeters Measurement: About .... centimeters	Estimate: About .... inches Measurement: About .... inches
Waist	Estimate: About .... centimeters Measurement: About .... centimeters	Estimate: About .... inches Measurement: About .... inches
Height	Estimate: About .... centimeters Measurement: About .... centimeters	Estimate: About .... inches Measurement: About .... inches

2.MD.1, 2.MD.2, 2.MD.3, 2.MD.9, SMP5   one hundred ninety-three   193

Math Masters, p. TA32

Line Plot

NAME   DATE

TA32

✓ Assessment Check-In  2.NBT.1a, 2.NBT.5, 2.NBT.7

Math Journal 2, p. 232

Circulate and observe as children complete journal page 232. Expect most children to be able to determine whether a trade is needed. Do not expect that all children will correctly solve the problems. In Grade 2 children are expected to fluently subtract within 100 using strategies based on place value, properties of operations, and/or the relationship between addition and subtraction. If children struggle to use base-10 blocks to subtract, suggest that they first solve each problem using any method they please and then use base-10 blocks to solve it again.

It is important for children to be exposed to and try different strategies so that, for any given problem, they can use the strategy that is most efficient for them. Support children by helping them model the steps with their base-10 blocks and record them using base-10 shorthand.

✓ Assessment and Reporting   〈Go Online〉 to record student progress and to see trajectories toward mastery for these standards.

Summarize Have children discuss how they know whether or not they need to make a trade. Sample answer: We need to make a trade if there are not enough blocks to remove what we need to take away.

③ Practice   10–15 min   〈Go Online〉 ePresentations eToolkit Home Connections

▶ Drawing a Line Plot

Math Journal 2, p. 193; *Math Masters,* p. TA32

| WHOLE CLASS | SMALL GROUP | PARTNER | INDEPENDENT |

Have children turn to journal page 193 and record their head-size measurements in centimeters on a stick-on note. Remind them to write large. Help children display the stick-on notes in order from smallest to largest. Tell children that they will draw line plots to show class head-size measurements.

Distribute *Math Masters,* page TA32. Display *Math Masters,* page TA32 or draw a line for the line plot. Have children suggest a label for the horizontal axis and write it below the line. Sample answer: Head Size in Centimeters Then have them suggest a title for the line plot and record it. Sample answer: Children's Head Sizes

Next discuss the horizontal scale for the line plot. The head-size data include measurements to the nearest centimeter. The scale should begin with the smallest head size in the class, increase in 1-centimeter increments, and end with the largest head size in the class. Model writing the scale while children do the same. Have children draw Xs to represent the class data on their line plots.

▶ Math Boxes 9-6

Math Journal 2, p. 233

WHOLE CLASS	SMALL GROUP	PARTNER	INDEPENDENT

Mixed Practice Math Boxes 9-6 are paired with Math Boxes 9-8.

▶ Home Link 9-6

Math Masters, p. 265

Homework Children solve subtraction problems. They show their work with base-10 shorthand.

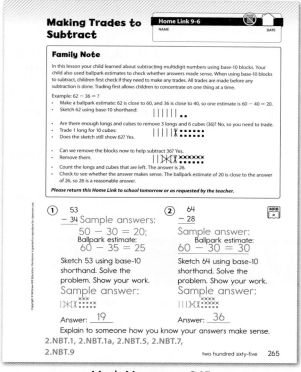

Math Masters, p. 265

Math Journal 2, p. 233

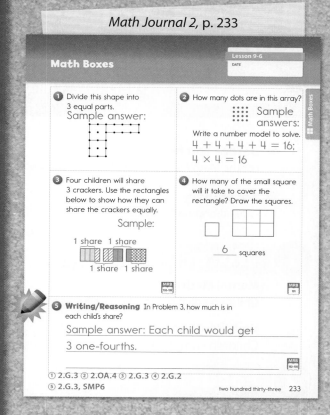

Expand-and-Trade Subtraction, Part 2

Overview Children use expand-and-trade subtraction to subtract multidigit numbers.

▶ **Before You Begin**
For the Math Message, provide each child with base-10 blocks (10 longs and 10 cubes).

▶ **Vocabulary**
expand-and-trade subtraction

Common Core State Standards

Focus Clusters
• Understand place value.
• Use place value understanding and properties of operations to add and subtract.

1 Warm Up 15–20 min	**Materials**	
Mental Math and Fluency Children solve subtraction problems involving multiples of 10.	slate	2.NBT.5
Daily Routines Children complete daily routines.	See pages 4–43.	See pages xiv–xvii.

2 Focus 35–40 min		
Math Message Children solve a subtraction problem using base-10 blocks.	base-10 blocks (10 longs and 10 cubes)	2.NBT.1, 2.NBT.1a, 2.NBT.5, 2.NBT.7
Introducing Expand-and-Trade Subtraction Children use expanded form to help them subtract.	base-10 blocks	2.NBT.1, 2.NBT.1a, 2.NBT.3, 2.NBT.5, 2.NBT.7, SMP2
Practicing Expand-and-Trade Subtraction Children practice solving problems using expand-and-trade subtraction.	*Math Journal 2*, p. 234–235; *My Reference Book*, pp. 87–89 (optional)	2.NBT.1, 2.NBT.1a, 2.NBT.3, 2.NBT.5, 2.NBT.7, 2.NBT.9
✓ **Assessment Check-In** See page 823.	*Math Journal 2*, p. 234–235; *My Reference Book*, pp. 87–89 (optional)	2.NBT.1a, 2.NBT.3, 2.NBT.5
Comparing Subtraction Strategies Children compare various subtraction strategies.	*Math Journal 2*, p. 234–235; slate	2.NBT.1a, 2.NBT.5, 2.NBT.7, 2.NBT.9, SMP1

CCSS 2.NBT.7 Spiral Snapshot

GMC Subtract multidigit numbers using models or strategies.

7-3 Warm Up	7-7 Practice	7-9 Practice	8-3 Practice	8-6 Practice	8-10 Practice	9-6 Focus Practice	9-7 Focus Practice

Spiral Tracker   **Go Online** to see how mastery develops for all standards within the grade.

3 Practice 10–15 min		
Playing *Beat the Calculator* **Game** Children practice addition facts.	*Assessment Handbook*, pp. 98–99 (optional); per group: 4 each of number cards 0–9, calculator, large paper triangle	2.OA.2
Math Boxes 9-7 Children practice and maintain skills.	*Math Journal 2*, p. 236	See page 825.
Home Link 9-7 **Homework** Children use expand-and-trade subtraction.	*Math Masters*, p. 266	2.NBT.1, 2.NBT.1a, 2.NBT.3, 2.NBT.5, 2.NBT.7

connectED.mheducation.com

Plan your lessons online with these tools.

 ePresentations   Student Learning Center   Facts Workshop Game   eToolkit   Professional Development   Home Connections   Spiral Tracker   Assessment and Reporting   English Learners Support   Differentiation Support

Differentiation Options

RtI

CCSS 2.NBT.1a, SMP2

Readiness
10–15 min

| WHOLE CLASS |
| SMALL GROUP |
| **PARTNER** |
| INDEPENDENT |

Trading with Base-10 Blocks

Math Masters, p. TA22; base-10 blocks

For experience with trading using a concrete model, children represent the same number in different ways with base-10 blocks. Children show their trades on the Place-Value Mat on *Math Masters,* page TA22. See the Readiness activity in Lesson 9-6 for details. **GMP2.2**

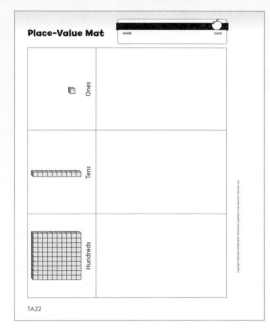

CCSS 2.NBT.1, 2.NBT.5, 2.NBT.7, SMP1

Enrichment
15–30 min

| WHOLE CLASS |
| SMALL GROUP |
| **PARTNER** |
| INDEPENDENT |

Exploring Subtraction Strategies

Activity Card 118; number grid (*Math Masters,* p. TA3); base-10 blocks; large poster paper

To further explore strategies for solving subtraction problems, children work together in groups or partnerships to write a 2- or 3-digit subtraction problem. They then solve the problem using as many different tools and strategies as they can. **GMP1.5** Children create posters to share with the class and compare their strategies with each other. **GMP1.6**

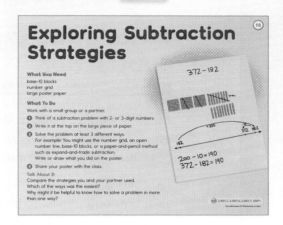

CCSS 2.NBT.1, 2.NBT.5, 2.NBT.7, 2.NBT.9

Extra Practice
5–15 min

| WHOLE CLASS |
| SMALL GROUP |
| **PARTNER** |
| INDEPENDENT |

Practicing Expand-and-Trade Subtraction

Activity Card 119, base-10 blocks, paper, 4 each of number cards 0–9

For additional practice subtracting 2-digit numbers, children use expand-and-trade subtraction to solve problems. They may use base-10 blocks if needed. Children explain how they solved the problems.

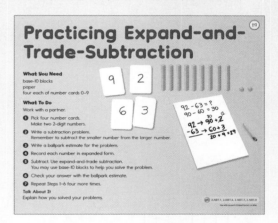

English Language Learners Support

Beginning ELL Use role play to review the term *trade.* Give a child 10 pennies and place a dime in front of you on the table. Say: *I need to trade. Please give me your 10 pennies, and I will give you my dime.* Have children practice with equivalent amounts in pennies, nickels, and dimes. As they trade, encourage them to use sentence frames like the following: *I need to trade. Please give me _____, and I will give you _____.*

 Go Online   **ELL** English Learners Support

Standards and Goals for
Mathematical Practice

SMP1 **Make sense of problems and persevere in solving them.**
GMP1.5 Solve problems in more than one way.
GMP1.6 Compare the strategies you and others use.

SMP2 **Reason abstractly and quantitatively.**
GMP2.2 Make sense of the representations you and others use.
GMP2.3 Make connections between representations.

1 Warm Up 15–20 min Go Online ePresentations eToolkit

▶ Mental Math and Fluency

Pose subtraction problems involving multiples of 10 for children to solve on their slates. *Leveled exercises:*

●○○ 20 − 10 10   ●●○ 70 − 20 50   ●●● 75 − 25 50
    50 − 20 30   65 − 10 55   82 − 42 40
    60 − 40 20   81 − 30 51   91 − 41 50

▶ Daily Routines

Have children complete daily routines.

2 Focus 35–40 min Go Online ePresentations eToolkit

▶ Math Message

Make a ballpark estimate for 93 − 68 = ? Sample answer: 90 − 70 = 20
Then solve using base-10 blocks. 25

▶ Introducing Expand-and-Trade Subtraction

| WHOLE CLASS | SMALL GROUP | PARTNER | INDEPENDENT |

Math Message Follow-Up Have children share their ballpark estimates and invite a volunteer to demonstrate his or her solution using base-10 shorthand. (*See margin.*)

Tell children that today they will use expanded form to help them think about making trades. To begin, model writing the numerical representations while children continue to use base-10 blocks. GMP2.3 As a class, solve the problem 79 − 34 by following these steps:

• Display the problem in vertical form and have children represent 79 with their base-10 blocks.
• Ask: *How can we write each number in expanded form?* 79 = 70 + 9 and 34 = 30 + 4 Draw arrows to the right of each number and write the expanded form as shown below. Point out that this way of writing the problem uses written numbers to show tens and ones, just like the blocks show tens and ones. GMP2.3

$$79 \rightarrow 70 + 9$$
$$\underline{-\ 34} \rightarrow 30 + 4$$

- Ask: *Do we need to make any trades?* No. We have enough to take away 3 longs and 4 cubes, or 30 and 4. Guide children through subtracting the tens and then the ones (or the ones and then the tens) as they take away their base-10 longs and cubes. Children subtract 30 from 70 to get 40 and 4 from 9 to get 5. Record these steps. (*See margin.*) Then demonstrate adding the tens and the ones to get the answer: 40 + 5 = 45.

- Write the following number sentence to summarize the problem: 79 − 34 = 45.

Repeat the process for other problems that do not require trades, such as 76 − 43 and 187 − 25. Be sure to model 187 − 25 because it involves a 3-digit number. (*See below.*)

$$
\begin{array}{r}
187 \quad \rightarrow \quad 100 + 80 + 7 \\
-\ 25 \quad \rightarrow \quad \underline{\hphantom{100 + 0} 20 + 5} \\
100 + 60 + 2 = 162
\end{array}
$$

Record the following number sentence to summarize: 187 − 25 = 162.

Next children will use expanded form to help them think about subtraction problems that require trades.

Tell children that you will once again illustrate this process with numbers while they model it using blocks. GMP2.3 Solve the problem 84 − 56 by following these steps:

- Display the problem in vertical form and ask children to represent 84 with their base-10 blocks.

- Ask: *How can we write the numbers in expanded form?* 84 = 80 + 4 and 56 = 50 + 6 Record each number in expanded form to the right of the problem as shown below.

$$
\begin{array}{r}
84 \quad \rightarrow \quad 80 + 4 \\
-\ 56 \quad \rightarrow \quad 50 + 6
\end{array}
$$

- Ask: *Do we need to make a trade?* Yes. Because 6 is larger than 4, we need to make a trade.

$$
\begin{array}{r}
79 \quad \rightarrow \quad 70 + 9 \\
-\ 34 \quad \rightarrow \quad \underline{30 + 4} \\
40 + 5 = 45
\end{array}
$$

Academic Language Development

Extend children's understanding of the term *expand* to other contexts by having them complete a 4-Square Graphic Organizer (*Math Masters,* page TA42). Ask children to complete the following sentence in the definitions quadrant of the organizer: "When we expand something, we _____." In the pictures quadrant, encourage children to draw pictures of objects before and after they expand.

- Have children trade 1 long for 10 cubes. Show children how to represent the trade by crossing out 80 and replacing it with 70 and crossing out 4 and replacing it with 14. Record these steps as shown below. Ask: *Is 70 + 14 still 84 all together?* Yes. *Do we have enough longs and cubes to remove 5 longs and 6 cubes?* Yes. Children subtract 50 from 70 to get 20 and 6 from 14 to get 8. Record these steps. Children then add the tens and the ones to get the answer: 20 + 8 = 28.

$$
\begin{array}{r}
84 \\
-\,56
\end{array}
\;\rightarrow\;
\begin{array}{r}
\overset{70}{\cancel{80}} + \overset{14}{\cancel{4}} \\
50 + 6 \\
\hline
20 + 8 = 28
\end{array}
$$

- Write the following number sentence to summarize: 84 − 56 = 28.

Tell children that this subtraction method is called **expand-and-trade subtraction** because children use expanded form to think about whether they need to make trades. Work as a class to use expand-and-trade subtraction to solve other problems requiring trades, such as 63 − 48 and 160 − 77. Be sure to model 160 − 77 because it involves two trades.

Display the vertical problem and write the expanded form of each number. Ask: *Do we need to make a trade?* Yes. Because 70 is larger than 60, we need to make a trade. Show how to trade 1 hundred for 10 tens as shown below.

$$
\begin{array}{r}
160 \\
-\,77
\end{array}
\;\rightarrow\;
\begin{array}{r}
\overset{0}{\cancel{100}} + \overset{160}{\cancel{60}} + 0 \\
70 + 7 \\
\hline
\end{array}
$$

Ask: *Do we need to make another trade?* Yes. Because 7 is larger than 0, we need to make a trade. Show how to trade 1 ten for 10 ones as shown below. Then demonstrate subtracting the hundreds, tens, and ones and adding to get the answer.

$$
\begin{array}{r}
160 \\
-\,77
\end{array}
\;\rightarrow\;
\begin{array}{r}
\overset{0}{\cancel{100}} + \overset{\overset{150}{\cancel{160}}}{\cancel{60}} + \overset{10}{\cancel{0}} \\
70 + 7 \\
\hline
80 + 3 = 83
\end{array}
$$

Record the following number sentence to summarize: 160 − 77 = 83. Encourage children to transition from using base-10 blocks to writing numbers to represent subtraction. GMP2.2

▶ Practicing Expand-and-Trade Subtraction

Math Journal 2, p. 234–235

| WHOLE CLASS | SMALL GROUP | PARTNER | INDEPENDENT |

Have partners work together to complete the problems on journal pages 234–235. As needed, children may refer to *My Reference Book,* pages 87–89. Remind children to check their solutions against their ballpark estimates. Circulate and ask: *Does this answer make sense?*

As children finish, remind them that number families, just like fact families, have four related number sentences. For example, the four number sentences for the problem $160 - 77 = 83$ relate 160, 77, and 83 by addition and subtraction: $77 + 83 = 160$, $83 + 77 = 160$, $160 - 77 = 83$, and $160 - 83 = 77$. For two of the problems on the journal page, have children use related addition number sentences to check their answers.

 Assessment Check-In (CCSS) **2.NBT.1a, 2.NBT.3, 2.NBT.5**

Math Journal 2, pp. 234–235

Expect most children to make reasonable estimates and write each number in expanded form for Problems 1–3 on journal pages 234–235. Because this lesson is an early exposure to expand-and-trade subtraction, don't expect all children to successfully solve Problems 1–3 using that method. In Grade 2 children are expected to fluently subtract within 100 using strategies based on place value, properties of operations, and/or the relationship between addition and subtraction.

If children struggle with expand-and-trade subtraction, suggest they first solve each problem using a different method and then use expand-and-trade subtraction to solve it again. It is important for children to be exposed to and try different strategies so that for any given problem they can use the strategy that is most efficient for them. Support children by helping them model each step with their base-10 blocks and record it on paper before moving to the next step. You may also wish to refer children to *My Reference Book,* pages 87–89.

 Assessment and Reporting   (Go Online)   to record student progress and to see trajectories toward mastery for these standards.

Math Journal 2, p. 234

Expand-and-Trade Subtraction   Lesson 9-7   DATE

For each problem do the following:

- Write a number model to show a ballpark estimate.
- Write each number in expanded form.   Sample ballpark
- Solve using expand-and-trade subtraction.   estimates given.

① Example: $45 - 27 = ?$

Ballpark estimate:
$45 - 20 = 25$

Solution:

$$
\begin{array}{r}
\overset{30}{} \quad \overset{15}{} \\
45 \rightarrow \cancel{40} + \cancel{5} \\
- 27 \rightarrow 20 + 7 \\
\hline
10 + 8 = 18
\end{array}
$$

$45 - 27 = \underline{18}$

② $31 - 17 = ?$

Ballpark estimate:
$30 - 15 = 15$

Solution:

$$
\begin{array}{r}
\overset{20}{} \quad \overset{11}{} \\
31 \rightarrow \cancel{30} + \cancel{1} \\
- 17 \rightarrow 10 + 7 \\
\hline
10 + 4 = 14
\end{array}
$$

$31 - 17 = \underline{14}$

234   two hundred thirty-four   2.NBT.1, 2.NBT.1a, 2.NBT.3, 2.NBT.5, 2.NBT.7

Math Journal 2, p. 235

Expand-and-Trade Subtraction (continued)   Lesson 9-7   DATE

③ $72 - 49 = ?$

Ballpark estimate: $70 - 50 = 20$

Solution:

$$
\begin{array}{r}
\overset{60}{} \quad \overset{12}{} \\
72 \rightarrow \cancel{70} + \cancel{2} \\
- 49 \rightarrow 40 + 9 \\
\hline
20 + 3 = 23
\end{array}
$$

$72 - 49 = \underline{23}$

Try This

④ $126 - 48 = ?$

Ballpark estimate: $130 - 50 = 80$

Solution:

$$
\begin{array}{r}
\quad\quad \overset{110}{} \\
0 \quad \overset{120}{} \quad \overset{16}{} \\
126 \rightarrow \cancel{100} + \cancel{20} + \cancel{6} \\
- 48 \rightarrow 40 + 8 \\
\hline
70 + 8 = 78
\end{array}
$$

$126 - 48 = \underline{78}$

2.NBT.1, 2.NBT.1a, 2.NBT.3, 2.NBT.5, 2.NBT.7   two hundred thirty-five   235

Math Journal 2, p. 234–235

Counting Up

Children count up by 10s from 17 to 27. Then they count up 4 more to 31 for a difference of 14.

Using Friendly Numbers

Children change 17 to 20 to make it easier to subtract from 31. Then they subtract: $31 - 20 = 11$. Because children need to subtract 17 from 31, not 20, they need to add 3 more: $11 + 3 = 14$.

▶ Comparing Subtraction Strategies

Math Journal 2, p. 234–235

WHOLE CLASS	SMALL GROUP	PARTNER	INDEPENDENT

Have children solve Problem 2 on journal page 234, $31 - 17$, by using a different strategy. **GMP1.5** As children work, circulate and note which strategies they are using. In particular, look for children who use the counting-up strategy or friendly numbers. (*See margin.*) Then bring the class together to discuss their strategies and compare them to expand-and-trade subtraction. Ask:

- *How is expand-and-trade subtraction different from counting up?* **GMP1.5** With counting up, we start with the smaller number. With expand-and-trade subtraction, we start by showing the numbers in expanded form.

- *Would you rather use a different method to solve some of the problems on journal pages 234–235? What method? Why do you think it would work well?* **GMP1.6** Answers vary.

- *What do you like about expand-and-trade subtraction? Do you think you could use it to solve any problem?* Answers vary.

Explain that although any strategy will work for any subtraction problem, no one strategy is the most efficient method for every problem. It is important for children to continue to reflect on and practice a variety of methods for solving subtraction problems.

Summarize Have children brainstorm a list of methods for solving multidigit subtraction problems. Expect their list to include strategies such as expand-and-trade subtraction, counting up, using friendly numbers, using a number line, and using a number grid.

3 Practice · 10–15 min

Go Online

ePresentations · eToolkit · Home Connections

▶ Playing *Beat the Calculator*

| WHOLE CLASS | **SMALL GROUP** | PARTNER | INDEPENDENT |

Have small groups play the game as introduced in Lesson 5-1. As you circulate and observe, consider using *Assessment Handbook*, pages 98–99 to monitor children's progress with addition facts. By the end of Grade 2, children are expected to know from memory all sums of two 1-digit numbers.

Observe

- Which facts do children know from memory?
- Which children need additional support to play the game?

Discuss

- *What strategies did you use to solve the facts you did not know?*
- *Why is it helpful to know addition facts?*

▶ Math Boxes 9-7

Math Journal 2, p. 236

| WHOLE CLASS | SMALL GROUP | PARTNER | **INDEPENDENT** |

Mixed Practice Math Boxes 9-7 are paired with Math Boxes 9-5.

▶ Home Link 9-7

Math Masters, p. 266

Homework Children practice expand-and-trade subtraction.

Math Journal 2, p. 236

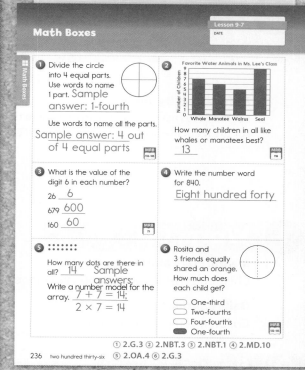

Math Masters, p. 266

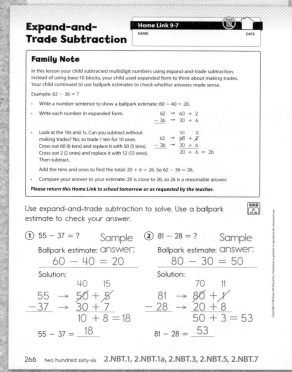

Equivalent Money Amounts

Overview Children practice finding coin and bill combinations with equivalent values and using cents and dollars-and-cents notation.

▶ **Before You Begin**
For the Focus activities, check to make sure that children's toolkits still contain coins and bills. For the Math Message, each child will need 2 quarters, 5 dimes, 4 nickels, and 7 pennies.

Common Core State Standards

Focus Clusters
• Understand place value.
• Use place value understanding and properties of operations to add and subtract.
• Work with time and money.

1 Warm Up 15–20 min

	Materials	
Mental Math and Fluency Children solve problems involving money.	slate	2.MD.8
Daily Routines Children complete daily routines.	See pages 4–43.	See pages xiv–xvii.

2 Focus 30–40 min

Math Message Children calculate the total value of a set of coins.	toolkit coins (2 quarters, 5 dimes, 4 nickels, 7 pennies); slate	2.NBT.2, 2.MD.8
Reviewing Values of Coins and Bills Children review values of coins.	slate	2.NBT.2, 2.MD.8
Using Dollars-and-Cents Notation Children practice using dollars-and-cents notation.	slate	2.MD.8 SMP2
Making Equivalent Amounts with Coins and Bills Children find two ways of paying for grocery items.	*Math Journal 2*, pp. 238–239; *Math Masters*, p. 270 (optional); toolkit coins and bills	2.NBT.2, 2.NBT.7, 2.MD.8 SMP1, SMP4
✓ **Assessment Check-In** See page 830.	*Math Journal 2*, p. 239	2.MD.8, SMP1, SMP4

CCSS 2.MD.8 **Spiral Snapshot**

GMC Solve problems involving coins and bills.

| 2-5 Practice | 2-8 Practice | 3-10 Practice | 3-11 Focus Practice | 5-2 through Warm Up Focus Practice | 5-4 | 5-11 Focus Practice | 9-8 Warm Up Focus Practice |

Spiral Tracker **Go Online** to see how mastery develops for all standards within the grade.

3 Practice 10–15 min

Playing *Hit the Target* **Game** Children find differences between numbers.	*Math Masters*, p. G25; per partnership: 1 calculator	2.NBT.5 SMP1, SMP3
Math Boxes 9-8 Children practice and maintain skills.	*Math Journal 2*, p. 237	See page 831.
Home Link 9-8 **Homework** Children draw coins to pay for items in ads.	*Math Masters*, p. 271	2.NBT.2, 2.MD.8

 ePresentations   Student Learning Center   Facts Workshop Game   eToolkit   Professional Development   Home Connections   Spiral Tracker   Assessment and Reporting   English Learners Support   Differentiation Support

Differentiation Options

RtI

Readiness 5–15 min

WHOLE CLASS
SMALL GROUP
PARTNER
INDEPENDENT

Finding Ways to Make a Dollar

Math Masters, p. 267;
toolkit coins

To explore the value of coins and coin combinations, children make various coin combinations that have a value of $1. Children record their results on *Math Masters*, page 267.

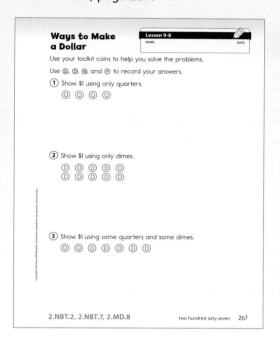

Enrichment 5–15 min

WHOLE CLASS
SMALL GROUP
PARTNER
INDEPENDENT

Planning a Picnic

Math Journal 2, p. 238;
Math Masters, p. 268

To apply their understanding of solving problems involving money, children use the Good Buys Poster on journal page 238 to plan and shop for food for a picnic. They should choose at least three different items to buy, find the total cost, and share the total cost equally among three people.

Remind children that money amounts that include dollars and cents can be written in different ways. For example, $1.49 and 149¢ both show the same amount. Children record their work on *Math Masters*, page 268.

Extra Practice 5–15 min

WHOLE CLASS
SMALL GROUP
PARTNER
INDEPENDENT

Practicing with Equivalent Money Amounts

Math Masters, p. 269;
toolkit coins and bills

For practice with equivalent money amounts, children complete *Math Masters*, page 269. They calculate the values of different bill and coin combinations and write the amounts in both cents notation and dollar-and-cents notation. They also show how to represent an amount of money two different ways. **GMP1.5**

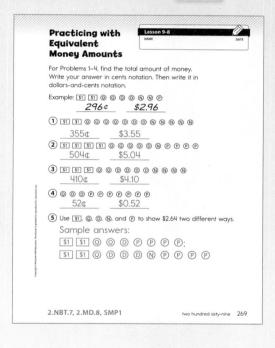

English Language Learners Support

Beginning ELL English language learners may confuse the homophones *good buy* and *good-bye*. Show children ads from a newspaper for familiar items, such as pencils, paper, or snacks. Point to the various items and model by saying the following: *I'm going to the store to buy pencils. They are a good buy. I am saving money.* Use the Good Buys Poster to model. Point to different items and say: *This is a good buy.* Then point to similar items that cost more in other ads in the newspaper and ask: *Is this a good buy or is this not a good buy?*

Go Online **ELL** English Learners Support

1 Warm Up 15–20 min

Go Online | ePresentations | eToolkit

▸ Mental Math and Fluency

Pose problems involving money. Children write the answers on their slates. *Leveled exercises:*

- ●○○ How much money is 2 dimes and 6 pennies? 26¢
- ●●○ How much money is 1 quarter, 1 dime, and 3 pennies? 38¢
- ●●● How much money is 2 quarters, 2 dimes, 1 nickel, and 2 pennies? 77¢

▸ Daily Routines

Have children complete daily routines.

2 Focus 30–40 min

Go Online | ePresentations | eToolkit

▸ Math Message

Take the following toolkit coins out of your toolkit: 2 Ⓠ, 5 Ⓓ, 4 Ⓝ, and 7 Ⓟ. How much money is this? Write the total on your slate. Possible answers: 127¢; $1.27

▸ Reviewing Values of Coins and Bills

| WHOLE CLASS | SMALL GROUP | PARTNER | INDEPENDENT |

Math Message Follow-Up Have volunteers share the strategies they used to find the total value of the coins they took out of their toolkits. To review the values of coins and bills, ask questions like the following and have children write their answers on their slates:

- *How many pennies are in a nickel?* 5 *In a dime?* 10
- *How many pennies are in a quarter?* 25 *In 50 cents?* 50
- *How many pennies are in a dollar?* 100 *In 2 dollars?* 200 *In 10 dollars?* 1,000
- *How many dimes are in a dollar?* 10 *In 60 cents?* 6
- *How many nickels are in a quarter?* 5 *In a dollar?* 20 *In half of a dollar?* 10
- *How many quarters are in a dollar?* 4 *In half of a dollar?* 2

Tell children that they will solve more problems involving money.

▶ Using Dollars-and-Cents Notation

WHOLE CLASS | **SMALL GROUP** | PARTNER | INDEPENDENT

Ask: *How many of you wrote 127¢ as your answer to the Math Message? Does anyone know a different way to write 127¢?* If no one suggests it, display the amount in dollars-and-cents notation: $1.27. **GMP2.3** Point out that 2 quarters and 5 dimes have the same value as 100 pennies, or 1 dollar.

Discuss how to interpret the dollars-and-cents notation:

- The symbol $ stands for the word *dollar*.
- The dot (or period) after the 1 is called a decimal point.
- The number before the decimal point is the number of dollars. The number after the decimal point is the number of cents.
- $1.27 is read as "1 dollar and 27 cents." Mention commonly used alternatives, such as "a dollar twenty-seven." Emphasize that $1.27 is more money than $1.

Tell children that an amount with a 0 before the decimal point, such as $0.74, is less than 1 dollar. It can be written with a cent symbol as 74¢. Have children practice writing the following in dollars-and-cents notation on their slates: **GMP2.3**

- 275¢ $2.75
- 305¢ $3.05
- 89¢ $0.89
- 1 dollar and 25 cents $1.25
- 2 dollars and 90 cents $2.90

Invite volunteers to discuss how they knew where to put the decimal point in 3-digit money amounts. Sample answers: I knew the decimal point went after the hundreds digit and before the tens digit, or after the word *dollar*.

Differentiate **Common Misconception**

Watch for children who write $2.75 as $2.75¢. Tell them that the dollar symbol and decimal point signal that the numbers after the decimal point are cents, so the ¢ symbol should not be used in this situation.

Go Online ⟩ Differentiation Support

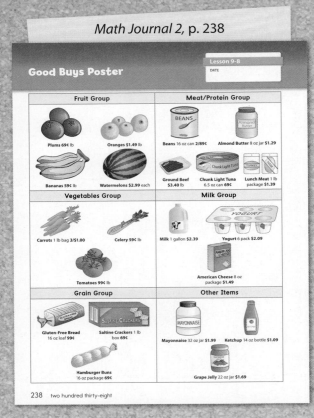

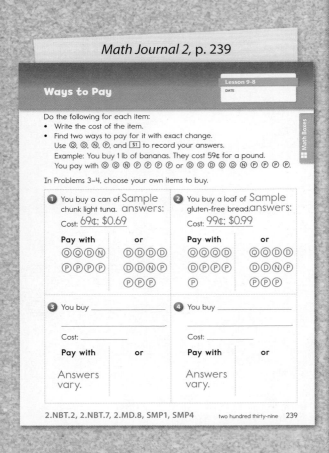

► Making Equivalent Amounts with Coins and Bills

Math Journal 2, pp. 238–239

WHOLE CLASS | SMALL GROUP | PARTNER | INDEPENDENT

Ask children to examine the Good Buys Poster on journal page 238. You may want to display the poster on *Math Masters,* page 239. Point out that *lb* and *oz* are abbreviations for *pounds* and *ounces*, that pounds and ounces are units of weight, and that there are 16 ounces in 1 pound. Check that children know how to read the prices of the items. Have them identify prices written in dollars-and-cents notation and instruct them to rewrite those prices in cents notation.

Academic Language Development Children may not be aware that the notation 2/89¢ means that they can buy two of that item for 89¢. Help them understand that the number before the slash (/) represents the quantity of items being bought, and the number after the slash represents the total cost for the items. Have them look at the Good Buys Poster and find examples. Sample answers: Carrots, 3/$1.00; Beans, 2/89¢

Partners complete Problems 1–2 on journal page 239 by showing two ways to pay for chunk light tuna and two ways to pay for gluten-free bread using exact change. For Problems 3–4 they choose two items from the poster to purchase and use their toolkit coins and bills to count out several ways to pay for each items with exact change. GMP4.1, GMP4.2 Children record two ways of paying for each item by drawing coins and bills on journal page 239. GMP1.5

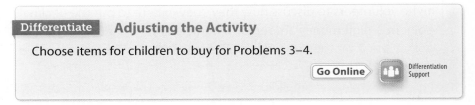

Differentiate **Adjusting the Activity**

Choose items for children to buy for Problems 3–4.

Go Online ▸ Differentiation Support

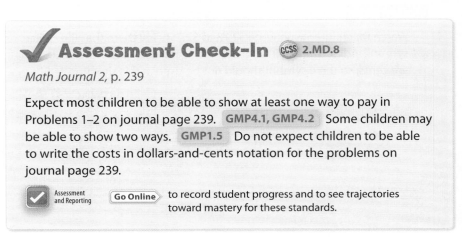

✓ **Assessment Check-In** CCSS 2.MD.8

Math Journal 2, p. 239

Expect most children to be able to show at least one way to pay in Problems 1–2 on journal page 239. GMP4.1, GMP4.2 Some children may be able to show two ways. GMP1.5 Do not expect children to be able to write the costs in dollars-and-cents notation for the problems on journal page 239.

Assessment and Reporting | Go Online ▸ to record student progress and to see trajectories toward mastery for these standards.

Summarize Have volunteers discuss the strategies they used to find the different ways to pay for the items on journal page 239.

③ Practice 10–15 min

Go Online

ePresentations eToolkit Home Connections

▶ Playing *Hit the Target*

Math Masters, p. G25

| WHOLE CLASS | SMALL GROUP | **PARTNER** | INDEPENDENT |

Have children play *Hit the Target* to practice mentally finding differences between multiples of 10 and smaller 2-digit numbers. They make conjectures about what number they need to add or subtract and then check their guesses on a calculator. **GMP3.1** See Lesson 7-1 for detailed directions.

Observe

- Which children seem to have a strategy for hitting the target number?
- Which children need additional support to understand and play the game?

Discuss

- *How did you decide which number to add or subtract?* **GMP3.1**
- *If you didn't hit the target number on your first try, how did you decide what to do next?* **GMP1.3**

▶ Math Boxes 9-8

Math Journal 2, p. 237

| WHOLE CLASS | SMALL GROUP | PARTNER | INDEPENDENT |

Mixed Practice Math Boxes 9-8 are paired with Math Boxes 9-6.

▶ Home Link 9-8

Math Masters, p. 271

Homework Children find the total value of 10 pennies, 10 nickels, 10 dimes, and 10 dollars. Then they cut out ads for items costing less than 99¢ and draw coin combinations they could use to pay for each item.

Math Journal 2, p. 237

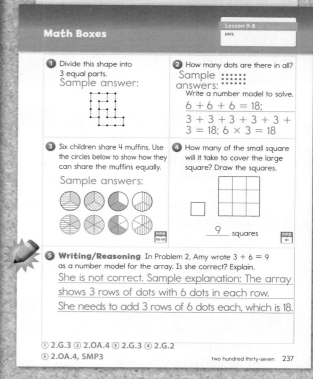

Math Masters, p. 271

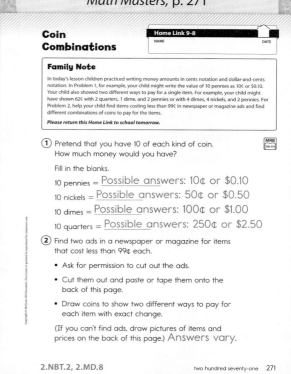

Estimating Costs

2-Day Lesson

Overview **Day 1:** Children select items from a store poster and use mental math to estimate the total cost. **Day 2:** The class discusses selected children's estimates, and children revise their work.

Day 1: Open Response

▶ **Before You Begin**
Solve the open response problem and consider which estimation and mental math strategies children may use. You may wish to ask children to read *My Reference Book,* page 18 to review GMP6.1. If possible, schedule time to review children's work and plan for Day 2 of this lesson with your grade-level team.

▶ **Vocabulary**
ballpark estimate · close-but-easier numbers · reasonable · precisely

CCSS **Common Core State Standards**

Focus Clusters
· Represent and solve problems involving addition and subtraction.
· Use place value understanding and properties of operations to add and subtract.

1 Warm Up   15–20 min

Materials

Mental Math and Fluency Children record number models and make a ballpark estimate.		2.NBT.5
Daily Routines Children complete daily routines.	See pages 4–43.	See pages xiv–xvii.

2a Focus   45–55 min

Math Message Children make sense of an estimation strategy for an addition problem and find a second estimation strategy for the same problem.	*Math Journal 2,* p. 240	2.NBT.5, 2.NBT.6, 2.NBT.8 SMP1, SMP6
Comparing Estimation Strategies Children discuss the estimation strategies.	*Math Journal 2,* p. 240; Standards for Mathematical Practice Poster	2.NBT.5, 2.NBT.6, 2.NBT.8 SMP1, SMP6
Solving the Open Response Problem Children use mental math and estimation strategies to decide what they can buy at a store for $100.	*Math Masters,* pp. 272– 274; scissors; glue	2.OA.1, 2.NBT.5, 2.NBT.6, 2.NBT.8 SMP1, SMP6

Getting Ready for Day 2 →

Review children's work and plan discussion for reengagement.   *Math Masters,* p. TA5; children's work from Day 1

CCSS 2.OA.1   **Spiral Snapshot**

GMC Use addition and subtraction to solve 2-step number stories.

Routines 2 and 6 | 6-7 Practice | 7-1 Practice | 7-2 Focus | 7-3 Practice | 7-5 Practice | 7-9 Practice | 9-9 Focus

III **Spiral Tracker**   **Go Online** to see how mastery develops for all standards within the grade.

connectED.mheducation.com

Plan your lessons online with these tools.

 ePresentations   Student Learning Center   Facts Workshop Game   eToolkit   Professional Development   Home Connections   Spiral Tracker   Assessment and Reporting   ELL English Learners Support   Differentiation Support

832   Unit 9 | Equal Shares and Whole Number Operations

1 Warm Up 15–20 min | Go Online | ePresentations eToolkit

▶ Mental Math and Fluency

Children make ballpark estimates and record them as number models on their slates.

● ○ ○ 48 + 46 = ? Sample answers: 50 + 50 = 100; 50 + 45 = 95
13 + 59 = ? Sample answers: 10 + 60 = 70; 13 + 60 = 73

● ● ○ 76 + 188 = ? Sample answer: 80 + 190 = 270; 80 + 200 = 280
85 + 165 = ? Sample answers: 80 + 160 = 240; 90 + 170 = 260

● ● ● 183 + 211 = ? Sample answers: 180 + 210 = 390; 200 + 200 = 400
296 + 373 = ? Sample answers: 300 + 370 = 670; 300 + 373 = 673

▶ Daily Routines

Have children complete daily routines.

2a Focus 45–55 min | Go Online | ePresentations eToolkit

▶ Math Message

Math Journal 2, p. 240

Complete journal page 240. Explain your thinking to your partner. **GMP1.4, GMP6.1, GMP6.2**

▶ Comparing Estimation Strategies

Math Journal 2, p. 240

| WHOLE CLASS | SMALL GROUP | **PARTNER** | INDEPENDENT |

Math Message Follow-Up Ask children to explain to their partners when they might use **ballpark estimates.** **GMP1.4, GMP6.2** Sample answers: when I need to solve problems in my head; when I do not need an exact answer; to check to see whether an answer to an addition or subtraction problem makes sense.

Have children explain Jayden's estimation strategy to their partners before asking volunteers to share with the class. Highlight the **close-but-easier numbers** Jayden uses, explaining that they are close enough to the numbers in the addition problem to estimate, but easier to add in one's head. Children may notice that Jayden rounded 48 to 50 (the nearest ten) and 24 to 25 (possibly thinking about money).

Have children explain their strategy for Problem 2 to their partners. Then ask: *When you shared your strategy with your partner, were you able to explain it so that your partner could solve a similar problem using your strategy?* Answers vary.

Professional Development

The focus of this lesson is GMP6.1. Children have been solving open response problems and explaining their solutions throughout the year. During the reengagement discussion, revisit what it means to "explain your mathematical thinking clearly and precisely." Precise mathematical language is accurate and detailed, so children need to use correct terms, describe all the steps in their strategies, and include details such as units with their answers.

| Go Online | for information about SMP6 in the *Implementation Guide*.

Math Journal 2, p. 240

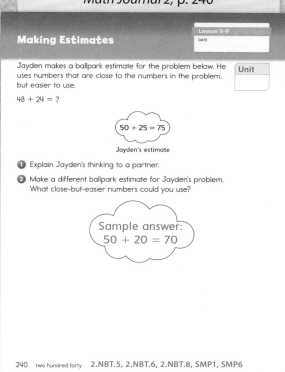

Making Estimates Lesson 9-9
DATE

Jayden makes a ballpark estimate for the problem below. He uses numbers that are close to the numbers in the problem, but easier to use.

Unit

48 + 24 = ?

50 + 25 = 75

Jayden's estimate

1 Explain Jayden's thinking to a partner.

2 Make a different ballpark estimate for Jayden's problem. What close-but-easier numbers could you use?

Sample answer:
50 + 20 = 70

240 two hundred forty   2.NBT.5, 2.NBT.6, 2.NBT.8, SMP1, SMP6

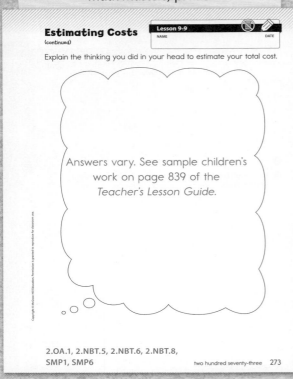
Discuss children's strategies. Ask: *What close-but-easier numbers did you use?* Sample answer: I rounded both numbers to the nearest ten. I rounded 48 to 50 and 24 to 20. *How did you find your estimate?* Sample answer: I added 50 + 20 = 70. *Can you do that addition in your head?* Yes.

Briefly discuss how both estimates are **reasonable**. The close-but-easier numbers you choose depend on how you add in your head or how close you want the estimate to be. Sometimes the strategy you choose depends on the situation in the problem. GMP6.1, GMP6.2

Tell children that today they will pretend they are in a store, and they have $100 to spend. They are going to make estimates in their head to decide what items they can buy and explain their thinking. GMP1.4, GMP6.2 A goal of their work is to explain their mathematical thinking clearly and precisely. Refer children to the Standards for Mathematical Practice Poster for **GMP6.1**.

English Language Learners Prior to the lesson, preview how to make a ballpark estimate with a simple problem, such as 21 + 47, and then explain why you chose the easier numbers for making the estimate. Model this process a few times before asking students to make their own ballpark estimates and explain how they decided what numbers to use. Be sure to use and describe vocabulary such as *ballpark estimate, making estimates in your head, doing a problem mentally,* and *explaining thinking clearly and precisely.*

▶ Solving the Open Response Problem

Math Masters, pp. 272–274

| WHOLE CLASS | SMALL GROUP | **PARTNER** | **INDEPENDENT** |

Distribute *Math Masters,* page 272 and Moran's Market Poster on *Math Masters,* page 274. Children will need scissors and glue, but they should put their pencils away for now. Read the problem with the class. Have partners discuss what they understand from the problem. Invite volunteers to explain the task, asking questions such as: *What do you need to figure out?* Sample answer: what I can buy from the poster *How many items should you buy?* 3 or more *How much money do you have?* $100 *Do you have to spend all the money?* Sample answer: No, but we should spend as much as possible. *Do you need an exact answer to decide what to buy?* No. *How will you show what items you plan to buy?* Sample answer: We will cut them out and glue them down. *Can you use a pencil?* No.

Review the prices on Moran's Market Poster. Encourage children to complete the first part of the problem mentally, or in their head.

Once children have chosen their items and glued them down, distribute *Math Masters,* page 273. In the thought bubble, children should show their mental math strategies and write a clear explanation of their thinking. Ask them to write down how they chose their items and how they know the total cost is close to $100, but not more than $100. GMP1.4, GMP6.1, GMP6.2

Allow children time to complete the page. Partners can talk about the task, but each child should write an explanation.

Circulate and assist. If children try to find an exact answer using paper and pencil, ask: *How can you find the total cost of the items in your head?*
GMP1.4, GMP6.2 Sample answers: I can use close-but-easier numbers that I can add mentally. I can look for numbers that are easy to add in my head. You may want to make notes about children's strategies.

> **Differentiate** **Adjusting the Activity**
>
> If children have trouble getting started, ask them to choose two items and estimate their total cost. Then discuss whether the total is close to $100 or not. Help them use these choices to understand how to make decisions for choosing more or different items. For those who have trouble writing their explanations, consider using sentence frames, such as "I chose my items because _____" and "I know that they are close to $100, but not more than $100 because _____." Ask them to describe their thinking orally to you or a partner and then have them write down or draw what they said or dictate their thinking to an adult.

Summarize Ask: *When is it helpful to estimate or use mental math?*
Sample answers: when you don't have a pencil or calculator or if you just want to see if an answer is reasonable

Collect children's work so that you can evaluate it and prepare for Day 2.

Getting Ready for Day 2

Math Masters, p. TA5

Planning a Follow-Up Discussion

Review children's work and your notes about their thinking. Use the Reengagement Planning Form (*Math Masters,* page TA5) and the rubric on page 837 to plan ways to help children meet expectations for both the content and practice standards. Look for work in which children used pencil and paper instead of estimation or mental math as well as work with interesting and efficient strategies children could complete in their head.

Organize the discussion in one of the ways below or in another way you choose. If children's work is unclear or if you prefer to show the work anonymously, rewrite the work for display.

Go Online for sample children's work that you can use in your discussion.

1. Display work from a child who used ballpark estimates to choose items with a total cost of just less than $100. Show the chosen items and written explanation. See Child A's work. Ask:

   • *Discuss Child A's thinking with your partner. Explain the strategy.*
   Sample answer: Child A chose the cell phone, calculator, headphones, and DVD. Child A found close-but-easier numbers for each one and added those numbers, which equaled $90.

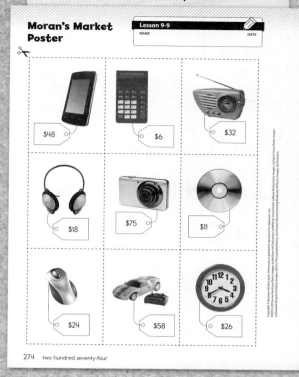

Moran's Market Poster Lesson 9-9

274 two hundred seventy-four

Sample child's work, Child A

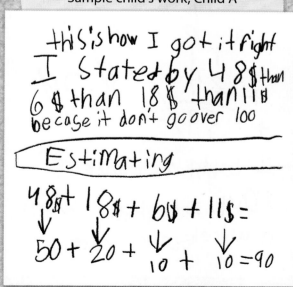

this is how I got it right
I started by 48$ than
6$ than 18$ than 11$
becage it don't go over 100

Estimating

48$ + 18$ + 6$ + 11$ =
50 + 20 + 10 + 10 = 90

26 is coser to 30
58 is cose to 60
6 is cose to 10
I kown it is less
than one obller
because the prise
is less.

$18
+ 6
+ 24
+ 59
/ $59

I bought a calculator
a DVD some Head phones
a wireless mouse

- *Is it clear what close-but-easier numbers Child A used?* GMP6.2 Yes. *What are they?* Sample answer: Child A used the closest ten for each: 50 for 48, 20 for 18, 10 for 6, and 10 for 11.
- *Is the total cost less than but close to $100? Why or why not?* GMP1.4, GMP6.2 Sample answer: Yes, $90 is close to $100.
- *What could Child A write or add to make the explanation clearer? Are any steps missing?* GMP6.1 Sample answers: Since Child A showed the prices with arrows pointing to the close-but-easier numbers in a number model, it was pretty easy to understand. The child could say more about deciding what to buy so that it costs less than $100. It would be good to add dollar signs.

2. Display work from a child who chose items correctly, but whose explanation is not clear or complete. See Child B's work. Ask:

- *Discuss Child B's thinking with your partner. Explain the strategy.* Sample answer: Child B chose the clock, calculator, and the car and showed close-but-easier numbers for each price. If you add those numbers, the total is $100. Child B said that the price was less than one dollar, but I think that he or she meant it was less than $100.
- *Do you agree that the items Child B chose cost less than but close to $100? Why or why not?* GMP1.4, GMP6.2 Sample answer: If you add 30 + 60 + 10 you get 100. Since each actual price is less than the close-but-easier number, the total will be less.
- *What could Child B write or add to make the explanation clearer? Are any steps missing?* Sample answer: Child B left out the steps of showing a number model or how to add 30 + 60 + 10 to equal $100. Child B should say that the total is less than $100 because each price is less. It would be clearer if the items were written with the prices.
- *Child A and Child B each chose different items. Can two different answers to the same question both be correct?* GMP6.1 Sample answer: Yes. The items are different, but both Child A and B chose items that cost close to but less than $100. They both added correctly.

3. Discuss a response from a child who didn't use mental math or estimation, or did not find a total close to $100. See Child C's work. Ask:

- *Discuss Child C's thinking with your partner. Explain the strategy.* Sample answer: Child C chose the headphones, calculator, DVD, and mouse and added up the prices to get an exact answer of $59.
- *Do you agree that the items Child C chose cost less than but close to $100? Why or why not?* GMP1.4, GMP6.2 Sample answer: They are less than $100, but $59 is not close to $100.
- *If you were in a store without a calculator or pencil, do you think the numbers are easy to add in your head?* No. *How could you choose items that total close to $100 just using your head?* GMP1.4, GMP6.2 Sample answer: You could use close-but-easier numbers to add in your head. Then it would be easy to pick another item to get closer to $100.

Planning for Revisions

Have copies of *Math Masters,* pages 272–274 or extra paper available for children to use in revisions. You might want to ask children to use colored pencils so you can see what they revised.

Estimating Costs

Day 2: Reengagement

▶ **Before You Begin**
Have extra copies available of *Math Masters,* pages 272–274 for children to revise their work.

▶ **Vocabulary**
precisely

Common Core State Standards

Focus Clusters
• Represent and solve problems involving addition and subtraction.
• Use place value understanding and properties of operations to add and subtract.

2b Focus   50–55 min

	Materials	
Setting Expectations Children review the open response problem and discuss what a good explanation would include. They also review how to discuss others' work respectfully.	Guidelines for Discussions Poster	2.OA.1, 2.NBT.5, 2.NBT.6, 2.NBT.8 SMP6
Reengaging in the Problem Children examine others' estimation strategies in a class discussion.	selected samples of children's work	2.OA.1, 2.NBT.5, 2.NBT.6, 2.NBT.8 SMP1, SMP6
Revising Work Children revise their explanations.	*Math Masters,* pp. 272–274 (optional); children's work from Day 1; scissors; glue; colored pencils (optional)	2.OA.1, 2.NBT.5, 2.NBT.6, 2.NBT.8 SMP1, SMP6

✓ **Assessment Check-In**   See page 839 and the rubric below.   2.OA.1, SMP6

Goal for Mathematical Practice **GMP6.1** Explain your mathematical thinking clearly and precisely.	**Not Meeting Expectations**	**Partially Meeting Expectations**	**Meeting Expectations**	**Exceeding Expectations**
	Does not show the selection of items, calculations, or any close-but-easier numbers for an estimate.	Shows or explains in words some steps such as selection of items, calculations, or the choice of any close-but-easier numbers used for an estimate.	Shows or explains in words all steps including selection of items, all calculations, and the choice of any close-but-easier numbers used for an estimate.	Meets expectations and explains the thinking for deciding whether the total is close to but less than $100 or shows a second strategy for a set of items that meet the criteria of the problem.

3 Practice   10–15 min

Math Boxes 9-9 Children practice and maintain skills.	*Math Journal 2,* p. 241	See page 838.
Home Link 9-9 Children practice making estimates for the cost of items in a store.	*Math Masters,* p. 275	2.OA.1, 2.NBT.5, 2.NBT.6, 2.NBT.8 SMP1, SMP6

2b Focus

50–55 min | Go Online |

ePresentations   eToolkit

▶ Setting Expectations

| **WHOLE CLASS** | SMALL GROUP | PARTNER | INDEPENDENT |

Briefly review the open response problem from Day 1. Remind children that their job was to find at least three items to buy so the total cost was close to but less than $100. They also needed to explain the strategies they used to estimate the total cost. Ask: *What do you think a good explanation would include?* Sample answers: We should include the items we decided to buy and the prices. We should tell how we added to find the total and any close-but-easier numbers we used for each item. We should tell how we know that the total is close to but less than $100. Remind children that a goal of their work is to explain their thinking clearly and precisely. Discuss the word **precisely**. Tell children that a precise explanation is one that gives details and is accurate and complete. GMP6.1, GMP6.2

Remind children that if they think someone else's work is unclear or incomplete, they should still be respectful when they explain why. Refer to your list of discussion guidelines and encourage children to use these sentence frames:

- I think this is a clear and complete explanation because _____.
- I think this explanation needs to include _____.

▶ Reengaging in the Problem

| **WHOLE CLASS** | SMALL GROUP | **PARTNER** | INDEPENDENT |

Children reengage in the problem by analyzing and critiquing other children's work in pairs and in a whole-group discussion. Have children discuss in partners before sharing with the whole group. Guide this discussion based on the decisions you made in Getting Ready for Day 2. GMP1.4, GMP6.1, GMP6.2

▶ Revising Work

| WHOLE CLASS | SMALL GROUP | **PARTNER** | **INDEPENDENT** |

Pass back children's work from Day 1. Before children revise anything, ask them to examine their explanations and decide how to improve them. Ask the following questions one at a time. Have partners discuss their responses and give a thumbs-up or thumbs-down based on their own work.

- *Did you choose at least three items and show the prices for each?*
- *Is your total close to but less than $100? Did you tell how you know?*
- *Did you show all the steps in your thinking? Did you show any close-but-easier numbers you chose?*
- *Did you show how you added the numbers?*

Math Journal 2, p. 241

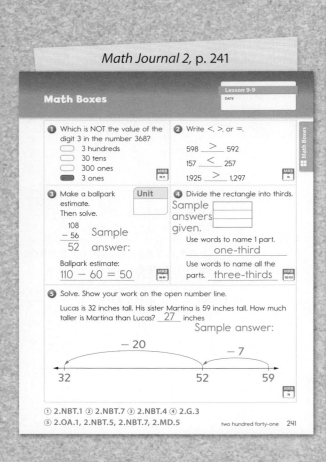

Tell children they now have a chance to revise their work. If their total was not close to $100 or was more than $100, they should use another copy of *Math Masters*, page 274 to add or change items. **GMP1.4, GMP6.2** Remind children that the strategies and explanations discussed are not the only ways to solve the problem. They should explain their strategies using their own number sentences and words. Tell children to add to their earlier work using colored pencils or another copy of *Math Masters*, page 273, instead of erasing their original work.

Summarize Ask children to reflect on their work and revisions. Ask: *How did you make your explanation clearer?* **GMP6.1** Answers vary.

✔ **Assessment Check-In** (CCSS) 2.OA.1

Collect and review children's revised work. Expect children to improve their work based on the class discussion. For the content standard, expect most children to choose at least three items, estimate or use mental math to find the total, and compare the total to $100. You can use the rubric on page 837 to evaluate children's revised work for **GMP6.1**.

☑ Assessment and Reporting | Go Online to record student progress and to see trajectories toward mastery for these standards.

Go Online for optional generic rubrics in the *Assessment Handbook* that can be used to assess any additional GMPs addressed in the lesson.

Sample Children's Work—Evaluated

See the sample in the margin. This work meets expectations for the content standard by showing selected items that total close to $100. Although the addition of the exact numbers is incorrect, with revision the child correctly found a reasonable estimate. The work meets expectations for the mathematical practice by showing the prices and, with revision, close-but-easier numbers for each item added together. **GMP6.1**

3 Practice 10–15 min | Go Online 🖥 ePresentations 🎚 eToolkit 🏠 Home Connections

▶ Math Boxes 9-9

Math Journal 2, p. 241

| WHOLE CLASS | SMALL GROUP | PARTNER | INDEPENDENT |

Mixed Practice Math Boxes 9-9 are paired with Math Boxes 9-11.

▶ Home Link 9-9

Math Masters, p. 275

Homework Children practice estimating the total cost of items in a store using mental math.

Sample child's work, "Meeting Expectations"

I got 92. I added 24 + 58 + 11 = 92, First I added 8 + 4 + 1 then 2 + 5 + 1 and got 92 But in a DIFFERENT Was 20 + 60 + 10 = 90

28
18
+ 60
90

added in my head. I got a remote control for 58$
wireless mouse 24$
and a DVD 11$

Go Online for other samples of evaluated children's work.

Math Masters, p. 275

Estimating Total Cost Home Link 9-9
NAME DATE

Family Note

In this lesson we worked on a problem in which your child pretended to be at a store and needed to estimate the total cost of selected items using mental math. When you are in a store together, choose two or three items and ask your child to try to estimate the total cost without using pencil and paper. Encourage the use of "close-but-easier" numbers for each item to make it easier to find the total cost using mental math.

Please return this Home Link to school tomorrow.

For each problem, pretend you are at a store and do not have a calculator or pencil and paper. Sample explanations given.

① You have $1. You want to buy a toy for 59¢ and an apple for 49¢. Do you have enough money? Tell why or why not.

No. 59¢ is almost 60¢, and 49¢ is almost 50¢. 60¢ + 50¢ is more than $1.

② You have $50. You want to buy a radio for $32, headphones for $18, and a calculator for $6. Do you have enough money? Tell why or why not. No. 30 + 10 = 40 and 2 and 8 make another 10, so the total for the radio and headphones is $50. I couldn't buy the calculator, too.

Practice Unit

Add or subtract.

③ 67
  − 29
  38

④ 35 + 56 = 91

⑤ 71
  − 46
  25

2.OA.1, 2.NBT.5, 2.NBT.6, 2.NBT.8 two hundred seventy-five 275

Lesson 9-10

Connecting Doubles Facts, Even Numbers, and Equal Groups

Overview Children solve number stories about 2 equal groups.

▶ **Before You Begin**
For the Focus activities, prepare a set of 20 centimeter cubes for each partnership.

Common Core State Standards

Focus Clusters
- Represent and solve problems involving addition and subtraction.
- Add and subtract within 20.
- Work with equal groups of objects to gain foundations for multiplication.

1 Warm Up 15–20 min

	Materials	
Mental Math and Fluency Children count by 2s to find the total number of objects in arrays.	slate	2.OA.4
Daily Routines Children complete daily routines.	See pages 4–43.	See pages xiv–xvii.

2 Focus 30–40 min

Math Message Children solve a number story about an array with 2 equal rows.	slate	2.OA.1, 2.OA.4 SMP4
Sharing Arrays and Number Models Children share and discuss answers to the Math Message problem.		2.OA.1, 2.OA.4 SMP4
Connecting Doubles and Equal Groups Children use doubles facts to solve equal-groups stories.	*Math Masters*, p. TA25; per partnership: 20 centimeter cubes	2.OA.1, 2.OA.2, 2.OA.4 SMP4, SMP7, SMP8
Connecting Even Numbers and Equal Groups Children connect even numbers to equal-groups stories.	*Math Journal 2*, p. 242; per partnership: 20 centimeter cubes	2.OA.1, 2.OA.2, 2.OA.3, 2.OA.4 SMP4, SMP7, SMP8
✓ **Assessment Check-In** See page 845.	*Math Journal 2*, p. 242	2.OA.1, 2.OA.3, 2.OA.4

CCSS 2.OA.3 Spiral Snapshot

GMC Express an even number as a sum of two equal addends.

2-9 Focus Practice | 2-10 Practice | 3-11 Practice | 4-2 Practice | 4-4 Practice | 9-10 Focus Practice

Spiral Tracker **Go Online** to see how mastery develops for all standards within the grade.

3 Practice 10–15 min

Equal Shares with Different Shapes Children divide figures into different-shaped equal parts.	*Math Journal 2*, p. 243	2.G.3
Math Boxes 9-10 Children practice and maintain skills.	*Math Journal 2*, p. 244	See page 845.
Home Link 9-10 **Homework** Children solve number stories about 2 equal groups.	*Math Masters*, p. 277	2.OA.1, 2.OA.2, 2.OA.3, 2.NBT.5

connectED.mheducation.com

Plan your lessons online with these tools.

ePresentations | Student Learning Center | Facts Workshop Game | eToolkit | Professional Development | Home Connections | Spiral Tracker | Assessment and Reporting | English Learners Support | Differentiation Support

Differentiation Options

 RtI

Readiness 10–15 min

Modeling 2 Equal Groups

WHOLE CLASS
SMALL GROUP
PARTNER
INDEPENDENT

Math Masters, p. TA6;
10 counters

For experience working with 2 equal groups using a concrete model, children represent the groups with counters on ten frames (*Math Masters,* page TA6). **GMP2.1** As children solve each problem, record their number sentences. *Suggestions:*

• Put 3 counters in each ten frame. How many do you have in all when you have 2 groups of 3 counters each? 6 How could we show this with a number sentence? 3 + 3 = 6

• Put 7 counters in each ten frame. How many do you have in all when you have 2 groups of 7 counters each? 14 How could we show this with a number sentence? 7 + 7 = 14

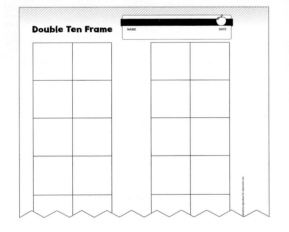

Enrichment 5–15 min

A Paper-Folding Problem

WHOLE CLASS
SMALL GROUP
PARTNER
INDEPENDENT

Math Masters, p. 276; poster paper (or other large paper)

To apply their understanding of doubling, children predict the number of equal parts that result from folding paper. **GMP4.1** Children follow the directions and record their work on *Math Masters,* page 276. It may be helpful for children to use poster paper or some other large piece of paper. Encourage them to look for patterns and think about doubles facts to help them solve the problem. **GMP7.1, GMP7.2**

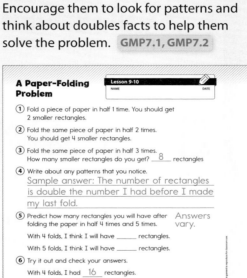

Extra Practice 5–15 min

Writing Number Stories with 2 Equal Groups

WHOLE CLASS
SMALL GROUP
PARTNER
INDEPENDENT

Activity Card 120;
Math Masters, p. TA7

To apply their understanding of equal groups, children write number stories about 2 equal groups. Children write number models for the stories. **GMP4.1** Partners discuss the strategies used to solve the problems.

English Language Learners Support

Beginning ELL Build on children's experiences with everyday examples of arrays, such as egg cartons and ice-cube trays, to review the convention of _____-by-_____ to describe arrays. For example, display a 2-by-6 ice-cube tray and use gestures going across and down along with a choral count-aloud to demonstrate that the tray is a 2-by-6 array. Give children pictures or other real examples of arrays and have them complete the following sentence frame: "This is a _____-by-_____ array."

Go Online   ELL English Learners Support

Standards and Goals for
Mathematical Practice

SMP4 Model with mathematics.
   GMP4.1 Model real-world situations using graphs, drawings, tables, symbols, numbers, diagrams, and other representations.

SMP7 Look for and make use of structure.
   GMP7.1 Look for mathematical structures such as categories, patterns, and properties.

SMP8 Look for and express regularity in repeated reasoning.
   GMP8.1 Create and justify rules, shortcuts, and generalizations.

× × × × × × × ×
× × × × × × × ×

$8 + 8 = 16$
2 [8s] is 16
$2 \times 8 = 16$

---

① Warm Up   15–20 min   Go Online   ePresentations   eToolkit

▶ Mental Math and Fluency

Display the following arrays. Have children count by 2s to find the total number of dots in each array. Then they write addition number models on their slates to represent the arrays. *Leveled exercises:*

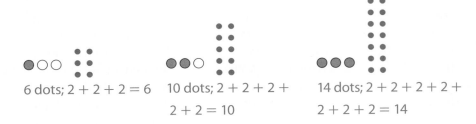

6 dots; $2 + 2 + 2 = 6$ 10 dots; $2 + 2 + 2 + 2 + 2 = 10$ 14 dots; $2 + 2 + 2 + 2 + 2 + 2 + 2 = 14$

▶ Daily Routines

Have children complete daily routines.

---

② Focus   30–40 min   Go Online   ePresentations   eToolkit

▶ Math Message

You have 2 rows of tomato plants with 8 plants in each row. How many plants do you have? 16 plants *Draw an array and write a number model.*   **GMP4.1**

▶ Sharing Arrays and Number Models

WHOLE CLASS   **SMALL GROUP**   PARTNER   **INDEPENDENT**

Math Message Follow-Up Invite a volunteer to sketch an array that matches the Math Message problem.   **GMP4.1**   (*See margin.*) Ask children to describe the array. Expect observations to include the following:

- It shows 2 rows of plants with 8 plants in each row.
- It is a 2-by-8 array.
- It has 2 rows and 8 columns.
- It has 16 objects in all.

Invite a child who wrote an addition number model to share it. $8 + 8 = 16$ Record the number model below the array. Remind children that this number model can be read as "2 rows of 8 is 16 all together." Write "2 [8s] is 16" and explain that this is a short way to record this idea. (*See margin.*)

Ask whether anyone wrote a multiplication number model and record $2 \times 8 = 16$. Tell children that the $\times$ is a *multiplication symbol* and that multiplication is an operation that involves finding the total number of things in equal groups. Explain that $2 \times 8 = 16$ can be read the same way as the addition number model: "2 groups of 8 is 16 in all." Have children read the model $2 \times 8 = 16$ aloud.

Tell children that they will solve more problems involving 2 equal groups and write number models.

▶ Connecting Doubles and Equal Groups

Math Masters, p. TA25

| WHOLE CLASS | SMALL GROUP | PARTNER | INDEPENDENT |

Distribute 20 centimeter cubes to each partnership. Returning to the Math Message context of tomato plants, tell children that there is space in the garden for only 2 rows of plants with up to 10 plants in each row. There should always be 2 equal rows, but each row can have fewer than 10 plants. Have children use their centimeter cubes to build at least three possible arrays of tomato plants that would fit in the garden. GMP4.1 Instruct them to record their arrays on centimeter grid paper (*Math Masters,* page TA25) and write addition or multiplication number models to match each array. When children are finished, have them share their arrays while you make an ordered list of the arrays and possible number models, as shown below.

Array	Addition Number Model	Multiplication Number Model
2-by-1 • •	1 + 1 = 2	2 × 1 = 2
2-by-2 • • • •	2 + 2 = 4	2 × 2 = 4
2-by-3 • • • • • •	3 + 3 = 6	2 × 3 = 6
⋮	⋮	⋮

After all 10 possible arrays have been recorded, have children examine the list. Ask: *What patterns do you notice?* GMP7.1 Sample answers: The total increases by 2 each time. The totals are all even numbers. The addition models are all doubles facts. Multiplying a number by 2 is the same as doubling the number. Referring to the two lists of possible number models, discuss the idea that when children need to find the total number of objects in 2 equal groups (or multiply by 2), they can use addition doubles. Ask: *How can we use doubles facts to help us solve number stories about 2 equal groups?* Sample answer: We just double the number in one of the groups. GMP8.1

> **NOTE** Some children might suggest repeatedly adding 2 or skip counting by 2, as opposed to thinking of 2 groups or 2 rows. If this is suggested, record those number models separately (for example, 2 + 2 + 2 = 6 and 3 × 2 = 6) and return to them at the end the discussion described above. Help children connect these alternative models to the idea of skip counting by 2s. For example, point to the 2s in 2 + 2 + 2 = 6 as you skip count by 2s to show how this number model represents the strategy of counting by 2s. Contrast this with the strategy of doubling 3, which is represented by the model 3 + 3 = 6.

Math Journal 2, p. 242

Two Equal Groups

Lesson 9-10

DATE

Solve the number stories. You can use cubes or draw pictures to help. Then write an addition number model for each problem. The number model should be a doubles fact.

1. There are 2 rows of paintings hanging on a wall. Each row has 3 paintings. How many paintings are on the wall?

   Answer: __6__ paintings

   Addition number model:

   $3 + 3 = 6$

2. A bookshelf has 2 shelves. Each shelf has 9 books on it. How many books are on the bookshelf?

   Answer: __18__ books

   Addition number model:

   $9 + 9 = 18$

3. You have 4 flowers in all and 2 vases. Each vase should have the same number of flowers. How many flowers should you put in each vase?

   Answer: __2__ flowers

   Addition number model:

   $2 + 2 = 4$

4. A total of 12 chairs are in 2 equal rows. How many chairs are in each row?

   Answer: __6__ chairs

   Addition number model:

   $6 + 6 = 12$

Try This

5. Write a multiplication number model for Problem 1.

   $2 \times 3 = 6$

6. Write a multiplication number model for Problem 2.

   $2 \times 9 = 18$

242 two hundred forty-two   2.OA.1, 2.OA.2, 2.OA.3, 2.OA.4, SMP4

Math Journal 2, p. 243

Equal Shares with Different Shapes

Lesson 9-10

DATE

Divide each shape into equal parts. Use the dots to help you draw the lines. Do not use any diagonal lines.

1. Divide this shape into 3 equal parts.
   Sample answer:

2. Divide this shape into 2 equal parts.

3. For Problem 2, are the parts the same shape? __No.__

   Are the parts the same size? __Yes.__

   How do you know?

   Sample answer: Each piece has 5 little squares in it.

Try This

4. Divide this shape into 4 equal parts.
   Sample answer:

2.G.3   two hundred forty-three   243

▶ # Connecting Even Numbers and Equal Groups

Math Journal 2, p. 242

WHOLE CLASS | **SMALL GROUP** | **PARTNER** | INDEPENDENT

Refer to the list of arrays and number models from the previous activity. If no one has mentioned it yet, ask children to look at the totals for each array and determine whether they are even or odd. **GMP7.1** They are all even. Ask: *Can the total number in 2 equal groups or rows be an odd number?* No. *How do you know?* **GMP8.1** Sample answer: If the 2 groups have the same number, you can match up one object from one group with one object from the other group, and every object will have a pair.

Ask: *If I have 14 cubes and I want to put them into 2 equal groups, what doubles fact could help me?* $7 + 7 = 14$ *Why?* Because it shows that 2 groups of 7 make 14 in all Guide children to see that they can also use doubles facts to help them put an even number of objects into 2 equal groups.

> **Differentiate**   **Adjusting the Activity**
>
> Display 14 cubes. Model moving two cubes at a time, one to the right and one to the left, until you have made 2 groups of 7. Display $7 + 7 = 14$ below the cubes to help children make the connection between doubles facts and 2 equal groups.
>
> **Go Online**   Differentiation Support

Pose number stories involving 2 equal groups or rows of objects. Encourage children to use their knowledge of doubles facts to help them solve the problems. They can use cubes or draw pictures to model the problems. **GMP4.1** Have them write addition and multiplication number models for the problems and share them with the class. *Suggestions:*

- You have 2 apples. Each apple is cut into 8 slices. How many slices are there in all? 16 slices; $8 + 8 = 16$; $2 \times 8 = 16$

- Your friend has 2 fish tanks with 6 fish in each tank. How many fish does your friend have in all? 12 fish; $6 + 6 = 12$; $2 \times 6 = 12$

- There are 10 pencils in all. You want to put an equal number of pencils in each of your 2 pencil cups. How many pencils should you put in each cup? 5 pencils; $5 + 5 = 10$; $2 \times 5 = 10$

> **Differentiate**   **Common Misconception**
>
> For problems where the total is given and children need to find the number in each group, some children may try to double the total. Sketch a parts-and-total diagram and write the given number in the total box. Point out that the total is what children know, and the number in each group (or part) is what they need to find. Ask: *Will the number in each group be larger or smaller than the total?* Smaller
>
> **Go Online**   Differentiation Support

After solving some problems as a class, have children work in partnerships to complete journal page 242.

✔ Assessment Check-In CCSS 2.OA.1, 2.OA.3, 2.OA.4

Math Journal 2, p. 242

Expect that most children will be able to correctly solve Problems 1–4 on journal page 242 and write addition number models for each problem. If children struggle with writing the number models in Problems 3–4, have them use cubes to model making 2 equal groups as described in the Adjusting the Activity note. Some children may be able to write multiplication number models for Problems 5–6.

☑ Assessment and Reporting   [Go Online] to record student progress and to see trajectories toward mastery for these standards.

Summarize Have children name something they know about number stories involving 2 equal groups. Sample answers: The total is always an even number. I can use doubles facts to help me solve them. The addition number models have equal addends.

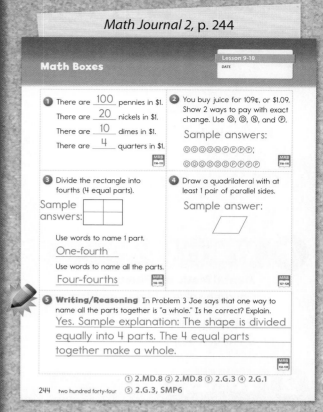

Math Journal 2, p. 244

Math Boxes

Lesson 9-10
DATE

1. There are __100__ pennies in $1.
There are __20__ nickels in $1.
There are __10__ dimes in $1.
There are __4__ quarters in $1.
MRB [116–117]

2. You buy juice for 109¢, or $1.09. Show 2 ways to pay with exact change. Use Ⓠ, Ⓓ, Ⓝ, and Ⓟ.
Sample answers:
ⓆⓆⓆⓆ ⓃⓅⓅⓅⓅ;
Ⓠ ⓀⓓⓓⓓⓅⓅⓅⓅ
MRB [116–117]

3. Divide the rectangle into fourths (4 equal parts).
Sample answers:
Use words to name 1 part.
One-fourth
Use words to name all the parts.
Four-fourths
MRB [122–123]

4. Draw a quadrilateral with at least 1 pair of parallel sides.
Sample answer:
MRB [127–128]

5. **Writing/Reasoning** In Problem 3 Joe says that one way to name all the parts together is "a whole." Is he correct? Explain.
Yes. Sample explanation: The shape is divided equally into 4 parts. The 4 equal parts together make a whole.
MRB [122–123]

244   two hundred forty-four   ① 2.MD.8 ② 2.MD.8 ③ 2.G.3 ④ 2.G.1
⑤ 2.G.3, SMP6

---

③ Practice   10–15 min

[Go Online]   ePresentations   eToolkit   Home Connections

▶ Equal Shares with Different Shapes

Math Journal 2, p. 243

| WHOLE CLASS | **SMALL GROUP** | PARTNER | INDEPENDENT |

Children partition rectilinear figures into same-size shares that are different shapes on journal page 243. Remind children to use the dots to help them figure out how to partition the figures. Encourage them to describe to partners how they know the shares are the same size. When they complete the page, have children share their answers.

▶ Math Boxes 9-10

Math Journal 2, p. 244

| WHOLE CLASS | **SMALL GROUP** | PARTNER | INDEPENDENT |

Mixed Practice Math Boxes 9-10 are paired with Math Boxes 9-12.

▶ Home Link 9-10

Math Masters, p. 277

Homework Children solve number stories about 2 equal groups.

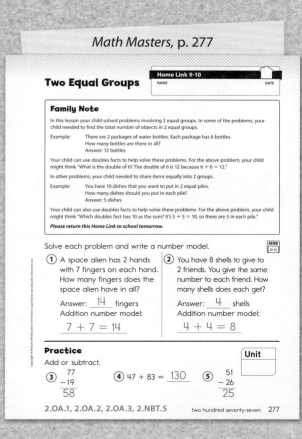

Math Masters, p. 277

Two Equal Groups

Home Link 9-10
NAME   DATE

Family Note

In this lesson your child solved problems involving 2 equal groups. In some of the problems, your child needed to find the total number of objects in 2 equal groups.

Example: There are 2 packages of water bottles. Each package has 6 bottles.
How many bottles are there in all?
Answer: 12 bottles

Your child can use doubles facts to help solve these problems. For the above problem, your child might think "What is the double of 6? The double of 6 is 12 because 6 + 6 = 12."

In other problems, your child needed to share items equally into 2 groups.

Example: You have 10 dishes that you want to put in 2 equal piles.
How many dishes should you put in each pile?
Answer: 5 dishes

Your child can also use doubles facts to help solve these problems. For the above problem, your child might think "Which doubles fact has 10 as the sum? It's 5 + 5 = 10, so there are 5 in each pile."

Please return this Home Link to school tomorrow.

Solve each problem and write a number model.   MRB [12–13]

1. A space alien has 2 hands with 7 fingers on each hand. How many fingers does the space alien have in all?
Answer: __14__ fingers
Addition number model:
7 + 7 = 14

2. You have 8 shells to give to 2 friends. You give the same number to each friend. How many shells does each get?
Answer: __4__ shells
Addition number model:
4 + 4 = 8

Practice
Add or subtract.

3. 77
  −19
  58

4. 47 + 83 = __130__

5. 51
  −26
  25

Unit

2.OA.1, 2.OA.2, 2.OA.3, 2.NBT.5   two hundred seventy-seven   277

Multiples of 10 and 5

Overview Children skip count and add to solve problems involving multiples of 10 and 5.

▶ **Vocabulary**
multiple

Common Core State Standards

Focus Clusters
- Understand place value.
- Use place value understanding and properties of operations to add and subtract.

1 Warm Up 15–20 min

	Materials	
Mental Math and Fluency Children solve addition problems involving multiples of 10.	slate	2.OA.2, 2.NBT.5, 2.NBT.7
Daily Routines Children complete daily routines.	See pages 4–43.	See pages xiv–xvii.

2 Focus 30–40 min

Math Message Children solve a number story involving groups of 10.	*Math Journal 2,* inside back cover; Class Number Line; base-10 blocks	2.OA.1, 2.NBT.2, 2.NBT.5 SMP5
Using Tools to Show Groups of 10 Children share and discuss their strategies for using tools.	*Math Journal 2,* number grid; Class Number Line; base-10 blocks	2.NBT.2, 2.NBT.5 SMP5, SMP7
Relating 10s and 5s Children discuss strategies for finding the total number of objects in groups of 10 and groups of 5.	*Math Journal 2,* p. 245; 1 dime and 1 nickel (for demonstration)	2.NBT.2, 2.NBT.5 SMP7, SMP8
Applying Strategies Children summarize and apply their strategies for solving problems involving 2s, 5s, and 10s.	*Math Journal 2,* p. 246	2.NBT.2, 2.NBT.5
✓ **Assessment Check-In** See page 851.	*Math Journal 2,* p. 246; *Math Masters,* p. TA3 (optional)	2.NBT.2, 2.NBT.5

CCSS 2.NBT.2 **Spiral Snapshot**

GMC Count by 5s, 10s, and 100s.

5-2 Focus Practice	5-6 Focus	8-8 through 8-10 Focus Practice	8-11 Practice	9-1 Practice	9-8 Focus Practice	9-11 Focus Practice

▓ Spiral Tracker   **Go Online**   to see how mastery develops for all standards within the grade.

3 Practice 10–15 min

Partitioning a Rectangle Children partition a rectangle into same-size squares.	*Math Journal 2,* p. 247	2.G.2 SMP3, SMP6
Math Boxes 9-11 Children practice and maintain skills.	*Math Journal 2,* p. 248	See page 851.
Home Link 9-11 **Homework** Children skip count to find multiples of 10 and 5.	*Math Masters,* p. 281	2.NBT.2, 2.NBT.5

connectED.mheducation.com

Plan your lessons online with these tools.

 ePresentations   Student Learning Center   Facts Workshop Game   eToolkit   Professional Development   Home Connections   Spiral Tracker   ✓ Assessment and Reporting   English Learners Support   Differentiation Support

Differentiation Options

RtI

Readiness
5–15 min

WHOLE CLASS
SMALL GROUP
PARTNER
INDEPENDENT

Counting on Calculators

Math Masters, p. 278; calculator

For experience skip counting by 10s, 5s, and 2s, children do calculator counts. See Lesson 1-6 for more information about calculator skip counting. Children record their results on *Math Masters*, page 278. Then they look for patterns in the counts. **GMP7.1**

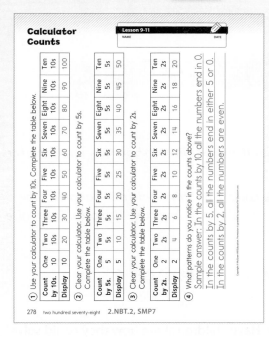

Enrichment
10–15 min

WHOLE CLASS
SMALL GROUP
PARTNER
INDEPENDENT

Making Multiples

Math Masters, p. 279; base-10 blocks

To apply their understanding of multiples, children solve function-machine problems involving equal groups of base-10 blocks. Children explain their strategies for solving the problems. **GMP6.1**

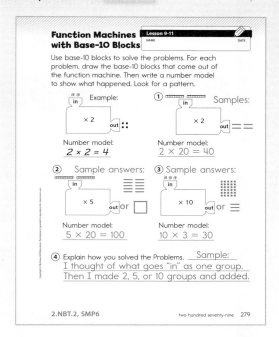

Extra Practice
10–15 min

WHOLE CLASS
SMALL GROUP
PARTNER
INDEPENDENT

Patterns in Multiples of 2, 5, and 10

Math Masters, p. 280; red, blue, and green colored pencils

For further practice with multiples of 2, 5, and 10, children circle the multiples of 2, 5, and 10 on a number grid, each with a different color. Prompt children to discuss the patterns they see and explain why they think the patterns occur. **GMP7.1** Expect children to notice that the multiples of 2, 5, and 10 each form vertical columns, and all multiples of 10 are also multiples of 2 and 5.

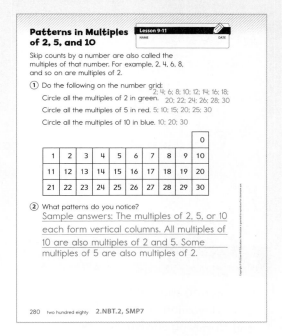

English Language Learners Support

Beginning ELL Children may confuse the terms *multiply* and *multiple*. Use teacher modeling to help children understand that *multiply* refers to an action, whereas *multiple* refers to a number. Introduce requests to *multiply* with action sentences, such as the following: *Here is what you do. Multiply _____ by _____.* For *multiple,* use number cards and show-me or find requests, such as these: *Show me a multiple of _____. Say the multiple. Find a multiple of _____. Say the multiple.*

Go Online — ELL English Learners Support

Professional Development

Lessons 9-10 and 9-11 focus on number stories involving multiples of 2, 5, and 10. Having efficient strategies for solving such stories by the end of second grade will help children develop fluency with the 2s, 5s, and 10s multiplication facts early in third grade. These key groups of facts will then be used as helper facts to help children learn the rest of the multiplication facts by the end of third grade.

Go Online Professional Development

1 Warm Up 15–20 min [Go Online]
ePresentations   eToolkit

▶ Mental Math and Fluency

Pose addition problems involving multiples of 10. *Leveled exercises:*

● ○ ○ 20 + 30 50
30 + 40 70

● ● ○ 50 + 50 100
50 + 60 110

● ● ● 60 + 70 130
90 + 80 170

▶ Daily Routines

Have children complete daily routines.

2 Focus 30–40 min [Go Online]
ePresentations   eToolkit

▶ Math Message

You have 6 boxes of markers with 10 markers in each box. How many markers do you have in all?

Talk to a partner about how you could solve this problem using each of the following tools: a number line, a number grid, and base-10 blocks. Then solve the problem. 60 markers **GMP5.2**

▶ Using Tools to Show Groups of 10

| WHOLE CLASS | SMALL GROUP | PARTNER | INDEPENDENT |

Math Message Follow-Up Have children share how they could solve the number story by using each tool suggested in the Math Message.
GMP5.2 *Sample strategies:*

• On a number line, start at 0 and make 6 hops of 10: 10, 20, 30, 40, 50, 60. You have 60 markers in all.

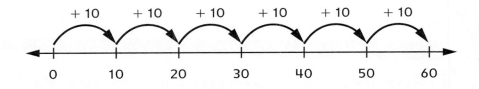

• On a number grid, start at 0 and move down 6 rows: 10, 20, 30, 40, 50, 60. You have 60 markers in all.

• With base-10 blocks, count out 6 longs for the 6 groups of 10. Count each long as 10: 10, 20, 30, 40, 50, 60. You have 60 markers in all.

Have children discuss how each tool illustrates the quantities in the problem. **GMP5.2** For example, on the number line, each hop represents a box of 10 markers. On the number grid, each row represents a box of 10 markers. With base-10 blocks, each long represents a box of 10 markers.

Ask: *What do all these strategies have in common?* **GMP7.1** Sample answer: They all use groups of 10. They involve either skip counting by 10 or adding groups of 10. Summarize the problem by displaying the following equal-groups notation: 6 [10s] is 60. Ask children to suggest a number model for the Math Message problem. Display both an addition model and a multiplication model. $10 + 10 + 10 + 10 + 10 + 10 = 60$; $6 \times 10 = 60$ Have children practice reading the number models aloud as "6 groups of 10 is 60 in all."

Tell children that they will solve more problems by skip counting and adding 10s.

▶ Relating 10s and 5s

Math Journal 2, p. 245

| WHOLE CLASS | SMALL GROUP | PARTNER | INDEPENDENT |

Display a dime and a nickel. Ask: *What is the value of each coin?* A dime is worth 10 cents. A nickel is worth 5 cents. Have children work as partners to determine the values of the sets of dimes and nickels in Problem 1 on journal page 245.

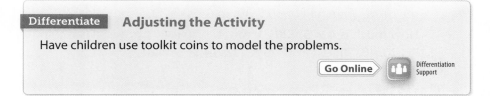

Differentiate **Adjusting the Activity**

Have children use toolkit coins to model the problems.

Go Online **Differentiation Support**

When most children have completed Problem 1, have each pair of children compare their answers with another partnership and resolve any discrepancies. Ask: *How did you find the values of the sets of dimes?* Sample answers: I skip counted by 10s. I added 10s in my head. *How did you find the values of the sets of nickels?* Sample answers: I skip counted by 5s. I added 5s in my head.

Remind children that *multiplication* is an operation that involves finding the total number of things that are in equal groups. Point out that a dime can be thought of as a "group" of 10 cents, so when children were finding the value of a set of dimes, they were multiplying the number of dimes in the set by 10. In each case the total value is a skip-count by 10s, or a **multiple** of 10. Display equal-groups notation for the sets of dimes in the table shown on the next page. Ask children to suggest a multiplication number model for each set of dimes and record them. (*See table on next page.*) Then repeat for the nickels, pointing out that finding the total value of a set of nickels is like multiplying the number of nickels by 5. In each case the total value is a skip-count by 5s, or a multiple of 5.

Academic Language Development

To further children's understanding of the term *multiple,* have them work in pairs to complete 4-Square Graphic Organizers (*Math Masters,* page TA42) showing pictures, examples, nonexamples, and their own definitions of the term.

Math Journal 2, p. 245

Dimes and Nickels   Lesson 9-11
                                         DATE

1. Find the value of each set of dimes and the value of each set of nickels.

Number of Dimes	Value	Number of Nickels	Value
1	10 cents	1	5 cents
2	20 cents	2	10 cents
4	40 cents	4	20 cents
5	50 cents	5	25 cents
8	80 cents	8	40 cents
10	100 cents	10	50 cents

2. Look at the value of the dimes and the value of the nickels in each row. What do you notice?
   Sample answer: The dimes are worth twice as much as the nickels.

2.NBT.2, 2.NBT.5, SMP7, SMP8   two hundred forty-five 245

Multiples of 10, 5, and 2

Lesson 9-11
DATE

1 How many is 3 [5s]? __15__

Number model:

$3 \times 5 =$ __15__

2 How many is 7 [10s]? __70__

Number model:

$7 \times 10 =$ __70__

3 How much money is 5 nickels? __25__ cents

Number model:

$5 \times 5 =$ __25__

4 How much money is 6 dimes? __60__ cents

Number model:

$6 \times 10 =$ __60__

5 How much money is 6 nickels? __30__ cents

Number model:

$6 \times 5 =$ __30__

6 A store display has 4 rows of oranges. There are 5 oranges in each row. Draw an array to show the oranges.

How many oranges are in the display? __20__ oranges

How many is 4 [5s]? __20__

Number model: $4 \times 5 =$ __20__

7 Jo puts pictures in an album. There are 3 rows of pictures on a page. There are 2 pictures in each row. Draw an array to show the pictures on one page.

How many pictures are on the page? __6__ pictures

How many is 3 [2s]? __6__

Number model: $3 \times 2 =$ __6__

246   two hundred forty-six   2.NBT.2, 2.NBT.5

Partitioning a Rectangle

Lesson 9-11
DATE

Paul's teacher asked him to partition the rectangle at right into 3 rows and 5 columns. Then she asked him to count the total number of small squares.

Paul's partitioned rectangle is shown at right.

How many small squares cover the rectangle? __24__

1 Do you agree with Paul's solution? __No.__ Why or why not?

Sample answer:
Paul drew one extra
column and one
extra row.

2 Show how you would partition the rectangle into 3 rows and 5 columns.

3 Describe how you partitioned the rectangle. Sample:
I drew 5 columns and
3 rows. I made all of
the squares the same
size.

4 How many small squares cover the rectangle? __15__ squares

2.G.2, SMP3, SMP6   two hundred forty-seven   247

850   Unit 9 | Equal Shares and Whole Number Operations

Dimes		Nickels	
1 [10] is 10	$1 \times 10 = 10$	1 [5] is 5	$1 \times 5 = 5$
2 [10s] is 20	$2 \times 10 = 20$	2 [5s] is 10	$2 \times 5 = 10$
4 [10s] is 40	$4 \times 10 = 40$	4 [5s] is 20	$4 \times 5 = 20$
5 [10s] is 50	$5 \times 10 = 50$	5 [5s] is 25	$5 \times 5 = 25$
8 [10s] is 80	$8 \times 10 = 80$	8 [5s] is 40	$8 \times 5 = 40$
10 [10s] is 100	$10 \times 10 = 100$	10 [5s] is 50	$10 \times 5 = 50$

Ask children to complete Problem 2 on the journal page by examining the values of the dimes and nickels in each row and looking for a pattern. **GMP7.1** Ask volunteers to share their thoughts with the class. Sample answer: Each set of dimes is worth twice as much as a set with the same number of nickels. Ask: *Can anyone think of a way we can use this pattern as a shortcut when we multiply a number by 5?* **GMP8.1** Sample answer: We can multiply the number by 10 and then take half of the answer.

Pose problems involving groups of 5 or multiplying by 5 and have volunteers describe how to use the shortcut to find the total. *Suggestions:*

- How many is 3 groups of 5? 15; I can skip count by 10s to find 3 groups of 10: 10, 20, 30. Then I can take half of that to find 3 groups of 5: 15.

- How many is 6 [5s]? 30; I know 6 [10s] is 60 because I added 6 tens in my head. Then half of 60 is 30, so 6 [5s] is 30.

- How much is 4×5? 20; I know 4×10 is 40 because I can think 10, 20, 30, 40. Then half of 40 is 20, so $4 \times 5 = 20$.

► Applying Strategies

Math Journal 2, p. 246

| WHOLE CLASS | SMALL GROUP | PARTNER | INDEPENDENT |

Display a table like the one shown below. Remind children that in the previous lesson, they multiplied by 2; in this lesson, they multiplied by 5 and 10. Have children share their strategies for solving each type of problem to create a class list of strategies. *Sample list:*

Multiplying by 2	Multiplying by 5	Multiplying by 10
Think of addition doubles.	Think of counting nickels.	Think of counting dimes.
Skip count by 2s.	Skip count by 5s.	Add 10s in your head.
	Find the multiple of 10 and then find half of that answer.	Skip count by 10s.
		Think of base-10 longs.

Have partners practice using the strategies as they complete journal page 246.

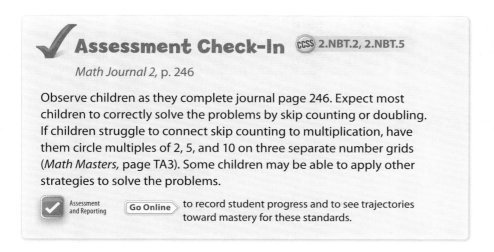

✓ Assessment Check-In CCSS 2.NBT.2, 2.NBT.5

Math Journal 2, p. 246

Observe children as they complete journal page 246. Expect most children to correctly solve the problems by skip counting or doubling. If children struggle to connect skip counting to multiplication, have them circle multiples of 2, 5, and 10 on three separate number grids (*Math Masters*, page TA3). Some children may be able to apply other strategies to solve the problems.

☑ Assessment and Reporting   Go Online ⟩ to record student progress and to see trajectories toward mastery for these standards.

Summarize Have partners talk about ways they could practice finding the total number of objects in groups of 2, 5, or 10 at home. Sample: Make up number stories; draw arrays with 2, 5, or 10 rows and find the total.

③ Practice   10–15 min

Go Online ⟩ ePresentations eToolkit Home Connections

▸ Partitioning a Rectangle

Math Journal 2, p. 247

| WHOLE CLASS | SMALL GROUP | PARTNER | INDEPENDENT |

Have children turn to journal page 247. Review the meanings of *rows* and *columns* and have children trace one of each with their fingers. Point out that there is a mistake in the way Paul partitioned the rectangle on journal page 247. (Have children count the rows and columns in Paul's rectangle and compare the results with his teacher's directions.) Then have them complete the page. When most children have finished, have volunteers explain Paul's mistake. GMP3.2 Have volunteers share their partitioned rectangles and their strategies for partitioning. GMP6.1

▸ Math Boxes 9-11

Math Journal 2, p. 248

| WHOLE CLASS | SMALL GROUP | PARTNER | INDEPENDENT |

Mixed Practice Math Boxes 9-11 are paired with Math Boxes 9-9.

▸ Home Link 9-11

Math Masters, p. 281

Homework Children skip count to find multiples of 5 and 10.

Math Journal 2, p. 248

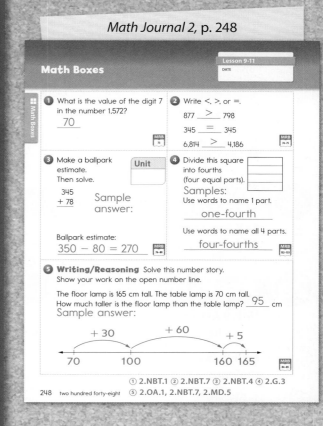

Math Masters, p. 281

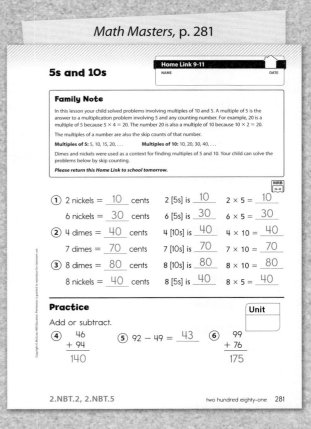

Unit 9 Progress Check

Overview **Day 1:** Administer the Unit Assessments.
Day 2: Administer the Open Response Assessment.

2-Day Lesson

 Student Learning Center
Students may take
assessments digitally.

 Assessment and Reporting
Record results and track
progress toward mastery.

Day 1: Unit Assessments

1 Warm Up 5–10 min

Materials

Self Assessment
Children complete the Self Assessment.

Assessment Handbook, p. 62

2a Assess 35–50 min

Unit 9 Assessment
These items reflect mastery expectations to this point.

Assessment Handbook, pp. 63–66

Unit 9 Challenge (Optional)
Children may demonstrate progress beyond expectations.

Assessment Handbook, p. 67

Common Core State Standards	Goals for Mathematical Content (GMC)*	Lessons	Self Assessment	Unit 9 Assessment	Unit 9 Challenge
2.OA.1	Use addition and subtraction to solve 1-step number stories.	9-10, 9-11		5, 6	1
	Model 1-step problems involving addition and subtraction.	9-10		5, 6	1
2.OA.2	Know all sums of two 1-digit numbers automatically.	9-10		5, 6	1
2.OA.3	Express an even number as a sum of two equal addends.	9-10	5	5, 6	
2.OA.4	Find the total number of objects in a rectangular array.	9-10	5	5	1
	Express the number of objects in an array as a sum of equal addends.	9-10	5	5	1
2.NBT.1	Understand 3-digit place value.	9-5, 9-6		4b	
	Represent whole numbers as hundreds, tens, and ones.	9-5 to 9-7		4b	
2.NBT.2	Count by 5s, 10s, and 100s.	9-8, 9-11		8, 9	
2.NBT.3	Read and write number names.			4a	
	Read and write numbers in expanded form.	9-5 to 9-7	3	4b	
2.NBT.4	Record comparisons using >, =, or <.	9-5	4	4c	
2.NBT.5	Add within 100 fluently.	9-9, 9-11		8, 9	
	Subtract within 100 fluently.	9-6, 9-7	6	7	
2.NBT.7	Add multidigit numbers using models or strategies.	9-8		10	
	Subtract multidigit numbers using models or strategies.	9-6, 9-7		7	
2.MD.1	Measure the length of an object.	9-4	2	3a	
2.MD.4	Measure to determine how much longer one object is than another.	9-4		3b	
2.MD.8	Solve problems involving coins and bills.	9-8		8–10	

Goals for Mathematical Content (GMC)*	Lessons	Self Assessment	Unit 1 Assessment	Unit 1 Challenge
2.G.3 Partition shapes into equal shares.	9-1, 9-2, 9-3	1	1a, 2a, 11	2
Describe equal shares using fraction words.	9-1 to 9-3		1b, 2b	
Describe the whole as a number of shares.	9-1 to 9-3		1c, 2c	
Recognize that equal shares of a shape need not have the same shape.	9-1, 9-2		11	2

Goals for Mathematical Practice (GMP)				
SMP1 Check whether your answer makes sense. GMP1.4	9-6, 9-9		7	
Solve problems in more than one way. GMP1.5	9-6 to 9-8		10	
SMP2 Create mathematical representations using numbers, words, pictures, symbols, gestures, tables, graphs, and concrete objects. GMP2.1	9-1, 9-2, 9-5, 9-6		4, 10	
Make sense of the representations you and others use. GMP2.2	9-1, 9-2, 9-5 to 9-7		4	1
SMP4 Model real-world situations using graphs, drawings, tables, symbols, numbers, diagrams, and other representations. GMP4.1	9-3, 9-8, 9-10			
Use mathematical models to solve problems and answer questions. GMP4.2	9-3, 9-8		10	

*All instruction and practice on the Grade 2 content is complete.

 Spiral Tracker 〈 **Go Online** 〉 to see how mastery develops for all standards within the grade.

1 Warm Up 5-10 min

▶ Self Assessment

Assessment Handbook, p. 62

| WHOLE CLASS | SMALL GROUP | PARTNER | **INDEPENDENT** |

Children complete the Self Assessment to reflect on their progress in Unit 9.

Assessment Handbook, p. 62

Put a check in the box that tells how you do each skill.

Skills		I can do this. I can explain how to do this.	I can do this by myself.	I can do this with help.
① Divide shapes into equal shares.	MJ2 220-221			
② Measure the lengths of objects.	MJ2 227			
③ Write 3-digit numbers in expanded form.	MJ2 230			
④ Use <, >, or = to compare 3-digit numbers.	MRB 74-75			
⑤ Solve number stories about equal groups.	MJ2 242			
⑥ Solve 2-digit subtraction problems.	MJ3 234			

Unit 9 Self Assessment   Lesson 9-12

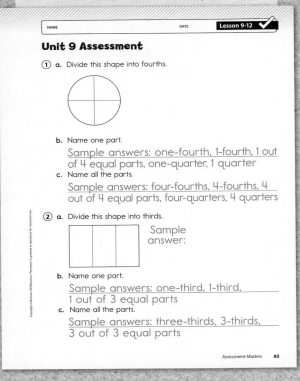

Unit 9 Assessment

① **a.** Divide this shape into fourths.

b. Name one part.
Sample answers: one-fourth, 1-fourth, 1 out of 4 equal parts, one-quarter, 1 quarter
c. Name all the parts.
Sample answers: four-fourths, 4-fourths, 4 out of 4 equal parts, four-quarters, 4 quarters

② **a.** Divide this shape into thirds.
Sample answer:

b. Name one part.
Sample answers: one-third, 1-third, 1 out of 3 equal parts
c. Name all the parts.
Sample answers: three-thirds, 3-thirds, 3 out of 3 equal parts

Assessment Masters   **63**

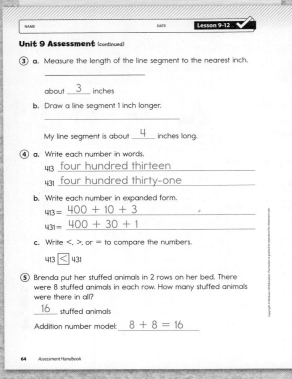

Unit 9 Assessment (continued)

③ **a.** Measure the length of the line segment to the nearest inch.

about __3__ inches

b. Draw a line segment 1 inch longer.

My line segment is about __4__ inches long.

④ **a.** Write each number in words.
413 four hundred thirteen
431 four hundred thirty-one

b. Write each number in expanded form.
413 = 400 + 10 + 3
431 = 400 + 30 + 1

c. Write <, >, or = to compare the numbers.
413 $<$ 431

⑤ Brenda put her stuffed animals in 2 rows on her bed. There were 8 stuffed animals in each row. How many stuffed animals were there in all?
__16__ stuffed animals
Addition number model: 8 + 8 = 16

64   Assessment Handbook

 **2a Assess**   35–50 min   Go Online

Assessment and Reporting   Differentiation Support

▶ Unit 9 Assessment

Assessment Handbotok, pp. 63–66

| WHOLE CLASS | SMALL GROUP | PARTNER | **INDEPENDENT** |

Children complete the Unit 9 Assessment to demonstrate their progress on the Common Core State Standards covered in this unit.

Go Online for generic rubrics in the *Assessment Handbook* that can be used to evaluate children's progress on the Mathematical Practices.

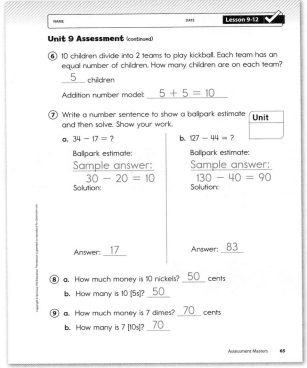

Unit 9 Assessment (continued)

⑥ 10 children divide into 2 teams to play kickball. Each team has an equal number of children. How many children are on each team?
__5__ children
Addition number model: 5 + 5 = 10

⑦ Write a number sentence to show a ballpark estimate and then solve. Show your work.   Unit

a. 34 − 17 = ?   **b.** 127 − 44 = ?

Ballpark estimate: Ballpark estimate:
Sample answer: Sample answer:
30 − 20 = 10   130 − 40 = 90
Solution: Solution:

Answer: __17__   Answer: __83__

⑧ **a.** How much money is 10 nickels? __50__ cents
b. How many is 10 [5s]? __50__

⑨ **a.** How much money is 7 dimes? __70__ cents
b. How many is 7 [10s]? __70__

Assessment Masters   **65**

Assessment Handbook, p. 65

Differentiate Adjusting the Assessment

Item(s)	Adjustments
1, 2	To scaffold items 1 and 2, direct children to the Equal Shares posters created in Lesson 9-1.
3	To extend item 3, have children measure the line segment to the nearest $\frac{1}{2}$ inch and explain their measurement.
4, 7	To scaffold items 4 and 7, provide children with base-10 blocks to represent the numbers.
5, 6	To scaffold items 5 and 6, provide children with counters.
8–10	To scaffold items 8, 9, and 10, provide children with coins.
10	To extend item 8, have children show all possible ways to pay for the orange juice.
11	To extend item 11, have children explain how they know their parts are equal.

Advice for Differentiation

All instruction in *Second Grade Everyday Mathematics* is complete. If you have concerns about children's progress on this content, see the online differentiation options for support.

> **NOTE** See the Unit Organizer on pages 768–769 or the online Spiral Tracker for details on Unit 9 focus topics and the spiral.

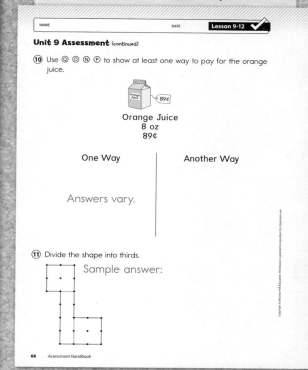

NAME   DATE   Lesson 9-12

Unit 9 Assessment (continued)

⑩ Use Ⓠ Ⓓ Ⓝ Ⓟ to show at least one way to pay for the orange juice.

Orange Juice
8 oz
89¢

One Way   Another Way

Answers vary.

⑪ Divide the shape into thirds.
Sample answer:

66   Assessment Handbook

▶ Unit 9 Challenge (Optional)

Assessment Handbook, p. 67

Children can complete the Unit 9 Challenge after they complete the Unit 9 Assessment.

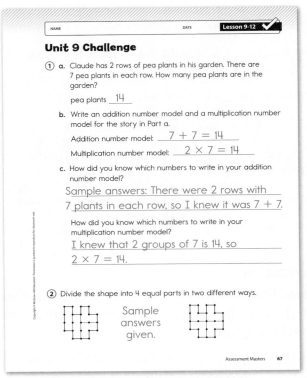

Assessment Handbook, p. 67

Day 2: Open Response Assessment

2b Assess   50–55 min

	Materials
Solving the Open Response Problem Children solve an open response problem organizing a set of digits to find the largest sum.	*Assessment Handbook*, pp. 68–69; scissors; base-10 blocks (optional); toolkit bills (optional); number grid (optional)
Discussing the Problem Children share how they know they found the largest sum.	*Assessment Handbook*, pp. 68–69; base-10 blocks (optional); toolkit bills (optional); number grid (optional)

CCSS Common Core State Standards

	Goal for Mathematical Content (GMC)	**Lessons**
2.NBT.5	Add within 100 fluently.	9-2, 9-5, 9-8 to 9-11
	Goal for Mathematical Practice (GMP)	
SMP3	Make mathematical conjectures and arguments. **GMP3.1**	9-2, 9-8

Spiral Tracker   **Go Online** ⟩ to see how mastery develops for all standards across the grade.

▶ **Evaluating Children's Responses**

Evaluate children's abilities to add two-digit numbers. Use the rubric below to evaluate their work based on **GMP3.1**.

Goal for Mathematical Practice	Not Meeting Expectations	Partially Meeting Expectations	Meeting Expectations	Exceeding Expectations
GMP3.1 Make mathematical conjectures and arguments.	Does not attempt an argument that the sum is maximized.	Does not maximize the sum, but argues that one addend should be in the 70s because the 7 in the tens place produces the greatest possible sum.	Argues that 74 + 53 or 73 + 54 produces the largest sum, referring to the need to put the two largest digits in the tens place **or** that 74 + 53 or 73 + 54 produces the largest sum using a guess-and-check strategy, testing the sum against at least one other pair of numbers.	Meets expectations and either identifies both pairs of digits that maximize the sum or gives an argument based both on place value and a guess-and-check strategy, each of which would represent an adequate argument on its own.

3 Look Ahead   10–15 min

	Materials
Math Boxes 9-12 Children practice and maintain skills.	*Math Journal 2*, p. 249
Home Link 9-12 Children take home the end-of-year Family Letter.	*Math Masters*, pp. 282–286

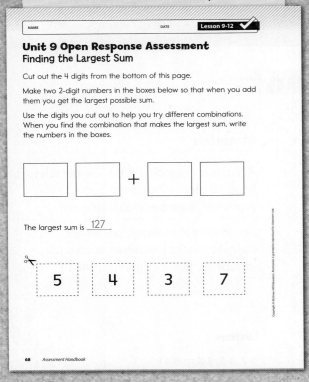

Assessment Handbook, p. 68

NAME   DATE   **Lesson 9-12** ✓

Unit 9 Open Response Assessment
Finding the Largest Sum

Cut out the 4 digits from the bottom of this page.

Make two 2-digit numbers in the boxes below so that when you add them you get the largest possible sum.

Use the digits you cut out to help you try different combinations. When you find the combination that makes the largest sum, write the numbers in the boxes.

☐ ☐ **+** ☐ ☐

The largest sum is __127__.

✂ ⌐ 5 ¬ ⌐ 4 ¬ ⌐ 3 ¬ ⌐ 7 ¬

68   *Assessment Handbook*

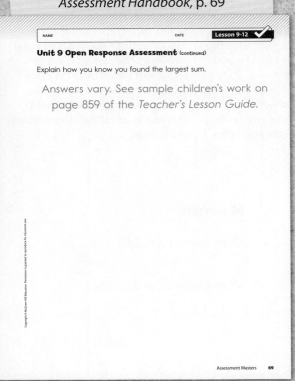

Assessment Handbook, p. 69

NAME   DATE   **Lesson 9-12** ✓

Unit 9 Open Response Assessment (continued)

Explain how you know you found the largest sum.

Answers vary. See sample children's work on page 859 of the *Teacher's Lesson Guide*.

Assessment Masters   **69**

Assessment
and Reporting

▶ Solving the Open Response Problem

Assessment Handbook, pp. 68–69

| WHOLE CLASS | SMALL GROUP | **PARTNER** | **INDEPENDENT** |

This open response problem requires children to apply skills and concepts from Unit 9 and earlier units to determine two 2-digit numbers that will produce the largest sum. The focus of this task is **GMP3.1:** Make mathematical conjectures and arguments.

Before children start the problem, tell them that today they will figure out how to organize a group of numbers to create the largest sum.

Distribute *Assessment Handbook,* pages 68–69. Read the directions aloud. Model how to lay and rearrange the cut-out numbers on the empty squares. Tell children to use whatever strategy they prefer to find the numbers that have the largest sum. Tell children to use the second page to show how they know they have found the numbers that have the largest sum. **GMP3.1**

> **Differentiate**   **Adjusting the Assessment**
>
> If children have difficulty, suggest tools they might use, such as base-10 blocks, drawings, or open number lines. Make base-10 blocks, toolkit bills ($1 and $10), and number grids available.

▶ Discussing the Problem

Assessment Handbook, pp. 68–69

| **WHOLE CLASS** | **SMALL GROUP** | PARTNER | INDEPENDENT |

After children complete their work, invite a few children to explain how they knew they found the largest sum.

Evaluating Children's Responses CCSS 2.NBT.5

Collect children's work. For the content standard, expect most children to determine the correct sum of the numbers they placed in the empty boxes. You can use the rubric on page 857 to evaluate children's work for **GMP3.1**.

See the sample below. This work meets expectations for the content standard because the child added 74 and 53 correctly. The work meets expectations for the mathematical practice because the child argues that the two largest digits should be placed in the tens column. **GMP3.1**

Go Online ✓ Assessment and Reporting

7 4 + 5 3

The largest sum is _____

Explain how you know you found the largest sum.

I know seven is the largest in all of the four and I knew 5 is next out of the three and it would be more if I made the other ten with the five and it would be better for the four to be after 70 so it would be 74 and 3 have to be with 5 so it is 12/7 is 74 + 53.

Sample child's work, "Meeting Expectations"

Math Journal 2, p. 249

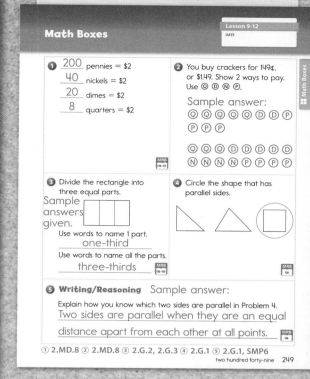

Math Boxes — Lesson 9-12 — DATE

1. 200 pennies = $2
   40 nickels = $2
   20 dimes = $2
   8 quarters = $2

2. You buy crackers for 149¢, or $1.49. Show 2 ways to pay. Use Q D N P.
   Sample answer:
   Q Q Q Q Q D D P P P P
   Q Q Q Q D D D D N N N N P P P P

3. Divide the rectangle into three equal parts.
   Sample answers given.
   Use words to name 1 part. one-third
   Use words to name all the parts. three-thirds

4. Circle the shape that has parallel sides.

5. **Writing/Reasoning** Sample answer:
   Explain how you know which two sides are parallel in Problem 4.
   Two sides are parallel when they are an equal distance apart from each other at all points.

① 2.MD.8 ② 2.MD.8 ③ 2.G.2, 2.G.3 ④ 2.G.1 ⑤ 2.G.1, SMP6

two hundred forty-nine 249

---

3 Look Ahead 10–15 min

Go Online Home Connections

▶ Math Boxes 9-12

Math Journal 2, p. 249

| WHOLE CLASS | SMALL GROUP | PARTNER | INDEPENDENT |

Mixed Practice Math Boxes 9-12 are paired with Math Boxes 9-10.

▶ Home Link 9-12

| WHOLE CLASS | SMALL GROUP | PARTNER | INDEPENDENT |

Math Masters, pp. 282–286

Homework Distribute copies of the end-of-year Family Letter for children to take home.

Math Masters, pp. 282–286

Unit 9: Family Letter — Home Link 9-12 — NAME — DATE

Congratulations!

By completing *Second Grade Everyday Mathematics*, your child has accomplished a great deal. Thank you for your support!

This Family Letter is provided as a resource for you to use throughout your child's vacation. It includes an extended list of Do-Anytime Activities, directions for games that can be played at home, and a sneak preview of what your child will be learning in *Third Grade Everyday Mathematics*. Enjoy your vacation!

Do-Anytime Activities

Mathematics concepts are more meaningful and easier to understand when they are rooted in real-life situations. To help your child review some of the concepts he or she has learned in second grade, we suggest the following activities for you and your child to do together over vacation. Doing so will help your child maintain and build on the skills learned this year and help prepare him or her for *Third Grade Everyday Mathematics*.

1. Pose addition and subtraction number stories about everyday life. For example, ask your child to count the number of grapes he or she has and then ask: *How many will you have if you eat 6 of them? How many will you have if you eat 2 of them and then I eat 3 more?* Here's another example: *If you have 1 quarter, 3 dimes, and 2 nickels, how many cents do you have?*

2. Review and practice addition and subtraction facts. Your child can use Fact Triangle cards to practice or play *Addition Top-It* or *Subtraction Top-It* as described on the second page of this letter.

3. Select everyday objects and have your child estimate their lengths and then measure to check the estimates. Your child could also measure objects to determine how much longer one thing is compared with another.

4. Ask your child to tell you the time to the nearest 5 minutes. Encourage your child to specify whether it is A.M. or P.M.

5. Encourage your child to identify and describe geometric shapes that can be seen in the world. For example: *I see rectangles in that bookcase. They all have 4 right angles.* You can also play I Spy to practice identifying and describing shapes. For example: *I spy a shape with 5 sides. All of the sides are the same length.*

6. Ask your child to share food items or other objects fairly with 1, 2, or 3 other people by dividing them into equal shares.

7. Count on or back by 10s and 100s from any given number.

282 two hundred eighty-two

Glossary

Everyday Mathematics strives to define terms clearly, especially when they can be defined in multiple ways.

This glossary focuses on terms and meanings for elementary school mathematics and omits details and complexities required at higher levels. The definitions here are phrased for teachers. Information for explaining terms and concepts to children can be found within the lessons themselves. Additional information is available online. In a definition, most terms in italics are defined elsewhere in this glossary.

0–9

0 fact (1) The *sum* of two 1-digit numbers when one of the *addends* is 0, as in $0 + 5 = 5$. If 0 is added to any number, there is no change in the number. See *Additive Identity*. (2) The *product* of two 1-digit numbers when one of the factors is 0, as in $4 \times 0 = 0$. The product of a number and 0 is always 0.

1-dimensional (1-D) (1) Having *length* but not area or volume; confined to a curve, such as an arc. (2) A figure whose points are all on one *line*. Line segments are 1-dimensional. Compare *2-dimensional* and *3-dimensional*.

2-dimensional (2-D) (1) Having *area* but not volume; confined to a *surface*. A 2-dimensional surface can be flat or curved, such as the surface of a sphere. (2) A figure whose points are all in one *plane* but not all on one line. Examples include polygons and circles. Compare *1-dimensional* and *3-dimensional*.

3-dimensional (3-D) Having *volume*. Solids such as cubes, cones, and spheres are 3-dimensional. Compare *1-dimensional* and *2-dimensional*.

A

accurate (1) As correct as possible for a given context. An answer can be accurate without being very *precise* if the units are large. For example, the driving time from Chicago to New York City is about 13 hours. See *approximate*. (2) Of a measurement or other quantity, having a high degree of correctness. A more accurate measurement is closer to the true value. Accurate answers must be reasonably *precise*.

add-to situation A situation in which there is a starting quantity, an additional quantity, and an ending quantity. Any of the three quantities may be unknown. See *addition/subtraction use class*.

addend Any one of a set of numbers that are added. For example, in $5 + 3 + 1 = 9$, the addends are 5, 3, and 1.

addition fact Two whole numbers from 0 through 10 and their sum, such as $9 + 7 = 16$. See *arithmetic facts*.

addition/subtraction use class A category of problem situations that can be solved using addition or subtraction or other methods, such as counting or direct modeling. *Everyday Mathematics* distinguishes four addition/subtraction use classes: *parts-and-total*, *change-to-more*, *change-to-less*, and *comparison* situations. The table below shows how these use classes correspond to those in the Common Core State Standards.

Everyday Mathematics	CCSS
change-to-more	add to
change-to-less	take from
parts-and-total	put together/take apart
comparison	compare

Additive Identity The number zero (0). The Additive Identity is the number that when added to any other number, yields that other number. See *additive inverses*.

additive inverses Two numbers whose sum is 0. Each number is called the additive inverse, or opposite, of the other. For example, 3 and -3 are additive inverses because $3 + (-3) = 0$. Zero is its own additive inverse: $0 + 0 = 0$. See *Additive Identity*.

adjacent angles Two nonoverlapping *angles* with a common side and *vertex*.

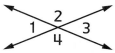

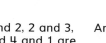

Angles 1 and 2, 2 and 3, 3 and 4, and 4 and 1 are pairs of adjacent angles.

Angle 5 is adjacent to angle 6.

adjacent sides (1) Two sides of a *polygon* with a common *vertex*. (2) Two faces of a *polyhedron* with a common *edge*.

A.M. The abbreviation for *ante meridiem,* meaning "before the middle of the day" in Latin. From midnight to noon.

analog clock (1) A clock that shows the time by the positions of the hour and minute hands. (2) Any device that shows time passing in a continuous manner, such as a sundial. Compare *digital clock.*

anchor chart A classroom display that is cocreated by a teacher and children and focuses on a central concept or skill.

angle (1) A figure formed by two rays or line segments with a common endpoint called the *vertex* of the angle. The rays or segments are called the sides of the angle. Angles can be named after their vertex point alone, as in ∠A; or by three points, one on each side and the vertex in the middle, as in ∠BCD. One side of an angle is *rotated* about the vertex from the other side through a number of *degrees.* (2) The measure of this rotation in degrees.

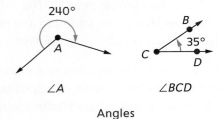

Angles

-angle A suffix meaning angle or corner, for example, triangle and rectangle.

apex (1) In a *pyramid,* the *vertex* opposite the *base.* All the nonbase faces meet at the apex. (2) The point at the tip of a *cone.*

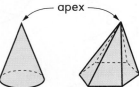

approximate Close to exact. In many situations it is not possible to get an exact answer, but it is important to be close to the exact answer. We might draw an angle that is approximately 60° with a protractor. In this case approximate suggests that the angle drawn is within a degree or two of 60°. Compare *precise.*

area The amount of *surface* inside a *2-dimensional* figure. The figure might be a triangle or a rectangle in a *plane,* the curved surface of a cylinder, or a state or a country on Earth's surface. Area can be measured in *square units,* such as square miles or square centimeters, or other units, such as acres.

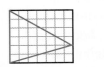   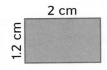

A triangle with area   A rectangle with area
  21 square units   1.2 cm * 2 cm = 2.4 square centimeters

arithmetic facts The *addition facts* (whole-number *addends* 10 or less); their inverse subtraction facts; multiplication facts (whole-number factors 10 or less); and their inverse division facts, except there is no division by zero. Facts and their corresponding inverses are organized into *fact families.*

arm span Same as *fathom.* See *Tables of Measures.*

array (1) An arrangement of objects in a regular *pattern,* usually *rows* and *columns.* (2) A rectangular array. In *Everyday Mathematics,* an array is a rectangular array unless specified otherwise.

arrow rule In *Everyday Mathematics,* a rule that determines the number that goes into the next *frame* in a *Frames-and-Arrows* diagram. There may be more than one arrow rule per diagram.

arrows In *Everyday Mathematics,* the links representing the *arrow rule*(s) in a *Frames-and-Arrows* diagram.

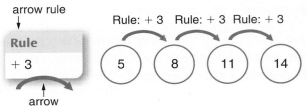

Associative Property of Addition A *property* of addition that for any numbers *a, b,* and *c,* (*a* + *b*) + *c* = *a* + (*b* + *c*). The grouping of the three *addends* can be changed without changing the *sum.* For example, (4 + 3) + 7 = 4 + (3 + 7) because 7 + 7 = 4 + 10. Subtraction is not associative. For example, (4 − 3) + 7 ≠ 4 − (3 + 7) because 8 ≠ −6. Compare *Commutative Property of Addition.*

attribute A characteristic or *property* of an object or a common characteristic of a set of objects. Size, shape, color, and number of sides are attributes.

attribute blocks   A set of blocks that vary in four *attributes:* color, size, thickness, and shape. The blocks are used for attribute identification and sorting activities. Compare *pattern blocks*.

automaticity   The ability to solve problems with great efficiency either by using recall or applying quick strategies. For example, one might "just know" that $8 + 7 = 15$ or quickly think $8 + 2 = 10$ and 5 more is 15. Compare *fluency*.

B

ballpark estimate   A rough *estimate*. A ballpark estimate can serve as a check of the reasonableness of an answer obtained through some other procedure, or it can be made when an exact value is unnecessary or impossible to obtain.

bar graph   A graph with horizontal or vertical bars that represent (typically categorical) *data*. The lengths of the bars may be *scaled*.

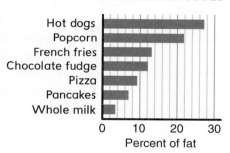

Fat Content of Foods

base angles of a trapezoid   Two *angles* that share a *base of a trapezoid*.

base of a number system   The foundation number for a *place-value* based numeration system. For example, our usual way of writing numbers uses a base-10 place-value system. In programming computers or other digital devices, bases of 2, 8, 16, or other powers of 2 are more common than base 10.

base of a prism or a cylinder   Either of the two *parallel* and *congruent faces* that define the shape of a *prism* or a *cylinder*.

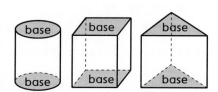

base of a pyramid or a cone   The *face* of a *pyramid* or a *cone* that is opposite its *apex*.

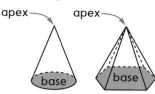

base of a trapezoid   (1) Either of a pair of *parallel* sides in a *trapezoid*. (2) The *length* of this side. The area of a trapezoid is the average of a pair of bases times the corresponding height.

base ten   (1) Related to powers of 10. (2) The most common system for writing numbers, which uses only 10 symbols 0, 1, 2, 3, 4, 5, 6, 7, 8, and 9, called digits. One can write any number using one or more of these 10 digits. Each *digit* has a value that depends on its place in the number (its *place value*). In the base-10 system, each place has a value 10 times that of the place to its right, and one-tenth the value of the place to its left.

base-10 blocks   A set of blocks to represent numbers. In *Everyday Mathematics,* the four standard sizes of blocks are called cubes, longs, flats, and big cubes from smallest to largest. Once one of these is designated as the unit whole, the values of the other blocks follow. For example, if the cube is the unit whole, then the long, flat, and big cube represent 10, 100, and 1,000, respectively. However, if the flat is the unit whole, then the cube, long, and big cube represent $\frac{1}{100}$, $\frac{1}{10}$, and 10, respectively. The cube is 1 cm on each edge. The long is 10 cm by 1 cm by 1 cm. The flat is 10 cm by 10 cm by 1 cm. The big cube looks like 1,000 cubes and is 10 cm by 10 cm by 10 cm. See *long, flat,* and *big cube* for photos of the blocks. See *base-10 shorthand*.

base-10 shorthand   In *Everyday Mathematics,* a written notation for base-10 blocks.

Name	Base-10 block	Base-10 shorthand
cube		.
long		|
flat		▢
big cube		▱

big cube In *Everyday Mathematics,* a *base-10 block* cube that measures 10 cm by 10 cm by 10 cm. A big cube is equivalent to one thousand 1-cm cubes.

billion By U.S. custom, 1 billion is 1,000,000,000, or 10^9. By British, French, and German custom, 1 billion is 1,000,000,000,000, or 10^{12}.

C

calendar (1) A *reference frame* to keep track of the passage of time over weeks, months, and years. (2) A practical model of the reference frame, such as the large, reusable Class Calendar in Kindergarten through Second Grade *Everyday Mathematics.* (3) A schedule or listing of events.

capacity (1) The amount of space contained by a *3-dimensional* figure. Capacity is often measured in liquid units such as cups, quarts, gallons, or liters. (2) The maximum weight a scale can measure.

cardinal number A number telling how many things are in a *set.* Compare *ordinal number.*

Celsius A *temperature scale* on which pure water at sea level freezes at 0° and boils at 100°. The Celsius scale is used in the *metric system.* A less common name for this scale is centigrade because there are 100 degrees between the freezing and boiling points of water. Compare *Fahrenheit.*

cent A penny; $\frac{1}{100}$ of a dollar. From the Latin word *centesimus,* which means "a hundredth part."

center of a circle The point in the *plane* of a *circle* that is equally distant from all points on the circle.

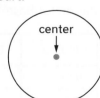

center of a sphere The point equally distant from all points on a *sphere.*

chance The possibility that an *outcome* or an event will occur. For example, in flipping a coin there is an equal chance of getting HEADS or TAILS.

change diagram In *Everyday Mathematics,* a diagram used to model situations in which quantities are increased or decreased. The diagram includes a starting quantity, an ending quantity, and an amount of change. See *situation diagram, change-to-less situation,* and *change-to-more situation.*

A change diagram for $14 - 5 = 9$

change-to-less situation A situation involving a starting quantity, a change, and an ending quantity that is less than the starting quantity. For example, a situation about spending money is a change-to-less situation. Compare *change-to-more situation.* See *addition/subtraction use class.*

change-to-more situation A situation involving a starting quantity, a change, and an ending quantity that is more than the starting quantity. For example, a situation about earning money is a change-to-more situation. Compare *change-to-less situation.* See *addition/subtraction use class.*

circle The set of all points in a *plane* that are equally distant from a fixed point in the plane called the *center.* The distance from the center to the circle is the radius of the circle. The diameter of a circle is twice the radius. A circle together with its interior is called a *disk* or a circular region.

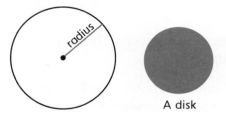

A disk

clockwise rotation A turn in the direction in which the hands move on a typical *analog clock;* a turn to the right.

column (1) A vertical arrangement of objects or numbers in an *array* or a table.

column

(2) A vertical selection of cells in a spreadsheet.

combinations of 10 Pairs of *whole numbers* from 0 to 10 that add to 10. Combinations of 10 are key *helper facts* for addition and subtraction fact strategies. For example, $4 + 6$ and $0 + 10$ are both combinations of 10.

Commutative Property of Addition A *property* of addition that for any numbers a and b, $a + b = b + a$. Two numbers can be added in either order without changing the *sum*. For example, $5 + 10 = 10 + 5$. In *Everyday Mathematics,* this and the Commutative Property of Multiplication are called *turn-around rules.* Subtraction is not commutative. For example, $8 - 5 \neq 5 - 8$ because $3 \neq -3$.

comparison diagram In *Everyday Mathematics,* a diagram used to model situations in which two quantities are compared. The diagram represents two quantities and their *difference.* See *situation diagram.*

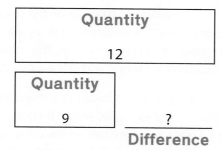

A comparison diagram for $12 = 9 + ?$

comparison situation A situation involving two quantities and the *difference* between them. See *addition/ subtraction use class.*

compose To make up or form a number or shape by putting together smaller numbers or shapes. For example, one can compose a 10 by putting together ten 1s: $1 + 1 + 1 + 1 + 1 + 1 + 1 + 1 + 1 + 1 = 10$. One can compose a pentagon by putting together an equilateral triangle and a square.

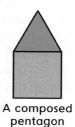

A composed pentagon

concave polygon A *polygon* on which there at least two points that can be connected with a *line segment* that passes outside the polygon. For example, segment *AD* is outside the hexagon between *B* and *C*. Informally, at least one *vertex* appears to be "pushed inward." At least one interior *angle* has a measure greater than 180°. Same as *nonconvex polygon.* Compare *convex polygon.*

A concave polygon

cone A *geometric solid* comprising a circular base, an *apex* not in the *plane* of the base, and all *line segments* with one endpoint at the apex and the other endpoint on the circumference of the *base.*

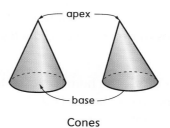

Cones

congruent figures Figures having the same size and shape. Two figures are congruent if they match exactly when one is placed on top of the other after a combination of isometry transformations (slides, flips, and/or turns). In diagrams of congruent figures, the corresponding congruent sides may be marked with the same number of hash marks. The symbol $\cong$ means "is congruent to."

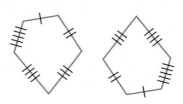

Congruent pentagons

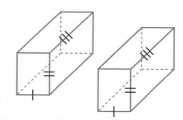

Congruent right rectangular prisms

conjecture A claim that has not been proved, at least by the person making the conjecture.

consecutive Following one after another in an uninterrupted order. For example, A, B, C, and D are four consecutive letters of the alphabet; 6, 7, 8, 9, and 10 are five consecutive whole numbers.

convex polygon A *polygon* on which no two points can be connected with a *line segment* that passes outside the polygon. Informally, all *vertices* appear to be "pushed outward." Each *angle* in the polygon measures less than 180°. Compare *concave polygon.*

A convex polygon

corner   Informal for *vertex* or *angle*.

count   (1) The number of objects in a *set*. (2) For use as a verb, see *rote counting*, *rational counting*, and *skip counting*.

counterclockwise rotation   Opposite the direction in which the hands move on a typical *analog clock;* a turn to the left.

counting numbers   The numbers used to *count* things. The set of counting numbers is {1, 2, 3, 4, . . .}. Sometimes 0 is included but not in *Everyday Mathematics*. Counting numbers are also known as *natural numbers*.

counting-up subtraction   A subtraction strategy in which a *difference* is found by counting or adding up from the smaller number to the larger number. For example, to calculate 87 − 49, one could start at 49, add 30 to reach 79, and then add 8 more to reach 87. The difference is 30 + 8 = 38.

cube   (1) A *regular polyhedron* with 6 square *faces*. A cube has 8 *vertices* and 12 *edges*.

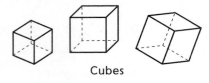

Cubes

(2) In *Everyday Mathematics*, the smaller cube of the *base-10 blocks,* measuring 1 cm on each edge.

cup   A U.S. customary unit of *volume* or *capacity* equal to 8 fluid ounces, or $\frac{1}{2}$ pint.

curved surface   A *2-dimensional surface* that does not lie in a *plane*. Spheres, cylinders, and cones have curved surfaces.

customary system of measurement   In *Everyday Mathematics,* same as *U.S. customary system* of measurement. See *Tables of Measures*.

cylinder   A *geometric solid* with two congruent, parallel circular regions for *bases* and a curved *face* formed by all the *segments* that have an endpoint on each circle and that are *parallel* to the segment with endpoints at the *centers* of the circles. Also called a circular cylinder.

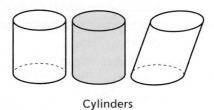

Cylinders

D

data   Information that is gathered by counting, measuring, questioning, or observing. Strictly, data is the plural of datum, but data is often used as a singular word.

decagon   A 10-sided *polygon*.

decimal   (1) A number written in standard *base-10* notation containing a *decimal point,* such as 2.54. (2) Any number written in standard base-10 notation. See *decimal fraction, repeating decimal,* and *terminating decimal*.

decimal fraction   (1) A *fraction* or mixed number written in standard *decimal notation*. (2) A fraction $\frac{a}{b}$, where b is a positive power of 10, such as $\frac{84}{100}$.

decimal notation   Same as *standard notation*.

decimal point   A mark used to separate the ones and tenths places in decimals. A decimal point separates dollars from cents in dollars-and-cents notation. The mark is a dot in the U.S. but a comma in Europe and some other countries.

decompose   To separate a number or a shape into smaller numbers or shapes. For example, 14 can be decomposed into 1 ten and 4 ones. A square can be decomposed into two isosceles right triangles. Any even number can be decomposed into two equal parts: $2n = n + n$.

degree (°)   (1) A unit of measure for *angles* based on dividing a *circle* into 360 equal parts. Latitude and longitude are measured in degrees based on angle measures. (2) A unit for measuring *temperature*. See *Celsius* and *Fahrenheit*. The symbol ° means degrees of any type.

diagonal   (1) A *line segment* joining two nonadjacent *vertices* of a *polygon*. (2) A segment joining two vertices not on the same *face* of a *polyhedron*.

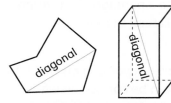

(3) A line of objects or numbers between opposite corners of an array or a table.

difference   (1) The distance between two numbers on a *number line*. The difference between 5 and 12 is 7. (2) The result of subtracting one number from another. For example, in 12 − 5 = 7 the difference is 7, and in 5 − 12 the difference is −7. Compare *minuend* and *subtrahend*.

digit (1) Any of the symbols 0, 1, 2, 3, 4, 5, 6, 7, 8, and 9 in the *base-10 numeration* system. For example, the *numeral* 145 is made up of the digits 1, 4, and 5. (2) Any one of the symbols in a *place-value* number system.

digital clock A clock that shows the time with numbers of hours and minutes, usually separated by a colon. This display is discrete, not continuous, meaning that the display jumps to a new time after a minute has elapsed. Compare *analog clock*.

dimension (1) A measurable extent such as *length*, width, or *height*. Having two measurable extents makes the measured thing 2-dimensional. See *1-*, *2-*, and *3-dimensional*. (2) The measures of those extents. For example, the dimensions of a box might be 24 cm by 20 cm by 10 cm. (3) The number of coordinates necessary to locate a point in a geometric space. A *plane* has two dimensions because an ordered pair of two coordinates uniquely locates any point in the plane.

direct comparison A measurement comparison made by aligning the objects. Compare *indirect comparison*.

disk A *circle* and its *interior* region.

double ten frame Two side-by-side *ten frames*. Double ten frames can be used to represent the numbers 0–20 as well as *addition facts* in a variety of

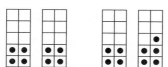
Sequence of double ten frames to elicit near doubles strategy

ways and are particularly useful for encouraging children to develop addition fact strategies, such as *near doubles* and *making 10*.

doubles fact The *sum* of a number 0 through 10 added to itself, such as 4 + 4 = 8, sometimes written as that number multiplied by 2: 2 × 4 = 8. These are key *helper facts*.

E

---

edge (1) Any side of a *polyhedron's* faces.

edges

(2) A line segment or curve where two *surfaces* of a *geometric solid* meet.

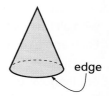
edge

elapsed time The amount of time that has passed from one point in time to the next. For example, between 12:45 P.M and 1:30 P.M, 45 minutes have elapsed.

equal (1) Identical in number or measure; neither more nor less. (2) *Equivalent*.

equal groups Sets with the same number of elements, such as cars with 5 passengers each, rows with 6 chairs each, and boxes containing 100 paper clips each.

equal-groups notation In *Everyday Mathematics*, a way to denote a number of equal-size groups. The size of each group is shown inside square brackets and the number of groups is written in front of the brackets. For example, 3 [6s] means 3 groups with 6 in each group. In general, *n* [*ks*] means *n* groups with *k* in each group.

equal parts *Equivalent* parts of a whole. For example, dividing a pizza into 4 equal parts means each part is $\frac{1}{4}$ of the pizza and is equal in size to each of the other 3 parts.

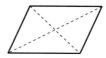

4 equal parts, each $\frac{1}{4}$ of a pizza   4 equal parts, each $\frac{1}{4}$ of the area

equal share One of several parts of a whole, each of which has the same amount of area, volume, mass, or other measurable or countable quantity. Sometimes called fair share. See *equal parts*.

equation A *number sentence* that contains an equals sign. For example, 5 + 10 = 15 and P = 2*l* + 2*w* are equations.

equilateral polygon A *polygon* in which all sides are the same length.

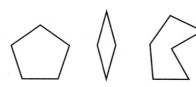

Equilateral polygons

equilateral triangle A *triangle* with all three sides equal in length. Each angle of an equilateral triangle measures 60°, so it is also called an equiangular triangle. All equilateral triangles are *isosceles triangles*.

An equilateral triangle

equivalent *Equal* in value but possibly in a different form. For example, $\frac{1}{2}$, 0.5, and 50% are all equivalent.

equivalent names Different ways of naming the same number. For example, $2 + 6$, $4 + 4$, $12 - 4$, $18 - 10$, $100 - 92$, $5 + 1 + 2$, eight, VIII, and ⊦⊦⊦ /// are all equivalent names for 8. See *name-collection box*.

estimate (1) An answer close to, or approximating, an exact answer. (2) To make an estimate.

evaluate a numerical expression To carry out the operations in a numerical *expression* to find a single value for the expression.

even number (1) A *counting number* that is divisible by 2: 2, 4, 6, 8 . . . .. (2) An *integer* that is divisible by 2. Compare *odd number*.

expand-and-trade subtraction A subtraction algorithm in which *expanded notation* is used to facilitate place-value exchanges.

expanded notation A way of writing a number as the *sum* of the values of each *digit*. For example, 356 is $300 + 50 + 6$ in expanded notation. Same as *expanded form*. Compare *standard notation* and *number-and-word notation*.

Explorations In First through Third Grade *Everyday Mathematics,* independent or small-group activities that focus on concept development, manipulatives, data collection, problem solving, games, and skill reviews.

expression (1) A mathematical phrase made up of numbers, *variables, operation symbols,* and/or grouping symbols. An expression does not contain *relation symbols* such as $=$, $>$, and $\leq$. (2) Either side of an *equation* or an *inequality*.

extended facts Variations of basic *arithmetic facts* involving *multiples* of 10, 100, and so on. For example, $30 + 70 = 100$, $40 \times 5 = 200$, and $560 \div 7 = 80$ are extended facts. See *fact extensions*.

---

face (1) A flat surface on a closed, *3-dimensional* figure. Some special faces are called *bases*. (2) More generally, any *2-dimensional surface* on a 3-dimensional figure.

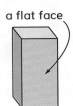 

a flat face   a curved face

fact extensions Calculations with larger numbers using knowledge of basic *arithmetic facts*. For example, knowing the addition fact $5 + 8 = 13$ makes it easier to solve problems such as $50 + 80 = ?$ and $65 + ? = 73$. Fact extensions apply to all four basic arithmetic operations. See *extended facts*.

fact family A set of related *arithmetic facts* linking two inverse operations. For example,

$$5 + 6 = 11 \qquad 6 + 5 = 11$$
$$11 - 5 = 6 \qquad 11 - 6 = 5$$

are an addition/subtraction fact family. Similarly,

$$5 \times 7 = 35 \qquad 7 \times 5 = 35$$
$$35 \div 7 = 5 \qquad 35 \div 5 = 7$$

are a multiplication/division fact family. Same as *number family*.

fact power In *Everyday Mathematics, automaticity* with basic *arithmetic facts*. Automatically knowing the facts is as important to arithmetic as knowing words by sight is to reading.

Fact Triangle In *Everyday Mathematics,* a triangular flash card labeled with the numbers of a *fact family* that children can use to practice addition/subtraction and multiplication/division facts. The two addends or factors and their *sum* or *product* (marked with a dot) appear in the corners of each triangle.

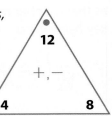

facts table A chart showing *arithmetic facts*. An addition/subtraction facts table shows addition and subtraction facts. A multiplication/division facts table shows multiplication and division facts.

Fahrenheit A *temperature* scale on which pure water at sea level freezes at 32° and boils at 212°. The Fahrenheit scale is widely used in the United States but in few other places. Compare *Celsius*.

false number sentence A *number sentence* that is not true. For example, 8 = 5 + 5 is a false number sentence. Compare *true number sentence*.

figurate numbers Numbers that can be illustrated by specific geometric patterns. Square numbers and triangular numbers are figurate numbers.

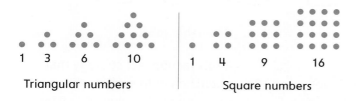

Triangular numbers   Square numbers

flat In *Everyday Mathematics*, the *base-10 block* that is equivalent to one hundred 1-cm cubes.

A flat

fluency The ability to compute using efficient, appropriate, and flexible strategies. Compare *automaticity*.

fraction (1) A number in the form $\frac{a}{b}$ or a/b, where a and b are *integers* and b is not 0. A fraction may be used to name part of an object or part of a collection of objects, to compare two quantities, or to represent division. For example, $\frac{12}{6}$ might mean 12 eggs divided in groups of 6, a ratio of 12 to 6, or 12 divided by 6. Also called a common fraction. (2) A fraction that satisfies the previous definition and includes a unit in both the numerator and denominator. For example, the rates $\frac{50\text{ miles}}{1\text{ gallon}}$ and $\frac{40\text{ pages}}{10\text{ minutes}}$ are fractions. (3) A number written using a fraction bar, where the fraction bar is used to indicate division. For example, $\frac{2.3}{6.5}$, $\frac{\frac{14}{5}}{12}$, $\frac{\pi}{4}$, and $\frac{\frac{3}{4}}{\frac{5}{8}}$. Compare *decimal*.

fractional part Part of a whole. Fractions represent fractional parts of numbers, sets, or objects.

frames In *Everyday Mathematics*, the empty shapes in which numbers are written in a Frames-and-Arrows diagram.

Frames and Arrows In *Everyday Mathematics*, diagrams consisting of *frames* connected by *arrows*. Frames-and-arrows diagrams are used to represent number sequences. Each frame contains a number, and each arrow represents a rule that determines which number goes in the next frame. There may be more than one rule, represented by different-color arrows. Frames-and-Arrows diagrams are also called chains.

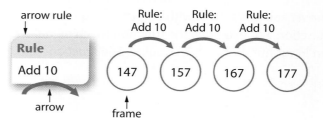

frequency (1) The number of times a value occurs in a set of *data*. (2) A number of repetitions per unit of time, such as the vibrations per second in a sound wave.

frequency graph A graph showing how often each value occurs in a *data* set.

Colors in a Bag of Marbles

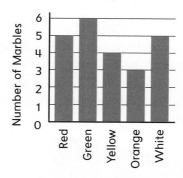

frequency table A table in which *data* are *tallied* and organized, often as a first step toward making a frequency graph.

Color	Number of Marbles
red	ЖЖ
green	ЖЖ I
yellow	////
orange	///
white	ЖЖ

friendly numbers An addition/subtraction strategy that uses a number that is easy to work from, typically 10 or a multiple of 10. For example, to solve 16 − 9, one might recognize that the friendly number 10 is 6 less than 16, then count down 1 more to 9 to find that the difference is 7. Compare *helper facts*.

function (1) A set of *ordered pairs* (*x*, *y*) in which each value of *x* is paired with exactly one value of *y*. A function is typically represented in a table, by points on a coordinate graph, or by a rule such as an *equation*. (2) A rule that pairs each *input* with exactly one *output*. For example, for a function with the rule "Double," 1 is paired with 2, 2 is paired with 4, 3 is paired with 6, and so on. See *"What's My Rule?"*

function machine An imaginary device that receives *inputs* and pairs them with *outputs* using a rule that is a function. For example, the function machine below pairs an input number with its double. See *function*.

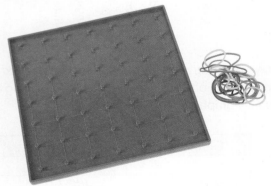

in	out
1	2
2	4
3	6
5	10
20	40
300	600

A function machine and function table

G

geoboard (1) A small wooden or plastic board with nails or other posts, usually arranged at equally spaced intervals in a *rectangular array*. Geoboards and rubber bands are useful for exploring basic concepts in plane geometry.

Geoboard and rubber bands

(2) A digital version of a geoboard.

geometric solid The *surface* or surfaces that make up a *3-dimensional* figure, such as a *prism, pyramid, cylinder, cone,* or *sphere*. Despite its name, a geometric solid is hollow; that is, it does not include the points in its *interior*. Informally, and in some dictionaries, a solid is defined as both the surface and its interior.

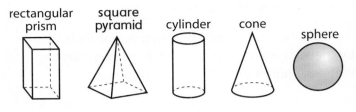

Geometric solids

going through 10 A subtraction fact strategy that involves using 10 as a benchmark to simplify the subtraction. For example, one might solve 16 − 9 by either going up through 10 and thinking 9 + 1 = 10 and 10 + 6 = 16, so the *difference* is 7, or by going back through 10 by thinking 16 take away 6 is 10, and 10 take away 3 more is 7, so the difference is 7. Compare *making 10*.

-gon A suffix meaning angle. For example, a *hexagon* is a plane figure with six *angles*.

grouping addends An addition strategy that involves adding three or more numbers in an order that makes the addition simpler, such as recognizing and adding a combination of 10 or a doubles fact first. See *Associative* and *Commutative Properties of Addition*.

H

height (1) The length of a perpendicular segment from one side of a geometric figure to a *parallel side* or from a vertex to the opposite *side*. (2) The line segment itself.

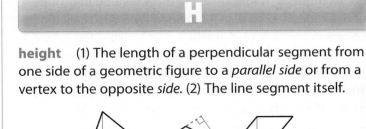

Height of 2-D figures are shown in red.

helper facts Well-known facts used to derive unknown facts. Doubles and combinations of 10 are key addition/subtraction helper facts. For example, knowing the doubles fact 6 + 6 can help one derive 6 + 7 by thinking 6 + 6 = 12 and 1 more makes 13.

heptagon A 7-sided *polygon*.

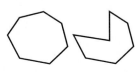

Heptagons

hexagon A 6-sided *polygon*.

A hexagon

hierarchy of shapes A classification in which shapes are organized into categories and subcategories. For each category, every defining *attribute* of a shape in that category is a defining attribute of all shapes in its subcategories. A hierarchy is often shown in a diagram with the most general category at the top and arrows or lines connecting categories to their subcategories. See *quadrilateral* and below for examples.

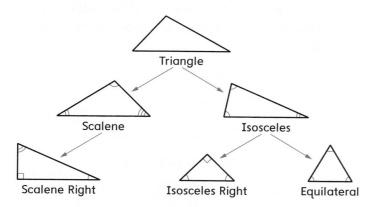

Hierarchy of triangles by angle size

Home Link In *Everyday Mathematics*, a suggested follow-up or enrichment activity to be done at home.

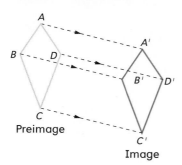

icon A small picture or diagram sometimes used to represent quantities. For example, an icon of a stadium might be used to represent 100,000 people on a *scaled picture graph*. Icons are also used to represent functions or objects in computer operating systems and applications.

image A figure that is produced by a transformation of another figure called the *preimage*.

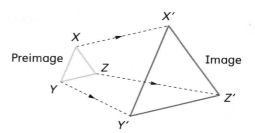

Preimage — Image

improper fraction A *fraction* with a numerator that is greater than or equal to its denominator. For example, $\frac{4}{3}, \frac{5}{2}, \frac{4}{4},$ and $\frac{24}{12}$ are improper fractions. In *Everyday Mathematics*, improper fractions are sometimes called "top-heavy" fractions.

indirect comparison A measurement comparison between two objects using a third object or unit. For example, if one knows that a wall is 8 feet tall and a person is 6 feet tall, and one knows that 8 > 6, then one knows that the wall is taller than the person by comparing their heights to the length of a foot. See *transitivity principle of indirect measurement*. Compare *direct comparison*.

inequality A *number sentence* with a *relation symbol* other than =, such as >, <, ≥, ≤, ≠, or ≈. Compare *equation*.

input (1) A number inserted into a *function machine*, which applies a rule to pair the input with an output. (2) Numbers or other information entered into a calculator or computer.

integer A number in the set {. . . , −4, −3, −2, −1, 0, 1, 2, 3, 4, . . .}. A *whole number* or its opposite, where 0 is its own opposite. Compare *rational numbers, irrational numbers,* and *real numbers*.

interior of a figure (1) The set of all points in a *plane* bounded by a closed *2-dimensional* figure, such as a polygon or a circle. (2) The set of all points in space bounded by a closed *3-dimensional* figure, such as a polyhedron or a sphere. The interior is usually not considered to be part of the figure.

irrational numbers Numbers that cannot be written as *fractions*, where both the numerator and denominator are *integers* and the denominator is not zero. For example, $\sqrt{2}$ and π are irrational numbers. In *standard notation*, an irrational number can only be written as a nonterminating, nonrepeating decimal. For example, $\pi = 3.141592653. . .$, continues forever without a repeating pattern. The number 1.10100100010000. . . is irrational because its pattern does not repeat. Compare *rational numbers*.

isosceles trapezoid   A *trapezoid* with a pair of *base angles* with the same measure. See *quadrilateral*.

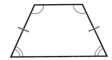
Isosceles trapezoids

isosceles triangle   A *triangle* with at least two *equal-length sides*. *Angles* opposite the equal-length sides are equal in measure.

Isosceles triangles

iterate units   To repeat a *unit* without gaps or overlaps while measuring. To measure *length*, units are placed end-to-end along a path. Unit iteration can be used to measure *area* (by tiling) or *volume* as well.

J

join situation   A *change-to-more situation*.

K

key sequence   The order in which calculator keys are pressed to perform a calculation.

kite   A *quadrilateral* that has two nonoverlapping pairs of *adjacent*, *equal*-length sides.

Kites

L

label   (1) A descriptive word or phrase used to put a number or numbers in context. Labels encourage children to associate numbers with real objects. (2) In a spreadsheet, table, or graph, words or numbers providing information, such as the title of the spreadsheet, the heading for a row or column, or the variable on an axis.

length   The distance between two points on a *1-dimensional* figure. For example, the figure might be a line segment, an arc, or a curve on a map modeling a hiking path. Length is measured in units, such as inches, kilometers, and miles.

line   A 1-dimensional straight path that extends forever in opposite directions. A line is named using two points on it or with a single, italicized lowercase letter, such as *l*. In formal Euclidean geometry, line is an undefined geometric term.

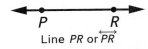
Line *PR* or $\overleftrightarrow{PR}$

line graph   A graph in which *data* points are connected by line segments. Also known as broken-line graph.

line plot   A sketch of *data* in which check marks, Xs, or other symbols above a labeled line show the *frequency* of each value.

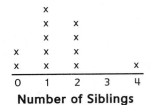
Number of Siblings

A line plot

line segment   A part of a line between and including two points called the endpoints of the segment. Same as *segment*. A line segment is often named by its endpoints.

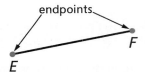
Segment *EF* or $\overline{EF}$

long   In *Everyday Mathematics*, the *base-10 block* that is equivalent to ten 1-cm cubes. Sometimes called a rod.

M

making 10   An addition fact strategy that involves *decomposing* one *addend* and adding part of it to another addend to make 10, then adding the remaining part to find the complete *sum*. For example, 9 + 5 can be solved by decomposing 5 into 1 and 4 and thinking 9 + 1 = 10 and 4 more makes 14. Compare *going through 10*.

Math Boxes   In *Everyday Mathematics,* a collection of problems to practice skills. Math Boxes for each lesson are in the *Math Journal*.

Math Message   In *Everyday Mathematics*, an introduction to the day's lesson designed for children to complete independently.

maximum   The largest amount; the greatest number in a set of *data*. Compare *minimum*.

measurement unit   The reference unit used when measuring. Examples of basic units include inches for length, grams for mass or weight, cubic inches for volume or capacity, seconds for elapsed time, and degrees Celsius for change of temperature. Compound units include square centimeters for area and miles per hour for speed.

memory in a calculator   Where numbers are stored in a calculator for use in later calculations. Most calculators have both short-term memory and long-term memory.

mental arithmetic   Computation done by people "in their heads," either in whole or in part. In *Everyday Mathematics,* children learn a variety of mental calculation strategies as they develop *automaticity* with basic facts and *fact power*.

Mental Math and Fluency   In *Everyday Mathematics,* short, leveled exercises presented at the beginning of lessons. Mental Math and Fluency problems prepare children to think about math, warm up skills they need for the lesson, and build mental-arithmetic skills. They also help teachers assess individual strengths and weaknesses.

meter (m)   The basic metric unit of length from which other metric units of length are derived. Originally, the meter was defined as $\frac{1}{10,000,000}$ of the distance from the North Pole to the equator along a meridian passing through Paris. From 1960 to 1983, the meter was redefined as 1,630,763.73 wavelengths of orange-red light from the element krypton. Today, the meter is defined as the distance light travels in a vacuum in $\frac{1}{299,792,458}$ second. One meter is equal to 10 decimeters, 100 centimeters, or 1,000 millimeters.

metric system   The measurement system used in most countries and by virtually all scientists around the world. *Units* within the metric system are related by powers of 10. Units for length include millimeter, centimeter, meter, and kilometer; units for mass and weight include gram and kilogram; units for volume and capacity include milliliter and liter; and the unit for temperature change is degrees Celsius. See *Tables of Measures*.

minimum   The smallest amount; the smallest number in a set of *data*. Compare *maximum*.

minuend   In subtraction, the number from which another number is subtracted. For example, in 19 − 5 = 14, the minuend is 19. Compare *subtrahend* and *difference*.

missing addend   An *addend* that is *unknown* within an addition *equation*. A subtraction problem can be represented by an addition number sentence with a missing addend. See *unknown*.

model   A mathematical representation or description of an object or a situation. For example, 60 × 3 can be a model for how much money is needed to buy 3 items that cost 60 cents each. A circle can be a model for the rim of a wheel. See *represent*.

multiple of a number n   (1) A product of n and a *counting number*. For example, the multiples of 7 are 7, 14, 21, 28, . . ., and the multiples of $\frac{1}{5}$ are $\frac{1}{5}, \frac{2}{5}, \frac{3}{15}, \ldots$. (2) A *product* of n and an *integer*. For example, the multiples of 7 are . . ., −21, −14, −7, 0, 7, 14, 21, . . ., and the multiples of π are −3π, −2π, −π, 0, π, 2π, 3π, ...

multiplication symbols   The number a multiplied by the number b is written in a variety of ways. Many mathematics textbooks and Second and Third Grade *Everyday Mathematics* use ×, as in $a \times b$. Beginning in fourth grade, *Everyday Mathematics* uses ∗, as in $a * b$. Other common ways to indicate multiplication are by a dot, as in $a \cdot b$, and by juxtaposition, as in ab, which is common in both formulas and algebra.

N

name-collection box In *Everyday Mathematics,* a diagram that is used for collecting *equivalent names* for a number.

25
37 − 12
20 + 5
~~HHT~~ ~~HHT~~ ~~HHT~~ ~~HHT~~ ~~HHT~~
twenty-five
veinticinco

natural numbers Same as *counting numbers.*

near doubles An addition fact strategy that involves relating a given fact to a nearby *doubles fact* to help solve the fact. For example, 7 + 8 can be solved by thinking 7 + 7 = 14, then 14 + 1 = 15. It can also be solved by thinking 8 + 8 = 16, then 16 − 1 = 15.

negative numbers Numbers less than 0; the opposites of the *positive numbers,* commonly written as a positive number preceded by a −. Negative numbers are plotted left of 0 on a horizontal number line or below 0 on a vertical number line.

n-gon A *polygon,* where *n* is the number of sides. Polygons that do not have special names are usually named using *n*-gon notation, such as 13-gon or 100-gon.

nonagon A 9-sided *polygon.*

nonconvex polygon Same as *concave polygon.*

number-and-word notation A notation consisting of the significant *digits* of a number and words for the *place value.* For example, 27 billion is number-and-word notation for 27,000,000,000. Compare *standard notation.*

number family Same as *fact family.*

number grid A table in which *consecutive* numbers are arranged in rows, usually 10 *columns* per *row.* A move from one number to the next within a row is a change of 1; a move from one number to the next within a column is a change of 10.

−9	−8	−7	−6	−5	−4	−3	−2	−1	0
1	2	3	4	5	6	7	8	9	10
11	12	13	14	15	16	17	18	19	20
21	22	23	24	25	26	27	28	29	30
31	32	33	34	35	36	37	38	39	40
41	42	43	44	45	46	47	48	49	50
51	52	53	54	55	56	57	58	59	60
61	62	63	64	65	66	67	68	69	70
71	72	73	74	75	76	77	78	79	80
81	82	83	84	85	86	87	88	89	90
91	92	93	94	95	96	97	98	99	100
101	102	103	104	105	106	107	108	109	110

A number grid

number-grid puzzle In *Everyday Mathematics,* a piece of a number grid in which some but not all of the numbers are missing. Children use number-grid puzzles to practice *place-value* concepts.

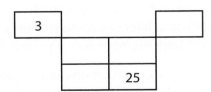

A number-grid puzzle

number line A *line* on which points are indicated by *tick marks* that are usually at regularly spaced intervals from a starting point called the *origin,* the zero point, or simply 0. Numbers are associated with the tick marks on a *scale* defined by the unit interval from 0 to 1. Every real number locates a point on the line, and every point corresponds to a real number. See *real numbers.*

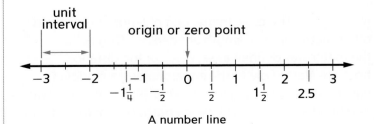

A number line

number model   A *number sentence, expression,* or other representation that fits a *number story* or situation. For example, the *number story* "Sally had $5, and then she earned $8." can be *modeled* as the number sentence $5 + 8 = 13$, as the expression $5 + 8$, or by

$$\begin{array}{r} 5 \\ + 8 \\ \hline 13 \end{array}$$

number scroll   In *Everyday Mathematics,* a series of number grids taped together.

number sentence   Two *expressions* linked by a *relation symbol,* such as $=$, $<$, or $>$.

$$5 + 5 = 10 \qquad 16 \le a \times b$$
$$2 - ? = 8 \qquad a^2 + b^2 = c^2$$

Number sentences

number sequence   A list of numbers, often generated by a rule. In *Everyday Mathematics,* children explore number sequences using *Frames-and-Arrows* diagrams.

$$1, 2, 3, 4, 5, \ldots \quad 1, 4, 9, 16, 25, \ldots$$
$$1, 2, 1, 2, 1, \ldots \quad 1, 3, 5, 7, 9, \ldots$$

Number sequences

number story   A story that involves numbers and one or more explicit or implicit questions. For example, "I have 7 crayons in my desk. Carrie gave me 8 more crayons." is a number story.

numeral   (1) A combination of *base-10 digits* used to express a number. (2) A word, symbol, or figure that represents a number. For example, six, VI, ̶H̶t̶̶ /, and 6 are all numerals that represent the same number.

numeration   A method of numbering or of reading and writing numbers. In *Everyday Mathematics,* numeration activities include counting, writing numbers, identifying *equivalent names* for numbers in *name-collection boxes,* exchanging coins such as 5 pennies for 1 nickel, and renaming numbers in computation.

octagon   An 8-sided *polygon.*

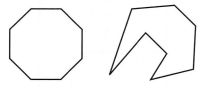

Octagons

octahedron   A *polyhedron* with 8 *faces.* An octahedron with 8 *equilateral triangle* faces is one of the five *regular polyhedrons.*

odd number   (1) A *counting number* that is not divisible by 2. (2) An *integer* that is not divisible by 2. Compare *even number.*

open number line   A line on which one can indicate points by *tick marks* and *labels* that are not spaced at regular intervals. Like *number lines,* there is an order and an implied *origin,* but unlike number lines, there is no *scale* or unit interval. Open number lines are useful *tools* for solving problems.

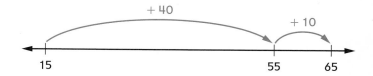

Open number line for solving $15 + ? = 65$

open sentence   A *number sentence* with one or more variables that is neither true nor false. For example, $9 + \underline{\quad} = 15$, $? - 24 < 10$, and $7 = x + y$ are open sentences. See *variable* and *unknown.*

operation   An action performed on one or more mathematical objects, such as numbers, *variables,* or *expressions,* to produce another mathematical object. Addition, subtraction, multiplication, and division are the four basic arithmetic operations. Taking a square root, squaring a number, and multiplying both sides of an equation by the same number are also operations. In *Everyday Mathematics,* children learn about many operations along with procedures, or algorithms, for carrying them out.

operation symbol   A symbol used in *expressions* and *number sentences* to stand for a particular mathematical operation. Symbols for common arithmetic operations are addition $+$; subtraction $-$; multiplication $\times$, $*$, $\bullet$; division $\div$, $/$; powering $\wedge$. See *General Reference.*

order To arrange according to a specific rule; for example, from smallest to largest, or from largest to smallest. See *sequence*.

ordered pair (1) Two numbers, or coordinates, used to locate a point on a rectangular coordinate grid. The first coordinate *x* gives the position along the horizontal axis of the grid, and the second coordinate *y* gives the position along the vertical axis. The pair is written (*x*, *y*). (2) Any pair of objects or numbers in a particular order, as in letter-number spreadsheet cell names, map coordinates, or functions given as sets of pairs of numbers.

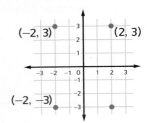

Ordered pairs

ordinal number A number describing the position or order of something in a *sequence*, such as first, third, or tenth. Ordinal numbers are commonly used in dates, as in "May fifth." Compare *cardinal number*.

origin The zero point in a coordinate system. On a *number line*, the origin is the point at 0. On a coordinate grid, the origin is the point (0, 0) where the two axes intersect.

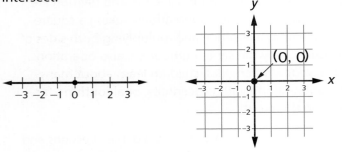

The points at 0 and (0, 0) are origins.

outcome A possible result of a *chance* experiment or situation. For example, HEADS and TAILS are the two possible outcomes of flipping a coin.

output (1) A number paired to an *input* by an imaginary *function machine* applying a rule. (2) The values for *y* in a *function* consisting of *ordered pairs* (*x*, *y*). (3) Numbers or other information displayed or produced by a calculator or computer.

P

pan balance A measuring device used to weigh objects or compare weights or masses. Simple pan balances have two pans suspended at opposite ends of a bar resting on a fulcrum at its midpoint. When the weights or masses of the objects in the pans are *equal*, the bar is level.

Pan balance

parallel *Lines, line segments,* or rays in the same *plane* are parallel if they never cross or meet, no matter how far they are extended. Two planes are parallel if they never cross or meet. A line and a plane are parallel if they never cross or meet. The symbol ∥ means is parallel to.

parallelogram A *trapezoid* that has two pairs of *parallel* sides. See *quadrilateral*.

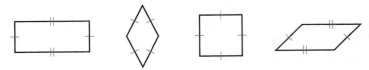

Parallelograms

partial-sums addition An addition algorithm in which separate *sums* are computed for each *place value* of the numbers and then added to get a final sum.

partition In geometry, to divide a shape into smaller shapes. For example, a *polygon* can be partitioned into *triangles*. Shapes can be partitioned into *equal shares* to represent *fractions*. Partitioning can also be used to find *length, area,* or *volume*.

parts-and-total diagram In *Everyday Mathematics*, a diagram to *model* situations in which two or more quantities (parts) are combined to make a total quantity. See *situation diagram*.

Total	
13	
Part	Part
8	**?**

Parts-and-total diagram for 13 = 8 + ?

parts-and-total situation   A situation in which a quantity is made up of two or more distinct parts. For example, the following is a parts-and-total situation: "There are 15 girls and 12 boys in Mrs. Dorn's class. How many children are there in all?" See *addition/subtraction use class*.

pattern   A repetitive order or arrangement. In *Everyday Mathematics*, children mainly explore visual and number patterns in which elements are arranged so that what comes next can be predicted.

pattern blocks   A standard set of *polygon*-shaped blocks used in geometry activities. Compare *attribute blocks*.

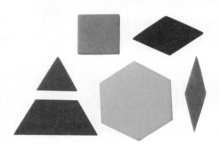

Pattern blocks

pentagon   A 5-sided *polygon*.

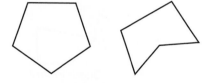

Pentagons

per   For each, as in ten chairs per row or six tickets per family.

pictograph   See *picture graph*.

picture graph   A graph constructed with icons representing *data* points. They are sometimes called *scaled picture graphs* when each icon represents more than 1 data point.

Trees Planted in Park

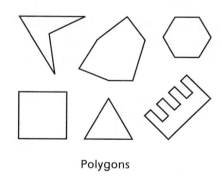

A picture graph

place value   A system that gives a *digit* a value according to its position, or place, in a number. In the standard *base-ten* (decimal) system for writing numbers, each place has a value 10 times that of the place to its right and one-tenth the value of the place to its left.

thousands	hundreds	tens	ones	.	tenths	hundredths

A place-value chart

plane   A *2-dimensional* flat *surface* that extends forever in all directions. In formal Euclidean geometry, plane is an undefined geometric term.

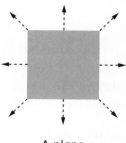

A plane

plane figure   A *set* of points that is entirely contained in a single plane. For example, squares, pentagons, circles, parabolas, lines, and rays are plane figures; cones, cubes, and prisms are not.

P.M.   The abbreviation for *post meridiem*, meaning "after the middle of the day" in Latin. From noon to midnight.

poly-   A prefix meaning many. See *General Reference, Prefixes* for specific numerical prefixes.

polygon   A plane figure formed by three or more *line segments* (sides) that meet only at their endpoints (*vertices*) to make a closed path. The sides may not cross one another.

Polygons

polyhedron A closed *3-dimensional* figure formed by polygons with their *interiors* (*faces*) that do not cross. The plural is polyhedrons or polyhedra.

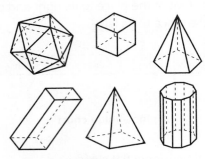

Polyhedrons

positive numbers Numbers greater than 0; the opposites of the *negative numbers*. Positive numbers are plotted to the right of 0 on a horizontal *number line* or above 0 on a vertical number line.

poster In *Everyday Mathematics,* a page displaying a collection of illustrated numerical *data*. A poster may be used as a source of data for developing *number stories*.

precise Of a measurement or other quantity, having a high degree of exactness. A measurement to the nearest inch is more precise than a measurement to the nearest foot. A measurement's precision depends on the *unit* scale of the *tool* used to obtain it. The smaller the unit is, the more precise a measure can be. For instance, a ruler with $\frac{1}{8}$-inch markings can give a more precise measurement than a ruler with $\frac{1}{2}$-inch markings. Compare *accurate*.

preimage The original figure in a transformation. Compare *image*.

prism A polyhedron with two *parallel* and *congruent polygonal* bases and lateral *faces* shaped like *parallelograms*. Right prisms have rectangular lateral faces. Prisms get their names from the shapes of their *bases*.

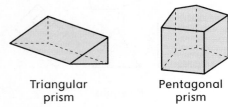

Triangular prism   Pentagonal prism

product The result of multiplying two or more numbers, called factors. For example, in $4 \times 3 = 12$, the product is 12.

property (1) A generalized statement about a mathematical relationship, such as the Distributive Property of Multiplication over Addition. (2) A feature of an object or a common feature of a set of objects. Same as *attribute*.

put-together/take-apart situation A situation in which a quantity is made up of two or more distinct parts. See *addition/subtraction use class*.

pyramid A *polyhedron* with a polygonal *base* and *triangular* other *faces* that meet at a common *vertex* called the *apex*. Pyramids get their names from the shapes of their bases.

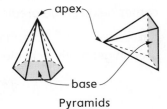

Pyramids

Q

quadrangle Same as *quadrilateral*.

quadrilateral A 4-sided *polygon*. Squares, rectangles, parallelograms, rhombuses, kites, and trapezoids are organized by defining attributes into a *hierarchy of shapes*.

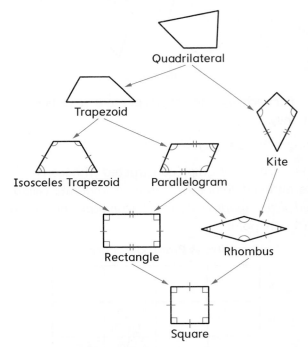

Hierarchy of quadrilaterals

Quick Looks An *Everyday Mathematics* routine in which an image of a quantity is displayed for 2–3 seconds and then removed. Quick Looks encourage children to *subitize, compose,* and *decompose* numbers in flexible ways and also develop strategies for *addition facts*.

R

range (1) The *difference* between the *maximum* and the *minimum* in a *set* of *data*. Used as a measure of the spread of the data. (2) The interval between the maximum and the *minimum* in a set of data.

rational counting Counting using one-to-one matching. For example, counting a number of chairs, people, or crackers. In rational counting, the last number gives the cardinality of the *set*.

rational numbers Numbers that can be written in the form $\frac{a}{b}$, where a and b are *integers* and $b \neq 0$. The *decimal* form of a rational number either terminates or repeats. For example, $\frac{2}{3}$, $-\frac{2}{3}$, 0.5, 20.5, and 0.333. . . are rational numbers.

r-by-c array A rectangular arrangement of elements with r rows and c elements per row. Among other things, an r-by-c array *models r sets* with c objects per *set*. Although listing *rows* before *columns* is arbitrary, it is in keeping with the order used in matrix notation, which children will study later in school.

real numbers All *rational* and *irrational numbers;* all numbers that can be written as *decimals*. For every real number there is a corresponding point on a *number line,* and for every point on the number line, there is a real number.

rectangle A *parallelogram* with four *right angles*. All rectangles are both parallelograms and *isosceles trapezoids*. See *quadrilateral*.

A rectangle

rectangular array An arrangement of objects in *rows* and *columns* that form a *rectangular* shape. All rows have the same number of objects, and all columns have the same number of objects. See r-*by*-c *array*.

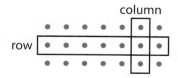

A rectangular array

rectangular prism A prism with rectangular *bases*. The four *faces* that are not bases are formed by *rectangles* or *parallelograms*. For example, a rectangular *prism* in which all sides are rectangular models a shoebox.

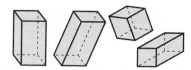

Rectangular prisms

rectangular pyramid A *pyramid* with a rectangular *base*.

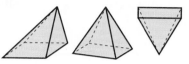

Rectangular pyramids

reference frame A system for locating numbers within a given context, usually with reference to an *origin* or a zero point. For example, *number lines,* clocks, *calendars,* temperature scales, and maps are reference frames.

regular polygon A *polygon* in which all sides are the same *length* and all interior *angles* have the same measure.

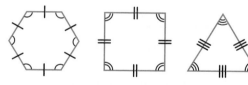

Regular polygons

regular polyhedron A *polyhedron* whose *faces* are all formed by *congruent regular polygons* and in which the same number of faces meet at each *vertex*. There are only five. They are called the Platonic solids.

Tetrahedron	Cube	Octahedron
(4 equilateral triangles)	(6 squares)	(8 equilateral triangles)

Dodecahedron	Icosahedron
(12 regular pentagons)	(20 equilateral triangles)

relation symbol   A symbol used to express a relationship between two quantities, figures, or *sets,* such as ≤, ∥, or ⊂. See *General Reference, Symbols.*

repeating decimal   A *decimal* in which one digit or block of digits is repeated without end. For example, 0.3333. . . and 0.$\overline{147}$ are repeating decimals. Compare *terminating decimal.*

represent   To show, symbolize, or stand for something. For example, numbers can be represented using base-10 blocks, spoken words, or written numerals. See *model.*

rhombus   A *parallelogram* with four sides of the same length. All rhombuses are both parallelograms and *kites.* See *quadrilateral.*

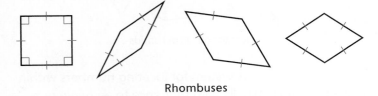

Rhombuses

right angle   A 90° *angle.*

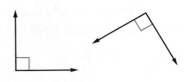

Right angles

rotation   (1) A turn about an axis or point. (2) Point *P'* is a rotation *image* of point *P* around a center of rotation *C* if *P'* is on the *circle* with center *C* and radius *CP.* If all the points in one figure are rotation images of all the points in another figure around the same center of rotation and with the same *angle* of rotation, then the figures are rotation images. The center can be inside or outside of the original image. Reflections, rotations, and translations are types of isometry transformations. (3) If all points on the image of a 3-dimensional figure are rotation images through the same angle around a point or a line called the axis of rotation, then the image is a rotation image of the original figure.

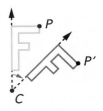

Rotation preimage and image

rote counting   Reciting a string of number words by rote, without necessarily understanding their significance. See *skip counting.*

round   (1) To approximate a number to make it easier to use or make it better reflect the *precision* of the *data.* "Rounding up" means to approximate larger than the actual value. "Rounding down" means to approximate smaller than the actual value. (2) Circular in shape.

row   (1) A horizontal arrangement of objects or numbers in an *array* or a table. (2) A horizontal section of cells in a spreadsheet.

ruler   (1) Traditionally, a wood, metal, or plastic strip *partitioned* into same-size standard *units,* such as inches or centimeters. (2) In *Everyday Mathematics,* a tool for measuring length comprising an end-to-end collection of same-size units.

S

same-change rule for subtraction   A subtraction algorithm in which the same number is added to or subtracted from both numbers.

scale   (1) A multiplicative comparison between the relative sizes or numbers of things. (2) Same as scale factor. (3) A tool for measuring weight and mass.

scale of a number line   The unit interval on a *number line* or a measuring device. The scales on this ruler are 1 millimeter on the left side and $\frac{1}{16}$ inch on the right side.

Scale of a number line

scaled picture graph   A graph constructed with icons each representing the same number of multiple data points. For example, each icon of a car on a graph may stand for 1,000 cars. See *picture graph.*

scalene triangle   A *triangle* with sides of three different lengths. The three *angles* of a scalene triangle have different measures.

segment   Same as *line segment.*

separate situation   A *change-to-less situation* or a *parts-and-total situation.*

sequence An ordered list of numbers, often with an underlying rule that may be used to generate numbers in the list. *Frames-and-Arrows* diagrams can be used to represent sequences. See *order*.

set A collection or group of objects, numbers, or other items.

short-term memory Memory in a calculator used to store values for immediate calculation. Short-term memory is usually cleared with a ⓒ, ⒶⒸ, Ⓒⓛⓔⓐⓡ, or a similar key.

similar figures Figures that have the same shape but not necessarily the same size. In similar figures, corresponding sides are proportional and corresponding *angles* are congruent. Compare *congruent figures*.

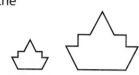

Similar polygons

situation diagram In *Everyday Mathematics,* a diagram used to organize information in a problem situation in one of the addition/subtraction or multiplication/division use classes.

Total
7

Part	Part
2	**5**

Susie has 2 pink balloons and 5 yellow balloons.
She has 7 balloons in all.

skip counting Counting by intervals, such as 2s, 5s, or 10s. See *rote counting*.

slate In *Everyday Mathematics,* a lap-size (about 8-inch by 11-inch) chalkboard or whiteboard that children use for recording responses during group exercises and informal group assessments.

solution of a problem (1) The answer to a problem. (2) The answer to a problem together with the method by which that answer was obtained.

solution of an open sentence A value or values for the *variable*(s) in an open sentence that make the sentence true. For example, 7 is a solution of $5 + n = 12$. Although *equations* are not necessarily open sentences, the solution of an open sentence is commonly referred to as a solution of an equation.

sphere The set of all points in space that are an equal distance from a fixed point called the center of the sphere. The distance from the center to the sphere is the radius of the sphere. The diameter of a sphere is twice its radius. Points inside a sphere are not part of the sphere.

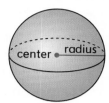

A sphere

Spiral Snapshot In *Everyday Mathematics,* an overview of nearby lessons that addresses one of the Goals for Mathematical Content in the Focus part of the lesson. It appears in the Lesson Opener.

Spiral Trace In *Everyday Mathematics,* an overview of work in the current unit and nearby units on selected Standards for Mathematical Content. It appears in the Unit Organizer.

Spiral Tracker In *Everyday Mathematics,* an online database that shows complete details about learning trajectories for all goals and standards.

square A rectangle with four sides of equal length. All squares are both *rectangles* and *rhombuses*. See *quadrilateral*.

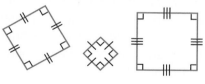

Squares

square array A *rectangular array* with the same number of rows as columns. For example, 16 objects will form a square array with 4 objects in each row and 4 objects in each column.

A square array

square corner Same as *right angle*.

square numbers Figurate numbers that are the *product* of a counting number and itself. For example, 25 is a square number because $25 = 5 \times 5$. A square number can be represented by a square array and as a number squared, such as $25 = 5^2$.

square pyramid A *pyramid* with a square *base*.

square root of a number n A number that multiplied by itself is n, commonly written $\sqrt{n}$. For example, 4 is a square root of 16, because $4 \times 4 = 16$. Normally, square root refers to the positive square root, but the opposite of a positive square root is also a square root. For example, -4 is also a square root of 16 because $-4 \times -4 = 16$.

square unit   A *unit* to measure *area*. A model of a square unit is a square with each side a related unit of length. For example, a square inch is the area of a square with 1-inch sides. Square units are often labeled as the length unit squared. For example, 1 cm^2 is read "1 square centimeter" or "1 centimeter squared."

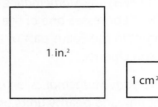

Square units

standard notation   The most common way of representing whole numbers, integers, and decimals. Standard notation for real numbers is *base-ten place-value* numeration. For example, standard notation for three hundred fifty-six is 356. Same as decimal notation. Compare *number-and-word notation*.

standard unit   A *unit* of measure that has been defined by a recognized authority, such as a government or the National Institute of Standards and Technology. For example, inches, meters, miles, seconds, pounds, grams, and acres are all standard units.

subitize   To recognize a quantity without needing to count. For example, many children can instantly recognize the number of dots on a rolled die without needing to count them. In *Everyday Mathematics, Quick Looks* help children develop the ability to subitize and use relationships between quantities to find more complex totals.

subtrahend   The number being taken away in a subtraction problem. For example, in $15 - 5 = 10$, the subtrahend is 5. Compare *difference* and *minuend*.

sum   The result of adding two or more numbers. For example, in $5 + 3 = 8$, the sum is 8.

surface   (1) The boundary of a *3-dimensional* object. (2) Any *2-dimensional* layer, such as a *plane* or a *face* of a *polyhedron*.

survey   (1) A study that collects *data* by asking people questions. (2) Any study that collects data. Surveys are commonly used to study demographics, such as people's characteristics, behaviors, interests, and opinions.

T

take-apart situation   See *put-together/take-apart situation*.

take-from situation   A *change-to-less situation*.

tally   (1) To keep a record of a count by making a mark for each item as it is counted. (2) The mark used in a count. Also called tally mark.

tally chart   A table to keep track of tallies, typically showing how many times each value appears in a set of data.

Number of Pull-Ups	Number of Children
0	卌 𝆑
1	卌
2	////
3	//

A tally chart

teen number   The teen numbers are 11, 12, 13, 14, 15, 16, 17, 18, and 19. Sometimes 10 is considered a teen number.

temperature   How hot or cold something is relative to another object or as measured on a standardized scale, such as *degrees Celsius* or *degrees Fahrenheit*.

template   In *Everyday Mathematics,* a sheet of plastic with geometric shapes cut out of it, used to draw patterns and designs.

ten frame   A 5-by-2 grid of squares that can be used to represent the numbers 0–10 in a variety of ways. Ten frames are particularly useful for encouraging children to relate given representations to 5 or 10.

Two representations of 6 on ten frames

terminating decimal   A *decimal* that ends. For example, 0.5 and 0.125 are terminating decimals.

thermometer   A tool used to measure *temperature* in *degrees* according to a fixed *scale*. The most common scales are *Celsius* and *Fahrenheit*.

think addition   A subtraction-fact strategy that involves using an addition fact in the same *fact family* to help solve the given subtraction problem. For example, for $10 - 4 = ?$, one might think $4 + ? = 10$ and use $4 + 6 = 10$ to determine that the *difference* is 6. This strategy is particularly useful with doubles or combinations of 10 as *helper facts*.

tick marks   (1) Marks showing the *scale of a number line* or ruler. (2) Marks indicating that two *line segments* have the same *length*. (3) Same as *tally*.

tile (verb)   To cover a *surface* completely with shapes without overlaps or gaps. Tiling with same-size squares is a way to measure *area*. See *iterate units*.

timeline   A number line showing when events took place. In some timelines the origin is based on the context of the events being graphed, such as the birth date of the child's life graphed below. The origin can also come from another reference system, such as the year CE, in which case the scale below might cover the years 2015 through 2020.

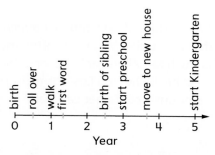

A timeline of a child's milestones

tool   Anything, physical or abstract, that serves as an instrument for performing a task. Physical tools include hammers for hammering, calculators for calculating, and rulers for measuring. Abstract tools include computational algorithms such as *partial-sums addition*, problem-solving strategies such as "guess and check," and technical drawings such as *situation diagrams*.

Total Physical Response   A teaching technique that facilitates beginning English language learners' acquisition of new English vocabulary through modeling of physical actions or display of visuals as target objects that are named aloud.

transitivity principle of indirect measurement   The property that provides for *indirect comparison* of measures of objects, such as length or weight. Given three measures A, B, and C, if $A > B$ and $B > C$, then $A > C$. See *indirect comparison*.

trapezoid   A *quadrilateral* that has at least one pair of *parallel* sides.

Trapezoids

trial-and-error method   A systematic method for finding the solution of an equation by trying a sequence of test numbers.

triangle   A 3-sided *polygon*.

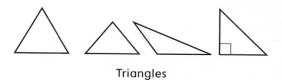

Triangles

triangular numbers   *Figurate numbers* that can be represented by triangular arrangements of dots. The triangular numbers are {1, 3, 6, 10, 15, 21, 28, 36, 45, . . .}.

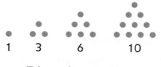

Triangular numbers

triangular prism   A *prism* whose *bases* are triangular.

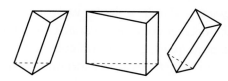

Triangular prisms

triangular pyramid   A *pyramid* in which all *faces* are *triangular*, any one of which is the *base*; also known as a tetrahedron. A regular tetrahedron has four faces formed by *equilateral triangles* and is one of the five regular *polyhedrons*.

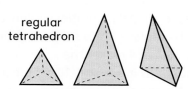

Triangular pyramids

true number sentence   A *number sentence* stating a correct fact. For example, $75 = 25 + 50$ is a true number sentence. Compare *false number sentence*.

turn-around rule   A rule for solving addition and multiplication problems based on the *Commutative Properties of Addition and Multiplication.* For example, if one knows that $6 \times 8 = 48$, then, by the turn-around rule, one also knows that $8 \times 6 = 48$.

unit   A label used to put a number in context. In measuring length, for example, inches and centimeters are units. In a problem about 5 apples, apple is the unit. In *Everyday Mathematics,* children keep track of units in unit boxes.

unit box   In *Everyday Mathematics,* a box displaying the unit for the numbers in the problems at hand.

A unit box

unknown   A quantity whose value is not known. An unknown is sometimes represented by a _____, a ?, or a letter. See *open sentence* and *variable.*

U.S. customary system   The measuring system used most often in the United States. Units for length include inch, foot, yard, and mile; units for weight include ounce and pound; units for volume or capacity include fluid ounce, cup, pint, quart, gallon, and cubic units; and the unit for temperature is degrees Fahrenheit. See *Tables of Measures.*

variable   A letter or other symbol that can be replaced by any value from a set of possible values. Some values replacing variables in *number sentences* may make them true. For example, to make number sentences true, variables may be replaced by a single number, as in $5 + n = 9$, where $n = 4$ makes the sentence true; many different numbers, as in $x + 2 < 10$, where any number less than 8 makes the sentence true; or any number, as in $a + 3 = 3 + a$, which is true for all numbers. See *open sentence* and *unknown.*

Venn diagram   A picture that uses circles or rings to show relationships between *sets.* In this diagram, $22 + 8 = 30$ girls are on the track team, and 8 are on both the track and the basketball teams.

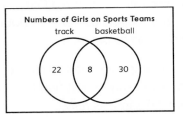

A Venn diagram

vertex   The point at which the sides of an *angle* or a *polygon,* or the *edges* of a *polyhedron,* meet. The plural is vertexes or vertices. See *corner.*

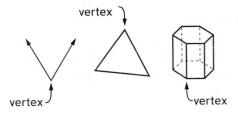

volume   (1) A measure of how much *3-dimensional* space something occupies. Volume is often measured in liquid units, such as gallons or liters, or cubic units, such as cm³ or cubic inches. (2) Less formally, the same as capacity: the amount a container can hold.

"What's My Rule?"   A problem in which two of the three parts of a *function* (*input, output,* and rule) are known and the third is to be found out. See *function.*

whole   An entire object, collection of objects, or quantity being considered in a problem situation; 100%. Can also be called ONE or unit whole.

whole numbers   The *counting numbers* and 0. The set of whole numbers is $\{0, 1, 2, 3, \ldots\}$.

zero fact   See *0 fact.*

General Reference

Symbols

$+$	plus or positive
$-$	minus or negative
$*$, $\times$	multiplied by
$\div$, $/$	divided by
$=$	is equal to
$\neq$	is not equal to
$<$	is less than
$>$	is greater than
$\leq$	is less than or equal to
$\geq$	is greater than or equal to
$\approx$	is approximately equal to
x^n, $x^{\wedge}n$	nth power of x
$\sqrt{x}$	square root of x
$\%$	percent
$a{:}b$, a/b, $\frac{a}{b}$	ratio of a to b or a divided by b or the fraction $\frac{a}{b}$
a [bs]	a groups, b in each group
$n \div d \longrightarrow a\ \mathrm{R}b$	n divided by d is a with remainder b
$\{\}, (\,), [\,]$	grouping symbols
∞	infinity
$n!$	n factorial
$\circ$	degree
(a, b)	ordered pair
$\overleftrightarrow{AS}$	line AS
$\overline{AS}$	line segment AS
$\overrightarrow{AS}$	ray AS
$\llcorner$	right angle
$\perp$	is perpendicular to
$\parallel$	is parallel to
$\triangle ABC$	triangle ABC
$\angle ABC$	angle ABC
$\angle B$	angle B
$\cong$	is congruent to
$\sim$	is similar to
$\equiv$	is equivalent to
$|n|$	absolute value of n

Prefixes

Prefix	Meaning	Prefix	Meaning
uni-	one	tera-	trillion (10^{12})
bi-	two	giga-	billion (10^{9})
tri-	three	mega-	million (10^{6})
quad-	four	kilo-	thousand (10^{3})
penta-	five	hecto-	hundred (10^{2})
hexa-	six	deca-	ten (10^{1})
hepta-	seven	uni-	one (10^{0})
octa-	eight	deci-	tenth (10^{-1})
nona-	nine	centi-	hundredth (10^{-2})
deca-	ten	milli-	thousandth (10^{-3})
dodeca-	twelve	micro-	millionth (10^{-6})
icosa-	twenty	nano-	billionth (10^{-9})

Constants

Constants	Approximations
Pi (π)	3.14159 26535 89793
Golden Ratio (ϕ)	1.61803 39887 49894
Radius of Earth at equator	6,378.388 kilometers 3,963.34 miles
Circumference of Earth at equator	40,076.59 kilometers 24,902.44 miles
Velocity of sound in dry air at 0°C	331.36 m/sec 1087.1 ft/sec
Velocity of light in a vacuum	2.997925×10^{10} cm/sec

The Order of Operations

1. Do operations inside grouping symbols following Rules 2–4. Work from the innermost set of grouping symbols outward.
2. Calculate all expressions with exponents or roots.
3. Multiply and divide in order from left to right.
4. Add and subtract in order from left to right.

Tables of Measures

Metric System

Units of Length

1 kilometer (km)	= 1,000 meters (m)
1 meter	= 10 decimeters (dm)
	= 100 centimeters (cm)
	= 1,000 millimeters (mm)
1 decimeter	= 10 centimeters
1 centimeter	= 10 millimeters

Units of Area

1 square meter (m^2)	= 100 square decimeters (dm^2)
	= 10,000 square centimeters (cm^2)
1 square decimeter	= 100 square centimeters
1 are (a)	= 100 square meters
1 hectare (ha)	= 100 ares
1 square kilometer (km^2)	= 100 hectares

Units of Volume and Capacity

1 cubic meter (m^3)	= 1,000 cubic decimeters (dm^3)
	= 1,000,000 cubic centimeters (cm^3)
1 cubic centimeter	= 1,000 cubic millimeters (mm^3)
1 kiloliter (kL)	= 1,000 liters (L)
1 liter	= 1,000 milliliters (mL)

Units of Mass and Weight

1 metric ton (t)	= 1,000 kilograms (kg)
1 kilogram	= 1,000 grams (g)
1 gram	= 1,000 milligrams (mg)

U.S. Customary System

Units of Length

1 mile (mi)	= 1,760 yards (yd)
	= 5,280 feet (ft)
1 yard	= 3 feet
	= 36 inches (in.)
1 foot	= 12 inches

Units of Area

1 square yard (yd^2)	= 9 square feet (ft^2)
	= 1,296 square inches ($in.^2$)
1 square foot	= 144 square inches
1 acre	= 43,560 square feet
1 square mile (mi^2)	= 640 acres

Units of Volume and Capacity

1 cubic yard (yd^3)	= 27 cubic feet (ft^3)
1 cubic foot	= 1,728 cubic inches ($in.^3$)
1 gallon (gal)	= 4 quarts (qt)
1 quart	= 2 pints (pt)
1 pint	= 2 cups (c)
1 cup	= 8 fluid ounces (fl oz)
1 fluid ounce	= 2 tablespoons (tbs)
1 tablespoon	= 3 teaspoons (tsp)

Units of Mass and Weight

1 ton (T)	= 2,000 pounds (lb)
1 pound	= 16 ounces (oz)

System Equivalents (Conversion Factors)

1 inch ≈ 2.5 cm (2.54)	1 liter ≈ 1.1 quarts (1.057)
1 kilometer ≈ 0.6 mile (0.621)	1 ounce ≈ 28 grams (28.350)
1 mile ≈ 1.6 kilometers (1.609)	1 kilogram ≈ 2.2 pounds (2.21)
1 meter ≈ 39 inches (39.37)	1 hectare ≈ 2.5 acres (2.47)

Body Measures

1 *digit* is about the width of a finger.

1 *hand* is about the width of the palm and thumb.

1 *span* is about the distance from the tip of the thumb to the tip of the first (index) finger of an outstretched hand.

1 *cubit* is about the length from the elbow to the tip of the extended middle finger.

1 *yard* is about the distance from the center of the chest to the tip of the extended middle finger of an outstretched arm.

1 *fathom* is about the length from fingertip to fingertip of outstretched arms. Also called an arm span.

Units of Time

1 century	= 100 years
1 decade	= 10 years
1 year (yr)	= 12 months
	= 52 weeks (plus one or two days)
	= 365 days (366 days in a leap year)
1 month (mo)	= 28, 29, 30, or 31 days
1 week (wk)	= 7 days
1 day (d)	= 24 hours
1 hour (hr)	= 60 minutes
1 minute (min)	= 60 seconds (s or sec)

Unpacking the Common Core State Standards

The **Common Core State Standards** include two groups of standards: Standards for Mathematical Content and Standards for Mathematical Practice. The Content Standards define the mathematical content to be mastered at each grade. The Practice Standards define the processes and habits of mind students need to develop as they learn the content for their grade level.

The Content Standards are organized into **Domains,** large groups of related standards. Within each Domain are **Clusters,** smaller groups of related standards. The **Standards** themselves define what students should understand and be able to do by the end of that grade.

The summary page from the Common Core State Standards for Grade 2 is on the following page. This page summarizes the mathematical content that children should learn in Grade 2.

The chart beginning on page EM3 lists the Common Core State Standards for Mathematical Content with corresponding *Everyday Mathematics* **Goals for Mathematical Content (GMC).**

The chart beginning on page EM6 lists the Common Core State Standards for Mathematical Practice with corresponding *Everyday Mathematics* **Goals for Mathematical Practice (GMP).**

Common Core State Standards

Standards for Mathematical Content

Domain Operations and Algebraic Thinking 2.OA	*Everyday Mathematics* Goals for Mathematical Content
Cluster Represent and solve problems involving addition and subtraction.	
2.OA.1 Use addition and subtraction within 100 to solve one- and two-step word problems involving situations of adding to, taking from, putting together, taking apart, and comparing, with unknowns in all positions, e.g., by using drawings and equations with a symbol for the unknown number to represent the problem.	**GMC** Model 1-step problems involving addition and subtraction. **GMC** Use addition and subtraction to solve 1-step number stories. **GMC** Model 2-step problems involving addition and subtraction. **GMC** Use addition and subtraction to solve 2-step number stories.

"…larger groups of related standards. Standards from different domains may…be… related."

"…groups of related standards… Standards from different clusters may…be… related, because mathematics is a connected subject."

the first standard under this Domain

program goals for finer-grained tracking of student progress

In the table on pages EM6–EM9, you may trace the **Goals for Mathematical Practice** as they unpack the Standards for Mathematical Practice.

The Grade 2 Content Standards are introduced in the CCSS document on page 17, as follows:

In Grade 2, instructional time should focus on four critical areas: (1) extending understanding of base-ten notation; (2) building fluency with addition and subtraction; (3) using standard units of measure; and (4) describing and analyzing shapes.

(1) Students extend their understanding of the base-ten system. This includes ideas of counting in fives, tens, and multiples of hundreds, tens, and ones, as well as number relationships involving these units, including comparing. Students understand multi-digit numbers (up to 1000) written in base-ten notation, recognizing that the digits in each place represent amounts of thousands, hundreds, tens, or ones (e.g., 853 is 8 hundreds + 5 tens + 3 ones).

(2) Students use their understanding of addition to develop fluency with addition and subtraction within 100. They solve problems within 1000 by applying their understanding of models for addition and subtraction, and they develop, discuss, and use efficient, accurate, and generalizable methods to compute sums and differences of whole numbers in base-ten notation, using their understanding of place value and the properties of operations. They select and accurately apply methods that are appropriate for the context and the numbers involved to mentally calculate sums and differences for numbers with only tens or only hundreds.

(3) Students recognize the need for standard units of measure (centimeter and inch) and they use rulers and other measurement tools with the understanding that linear measure involves an iteration of units. They recognize that the smaller the unit, the more iterations they need to cover a given length.

(4) Students describe and analyze shapes by examining their sides and angles. Students investigate, describe, and reason about decomposing and combining shapes to make other shapes. Through building, drawing, and analyzing two- and three-dimensional shapes, students develop a foundation for understanding area, volume, congruence, similarity, and symmetry in later grades.

Common Core State Standards

Standards for Mathematical Content

Domain Operations and Algebraic Thinking 2.OA	*Everyday Mathematics* Goals for Mathematical Content

Cluster Represent and solve problems involving addition and subtraction.

2.OA.1 Use addition and subtraction within 100 to solve one- and two-step word problems involving situations of adding to, taking from, putting together, taking apart, and comparing, with unknowns in all positions, e.g., by using drawings and equations with a symbol for the unknown number to represent the problem.[1]	**GMC** Model 1-step problems involving addition and subtraction.
	GMC Use addition and subtraction to solve 1-step number stories.
	GMC Model 2-step problems involving addition and subtraction.
	GMC Use addition and subtraction to solve 2-step number stories.

Cluster Add and subtract within 20.

2.OA.2 Fluently add and subtract within 20 using mental strategies.[2] By end of Grade 2, know from memory all sums of two one-digit numbers.	**GMC** Add within 20 fluently.
	GMC Subtract within 20 fluently.
	GMC Know all sums of two 1-digit numbers automatically.

Cluster Work with equal groups of objects to gain foundations for multiplication.

2.OA.3 Determine whether a group of objects (up to 20) has an odd or even number of members, e.g., by pairing objects or counting them by 2s; write an equation to express an even number as a sum of two equal addends.	**GMC** Determine whether the number of objects in a group is odd or even.
	GMC Express an even number as a sum of two equal addends.
2.OA.4 Use addition to find the total number of objects arranged in rectangular arrays with up to 5 rows and up to 5 columns; write an equation to express the total as a sum of equal addends.	**GMC** Find the total number of objects in a rectangular array.
	GMC Express the number of objects in an array as a sum of equal addends.

Domain Number and Operations in Base Ten 2.NBT

Cluster Understand place value.

2.NBT.1 Understand that the three digits of a three-digit number represent amounts of hundreds, tens, and ones; e.g., 706 equals 7 hundreds, 0 tens, and 6 ones. Understand the following as special cases:	**GMC** Understand 3-digit place value.
	GMC Represent whole numbers as hundreds, tens, and ones.
2.NBT.1a 100 can be thought of as a bundle of ten tens— called a "hundred."	**GMC** Understand exchanging tens and hundreds.
2.NBT.1b The numbers 100, 200, 300, 400, 500, 600, 700, 800, 900 refer to one, two, three, four, five, six, seven, eight, or nine hundreds (and 0 tens and 0 ones).	**GMC** Understand *100, 200, …, 900* as some hundreds, no tens, and no ones.
2. NBT.2 Count within 1000; skip-count by 5s, 10s, and 100s.	**GMC** Count by 1s.
	GMC Count by 5s, 10s, and 100s.
2. NBT.3 Read and write numbers to 1000 using base-ten numerals, number names, and expanded form.	**GMC** Read and write numbers.
	GMC Read and write number names.
	GMC Read and write numbers in expanded form.

---

[1] See Glossary, Table 1. http://www.corestandards.org/assets/CCSSI_Math Standards.pdf
[2] See standard 1.OA.6 for a list of mental strategies.

Common Core State Standards

Standards for Mathematical Content

Cluster Use place value understanding and properties of operations to add and subtract.

2. NBT.4 Compare two three-digit numbers based on meanings of the hundreds, tens, and ones digits, using >, =, and < symbols to record the results of comparisons.	**GMC** Compare and order numbers. **GMC** Record comparisons using >, =, or <.
2.NBT.5 Fluently add and subtract within 100 using strategies based on place value, properties of operations, and/or the relationship between addition and subtraction.	**GMC** Add within 100 fluently. **GMC** Subtract within 100 fluently.
2.NBT.6 Add up to four two-digit numbers using strategies based on place value and properties of operations.	**GMC** Add up to four 2-digit numbers.
2.NBT.7 Add and subtract within 1000, using concrete models or drawings and strategies based on place value, properties of operations, and/or the relationship between addition and subtraction; relate the strategy to a written method. Understand that in adding or subtracting three-digit numbers, one adds or subtracts hundreds and hundreds, tens and tens, ones and ones; and sometimes it is necessary to compose or decompose tens or hundreds.	**GMC** Add multidigit numbers using models or strategies. **GMC** Subtract multidigit numbers using models or strategies.
2.NBT.8 Mentally add 10 or 100 to a given number 100–900, and mentally subtract 10 or 100 from a given number 100–900.	**GMC** Mentally add 10 to and subtract 10 from a given number. **GMC** Mentally add 100 to and subtract 100 from a given number.
2.NBT.9 Explain why addition and subtraction strategies work, using place value and the properties of operations	**GMC** Explain why addition and subtraction strategies work.

Domain Measurement and Data   2.MD

Everyday Mathematics
Goals for Mathematical Content

Cluster Measure and estimate lengths in standard units.

2.MD.1 Measure the length of an object by selecting and using appropriate tools such as rulers, yardsticks, meter sticks, and measuring tapes.	**GMC** Measure the length of an object. **GMC** Select appropriate tools to measure length.
2.MD.2 Measure the length of an object twice, using length units of different lengths for the two measurements; describe how the two measurements relate to the size of the unit chosen.	**GMC** Measure an object using 2 different units of length. **GMC** Describe how length measurements relate to the size of the unit.
2.MD.3 Estimate lengths using units of inches, feet, centimeters, and meters.	**GMC** Estimate lengths.
2.MD.4 Measure to determine how much longer one object is than another, expressing the length difference in terms of a standard length unit.	**GMC** Measure to determine how much longer one object is than another.

Cluster Relate addition and subtraction to length.

2.MD.5 Use addition and subtraction within 100 to solve word problems involving lengths that are given in the same units, e.g., by using drawings (such as drawings of rulers) and equations with a symbol for the unknown number to represent the problem.	**GMC** Solve number stories involving length by adding or subtracting. **GMC** Model number stories involving length.
2.MD.6 Represent whole numbers as lengths from 0 on a number line diagram with equally spaced points corresponding to the numbers 0, 1, 2, ..., and represent whole-number sums and differences within 100 on a number line diagram.	**GMC** Represent whole numbers as lengths from 0 on a number-line diagram. **GMC** Represent sums and differences on a number-line diagram.

Common Core State Standards

Standards for Mathematical Content

Cluster **Work with time and money.**

2.MD.7 Tell and write time from analog and digital clocks to the nearest five minutes, using A.M. and P.M.	**GMC** Tell and write time using analog and digital clocks **GMC** Use A.M. and P.M.
2.MD.8 Solve word problems involving dollar bills, quarters, dimes, nickels, and pennies, using $ and ¢ symbols appropriately. Example: If you have 2 dimes and 3 pennies, how many cents do you have?	**GMC** Solve problems involving coins and bills. **GMC** Read and write monetary amounts.

Cluster **Represent and interpret data.**

2.MD.9 Generate measurement data by measuring lengths of several objects to the nearest whole unit, or by making repeated measurements of the same object. Show the measurements by making a line plot, where the horizontal scale is marked off in whole-number units.	**GMC** Generate measurement data. **GMC** Represent measurement data on a line plot.
2.MD.10 Draw a picture graph and a bar graph (with single-unit scale) to represent a data set with up to four categories. Solve simple put-together, take-apart, and compare problems[3] using information presented in a bar graph.	**GMC** Organize and represent data on bar and picture graphs. **GMC** Answer questions using information in graphs.

Domain Geometry   2.G

Everyday Mathematics
Goals for Mathematical Content

Cluster **Reason with shapes and their attributes.**

2.G.1 Recognize and draw shapes having specified attributes, such as a given number of angles or a given number of equal faces.[4] Identify triangles, quadrilaterals, pentagons, hexagons, and cubes.	**GMC** Recognize and draw shapes with specified attributes. **GMC** Identify 2- and 3-dimensional shapes.
2.G.2 Partition a rectangle into rows and columns of same-size squares and count to find the total number of them.	**GMC** Partition a rectangle into rows and columns of same-size squares and count to find the total number of squares.
2.G.3 Partition circles and rectangles into two, three, or four equal shares, describe the shares using the words *halves, thirds, half of, a third of,* etc., and describe the whole as two halves, three thirds, four fourths. Recognize that equal shares of identical wholes need not have the same shape.	**GMC** Partition shapes into equal shares. **GMC** Describe equal shares using fraction words. **GMC** Describe the whole as a number of shares. **GMC** Recognize that equal shares of a shape need not have the same shape.

[3] See Glossary, Table 1. http://www.corestandards.org/assets/CCSSI_Math Standards.pdf

[4] Sizes are compared directly or visually, not compared by measuring.

Common Core State Standards

Standards for Mathematical Practice	*Everyday Mathematics* Goals for Mathematical Practice

1 Make sense of problems and persevere in solving them.

Mathematically proficient students start by explaining to themselves the meaning of a problem and looking for entry points to its solution. They analyze givens, constraints, relationships, and goals. They make conjectures about the form and meaning of the solution and plan a solution pathway rather than simply jumping into a solution attempt. They consider analogous problems, and try special cases and simpler forms of the original problem in order to gain insight into its solution. They monitor and evaluate their progress and change course if necessary. Older students might, depending on the context of the problem, transform algebraic expressions or change the viewing window on their graphing calculator to get the information they need. Mathematically proficient students can explain correspondences between equations, verbal descriptions, tables, and graphs or draw diagrams of important features and relationships, graph data, and search for regularity or trends. Younger students might rely on using concrete objects or pictures to help conceptualize and solve a problem. Mathematically proficient students check their answers to problems using a different method, and they continually ask themselves, "Does this make sense?" They can understand the approaches of others to solving complex problems and identify correspondences between different approaches.

GMP1.1 Make sense of your problem.

GMP1.2 Reflect on your thinking as you solve your problem.

GMP1.3 Keep trying when your problem is hard.

GMP1.4 Check whether your answer makes sense.

GMP1.5 Solve problems in more than one way.

GMP1.6 Compare the strategies you and others use.

2 Reason abstractly and quantitatively.

Mathematically proficient students make sense of quantities and their relationships in problem situations. They bring two complementary abilities to bear on problems involving quantitative relationships: the ability to *decontextualize*—to abstract a given situation and represent it symbolically and manipulate the representing symbols as if they have a life of their own, without necessarily attending to their referents—and the ability to *contextualize*, to pause as needed during the manipulation process in order to probe into the referents for the symbols involved. Quantitative reasoning entails habits of creating a coherent representation of the problem at hand; considering the units involved; attending to the meaning of quantities, not just how to compute them; and knowing and flexibly using different properties of operations and objects.

GMP2.1 Create mathematical representations using numbers, words, pictures, symbols, gestures, tables, graphs, and concrete objects.

GMP2.2 Make sense of the representations you and others use.

GMP2.3 Make connections between representations.

Common Core State Standards

Standards for Mathematical Practice	*Everyday Mathematics* Goals for Mathematical Practice

3 Construct viable arguments and critique the reasoning of others.

Mathematically proficient students understand and use stated assumptions, definitions, and previously established results in constructing arguments. They make conjectures and build a logical progression of statements to explore the truth of their conjectures. They are able to analyze situations by breaking them into cases, and can recognize and use counterexamples. They justify their conclusions, communicate them to others, and respond to the arguments of others. They reason inductively about data, making plausible arguments that take into account the context from which the data arose. Mathematically proficient students are also able to compare the effectiveness of two plausible arguments, distinguish correct logic or reasoning from that which is flawed, and—if there is a flaw in an argument—explain what it is. Elementary students can construct arguments using concrete referents such as objects, drawings, diagrams, and actions. Such arguments can make sense and be correct, even though they are not generalized or made formal until later grades. Later, students learn to determine domains to which an argument applies. Students at all grades can listen or read the arguments of others, decide whether they make sense, and ask useful questions to clarify or improve the arguments.

GMP3.1 Make mathematical conjectures and arguments.

GMP3.2 Make sense of others' mathematical thinking.

4 Model with mathematics.

Mathematically proficient students can apply the mathematics they know to solve problems arising in everyday life, society, and the workplace. In early grades, this might be as simple as writing an addition equation to describe a situation. In middle grades, a student might apply proportional reasoning to plan a school event or analyze a problem in the community. By high school, a student might use geometry to solve a design problem or use a function to describe how one quantity of interest depends on another. Mathematically proficient students who can apply what they know are comfortable making assumptions and approximations to simplify a complicated situation, realizing that these may need revision later. They are able to identify important quantities in a practical situation and map their relationships using such tools as diagrams, two-way tables, graphs, flowcharts and formulas. They can analyze those relationships mathematically to draw conclusions. They routinely interpret their mathematical results in the context of the situation and reflect on whether the results make sense, possibly improving the model if it has not served its purpose

GMP4.1 Model real-world situations using graphs, drawings, tables, symbols, numbers, diagrams, and other representations.

GMP4.2 Use mathematical models to solve problems and answer questions.

Common Core State Standards

Standards for Mathematical Practice	*Everyday Mathematics* Goals for Mathematical Practice

5 Use appropriate tools strategically.

Mathematically proficient students consider the available tools when solving a mathematical problem. These tools might include pencil and paper, concrete models, a ruler, a protractor, a calculator, a spreadsheet, a computer algebra system, a statistical package, or dynamic geometry software. Proficient students are sufficiently familiar with tools appropriate for their grade or course to make sound decisions about when each of these tools might be helpful, recognizing both the insight to be gained and their limitations. For example, mathematically proficient high school students analyze graphs of functions and solutions generated using a graphing calculator. They detect possible errors by strategically using estimation and other mathematical knowledge. When making mathematical models, they know that technology can enable them to visualize the results of varying assumptions, explore consequences, and compare predictions with data. Mathematically proficient students at various grade levels are able to identify relevant external mathematical resources, such as digital content located on a website, and use them to pose or solve problems. They are able to use technological tools to explore and deepen their understanding of concepts.

GMP5.1 Choose appropriate tools.

GMP5.2 Use tools effectively and make sense of your results.

6 Attend to precision.

Mathematically proficient students try to communicate precisely to others. They try to use clear definitions in discussion with others and in their own reasoning. They state the meaning of the symbols they choose, including using the equal sign consistently and appropriately. They are careful about specifying units of measure, and labeling axes to clarify the correspondence with quantities in a problem. They calculate accurately and efficiently, express numerical answers with a degree of precision appropriate for the problem context. In the elementary grades, students give carefully formulated explanations to each other. By the time they reach high school they have learned to examine claims and make explicit use of definitions.

GMP6.1 Explain your mathematical thinking clearly and precisely.

GMP6.2 Use an appropriate level of precision for your problem.

GMP6.3 Use clear labels, units, and mathematical language.

GMP6.4 Think about accuracy and efficiency when you count, measure, and calculate.

Common Core State Standards

Standards for Mathematical Practice	*Everyday Mathematics* Goals for Mathematical Practice

7 Look for and make use of structure.

Mathematically proficient students look closely to discern a pattern or structure. Young students, for example, might notice that three and seven more is the same amount as seven and three more, or they may sort a collection of shapes according to how many sides the shapes have. Later, students will see 7×8 equals the well remembered $7 \times 5 + 7 \times 3$, in preparation for learning about the distributive property. In the expression $x^2 + 9x + 14$, older students can see the 14 as 2×7 and the 9 as $2 + 7$. They recognize the significance of an existing line in a geometric figure and can use the strategy of drawing an auxiliary line for solving problems. They also can step back for an overview and shift perspective. They can see complicated things, such as some algebraic expressions, as single objects or as being composed of several objects. For example, they can see $5 - 3(x - y)^2$ as 5 minus a positive number times a square and use that to realize that its value cannot be more than 5 for any real numbers x and y.

GMP7.1 Look for mathematical structures such as categories, patterns, and properties.

GMP7.2 Use structures to solve problems and answer questions.

8 Look for and express regularity in repeated reasoning.

Mathematically proficient students notice if calculations are repeated, and look both for general methods and for shortcuts. Upper elementary students might notice when dividing 25 by 11 that they are repeating the same calculations over and over again, and conclude they have a repeating decimal. By paying attention to the calculation of slope as they repeatedly check whether points are on the line through (1, 2) with slope 3, middle school students might abstract the equation $(y - 2)/(x - 1) = 3$. Noticing the regularity in the way terms cancel when expanding $(x - 1)(x + 1)$, $(x - 1)(x^2 + x + 1)$, and $(x - 1)(x^3 + x2 + x + 1)$ might lead them to the general formula for the sum of a geometric series. As they work to solve a problem, mathematically proficient students maintain oversight of the process, while attending to the details. They continually evaluate the reasonableness of their intermediate results.

GMP8.1 Create and justify rules, shortcuts, and generalizations.

K-2 Games Correlation

Game	Grade K Lesson	Grade 1 Lesson	Grade 2 Lesson	Counting and Cardinality***	Operations and Algebraic Thinking	Number and Operations in Base Ten	Measurement and Data	Geometry
				CCSS Domain				
Addition Flip-It	9-11				•			
Addition/Subtraction Spin			5-6		•			
Addition Top-It	8-11	5-5	4-3	•	•	•*		
Addition Top-It with Dot Cards	2-2			•	•			
Animal Weight Top-It		5-11				•		
Array Bingo			8-10		•			
Array Concentration			8-10		•	•		
Attribute Spinner	6-10						•	•
Attribute Train Game		7-6						•
Base-10 Exchange (See also The Exchange Game)		5-8				•		
Basketball Addition			7-3			•		
Beat the Calculator		7-2	5-1		•			
Beat the Calculator (Extended Facts)			5-1		•	•		
Beat the Timer	3-9			•				
Before and After		5-6				•		
Bunny Hop		1-5			•	•		
Car Race	8-8				•			
Clear the Board	7-12			•	•			
Count and Sit	1-6			•				
Count and Sit (by Tens)	4-12			•				
Count and Sit (Counting On)	4-12			•				
Dice Addition	7-12			•	•			
Dice Race	3-11			•				
Dice Subtraction	8-5			•	•			
The Difference Game		5-10	3-5		•	•**		
Digit Game		5-1				•		
Dime-Nickel-Penny Grab			5-2				•	
Disappearing Train	6-9			•	•			
Domino Concentration	7-2				•			
Domino Top-It		3-1			•	•		

*This standard is not covered in the Kindergarten version of the game.
**This standard is not covered in the Grade 1 version of the game.
***Counting and Cardinality is not covered in Grades 1 and 2.

Game	Grade K Lesson	Grade 1 Lesson	Grade 2 Lesson	Counting and Cardinality***	Operations and Algebraic Thinking	Number and Operations in Base Ten	Measurement and Data	Geometry
Evens and Odds			2-9		•			
Find the Block	6-10						•	•
Fishing for 10	9-11	4-9	1-7		•			
Fishing for 100			1-7		•	•		
Frog Hop	7-1			•	•			
Gotcha	1-3			•				
Growing and Disappearing Train	6-12			•	•			
Growing Train	5-11			•	•			
Guess My Number	6-12			•	•	•		
Guess My Shape	6-4							•
Hiding Bears	6-11			•	•			
Hiding Bears (3 Addends)	6-11			•	•			
High Roller		2-6			•	•		
Hit the Target			7-1			•		
How Many Now?	2-6			•				
I Spy (with 2-dimensional shapes)	5-5	8-1						•
I Spy (with 3-dimensional shapes)	6-8	8-6						•
Make My Design	9-1	8-5						•
Making Five	8-11				•			
Match Up with Dot and Number Cards	3-1			•				
Match Up with Dot Cards	2-1			•				
Match Up with Ten Frames and Numbers	4-5			•	•			
Mini Monster Squeeze	3-12			•				
Monster Squeeze	3-12	1-2		•		•*		
Mystery Block	7-13						•	•
Mystery Change	2-5				•			
Name That Number			2-11		•	•		
Number-Grid Cover-Up	4-13			•				
Number-Line Squeeze			1-2			•		
Number Top-It			1-11			•		
Penny-Dice		1-3			•	•		

Game	Grade K Lesson	Grade 1 Lesson	Grade 2 Lesson	Counting and Cardinality***	Operations and Algebraic Thinking	Number and Operations in Base Ten	Measurement and Data	Geometry
Penny-Dime-Dollar Exchange		6-11				•		
Penny-Dime Exchange		5-3				•		
Penny Plate	6-11	2-3	1-7		•	• * **		
Quarter-Dime-Nickel-Penny Grab			1-8				•	
Racing Two Trains	5-11			•	•			
Rock, Paper, Scissors		1-8					•	
Rock, Paper, Scissors, Pencil		1-8					•	
Roll and Record	3-11			•				
Roll and Record with Dot Dice	5-2			•	•			
Roll and Record with Doubles	9-10	4-7	2-5		•			
Roll and Record with Numeral Dice	9-6				•			
Roll and Total		2-1			•	•		
Rolling for 50		1-11			•	•		
Salute!		7-3	3-4		•			
Shaker Addition Top-It		7-4			•	•		
Shape Capture			8-2					•
Solid Shapes Match Up	7-4							•
Spin a Number	3-8			•				
Spinning for Money			2-1				•	
Stand Up If…	6-4							•
Stop and Go		5-11				•		
Subtraction Bingo		2-4			•			
Subtraction Roll and Record	9-6				•			
Subtraction Top-It	9-2			•	•			
Target (to 50)			4-7			•		
Target (to 200)			4-7			•		
Teens on Double Ten Frames	5-8			•		•		
Ten Bears on a Bus	5-3			•	•			
Ten-Frame Top-It		2-2			•	•		
The Digit Game (with Symbols)			4-5			•		

*This standard is not covered in the Kindergarten version of the game.

**This standard is not covered in the Grade 1 version of the game.

***Counting and Cardinality is not covered in Grades 1 and 2.

Game	Grade K Lesson	Grade 1 Lesson	Grade 2 Lesson	CCSS Domain				
				Counting and Cardinality***	Operations and Algebraic Thinking	Number and Operations in Base Ten	Measurement and Data	Geometry
The Exchange Game (with Pennies and Nickels)			2-1				•	
The Exchange Game (with Pennies, Nickels, and Dimes)			1-8				•	
The Exchange Game (with base-10 blocks)			4-10			•		
The Exchange Game (with money)			2-1			•	•	
The Number-Grid Difference Game			3-5			•		
The Number-Grid Game			1-4			•		
Time Match		7-11					•	
Top-It		1-6				•		
Top-It with Dot Cards	2-2			•				
Top-It with Multiple Cards	4-12			•				
Top-It with Number Cards	4-12			•				
Tric-Trac		7-10			•			
Turning Over 10			1-7		•	•		
Two-Fisted Penny Addition			1-6		•	•		
What Changed? Train Games	6-12				•			
"What's My Rule?" Fishing	6-6			•			•	
"What's My Rule?" with Attribute Blocks	7-13						•	•
"What's My Rule?" with Numbers	9-3				•			
"What's My Rule?" with Patterns	8-11							•
What's Your Way		4-11				•		
Which Number Doesn't Belong?	3-7			•				
Who Am I Thinking Of?	6-6						•	

F

G

Notes

Notes

Notes